www.brookscole.com/astronomy

Discover the Brooks/Cole Galaxy is the heart of our Astronomy Resource Center. It includes a wealth of information and links to all Brooks/Cole astronomy titles. At brookscole.com/astronomy you can find out about supplements, demonstration software, and student resources. You can also send e-mail to authors and preview new publications and exciting new technologies.

Horizons

About the Author

Mike Seeds is the John W. Wetzel Professor of Astronomy at Franklin and Marshall College as well as Director of the College's Joseph R. Grundy Observatory. His research interests focus on peculiar variable stars and the automation of astronomical photometry. He is the Principal Astronomer in charge of the Phoenix 10, the first fully robotic telescope, located in southern Arizona. In 1989, he received the Christian R. and Mary F. Lindback Award for Distinguished Teaching. In addition to writing textbooks, Seeds pursues an active interest in astronomical research, creates educational tools for use in computer-smart classrooms, and continues to develop his upper-level courses in Archaeoastronomy and in the History of Astronomy. He has also published educational software for preliterate toddlers! Seeds was Senior Consultant in the creation of the 26-episode telecourse *Universe: The Infinite Frontier*. He is the author of *Foundations of Astronomy*, Fifth Edition (1999) and *Astronomy: The Solar System and Beyond* (1999), published by Brooks/Cole.

Horizons
EXPLORING THE UNIVERSE
Sixth Edition

Michael A. Seeds

Joseph R. Grundy Observatory
Franklin and Marshall College

Brooks/Cole
Thomson Learning

Pacific Grove • Albany • Belmont • Boston • Cincinnati • Johannesburg • London • Madrid
Melbourne • Mexico City • New York • Scottsdale • Singapore • Tokyo • Toronto

Sponsoring Editor: *Gary Carlson*
Marketing Manager: *Steve Catalano*
Marketing Assistant: *Christina De Veto*
Editorial Assistant: *Larisa Lieberman*
Advertising Communications: *Laura Hubrich*
Senior Developmental Editor: *Heather Dutton*
Developmental Editor: *Elmarie Hutchinson*
Project Development Editor: *Marie Carigma-Sambilay*

Production Coordinator: *Kirk Bomont*
Manuscript Editor: *Margaret Pinette*
Permissions Editor: *Mary Kay Hancharick*
Interior Design: *Jeanne Calabrese*
Cover Design: *Vernon T. Boes and Bob Western*
Typesetting: *Thompson Type*
Printing and Binding: *Transcontinental Printing Inc.*

For more information, contact:

BROOKS/COLE
511 Forest Lodge Road
Pacific Grove, CA 93950
USA
www.brookscole.com

Printed in Canada

10   9   8   7   6   5   4   3   2

Library of Congress Cataloging-in-Publication Data
Seeds, Michael A.
  Horizons : exploring the universe / Michael A. Seeds. — 6th ed.
    p.   cm.
  Includes index.
  ISBN 0-534-57258-8
    1. Astronomy.   I. Title
  QB45.S44   2000
  520—dc21   99-30934
                                                 CIP

Brief Contents

Contents

Manuscript Reviewers

J. Wayne Wooten, *Pensacola Junior College*

Steve Herbert, *Xavier University (OH)*

John Silva, *University of Massachusetts, Dartmouth*

Francette Fey, *Macomb County Community College (MI)*

Jon U. Bell, *Indian River Community College (FL)*

John D. Eggert, *Daytona Beach Community College*

Jorge F. Cossio, *Miami Dade Community College*

Gordon Baird, *University of Mississippi*

Henry Bass, *University of Mississippi*

William Bittle, *Seminole Community College*

Jeff Conn, *Lansing Community College*

Eugene Capriotti, *Michigan State University*

Charles Hager, *San Francisco State University*

Dennis Hibbert, *North Seattle Community College*

Fred Holmstom, *San Jose State University*

Douglas Ingram, *Texas Christian University*

Kenneth Janes, *Boston University*

Robert Kimble, *Feather River College*

Michael LoPresto, *Henry Ford Community College*

J. Scott Miller, *University of Louisville*

Gerald Newsom, *Ohio State University*

Jim Peters, *San Francisco State University*

Harold Pray, *Southern Illinois University*

James Roberts, *University of North Texas*

Cecil Shugart, *University of Memphis*

Joseph Silk, *University of California at Berkeley*

Alex Smith, *University of Florida*

Don Summers, *University of Mississippi*

Richard Wade, *Pennsylvania State University*

A Note to the Student

From Mike Seeds

Each of us faces the ultimate question that has intrigued humans since the beginning of awareness. Great minds like Plato, Beethoven, and Hemingway have tried to give their personal answers, but it continues to perplex every one of us. The ultimate question is simply "What are we?"

Answering in a physical sense, we are the product of nuclear and atomic processes that began in the big bang and continued through the births, lives, and deaths of stars. That physical sense of what we are is the heart of modern astronomy.

Certainly part of the answer goes beyond the physical universe to include philosophical and cultural issues. We define ourselves by what we create, what we worship, what we admire, and what we demand of each other. But a major part of the answer is embedded in the physical universe, and for this we turn to astronomy. No other science speaks so clearly about our place in the universe. We cannot hope to understand what we are until we understand where we are in a universe filled with worlds, stars, and galaxies.

While astronomy is exciting because it tells us about our role in the universe, it is also exciting because astronomers are discovering new things every day. That continuing adventure is woven into this book in the form of new images, new discoveries, and new understandings, including the possible discovery of ancient life on Mars, planets orbiting other stars, planetary disks in Orion, evidence of supermassive black holes in galaxies, planets orbiting pulsars, and more. Astronomers love photographs, and these pages include the latest Hubble Space Telescope images (including the Eagle Nebula and the Hubble Deep Field), Galileo spacecraft images of Jupiter and its moons, the Crab Nebula pulsar, comet impacts on Jupiter, and much more. Many ground-based photos in this edition have never been published before, and some were custom made for this book.

As a teacher, my quest is simple. By the end of this book, I want you to know a few basic things: that you live in a very big universe that is described by a small set of rules; that those rules are knowable; and that the human race has found a logical method to figure out the rules—a method called science. The most important concept in introductory astronomy is the process of science.

Moreover, the interplay of evidence and hypothesis is the principal organizing theme for this book at every level from individual paragraphs to the order of chapters. I introduce basic observational facts, synthesize hypotheses, and discuss further observations as evidence supporting or contradicting those hypotheses. A good example of how the ordering of chapters also follows this theme can be seen with Chapter 16, "The Origin of the Solar System." Its placement at the beginning of the section on the solar system allows you to use the following chapters on the planets as further tests of the "origins" hypothesis. Furthermore, placing the solar system section after the stars and galaxies sections allows you to see how the solar nebula hypothesis develops from our understanding of stellar evolution, star formation, and nucleosynthesis.

One big advantage of emphasizing evidence and hypothesis is that it unifies the mass of facts that discourage so many nonscience readers. I want you to understand logical arguments. My approach makes the facts easier to remember, but, more importantly, the facts make sense because they fit into a logical framework. Understanding nature is the heart of science, and the interplay of evidence and hypothesis is the tool science uses to unlock the secrets of the universe. Once you begin to understand that exciting story, the facts will fall into place.

To help you more clearly grasp the subject, I have created a number of features within the book.

- **Windows on Science** provide a parallel commentary on how all of science works. For example, the Windows point out where we are using statistical evidence, where we are reasoning by analogy, and where we are building a scientific model in place of a scientific hypothesis.

- **Guideposts** at the opening page of each chapter help you see the organization of the book. The Guidepost connects the chapter with preceding and following chapters to provide an overall organizational guide.

- **Review/Critical Inquiry** at the end of each text section is a carefully designed question to help you review and synthesize concepts in the section and understand the scientific procedures being used. A short answer follows to show how scientists construct logical arguments from observations, evidence, theories, and the natural laws that lead to a conclusion. A further question then gives you a chance to construct your own argument on a related issue.

- *End-of-Chapter Review Questions* are designed to help you test your understanding of the material when you finish reading each chapter.

- *End-of-Chapter Discussion Questions* go beyond the text and invite you to think critically and creatively about scientific questions. You can think about these questions yourself or discuss them in class.

- *By the Numbers* are short math boxes that occasionally present an equation and an example or two. The mathematical aspects of science are integral to our understanding of the universe, so it is important for you to see these ideas. But because students in an introductory science course have different levels of math experience, arguments in the text do not depend entirely on mathematical reasoning.

As additional aids, be aware that this book also includes, free of charge, the following electronic enhancements:

- *The Sky Student Edition CD-ROM.* With this CD-ROM, a personal computer becomes a powerful personal planetarium. Loaded with data on 118,000 stars and 13,000 deep-sky objects with images, it allows you to view the universe at any point in time from 4000 years ago to 8000 years in the future, to see the sky in motion, to view constellations, to print star charts, and much more.

- *Voyage Through the Solar System CD-ROM.* Launch yourself into interplanetary space with close-up views of the planets, the sun, asteroids, and comets. This CD-ROM includes more than 70 full-motion sequences, narrated videos, and photos from NASA and Jet Propulsion Lab space missions.

- *InfoTrac® College Edition.* You receive free unlimited access for one academic term to this online database of complete articles and images from more than 700 popular and scholarly periodicals of the past four years, updated daily. An online student guide gives you tips for using InfoTrac and correlates each chapter in this book to InfoTrac articles. (Check with your instructor for availability outside North America.)

- *The Brooks/Cole Astronomy Resource Center (www.brookscole.com/astronomy).* Thousands of students and instructors visit this free Web site every week. The Astronomy Resource Center offers—among other things—online study aids, virtual field trips, and links to other sites.

- *Critical Inquiries for the Web.* At the end of each chapter, this feature offers you more opportunities to use the Internet to research questions.

This book focuses on the things that nonscientists should understand to function in the modern world and to appreciate their lives in this beautiful universe. That is why introductory astronomy is the ideal introductory science course. It focuses not on the facts of astronomy, but on an understanding of what we are, where we are, and how we know. Every page of this new edition reflects that ideal. I hope you enjoy exploring our universe.

Mike Seeds
m_seeds@acad.fandm.edu

Acknowledgments

In 1973, I started writing astronomy textbooks, and over the years I have had the guidance of a great many people who care about astronomy and teaching. I would like to thank all of the students and teachers who have responded so enthusiastically to *Horizons: Exploring the Universe*. Their comments and suggestions have been very helpful in shaping the book over the years. I would especially like to thank the many reviewers whose careful analysis and thoughtful suggestions have been invaluable in completing this most recent edition.

The following institutions have provided the many photos and diagrams in this book: Anglo-Australian Telescope Board, Arecibo Observatory, Association of Universities for Research in Astronomy, The Astronomical Journal, The Astrophysical Journal, AT&T Archives, Brookhaven National Laboratories, California Association for Research in Astronomy, California Institute of Technology, Carnegie Institution of Washington, Celestron International, Cornell University, Granger Collection, Grundy Observatory, Hobby–Eberly Telescope Project, Houghton Library, IBM Research Center, Infrared Processing and Analysis Center, Institute for Astronomy, Japan/U.S. Yohkoh Team, Joint Astronomy Center, JPL, Lick Observatory, Los Alamos National Laboratory, Lowell Observatory, Lunar and Planetary Institute, Max Planck Institute for Extraterrestrial Physics, NASA, National Geophysical Data Center, National Museum of Natural History, National Oceanic and Atmospheric Administration, National Optical Astronomy Observatory, National Radio Astronomy Observatory, New England Meteorical Services, Palomar Observatory, Penn State University, Pennsylvania Department of Tourism, Sac Peak Observatory, SETI Institute, Smithsonian Astrophysical Observatory, Space Environmental Laboratory, Space Telescope Science Institute, Stanford University, Steward Observatory, University of Goettingen, University of Hawaii, University of Munich, University of Texas, U.S. Geological Survey, USAF Air Weather Service, William Keck Observatory, and Yerkes Observatory.

I would especially like to thank the following individuals who went out of their way to help me locate materials for *Horizons:* H. A. Abt, Sam Barden, Allen Beechel, Roger Bell, Dana Berry, David Bradstreet, Robert Braun, Michael Briley, Jack Burns, Bruce Campbell, Chip Clark, Craig Clark, Wesley N. Colley, Coral Cooksley, Don Davis, Frank Drake, Richard Fluck, S. W. Fox, L. A. Frank, James Gelb, Margaret J. Geller, Dan Gezari, Owen Gingerich, Bradford Greeley, Cheryl Gundy, R. R. Harnden, William Hartmann, John Hayes, Jeff Hester, Paul Hintzen, George Jacoby, William Keel, Charles Keller, R. Kempton, Ivan King, Ruth Kneale, Stephen Larson, Tod Lauer, Deborah Levine, Kenneth Libbrecht, William Livingston, Kwok-Yung Lo, John W. Mackenty, David Malin, Jack Marling, Laurence Marschall, Ian McLean, David Miller, Stanley Miller, Reinhard Mundt, Harold Nations, C. R. O'Dell, Seven Ostro, Linda C. Owens, Andy Perala, C. M. Pieters, Carolyn C. Porco, William Romanishin, Vera Rubin, Laurence Rudnick, Anneila Sargent, Rudolph Schild, Maartin Schmidt, J. William Schopf, Francois Schweizer, M. Seldner, Anna Sergi, Nigel Sharp, L. D. Simon, Damon Simonelli, Patrica Smiley, Steve Snowden, Hyron Spinrad, John Stauffer, Alan Stern, William P. Stern, Jr., Alan Stockton, R. J. Terrile, L. Thompson, J. Trumper, Ed Turner, Tony Tyson, Russ Underwood, Joseph Ververka, Steven Vogt, Richard Wainscoat, Matt Weinberg, Michael J. West, Richard E. White, Simon D. M. White, David Wittman, Dennis Zaritsky, and Ben Zellner.

Certain diagrams in Chapters 9 and 10 are based on figures I designed that first appeared in my article "Stellar Evolution," *Astronomy,* February 1979.

I would like to thank my developmental editor for this edition, Elmarie Hutchinson, for her expert and detailed analysis of the informational structure of the text, figures, and tables. Her advice will help countless students understand astronomy more easily and more clearly.

It has been a great pleasure to work with and learn from the Brooks/Cole team. Special thanks go to all of the people who have contributed to the project, including Marie Carigma-Sambilay, Vernon Boes, Kirk Bomont, Jeanne Calabrese, Steve Catalano, Christina De Veto, Heather Dutton, Sue Ewing, Laura Hubrich, Larisa Lieberman, Claire Masson, and Pat Waldo. I have enjoyed working with Margaret Pinette of Heckman & Pinette on production, and I appreciate her understanding of my multifaceted failings. I would especially like to thank my editor, Gary Carlson, for all his help and advice.

Most of all, I would like to thank my wife, Janet, and my daughter, Kate, for putting up with "the books." They know all too well that textbooks are made of time.

Mike Seeds

Horizons

CHAPTER 1

The Scale of the Cosmos

The longest journey begins with

a single step.

Confucius

Guidepost

How can we study something so big it includes everything, even us? The cosmos, also called the universe, is our subject in astronomy. Perhaps the best way to begin our study is to grab a quick impression as we zoom from things our own size up to the largest things in the universe.

That cosmic zoom, the subject of this chapter, gives us our first glimpse of the objects we will study in the rest of this book. In the next chapter, we will return to Earth to think about the appearance of the sky, and subsequent chapters will discuss the stars, galaxies, and worlds that fill our universe.

While we study the cosmos, we may observe the process by which we learn. That process, science, gives us a powerful way to understand not only the universe but also ourselves.

We are going on a voyage out to the end of the universe. Marco Polo journeyed east, Columbus west, but we will travel away from our home on Earth, out past our moon, the sun, and the other planets, past the stars we see in the sky, and past billions more that we cannot see without the aid of the largest telescopes. We will journey through great whirlpools of stars to the most distant galaxies visible from Earth—and then we will continue on, carried only by experience and imagination—looking for the structure of the universe itself.

Besides journeying through space, we will also travel in time. We will explore the past to see the sun and planets form and search for the formation of the first stars and the origin of the universe. We will also explore forward in time to watch the sun die and Earth wither. Our imagination will become a scientific time machine searching for the ultimate end of the universe.

Though we may find an end to the universe, a time when it will cease to exist, we will not discover an edge. It is possible that our universe is infinite and extends in all directions without limit. Such vastness dwarfs our human dimensions, but not our intelligence or imagination.

Astronomy—this imagined voyage—is more than the study of stars and planets. It is the study of the universe in which we exist. Our personal lives are confined to a small part of a planet circling a small sun drifting through the universe, but astonomy can take us out of ourselves and thus help us understand what we are.

We live small lives on our planet, but our study of astronomy introduces us to sizes, distances, and times far beyond our usual experience. Our task in this chapter is to grasp the meaning of these unfamiliar sizes, distances, and times. The solution lies in a single word, *scale.* We will compare objects of different sizes to grasp the scale of the universe.

We begin with a human figure whose size we recognize. Figure 1-1 shows a region about 52 feet across occupied by a human being, a sidewalk, and a few trees—all objects whose size we can understand. Each successive picture in this chapter will show a region of the universe that is 100 times wider than the preceding picture. That is, each step will widen our view by a factor of 100.

In Figure 1-2 we widen our field of view by a factor of 100 and see an area 1 mile in diameter. The arrow points to the scene shown in the preceding photograph. People, trees, and sidewalks have vanished, but now we can see a college campus and the surrounding streets and houses. The dimensions of houses and streets are familiar to us. We have been in houses, crossed streets, and walked or run a mile. This is the familiar world around us, and we can relate such objects to the scale of our bodies.

Figure 1-1
(Michael A. Seeds)

Figure 1-2
(Pennsylvania Department of Tourism, Bureau of Design)

Figure 1-3
(NASA infrared photograph)

ing material down the rivers and into the sea. The mountains and valleys that we know are only temporary features; they are constantly changing. As we explore the universe, we will see that it, like Earth's surface, is evolving. Change is the norm in the universe.

Look again at Figure 1-3, and note the red color. This photo is an infrared photograph in which healthy green leaves and crops show up as red. Our eyes are only sensitive to a narrow range of colors. As we explore the universe, we must learn to use a wide range of "colors," from X rays to radio waves, to reveal sights invisible to our unaided eyes.

We have begun our adventure using feet and miles, but we should use the metric system of units. Not only is it used by all scientists around the world, but it makes calculations much easier. For example, just calculating the number of inches in a mile is a chore. We must multiply 5280 feet per mile by 12 inches per foot. It is much easier to calculate the number of centimeters in 1 kilometer. We just move the decimal place three places to the right to get 1000 meters per kilometer and then move it two more places to get 100,000 centimeters per kilometer. If you are not already familiar with the metric system, or need a review, study Appendix A before reading on.

The photo in Figure 1-2 is 1 mile in diameter. A mile equals 1.609 kilometers, so we can see in the photo that a kilometer is a bit over two thirds of a mile—a short walk across a neighborhood.

Our view in Figure 1-3 spans 160 kilometers. Green foliage of different kinds appears as shades of red in this infrared photo. The college campus is now invisible. The gray blotches are small cities, and the suburbs of Philadelphia are visible at the lower right. At this scale, we see the natural features of Earth's surface. The Allegheny Mountains of southern Pennsylvania cross the image in the upper left, and the Susquehanna River flows southeast into Chesapeake Bay. What appear as white bumps are a few puffs of clouds.

These features remind us that we live on the surface of a changing planet. Forces in Earth's crust pushed the mountain ranges up into parallel folds, like a rug wrinkled on a polished floor. The clouds remind us that Earth's atmosphere is rich in water, which falls as rain and erodes the mountains, wash-

Figure 1-4
(NASA)

At the next step in our journey, we see our entire planet (Figure 1-4), which is 12,756 km in diameter. Earth rotates on its axis once a day, exposing half of its surface to daylight at any particular moment. The photo shows most of the daylight side of the planet. The blurriness at the extreme right is the sunset line. The rotation of Earth carries us eastward, and as we cross the sunset line into darkness, we say the sun has set. Thus, the rotation of our planet causes the cycle of day and night.

We know that Earth's interior is made mostly of iron and nickel and that its crust is mostly silicate rocks. Only a thin layer of water makes up the oceans, and the atmosphere is only a few hundred kilometers deep. On the scale of this photograph, the depth of the atmosphere on which our lives depend is less than the thickness of a piece of thread.

Again we enlarge our field of view by a factor of 100, and we see a region 1,600,000 km wide (Figure 1-5). Earth is the small white dot in the center, and the moon, whose diameter is only one-fourth that of Earth, is an even smaller dot along its orbit 380,000 km from Earth.

These numbers are so large that it is inconvenient to write them out. Astronomy is the science of big numbers, and we will use numbers much larger than these to discuss the universe. Rather than writing out these numbers as in the previous paragraph, it is convenient to write them in **scientific notation.** This is nothing more than a simple way to write numbers without writing a great many zeros. In scientific notation, we write 380,000 as 3.8×10^5. The 5 tells us to begin with 3.8 and move the decimal point 5 places to the right to get 380,000. In the same way, we write 1,600,000 as 1.6×10^6. Notice that we could also write this as 16×10^5 or 0.16×10^7. All represent the same quantity. If you are not familiar with scientific notation, read the section on "Powers of 10 Notation" in Appendix A. The universe is too big to discuss without using scientific notation.

When we once again enlarge our field of view by a factor of 100 (Figure 1-6), Earth, its moon, and the moon's orbit all lie in the small red box at lower left. But now we see the sun and two other planets that are part of our solar system. Our **solar system** consists of the sun, its family of planets, and some smaller bodies such as moons and comets.

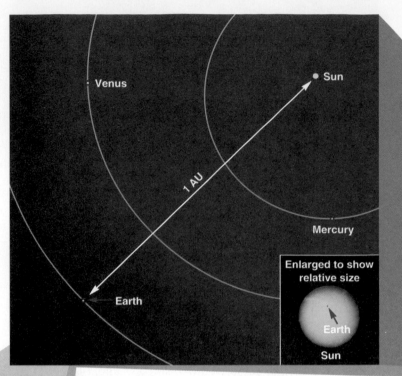

Figure 1-6

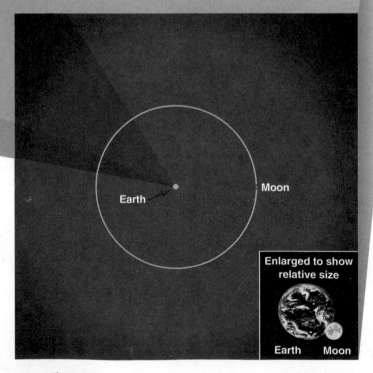

Figure 1-5

Like Earth, Venus and Mercury are **planets,** small, nonluminous bodies that shine by reflected light. Venus is about the size of Earth, and Mercury is a bit larger than our moon. On this diagram, they are both too small to be seen as anything but tiny dots. The sun is a **star,** a self-luminous ball of hot gas that generates its own energy. The sun is 10^9 times larger in diameter than Earth (inset), but it too is nothing more than a dot in this diagram.

This diagram has a diameter of 1.6×10^8 km. One way astronomers deal with large numbers is to define new units. The average distance from Earth to the sun is a unit of distance called the **astronomical unit (AU),** a distance of 1.5×10^{11} m. Using this unit we can say that the average distance from Venus to the sun is about 0.7 AU. The average distance from Mercury to the sun is about 0.39 AU.

The orbits of the planets are not perfect circles, and this is particularly apparent for Mercury. Its orbit carries it as close to the sun as 0.307 AU and as far away as 0.467 AU. Earth's orbit is more circular, and its distance from the sun varies by only a few percent.

Our first field of view was only 52 feet (about 16 m) in width. After only six steps of enlarging by a factor of 100, we now see the entire solar system (Figure 1-7). Our field of view is 1 trillion (10^{12}) times wider than in our first view. The details of the preceding figure are now lost in the red square at the center of this diagram. We see only the brighter, more widely separated objects as we back away. The sun, Mercury,

Venus, and Earth lie so close together that we cannot separate them at this scale.

Mars, the next outward planet, lies only 1.5 AU from the sun. In contrast, Jupiter, Saturn, Uranus, Neptune, and Pluto are so far from the sun that they are easy to place in this diagram. These are cold worlds far from the sun's warmth. Light from the sun reaches Earth in only 8 minutes, but it takes over 4 hours to reach Neptune. Notice that Pluto's orbit is so elliptical that Pluto can come closer to the sun than Neptune does, as Pluto did between 1979 and 1999.

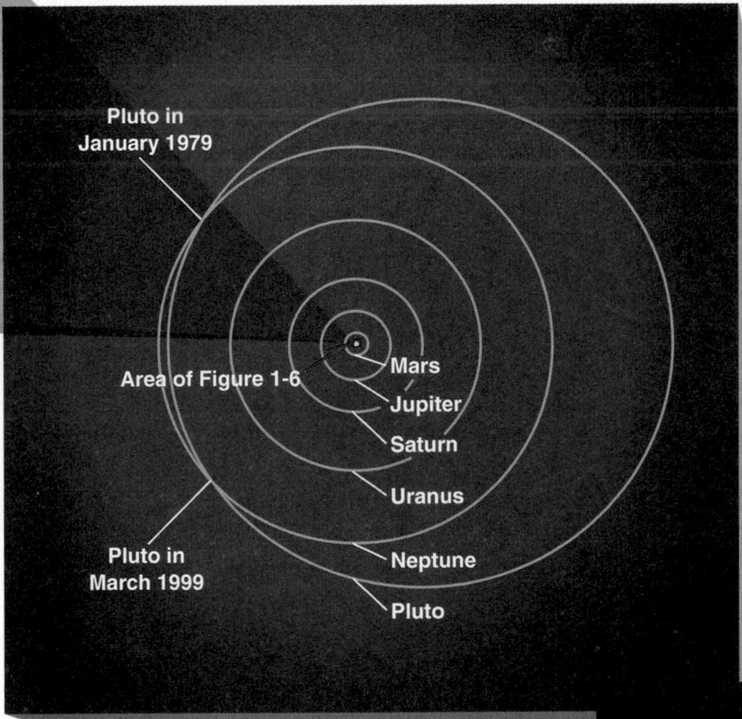

It is difficult to grasp the isolation of the stars. If the sun were represented by a golf ball in New York City, the nearest star would be another golf ball in Chicago. Except for the widely scattered stars and a few atoms of gas drifting between the stars, the universe is nearly empty.

In Figure 1-9, our field of view has expanded to a diameter a bit over 1 million AU. The sun is at the center, and we see a few of the nearest stars. These stars are so distant that it is not reasonable to give their distances in astronomical units. We must define a new unit of distance, the light-year. One **light-year (ly)** is the distance that light travels in 1 year, roughly 10^{13} km or 63,000 AU. The diameter of our field of view is, in our new unit, 17 ly. The nearest star to the sun, Proxima Centauri, is 4.2 ly from Earth. In other words, light from Proxima Centauri takes 4.2 years to reach Earth.

Although these stars are roughly the same size as the sun, they are so far away that we cannot see them as anything but points of light. Even with the largest telescopes on Earth, we still see only points of light when we look at stars. In Figure 1-9, the sizes of the dots represent not the size of the stars but their brightness. This is the custom in astronomical diagrams, and it is also how star images are recorded on photographs. Bright stars make larger spots on a photograph than faint stars. Thus, the size of a star image in a photograph tells us not how big the star is but only how bright it looks.

Figure 1-7

When we again enlarge our field of view by a factor of 100, our solar system vanishes (Figure 1-8). The sun is only a point of light, and all the planets and their orbits are now crowded into the small, red square at the center. The planets are too small and reflect too little light to be visible so near the brilliance of the sun.

Nor are any stars visible except for the sun. The sun is a fairly typical star, and it seems to be located in a fairly average neighborhood in the universe. Although there are many billions of stars like the sun, none are close enough to be visible in this diagram, which shows an area only 11,000 AU in diameter. The stars are typically separated by distance about 10 times larger than the distance represented in this diagram. We will see stars in our next field of view, but, except for the sun at the center, this diagram is empty.

Figure 1-8

Figure 1-9

enough into space to look back and photograph our galaxy, so the photo in Figure 1-11 shows a galaxy similar to our own. Our sun would be invisible in such a photo, but if we could see it, we would find it about two-thirds of the way from the center to the edge.

Ours is a fairly large galaxy. Only a century ago astronomers thought it was the entire universe—an island universe of stars in an otherwise empty vastness. Now we know that our galaxy is not unique. Indeed ours is only one of many billions of galaxies scattered throughout the universe.

As we expand our field of view by another factor of 100, our galaxy appears as a tiny luminous speck surrounded by other specks (Figure

In Figure 1-10, we expand our field of view by another factor of 100, and the sun and its neighboring stars vanish into the background of thousands of stars. The field of view is now 1700 ly in diameter. Of course, no one has ever journeyed thousands of light-years from Earth to look back and photograph the solar neighborhood, so we use a representative photo of the sky. The sun is a relatively faint star that would not be easily located in a photo at this scale. Looking at this photograph, we notice a loose cluster of stars in the lower left quarter of the image. We will discover that many stars are born in clusters and that both old and young clusters exist in the sky. Star clusters are forming right now.

What we do not see is critically important. We do not see the thin gas that fills the spaces between the stars. Although those clouds of gas are thinner than the best vacuum on Earth, it is those clouds that give birth to new stars. Our sun formed from such a cloud about 5 billion years ago. We will see more star formation in our next view.

If we expand our field of view by a factor of 100, we see our galaxy (Figure 1-11). A **galaxy** is a great cloud of stars, gas, and dust bound together by the combined gravity of all the matter. Galaxies range from 1500 to over 300,000 ly in diameter and can contain over 100 billion stars. In the night sky, we see our galaxy as a great, cloudy wheel of stars ringing the sky as the **Milky Way,** and we refer to our galaxy as the Milky Way Galaxy. Of course, no one can journey far

Figure 1-10
This box ■ represents the relative size of the previous frame.
(NOAO)

1-12). This diagram includes a region 17 million ly in diameter, and each of the dots represents a galaxy. Notice that our galaxy is part of a cluster of a few dozen galaxies. We will find that galaxies are commonly grouped together in clusters. Some of these galaxies have beautiful spiral patterns like our own galaxy, but others do not. Some are strangely distorted. One of the mysteries we will try to solve is what produces these differences among the galaxies.

If we again expand our field of view, we see that the clusters of galaxies are connected in a vast network (Figure 1-13). Clusters are grouped into superclusters—clusters of clusters—and the superclusters

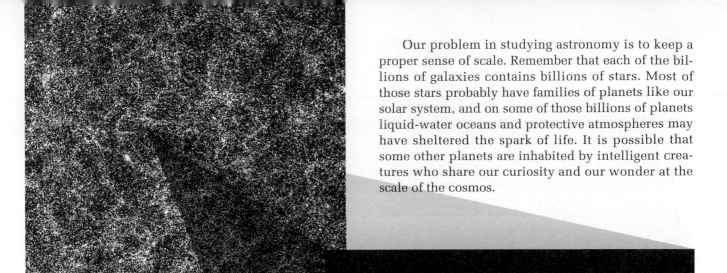

Our problem in studying astronomy is to keep a proper sense of scale. Remember that each of the billions of galaxies contains billions of stars. Most of those stars probably have families of planets like our solar system, and on some of those billions of planets liquid-water oceans and protective atmospheres may have sheltered the spark of life. It is possible that some other planets are inhabited by intelligent creatures who share our curiosity and our wonder at the scale of the cosmos.

Figure 1-13
This box ■ represents the relative size of the previous frame. *(Detail from galaxy map from M. Seldner, B. L. Siebers, E. J. Groth, and P. J. E. Peebles,* Astronomical Journal *82 [1977])*

are linked to form long filaments and walls outlining voids that seem nearly empty of galaxies. These appear to be the largest structures in the universe. Were we to expand our field of view another time, we would probably see a uniform fog of filaments and voids. When we puzzle over the origin of these structures, we are at the frontier of human knowledge.

Milky Way Galaxy →

Figure 1-12

Figure 1-11
(© Anglo-Australian Telescope Board)

Summary

Our goal in this chapter is to preview the scale of astronomical objects. To do so, we journey outward from a familiar campus scene by expanding our field of view by factors of 100. Only 12 such steps take us to the largest structures in the universe.

The numbers in astronomy are so large that it is not convenient to express them in the usual way. Instead, we use the metric system to simplify our calculations and scientific notation to simplify the writing of big numbers. The metric system and scientific notation are discussed in Appendix A.

We live on the rotating planet Earth, which orbits a rather typical star we call the sun. A unit of distance convenient for discussing our solar system, the astronomical unit, equals the average distance from Earth to the sun. Of the other planets in our solar system, Mercury is closest to the sun, only 0.39 AU away, and Neptune is about 30 AU.

The sun, like most stars, is very far from its neighboring stars, and this leads us to use another unit of distance, the light-year. A light-year is the distance light travels in 1 year. The star nearest to the sun is Proxima Centauri, at a distance of 4.2 ly.

As we enlarge our field of view, we discover that the sun is only one of about 100 billion stars in our galaxy and that our galaxy is only one of many billions of galaxies in the universe. Galaxies appear to be grouped together in clusters, superclusters, and filaments, the largest structures known.

As we explore, we note that the universe is evolving. Earth's surface is evolving, and so are stars. Stars form from the gas in space, grow old, and eventually die. We do not yet understand how galaxies form or evolve.

Among the billions of stars in each of the billions of galaxies, many probably have planets, but detecting such planets is very difficult, and only a few have been found. We suppose that among the many planets in the universe, there must be some that are like Earth. We wonder if a few are inhabited by intelligent beings like ourselves.

New Terms

scientific notation	astronomical unit (AU)
solar system	light-year (ly)
planet	galaxy
star	Milky Way

Review Questions

1. What is the largest dimension you have personal knowledge of? Have you run a mile? hiked 10 miles? run a marathon?

2. Why are astronomical units more convenient than miles or kilometers for measuring some astronomical distances?

3. In what ways is our planet changing?

4. What is the difference between our solar system, our galaxy, and the universe?

5. Why do all stars, except for the sun, look like points of light as seen from Earth?

6. Why are light-years—rather than miles or kilometers—more convenient units for measuring some astronomical distances?

7. Why is it difficult to see planets orbiting other stars?

8. Why can't we measure the diameters of stars from the size of the star images on photographs? What does the diameter of a star image really tell us about the star?

9. How long does it take light to cross the diameter of our solar system? of our galaxy?

10. What are the largest known structures in the universe?

11. How many planets inhabited by intelligent life do you think the universe contains? Explain your answer.

Problems

1. If 1 mile equals 1.609 km and the moon is 2160 miles in diameter, what is its diameter in kilometers? in meters?

2. If sunlight takes 8 minutes to reach Earth, how long does moonlight take?

3. How many suns would it take, laid edge to edge, to reach the nearest star?

4. How many kilometers are there in a light-minute? (Hint: The speed of light is 3×10^5 km/sec.)

5. How many galaxies like our own, laid edge to edge, would it take to reach the nearest large galaxy (which is 2×10^6 ly away)?

CHAPTER 2

The Sky

The Southern Cross I saw every

night abeam. The sun every morning

came up astern; every evening it

went down ahead. I wished for no

other compass to guide me, for these

were true.

Captain Joshua Slocum
Sailing Alone Around the World

Guidepost

Astronomy is about us. As we learn about astronomy, we learn about ourselves. We search for an answer to the question "What are we?" The quick answer is that we are thinking creatures living on a planet that circles a star we call the sun. In this chapter, we begin trying to understand that answer. What does it mean to live on a planet?

The preceding chapter gave us a quick overview of the universe, and chapters later in the book will discuss the details. This chapter and the next help us understand what the universe looks like seen from the surface of our spinning planet.

But appearances are deceiving. We will see in Chapter 4 how difficult it has been for humanity to understand what we see in the sky every day. In fact, we will discover that modern science was born when people tried to understand the appearance of the sky.

The night sky is the rest of the universe as seen from our planet. When we look up at the stars, we look out through a layer of air only a few hundred kilometers deep. Beyond that, space is nearly empty, with the stars scattered light-years apart. We begin our search for the natural laws that govern the universe by trying to understand what that universe looks like.

As you read this chapter, keep in mind that we live on a planet. The stars are scattered into the void all around us, most very distant and some closer. Our planet rotates on its axis once a day, so from our viewpoint the sky appears to rotate around us each day. Not only does the sun rise in the east and set in the west, but so also do the stars. That apparent daily motion is caused by the rotation of our planet.

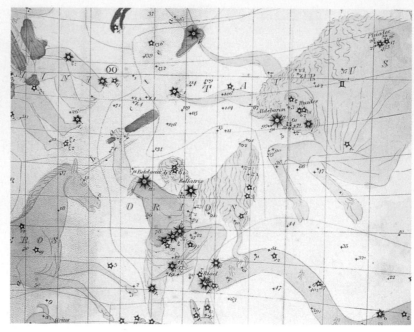

Figure 2-1
The ancient constellations Orion and Taurus celebrate a great hunter and a powerful bull. *(From Duncan Bradford,* Wonders of the Heavens, *Boston: John B. Russell, 1837.)*

2-1 The Stars

On a dark night far from city lights, we can see a few thousand stars in the sky. Like ancient astronomers, we will try to organize what we see by naming groups of stars and individual stars and by specifying the brightness of individual stars.

Constellations

All around the world, ancient cultures celebrated heroes, gods, and mythical beasts by naming groups of stars—**constellations** (Figure 2-1). We should not be surprised that the star patterns do not look like the creatures they represent any more than Columbus, Ohio, looks like Christopher Columbus. The constellations "celebrate" the most important mythical figures in each culture. The constellations we know within western culture originated in Mesopotamia over 5000 years ago, with other constellations added by Babylonian, Egyptian, and Greek astronomers during the classical age. Of these ancient constellations, 48 are still in use.

To the ancients, a constellation was a loose grouping of stars. Many of the fainter stars were not included in any constellation, and regions of the southern sky not visible to the ancient astronomers of northern latitudes were not identified with constellations. Constellation boundaries, when they were defined at all, were only approximate (Figure 2-2a), so a star like

Figure 2-2
(a) In antiquity, constellation boundaries were poorly defined, as shown on this map by the curving dotted lines that separate Pegasus from Andromeda. *(From Duncan Bradford,* Wonders of the Heavens, *Boston: John B. Russell, 1837.)* (b) Modern constellation boundaries are precisely defined by international agreement.

a

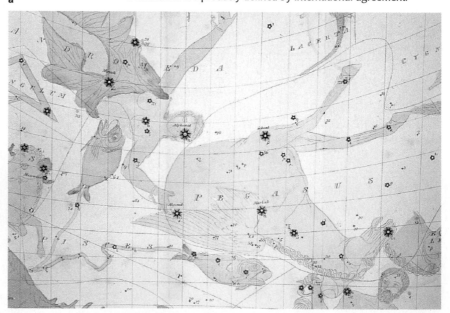

Alpheratz could be thought of as part of Pegasus or part of Andromeda. In recent centuries astronomers have added 40 modern constellations to fill gaps, and in 1928 the International Astronomical Union established 88 official constellations with clearly defined boundaries (Figure 2-2b). Thus a constellation now represents not a group of stars but an area of the sky, and any star within the region belongs to one and only one constellation.

In addition to the 88 official constellations, the sky contains a number of less formally defined groupings called **asterisms**. The Big Dipper, for example, is a well-known asterism that is part of the constellation Ursa Major (the Great Bear). Another asterism is the Great Square of Pegasus (Figure 2-2b), which includes three stars from Pegasus and one from Andromeda. The star charts at the end of this book will introduce you to the brighter constellations and asterisms.

Although we name constellations and asterisms, most are made up of stars that are not physically associated with one another. Some stars may be many times farther away than others and moving through space in different directions. The only thing they have in common is that they lie in approximately the same direction from Earth (Figure 2-3).

The Names of the Stars

In addition to naming groups of stars, ancient astronomers named the brighter stars, and modern astronomers still use many of those names. Whereas the names of the constellations are in Latin, the common language of science in Renaissance Europe, most star names derive from ancient Arabic, though much altered by the passing centuries. The name of Betelgeuse, the bright red star in Orion, for example, comes

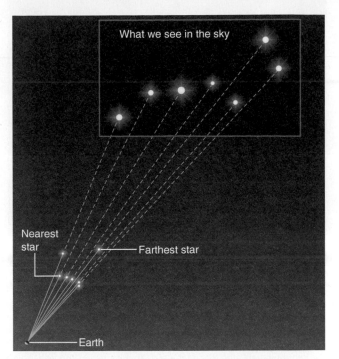

Figure 2-3
The stars we see in the Big Dipper—the brighter stars of the constellation Ursa Major, the Great Bear—are not at the same distance from Earth. We see the stars in a group in the sky because they lie in the same general direction as seen from Earth. The sizes of the star dots in the star chart represent the apparent brightness of the stars.

from the Arabic *yad al-jawza,* meaning "armpit of Jawza [Orion]." Aldebaran, the bright red eye of Taurus the bull, comes from the Arabic *al-dabar an,* meaning "the follower."

Naming individual stars is not very helpful because we can see thousands of them, and names do not help us locate stars in the sky. Another way to identify stars is to assign Greek letters to the bright stars in a constellation in approximate order of brightness. Thus the brightest star is usually designated α (alpha), the second brightest β (beta), and so on. For many constellations, the letters follow the order of brightness, but some constellations, by tradition, mistake, or the personal preferences of early chartmakers, are exceptions (Figure 2-4).

To identify a star by its Greek letter designation, we give the Greek letter followed by the genitive (possessive) form of the constellation name; for example, the brightest star in the constellation Canis Major is α Canis Majoris. This both identifies the star and the constellation and gives us a clue to the relative brightness of the star. Compare this with the ancient name for this star, Sirius, which tells us nothing about location or brightness.

This method of identifying a star's brightness is only approximate. In order to

b

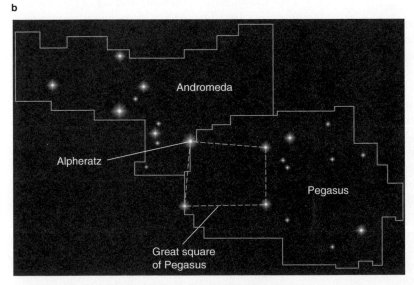

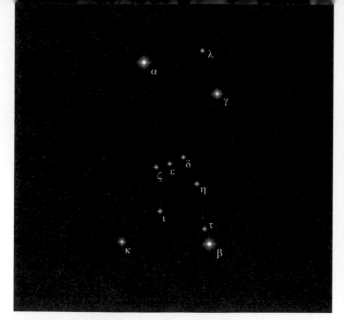

a

Figure 2-4
The brighter stars in the constellation Orion are designated by Greek letters. Although most of the stars are lettered in order of brightness, β is brighter than α and κ is brighter than η. (b) A long-exposure photograph reveals the many faint stars that lie within Orion's boundaries. These are members of the constellation, but they do not have Greek letter designations. Note the differences in the colors of the stars. The crosses through the bright stars are caused by an optical effect in the telescope. *(Photo courtesy William Hartmann.)*

b

discuss the sky with precision, we must have an accurate way of referring to the brightness of stars, and for that we must consult one of the first great astronomers.

The Brightness of Stars

Hipparchus, a Greek astronomer who lived about 2100 years ago (Figure 2-5), divided the stars into six classes. The brightest were first-class stars, and those slightly fainter were second-class stars. Continuing down to the faintest stars he could see, the sixth-class stars, he recorded his classifications in a great star catalogue that became a basic reference in ancient astronomy. His system, slightly modified, is still in use today.

In spite of its value, Hipparchus's system may seem a bit confusing. First, when early astronomers translated the catalogue into Latin, they used the word *magnitudo,* meaning "size." In English, this became *magnitude,* even though it refers to the brightness of the stars and not to their size. Thus the **magnitude scale** is the astronomer's brightness scale.

The second source of confusion is that the fainter the star, the larger the magnitude number. For example, 6th-magnitude stars are fainter than 1st-magnitude stars. This may seem backwards at first, but think of it as Hipparchus did. The brightest stars are first-class stars, the fainter stars are second- and third-class, and so on.

Modern astronomers have made a major improvement in Hipparchus's magnitude system by measuring stellar brightness with sensitive instruments and developing a numerical scale that gives finer gradations of the six magnitudes. For example, instead of merely saying that θ (theta) Leonis is a 3rd-magnitude star, they can say specifically that its magnitude is 3.34.

If we measure the brightness of all of the stars in Hipparchus's first brightness class, we discover that

Figure 2-5
Hipparchus (2nd century B.C.) was the first great observational astronomer. Among other things, he constructed a catalogue listing 1080 of the brightest stars. He is honored here on a Greek stamp that also shows one of his observing instruments.

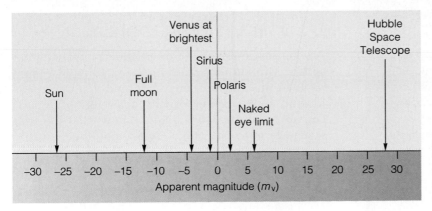

Figure 2-6
The scale of apparent visual magnitudes extends into negative numbers to represent the brightest objects and to positive numbers larger than 6 to represent objects fainter than the human eye can see.

some are brighter than 1.0. For example, Vega (α Lyrae) is so bright that its magnitude, 0.04, is almost zero. A few are so bright the magnitude scale must extend into negative numbers (Figure 2-6). On this scale, Sirius, the brightest star in the sky, has a magnitude of −1.47. If we use a telescope to search for very faint stars, we can find stars much fainter than the limit for the unaided eye. Thus the magnitude system has been extended to numbers larger than 6th magnitude to include fainter stars.

These magnitudes are known as **apparent visual magnitudes.** They refer to how bright the stars look and do not compensate for their distance from Earth. A star that emits a million times more light than the sun might appear very faint if it were very far away, and a star that is much fainter than the sun might look bright if it were nearby. In Chapter 8, we will develop a magnitude scale that takes distance into account and tells us how bright the stars really are. Apparent visual magnitude only tells us how bright they *appear*.

Hipparchus used apparent visual magnitudes to describe how bright a star looks, but brightness is subjective. How bright a star looks depends on such things as the physiology of the eye and the psychology of perception. To be accurate, we should use the more precise term *intensity*—a measure of the light energy from a star that hits 1 square meter in 1 second. If we compare the intensity of the light coming from two stars, then we can be precise in describing their relative brightness.

Nevertheless, nearly all star catalogs from today back to the time of Hipparchus use magnitudes, so astronomers need a way to calculate intensity from magnitude and vice versa. By the Numbers 2-1 (on page 14) shows how we can convert from one to the other.

The nearest star in the Big Dipper is ζ (zeta) Ursa Majoris, with apparent magnitude of 2.40. The most distant is α Ursa Majoris with apparent magnitude 1.79. What does that tell us?

The smaller the magnitude number, the brighter the star, so we can conclude immediately that α looks brighter in the sky than ζ. We can go further with our analysis by noting that the difference in magnitude tells us the intensity ratio. In this case the magnitude difference is $2.4 - 1.79$, which is 0.61 magnitudes, and the ratio of the brightness is $(2.512)^{0.61}$. A calculator tells us that the intensity ratio is 1.75. So light from α is 1.75 times more intense than light from ζ. We might say that α looks 1.75 times brighter than ζ, but our judgment of brightness is subjective and depends on the psychology of vision. We would be right to say α looks about 1.75 times brighter, but for precision, we should refer to the intensity of the light and not to the subjective brightness.

These two numbers can tell us even more. The nearest star, ζ, is one of the fainter stars in the Big Dipper. We might expect nearer stars to be brighter. The most distant star, α, is very nearly the brightest star in the Big Dipper. (ε Ursa Majoris is 0.03 magnitudes brighter than α.) We would expect the most distant star to be the faintest. This must mean that stars don't all emit the same amount of light. Clearly, α must be emitting much more light than ζ to be so bright even though it is farther away. We discuss this in more detail in Chapter 9.

The sun is very close to us and so looks very bright. Compare the intensity of sunlight (apparent magnitude −26.5) with the intensity of light from the brightest star, Sirius (apparent magnitude −1.42).

We have found a way to identify stars by name and by brightness. Next we must look at the sky as a whole and notice its motion.

2-2 The Celestial Sphere

Ancient astronomers thought the sky was a great sphere with the stars stuck on the inside like thumbtacks in the ceiling. They imagined that Earth was located at the center of this **celestial sphere.** Of course, modern astronomers know the celestial sphere is not real. The stars are not attached to a sphere but are scattered at different distances through space. Nevertheless, the celestial sphere is still a useful model to help us think about the motions of the sky (see Window on Science 2-1).

A Model of the Sky

The sky we see above us at any moment is the half of the celestial sphere above the **horizon,** the circular boundary between Earth and the sky. The other half of the sphere is hidden below the horizon. The very top of the sky, the **zenith,** is directly overhead (Figure 2-7).

If we watch the night sky for a few hours, we can see movement (Figure 2-8 on page 16). As the rotation of Earth carries us eastward, stars move across the sky and set in the west as new stars rise in the east.

Figure 2-8b shows how stars in the northern sky appear to revolve around a point called the **north celestial pole,** the point on the sky directly above Earth's north pole. The star Polaris happens to lie very near the north celestial pole and thus hardly moves as Earth rotates. At any time of night, in any season of the year, Polaris stands above our northern horizon and is consequently known as the North Star. (Locate Polaris on the star charts at the end of the book.)

Table 2-1 Magnitude and Intensity

Magnitude Difference	Intensity Ratio
0	1
1	2.5
2	6.3
3	16
4	40
5	100
6	250
7	630
8	1,600
9	4,000
10	10,000
⋮	⋮
15	1,000,000
20	100,000,000
25	10,000,000,000
⋮	⋮

In everyday language, we use the word *model* in various ways—fashion model, model airplane, model student—but scientists use the word in a very specific way. A **scientific model** is a carefully devised mental conception of how something works, a framework that helps scientists think about some aspect of nature. For example, astronomers use the celestial sphere as a way to think about the motions of the sky, sun, moon, and stars.

Although we will think of a scientific model as a mental conception, it can take many forms. Some models are quite abstract—the psychologist's model of how the human mind processes visual information into images, for instance. But other models are so specific that they can be expressed as a set of mathematical equations. For example, an astronomer might use a set of equations to describe in detail how gas falls into a black hole. We could refer to such a calculation as a model. Of course, we could use metal and plastic to build a celestial globe, but the thing we build wouldn't really be the model any more than the equations are a model. A scientific model is a mental conception, an idea in our heads, that helps us think about nature.

On the other hand, a model is not meant to be a statement of truth. The celestial sphere is not real; we know the stars are scattered through space at various distances, but we can imagine a celestial sphere and use it to help us think about the sky. A scientific model does not have to be true to be useful. Chemists, for example, think about the atoms in molecules by visualizing them as balls joined together by bonds. This model of a molecule is not fully correct, but it is a helpful way to think about molecules; it gives chemists a framework within which to organize their ideas.

Because scientific models are not meant to be totally correct, we must always remember the assumptions on which they are based. If we begin to think a model is true, it can mislead us instead of helping us. The celestial sphere, for instance, can help us think about the sky, but we must remember that it is only a mental crutch. The universe is much larger and much more interesting than this ancient scientific model of the heavens.

The **south celestial pole** is the corresponding point directly above Earth's south pole. No easily visible star marks the south celestial pole. The south celestial pole lies below the southern horizon as seen from the United States. We can see stars in our southern sky following curved paths westward as they circle the south celestial pole (Figure 2-8d).

Earth's rotation defines the celestial poles and the directions around the horizon. The **north point** on the horizon is the point directly below the north celestial pole, and the **south point** is directly opposite the north point. The **east point** and the **west point** are halfway between the north and south points. Thus the directions we use every day are defined astronomically.

The **celestial equator** is an imaginary line directly above Earth's equator. It touches the horizon at the east point and the west point and divides the celestial sphere into a two equal hemispheres (Figure 2-9).

Angles on the Sky

One way we might use the celestial sphere is to describe the location of objects. Just as we might tell a friend to meet us 10 kilometers north of Flagstaff, Arizona, we might want to tell our friend to look for an interesting star a certain distance north of the bright star Sirius. However, we can't measure distances in the sky in kilometers. Rather, we need to measure angles on the celestial sphere.

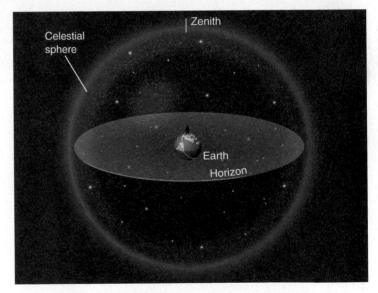

Figure 2-7
The horizon is the dividing line between the half of the celestial sphere we can see and the half we cannot see. The zenith is the point on the sky directly above our heads. Notice that people in different parts of the world have different horizons and zeniths.

Astronomers often use angles to describe distance across the sky. They might say, for instance, that the **angular distance** between the moon and the bright star Spica is 8°, meaning that if we point one arm at the moon and the other arm at Spica, the angle between our arms is 8°. We know the star is hundreds of light-years away, and we know that the moon is much

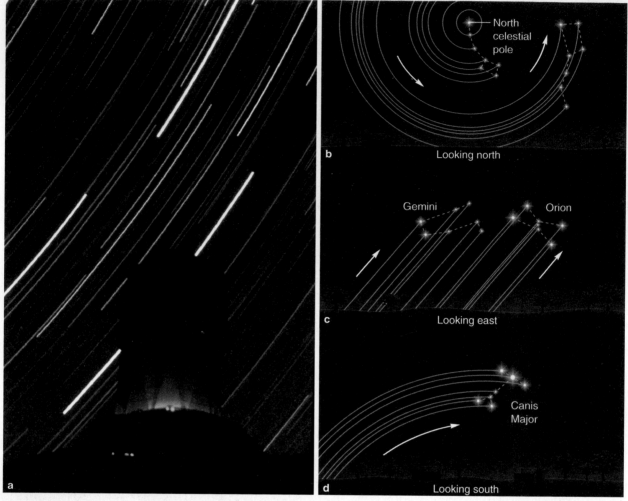

Figure 2-8

Motion in the sky. (a) Stars rising in the northeast make streaks in this time exposure photograph taken from a road near an observatory dome. *(NOAO.)* (b) Looking north, we would see stars circling the north celestial pole. (c) In the eastern sky, stars rise along straight lines, and in the southern sky, the stars follow arcs as they circle the south celestial pole hidden below the southern horizon.

closer, so the true distance between them is immense. But if we imagine them painted on the celestial sphere, we can think of their separation as an "angle *on* the sky." Thus we can discuss the angular separation of two objects even when we don't know their true distance from each other.

The **angular diameter** of an object is the angular distance across the object. For example, if we point at the left edge of the full moon with one arm and the right edge with the other arm, the angle between our arms is the angular diameter of the moon, about 0.5°. The sun also has an angular diameter of about 0.5°.

Being able to measure angles on the sky is a great help as we discuss the appearance of the sky as seen from different locations on Earth.

Latitude and the Sky

Our discussion so far has assumed we live at a latitude representative of North America, that is, about 40° north of Earth's equator. If we live at a different latitude, then our zenith and horizon will be quite different. Notice in Figure 2-7 how someone in southern Africa would have a different zenith and horizon.

What we see in the sky depends on our latitude. If we live in Australia, the south celestial pole will be above our horizon and the north celestial pole will be hidden below the northern horizon. To be precise, the angular distance from the horizon to the north celestial pole equals the latitude of the observer.

Figure 2-10 illustrates the effect of latitude on what we see in the sky by taking us on an imaginary journey southward from Earth's north pole. We begin in the ice and snow at Earth's north pole where our latitude is 90° N and the north celestial pole is directly overhead, 90° from the horizon. As we walk southward, our latitude decreases, and the north celestial pole sinks closer to the northern horizon. As we cross Earth's equator, as shown in Figure 2-10d, our latitude is 0°, and the north celestial pole lies on our northern horizon. If we continue our journey

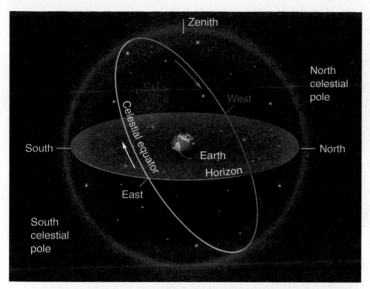

Figure 2-9
The celestial equator circles the celestial sphere directly above Earth's equator and touches the horizon at the east point and the west point (yellow). Earth's rotation carries us eastward and makes stars rise in the east and set in the west. Earth's axis of rotation (red) extended out to the celestial sphere defines the celestial poles (red), the pivots around which the sky appears to rotate.

south of Earth's equator, our latitude becomes negative, and the north celestial pole sinks below the northern horizon. That is, its angular distance from the horizon becomes negative.

This relationship between latitude and the distance from the horizon to the north celestial pole is a great aid to sailors. So long as we can see Polaris, the North Star, we can find our latitude by measuring its distance above the northern horizon. (For precision, we would want to use navigation tables to correct for the fact that Polaris is almost a degree from the north celestial pole.) Of course, if we were voyaging through southern oceans, we could never see Polaris, and then we would have to observe the south celestial pole, which is more difficult because of the lack of a bright star as a marker.

Our latitude also determines what constellations we can see. From northern latitudes, constellations near the north celestial pole never set below the horizon. These **north circumpolar constellations,** such as Ursa Minor and Cassiopeia, are always above the horizon (Figure 2-11). Constellations near the south celestial pole never rise above the horizon for an observer in Earth's northern hemisphere. These **south circumpolar constellations** are not visible from the United States and similar latitudes. How many constellations are circumpolar depends on your latitude; if you live on Earth's equator, you have no circumpolar constellations at all.

Many of the technical terms that describe the celestial sphere are not common in out daily lives, but we do not learn them here for their own sake. Merely

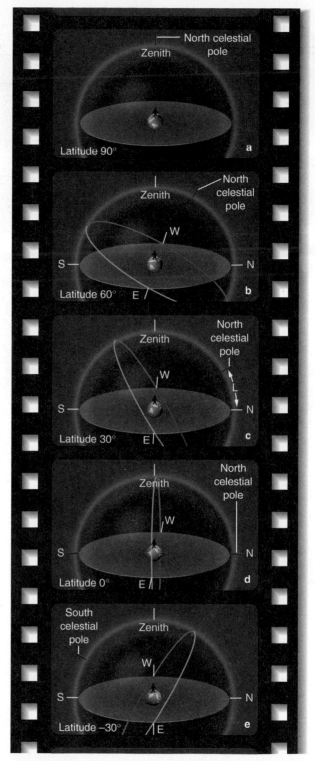

Figure 2-10
A journey beginning at Earth's north pole (a) shows how our latitude L determines what we see in the sky. As we travel south, our latitude decreases, and the north celestial pole moves down toward the northern horizon. The angular distance from the northern horizon up to the north celestial pole always equals our latitude. When we cross Earth's equator, our latitude becomes negative, the north celestial pole sinks below the horizon, and the south celestial pole rises above our southern horizon.

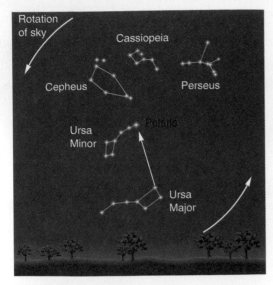

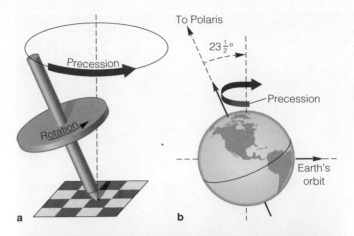

Figure 2-11
The brighter north circumpolar constellations visible from the latitude of the United States never set. The pointer stars at the front of the cup of the Big Dipper always point toward Polaris.

naming things is not the same as understanding (Window on Science 2-2). Our goal in science is to understand nature, so we can now use this powerful vocabulary of the sky to analyze the motions of Earth and the heavenly bodies.

Precession

Over 2000 years ago, Hipparchus compared a few of his star positions with those made nearly two centuries before and realized that the celestial poles and equator were slowly moving across the sky. Later astronomers understood that this motion is caused by the toplike motion of Earth.

If you have ever played with a gyroscope or top, you have seen how the spinning mass resists any change in the direction of its axis of rotation. The more massive the top and the more rapidly it spins, the more difficult it is to change the direction of its axis of rotation. But you probably recall that the axis of even the most rapidly spinning top sweeps around in a conical motion. The weight of the top tends to make it tip, and this combines with its rapid rotation to make its axis sweep around in a conical motion called **precession** (Figure 2-12a).

Earth spins like a giant top, but it does not spin upright in its orbit; it is tipped 23.5° from vertical. Earth's large mass and rapid rotation keep its axis of rotation pointed toward a spot, the star Polaris, and the axis would not wander if Earth were a perfect sphere. However, Earth, because of its rotation, has a slight bulge around its middle, and the gravity of the sun and moon pulls on this bulge, tending to twist Earth upright in its orbit. The combination of these

Figure 2-12
Precession. (a) A spinning top precesses in a conical motion around the perpendicular to the floor because its weight tends to make it fall over. (b) Earth precesses around the perpendicular to its orbit because the gravity of the sun and moon tend to twist it upright. (c) Precession causes the north celestial pole to drift among the stars, completing a circle in 26,000 years.

forces and Earth's rotation causes the Earth's axis to precess in a conical motion, taking about 26,000 years for one cycle (Figure 2-12b).

Because the celestial poles and equator are defined by Earth's rotational axis, precession moves these reference marks. We notice no change at all from night to night or year to year, but precise measurements reveal the precessional motion of the celestial poles and equator.

Over centuries, precession has dramatic effects. Egyption records show that 4800 years ago the north celestial pole was near the star Thuban (α Draconis). The pole is now approaching Polaris and will be closest to it in about 2100. In about 12,000 years, the pole

One of the fascinations of science is that it can reveal things we normally do not sense, such as the precessional motion of Earth's axis. Science can even take our imagination into realms beyond our experience, from the inside of an atom to the inside of a star. Because these experiences are so unfamiliar, we need a vocabulary of technical terms just to talk about them, and that leads to a common confusion between naming things and understanding things.

The first step in understanding something is naming the parts, but it is only the beginning. For example, we need to know terms such as *celestial equator* and *celestial pole* in order to discuss the apparent motion of the sky. That can lead us to understand precession, something we do not sense in our daily lives.

We must know the words, but real understanding comes from being able to use the vocabulary to discuss the way nature works. We must be able to tell the stories that scientists call theories and hypotheses and use the technical words as needed to cite the evidence that makes us think those stories are true. That goes far beyond merely memorizing the terminology; that is real understanding.

In folklore, naming a thing gives us power over it—recall the story of Rumpelstiltskin. In science, naming things is only the beginning. Our goal lies in understanding, not just naming.

will have moved to within 5° of Vega (α Lyrae). Figure 2-12c shows the path followed by the north celestial pole.

REVIEW Critical Inquiry

Does everyone see the same circumpolar constellations?

Here we can use the celestial sphere as a convenient model of the sky (Figure 2-9). A circumpolar constellation is one that does not set or rise. Which constellations are circumpolar depends on our latitude. If we live on Earth's equator, we see all the constellations rising and setting, and there are no circumpolar constellations at all. If we live at Earth's north pole, all the constellations north of the celestial equator never set, and all the constellations south of the celestial equator never rise. In that case, every constellation is circumpolar. At intermediate latitudes, the circumpolar regions are caps whose angular radius equals the latitude of the observer. If we live in Iceland, the caps are very large, and if we live in Egypt, near the equator, the caps are much smaller.

Locate Ursa Major and Orion on the star charts at the end of this book. For people in Canada, Ursa Major is circumpolar, but people in Mexico see most of this constellation slip below the horizon. From much of the United States, some of the stars of Ursa Major set, and some do not. In contrast, Orion rises and sets as seen from nearly everywhere on Earth. Explorers at Earth's poles, however, never see Orion rise or set. How would you improve the definition of a circumpolar constellation to clarify the status of Ursa Major? Would your definition help in the case of Orion?

We now have a model of the sky—the celestial sphere—complete with reference marks and angular measurements to guide us. We can locate constellations, name the stars, and express their brightness numerically. We are now ready to discuss the principal motions of Earth, its moon, and the sun—the subject of the next chapter.

Summary

Astronomers divide the sky into 88 areas called constellations. Although the constellations originated in Greek and Middle Eastern mythology, the names are Latin. Even the modern constellations, added to fill in the spaces between the ancient figures, have Latin names. The names of stars usually come from ancient Arabic, though modern astronomers often refer to a star by its constellation and a Greek letter assigned according to its brightness within the constellation.

The magnitude system is the astronomer's brightness scale. First-magnitude stars are brighter than 2nd-magnitude stars, which are brighter than 3rd-magnitude stars, and so on. The magnitude we see when we look at a star in the sky is its apparent visual magnitude, which does not take into account its distance from us.

The celestial sphere is a model of the sky, carrying the celestial objects around Earth. Because Earth rotates eastward, the celestial sphere appears to rotate westward on its axis. The northern and southern celestial poles are the pivots on which the sky appears to rotate. The celestial equator, an imaginary line around the sky above Earth's equator, divides the sky in half.

Astronomers often refer to angles "on" the sky as if the stars, sun, moon, and planets were equivalent to spots painted on a plaster ceiling. These angular distances are unrelated to the true distance between the objects in light-years.

The gravitational forces of the moon and sun act on the spinning Earth and cause it to precess like a top. Earth's axis of rotation sweeps around in a conical motion with a period of 26,000 years, and consequently the celestial poles and celestial equator move slowly against the background of the stars.

New Terms

constellation
asterism
magnitude scale
apparent visual magnitude
celestial sphere
horizon
zenith
scientific model
north and south celestial
  poles

north, south, east, and west
  points
celestial equator
angular distance
angular diameter
north and south circumpolar
  constellations
precession

Review Questions

1. Why have astronomers added modern constellations to the sky?

2. What is the difference between an asterism and a constellation? Give some examples. `Goup` `section of sky`

3. What characteristic do stars in a constellation or asterism share?

4. Do people from other cultures on Earth see the same stars, constellations, and asterisms that you see? What about people on planets circling other stars?

5. How does the Greek letter designation of a star give us a clue to its brightness?

6. Give two features of the magnitude scale that might seem confusing. What is the origin of these two features?

7. What does the word *apparent* mean in *apparent visual magnitude?*

8. In what ways is the celestial sphere a scientific model?

9. Why do astronomers use the word *on* to describe angles *on* the sky rather than angles *in* the sky?

10. If Earth did not rotate, could we define the celestial poles and celestial equator?

11. Where would you go on Earth if you wanted to be able to see both the north celestial pole and the south celestial pole at the same time?

12. Where would you go on Earth to place a celestial pole at your zenith?

13. Explain how to make a simple astronomical observation that would determine your latitude.

14. Why does the number of circumpolar constellations depend on the latitude of the observer?

15. How could we detect Earth's precession by examining star charts from ancient Egypt?

Discussion Questions

1. All cultures on Earth named constellations. Why do you suppose this was such a common practice?

2. If you were lost at sea, you could find your approximate latitude by measuring the altitude of Polaris. But Polaris isn't exactly at the celestial pole. What else would you need to know to measure your latitude more accurately?

Problems

1. If light from one star is 40 times more intense than light from another star, what is their difference in magnitudes?

2. If two stars differ by 8.6 magnitudes, what is their intensity ratio? $2.512(2nd) \gamma^x 8.6$ $2nd$ $x^2 = 2755$

3. Star A has a magnitude of 2.5; Star B, 5.5; and Star C, 9.5. Which is brightest? Which are visible to the unaided eye? Which pair of stars has an intensity ratio of 16?

4. By what factor is sunlight more intense than moonlight? (*Hint:* See Figure 2-6.)

5. If you are at a latitude of 35 degrees north of Earth's equator, what is the angular distance from the northern horizon up to the north celestial pole? from the southern horizon down to the south celestial pole?

Critical Inquiries for the Web

1. We've all heard of the "North Star" (Polaris, shown in Figures 2-11 and 2-12), but is there a "South Star" for southern hemisphere observers? As Earth precesses on its axis, Polaris will cease to be our pole star and others will eventually hold that title. Search for star maps available online and for information on precession, and determine what bright southern hemisphere stars might be future pole stars for people living south of the equator.

2. Would the stars of a familiar constellation like Pegasus or Orion look similar to our view from Earth if we lived on planets orbiting other stars? Find a table of distance data for the bright stars in a familiar constellation and construct a diagram that visualizes the positions of these stars in space (similar to Figure 2-3).

 Go to the Brooks/Cole Astronomy Resource Center (www.brookscole.com/astronomy) for critical thinking exercises, articles, and additional readings from InfoTrac College Edition, Brooks/Cole's online student library.

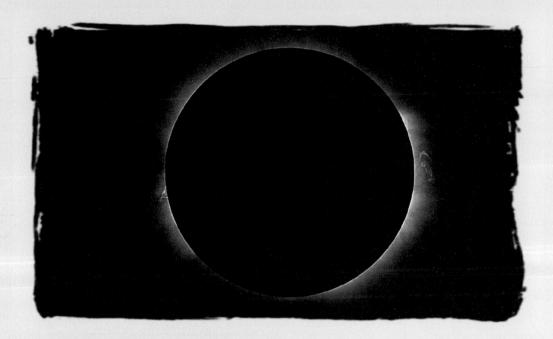

CHAPTER 3

Cycles of the Sky

Even a man who is pure in heart

and says his prayers by night

may become a wolf when the

wolfbane blooms and the moon

shines full and bright.

Proverb from old Wolfman movies

Guidepost

The preceding chapter showed how the daily apparent motion of the sky is caused by the rotation of our planet. Now we are ready to discuss other motions in the sky involving the sun and moon.

This chapter illustrates a powerful way to study certain kinds of problems. If a process includes events that repeat, we can analyze the process by searching for cycles. The motions of the sun and moon become simple and elegant when we see them as cycles.

As we try to understand the appearance of the sun and moon, we discover that what we see is a product of light and shadow. To understand the universe, we must understand light. Later chapters will show that much of astronomy hinges on understanding light.

In the next chapter, we will see how Renaissance astronomers analyzed the motions in the sky and came to a revolutionary conclusion—that we live on a planet.

The Aztecs sacrificed humans and offered their beating hearts to the brilliant object that appeared sometimes in the evening sky and sometimes in the morning sky—the object we know as the planet Venus. The Aztecs saw in the cycle of Venus a metaphor for life, death, and rebirth; and they associated Venus with their principal god, Quetzalcoatl (kwét-zal-ko-áh-tl), who had died, journeyed through the underworld, and returned to life. The Aztecs linked the cycles of their lives on Earth directly to the cycles they saw in the sky.

Even today our lives are linked to sky cycles. The rotation of Earth on its axis makes the sun appear to rise and set, and our bodies follow that cycle of day and night in sleeping and waking. Our moon revolves around Earth in its orbit—passing through a cycle of phases—and we divide our calendar into months of roughly one lunar cycle each. The orbital motion of Earth around the sun causes a year-long cycle of seasons by which we regulate our agricultural, political, and social lives.

The cycles in the sky are the cycles of the sun and moon. In this chapter, we will study those motions and see our Earth as a rotating planet circling its sun. While we may be studying the sky, we will be learning about our world.

3-1 The Cycle of the Sun

Perhaps the most obvious cycles in the sky are those of the sun, but those apparent motions are produced by the rotation and revolution of Earth. **Rotation** is the turning of a body about an axis through its center. Thus we say Earth rotates on its axis once a day. **Revolution** is the motion of a body around a point located outside the body, and we say Earth revolves around the sun once a year.

The cycle of day and night is caused by the rotation of Earth on its axis. In Figure 3-1 we see that four people in different places on Earth have different times, and that the rotation of Earth carries us across the daylight side and then across the night side of Earth, producing day and night.

The annual revolution of Earth around the sun produces a cycle in the sky that we experience as the seasons. To understand that motion, we must imagine that we can make the sun fainter.

The Annual Motion of the Sun

If the sun were fainter, it would not illuminate the atmosphere and prevent us from seeing the stars, and we would notice that day by day the sun was moving eastward against the background of stars. This motion is caused by the motion of Earth as it revolves around the sun. In Figure 3-2, we can see how the sun would

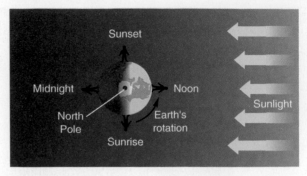

Figure 3-1
Looking down on Earth from above the North Pole shows how the time of day or night depends on our location on Earth.

appear in front of the stars of the constellation Sagittarius on January 1, and how the motion of Earth along its orbit would cause the sun to appear to move eastward against the constellations. By March 1 the sun appears to be in Aquarius.

Of course, it is not quite correct to say that the sun is "in Aquarius." The sun is only 93 million miles away, and the stars of Aquarius are at least a million times farther away. But in March of each year, we would see the sun against the background of the stars in Aquarius, and thus we could say, "The sun is in Aquarius."

If we continued watching the sun against the background of stars throughout the year, we could plot its path on a star chart. After one full year, we would see the sun begin to retrace this line as it continued its annual cycle of motion around the sky. This line, the apparent path of the sun around the sky, is called the **ecliptic.** Another way to define the ecliptic is to say it is the projection of Earth's orbit on the sky. If the sky were a great screen illuminated by the sun at the center, then the shadow cast by Earth's orbit would be the ecliptic. Yet a third way to define the ecliptic is to refer to it as the plane of Earth's orbit. These three definitions of the ecliptic are equivalent, and it is worth considering them all because the ecliptic is one of the most important reference lines on the sky. We will use it, for instance, to discuss the seasons.

The eastward motion of the sun along the ecliptic is a consequence of Earth's moving around its orbit. Earth completely circles the sun in 365.25 days, and consequently the sun circles the sky along the ecliptic in the same number of days. This means the sun, traveling 360° around the ecliptic in 365.25 days, travels about 1° eastward each day. The sun is about 0.5° in diameter, so it travels twice its own diameter per day.

The sun's annual motion around the sky is complicated by the fact that Earth rotates about an axis that is tipped 23.5° from the perpendicular to its orbit. Like a spinning top, it holds its axis fixed in space as it revolves around the sun (Figure 3-3). We saw in a previous chapter that Earth's axis always points toward a spot on the sky very near Polaris, the North Star. Although precession causes Earth's axis to drift, the

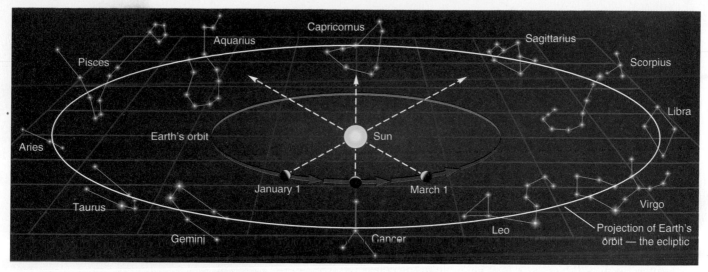

Figure 3-2
The motion of Earth around the sun makes the sun appear to move against the background of the stars.
The circular orbit of Earth is thus projected on the sky as the circular path of the sun, the ecliptic.

change is very slow and does not alter the angle between the axis of rotation and the perpendicular.

Because Earth is tipped 23.5° in its orbit, the ecliptic is tipped 23.5° from the celestial equator. Recall that the celestial equator is the projection of Earth's equator and that the ecliptic is the projection of Earth's orbit. Because Earth is tipped 23.5°, its equator is tipped 23.5° from the plane of its orbit. When we project these two lines onto the sky, we find that the ecliptic and celestial equator meet at an angle of 23.5°. (Find the ecliptic on the star charts at the end of this book.)

Adding the ecliptic to the celestial sphere gives us four more reference marks on the sky. As we see in Figure 3-4, the ecliptic and celestial equator cross at two places called equinoxes. (Notice that they cross at an angle of 23.5°.) The **vernal equinox** is the place where the sun crosses the celestial equator moving northward, and the **autumnal equinox** is the place where it crosses moving southward. The sun crosses the vernal equinox on or about

March 21, and it crosses the autumnal equinox on or about September 22. The exact dates of the equinoxes can vary by a day or two because of leap year and other factors. Figure 3-4 can help us identify two other reference marks on the ecliptic, the two points where the sun is farthest from the celestial equator. About June 22, the sun is farthest north at the point called the **summer solstice.** The **winter solstice** is the point where the sun is farthest south, about December 22.

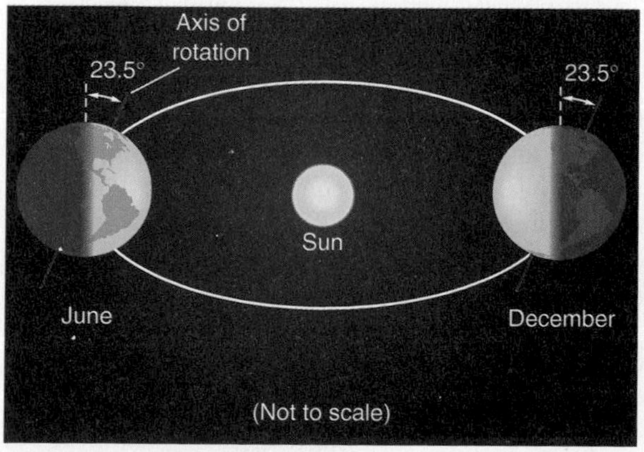

Figure 3-4
The ecliptic crosses the celestial equator at the equinoxes and reaches its most northerly point at the summer solstice and its most southerly point at the winter solstice.

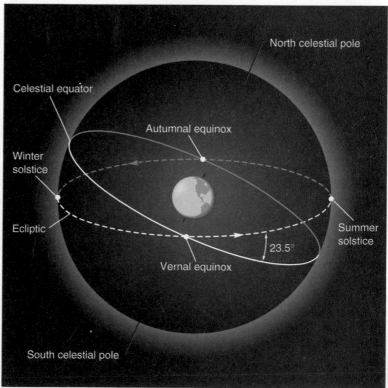

Figure 3-3
Earth, spinning like a top, holds its axis fixed as it orbits the sun. Earth's northern hemisphere is tipped toward the sun in June and away in December. The dotted line is perpendicular to Earth's orbit.

Note that the equinoxes and solstices are points on the sky, but the same words refer to the times when the sun crosses those points. You might hear someone say, "This year the vernal equinox occurs at 3:02 A.M. on March 21." Whether we think of them as places or times, the equinoxes and solstices are important because they mark the beginning of each of the seasons.

The Seasons

The seasonal temperature depends on the amount of heat we receive from the sun. The motion of the sun around the ecliptic tips the balance between the amount of heat a particular place on Earth gains and the amount that is radiated to space. The balance is tipped one way in summer and the opposite way in winter. Because the ecliptic is at an angle to the celestial equator (Figure 3-4), the sun spends half the year in the northern celestial hemisphere and half the year in the southern celestial hemisphere. When the sun is in the northern sky, Earth's northern half receives more direct sunlight—and therefore more heat—than the southern half (Figure 3-3). This makes North America, Europe, and Asia warmer.

The seasons are reversed in Earth's southern half. While the sun is in the northern celestial hemisphere warming North America, South America becomes cooler. Chile has warm weather on New Year's Day and cold in July.

To see how the sun can give us more heat in summer, think about the path the sun takes across the sky between sunrise and sunset. Figure 3-5 shows these paths when the sun is at the summer solstice and at the winter solstice as seen by a person living at latitude 40°, a good average latitude for the United States. Notice that at the summer solstice, the sun rises in the northeast, moves high across the sky, and sets in the northwest (Figure 3-5a). But at the winter solstice, the sun rises in the southeast, moves low across the sky, and sets in the southwest (Figure 3-5b). Two features of these paths tip the heat balance.

First, the summer sun is above the horizon for more hours of each day than the winter sun. Summer days are long, and winter days are short. Because the sun is above our horizon longer in summer, we receive more energy each day.

Second, the sun stands high in the sky at noon on a summer day. It shines almost straight down, as shown by our small shadows. On a winter day, however, the noon sun is low in the southern sky. Each square meter of the ground gains little heat from the winter sun because the sunlight strikes the ground at an angle and spreads out, as shown by our longer shadows (Figures 3-5 and 3-6). These two effects work together to tip the heat balance and produce the seasons.

We mark the beginning of the seasons by the position of the sun. Spring begins at the moment the sun

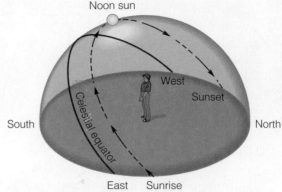

a At summer solstice

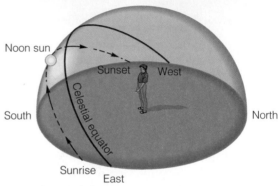

b At winter solstice

Figure 3-5
The daily paths of the sun at latitude 40°. Notice the person's shadows. (a) The summer sun is above the horizon for more than 12 hours and stands high in the sky at noon. (b) The winter sun is above the horizon for less than 12 hours and never stands high in the sky, even at noon.

crosses the celestial equator going north (the vernal equinox). Summer begins at the moment the sun reaches its most northerly point (the summer solstice), and autumn begins when the sun crosses the celestial equator going south (the autumnal equinox). We mark the official beginning of winter when the sun reaches its most southerly position (the winter solstice).

Of course, the weather does not turn warm the instant spring begins. The ground, air, and oceans are still cool from winter, and they take a while to warm up. Likewise, in the autumn, the ground, air, and oceans slowly release the heat stored through the summer. Due to this thermal lag, the average daily temperatures lag behind the solstices by about 1 month. Although the sun crosses the summer solstice on about June 22, the hottest months at northern latitudes are July and August. The coldest months are January and February, even though the sun passes the winter solstice earlier, about December 22.

The seasons are not related to variations in Earth–sun distance. The average distance from Earth to the sun is one astronomical unit (1 AU), 1.5×10^8 km. Earth's orbit is slightly elliptical, however, so its

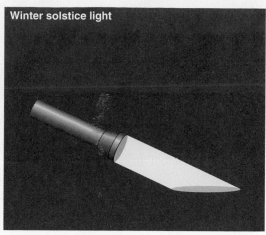

Summer solstice light

Winter solstice light

a

b

Figure 3-6

The spreading of light. (a) At noon on the day of the summer solstice, the sun shines form nearly overhead at the average latitude of the United States. Like the light from a flashlight shining nearly straight downward, the summer sunlight is not spread out very much. (b) On the day of the winter solstice, sunlight strikes the ground at a steep angle and spreads out. Thus, the ground receives less energy per square meter from the winter sun than from the summer sun.

distance varies. About January 4, Earth reaches **perihelion,** its closest point to the sun; it reaches **aphelion,** its farthest point, about July 4. The total variation is only about 3 percent, and it has only a tiny influence on the seasons.

The ecliptic is important to our daily lives because of its connection with the seasons, but it may also be familiar in a different guise. The ecliptic marks the center line of the zodiac, a band 18° wide that encircles the sky. The signs of the zodiac take their names from the 12 principal constellations along the ecliptic. Astrology was once an important part of astronomy, but the two are now almost exact opposites — astronomy is a science that depends on evidence, and astrology is a superstition that depends on faith (Window on Science 3-1). Thus the signs of the zodiac are no longer important in astronomy. The zodiac itself is of interest only because it is on the path followed by the planets as they move around the sky.

The Motion of the Planets

The planets of our solar system produce no visible light of their own; we see them by reflected sunlight. Mercury, Venus, Mars, Jupiter, and Saturn are all easily visible to the naked eye; but Uranus is usually too faint to be seen, and Neptune is never bright enough. Pluto is even fainter, and we need a large telescope to find it.

All the planets of our solar system move in nearly circular orbits around the sun. If we were looking down on the solar system from the north celestial pole, we would see the planets moving in the same counterclockwise direction around their orbits with

the planets farthest from the sun moving the slowest.

When we look for planets in the sky, we always find them near the ecliptic because their orbits lie in nearly the same plane as Earth's orbit. As they orbit the sun, they appear to move eastward along the ecliptic.* Mars moves completely around the ecliptic in slightly less than 2 years, but Saturn, being farther from the sun, takes nearly 30 years.

As seen from Earth, Venus and Mercury can never be seen far from the sun because their orbits are inside Earth's orbit. They appear at times above the western horizon just after sunset or above the eastern horizon just before sunrise. Venus is easier to locate because its larger orbit carries it higher above the horizon than does Mercury's (Figure 3-7). Mercury's orbit is so small that it can never get farther than 27°50' from the

*We will discuss occasional exceptions to this eastward motion in Chapter 4.

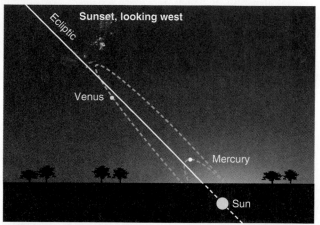

Sunset, looking west

Ecliptic

Venus

Mercury

Sun

a

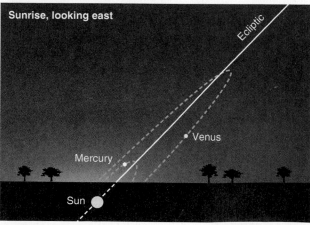

Sunrise, looking east

Ecliptic

Venus

Mercury

Sun

b

Figure 3-7

Mercury and Venus are sometimes visible in the western sky just after sunset (a) or in the eastern sky just before sunrise (b).

Astronomers have a low opinion of astrology not so much because it is groundless but because it pretends to be a science. It is a pseudoscience, from the prefix *pseudo*, meaning "false." There are many examples of pseudoscience, and it is illuminating to consider the difference between pseudoscience and science.

A pseudoscience is a set of beliefs that appear to be based on scientific ideas but that fail to obey the most basic rules of science. For example, some years ago a fad arose in which people placed objects under pyramids made of paper, plastic, wire, and so on. The claim was that the pyramidal shape would focus cosmic forces on anything inside and so preserve fruit, sharpen razor blades, and do other miraculous things. Many books promoted this idea, but simple experiments showed that any shape would protect a piece of fruit from airborne spores and allow it to dry without rotting. Likewise, any shape would allow oxidation to improve the cutting edge of a razor blade. In short, experimental evidence contradicted the claim. Nevertheless, supporters of the theory declined to abandon or revise their claims. Thus, the fad of pyramid power was a pseudoscience.

One characteristic of a pseudoscience is that it appeals to our needs and desires. Thus, some pseudoscientific claims are self-fulfilling. For example, some people bought pyramidal tents to put over their beds and thus improve their rest. While there is no logical mechanism by which such a tent could affect a sleeper, people slept more soundly because they wanted and expected the claim to be true. Many pseudoscientific claims involve medical cures, ranging from copper bracelets and crystals to focus spiritual power to astonishingly expensive and illegal treatments for cancer. Logic is a stranger to pseudoscience, but human fears and needs are not.

Astrology is a pseudoscience. Over the centuries, astrology has been tested repeatedly, and no correlation between the movement of the heavens and human lives has been found. But it survives, and its supporters disregard any evidence that it doesn't work. Like all pseudosciences, astrology is not open to revision in the face of contradictory evidence. Furthermore, astrology fulfills our human need to believe that there is order and meaning to our lives. It may comfort us to believe that our sweetheart has rejected us because of the motions of the sun, moon, and planets along the zodiac rather than to admit that we behaved badly on our last date. Comfort aside, astrology is a poor basis for life decisions.

Human nature and human needs probably ensure that pseudoscientific beliefs will continue to plague us like emotional viruses propagating from person to person. If we recognize them for what they are, we can more easily guide our lives by rational principles and not assign the stars blame for our failures and credit for our successes.

---

sun. Consequently, it is hard to see against the sun's glare and is often hidden in the clouds and haze near the horizon.

By tradition, any planet visible in the evening sky is called an **evening star,** even though planets are not stars. Similarly, any planet visible in the sky shortly before sunrise is called a **morning star.** Perhaps the most beautiful is Venus, which can become as bright as −4.7. As Venus moves around its orbit, it can dominate the western sky each evening for many weeks, but eventually its orbit carries it back toward the sun, and it is lost in the haze near the horizon. In a few weeks, it reappears in the dawn sky, a brilliant morning star.

REVIEW Critical Inquiry

How does the elliptical shape of Earth's orbit affect the seasons?

In the critical analysis of an idea it can be helpful to exaggerate the importance of a single factor. Doing so not only reveals the effect of that factor but can also reveal the inner workings of the process itself. In this case, we might try to imagine what life on Earth would be like if Earth's orbit were more elliptical than it really is. Earth now passes perihelion about January 4, and on that date we are slightly closer to the sun than average. If Earth's orbit were more elliptical, we would be significantly closer. This would make winters in the northern hemisphere much warmer than they are now; and the weather in the southern hemisphere, normally warm in January, would be dreadfully hot. Six months later, about July 4, Earth reaches aphelion, its farthest point from the sun. If Earth had a more elliptical orbit, it would be significantly farther from the sun on this date. Then the northern hemisphere would experience a much cooler summer than it now does; and in the southern hemisphere, the cool season would be bitter cold.

Of course, Earth's orbit is nearly circular, so its distance from the sun varies by only about 3 percent from nearest to farthest. That should make the northern seasons a little milder and the southern seasons a little more extreme, an effect that is hardly noticeable. Now use exaggeration to analyze the true cause of the seasons. Try to imagine what the seasons would be like if Earth's axis were inclined more than 23.5° or less than 23.5° from the perpendicular to its orbit.

The cycle of the sun around the ecliptic dominates our lives through the seasons, but there is another moving light in the sky. What the cycles of our moon lack in intensity, they make up in beauty and elegance.

3-2 The Cycles of the Moon

Among the changing aspects of the sky, none are more striking than the cycles of Earth's moon. Here we will study three kinds of lunar cycles, the phases of the moon, tides, and eclipses.

The Phases of the Moon

Our moon orbits eastward around Earth in 27.322 days. This is called the moon's **sidereal period.** The word *sidereal* means "relative to the stars." The moon takes 27.322 days to circle the sky once and return to the same place among the stars.

The moon rotates on its axis as it circles Earth, keeping the same side toward Earth. Thus we always see the same features on the moon's face. A mountain on the near side of the moon remains facing Earth throughout the moon's circuit around Earth.

The moon moves rapidly across the sky, traveling 13°, about 26 times its apparent diameter, in 24 hours. If you compare the position of the moon to the stars in the background, you will see it move a distance slightly greater than its own diameter in an hour. Because the moon's orbit is inclined 5°8′42″ to the plane of Earth's orbit, the moon never travels farther than 5°8′42″ north or south of the ecliptic.

Because the moon, like the planets, does not produce visible light of its own, it is visible only by the sunlight it reflects, and we can see only that portion illuminated by the sun. As the moon moves around the sky, the sun illuminates different amounts of the side of the moon facing Earth, and so the moon passes through a sequence of phases.

The top panel of Figure 3-8 shows the phases of our moon. We can understand these by looking at the location of Earth's moon relative to the sun. For example, the *new moon* in the second panel of Figure 3-8 is located roughly between Earth and the sun, and the side of the moon facing Earth is in darkness. We

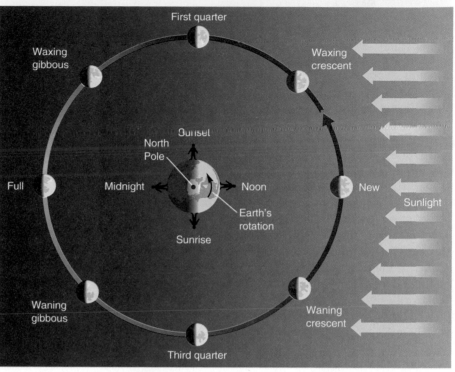

Figure 3-8

The phases of the moon are caused by the varying amounts of sunlight that reach the side of the moon facing Earth. At first quarter, for example, only half of the near side is illuminated. At full moon, all of the near side is illuminated. To locate a specific lunar phase in the sky at a given time of day or night, rotate the diagram until the corresponding human figure is at the top of Earth globe; the upper half of the of the moon's orbit will show the location of the lunar phases above the horizon with east at the left and west at the right. *(Lick Observatory photographs)*

see no moon at new moon. A few days after new moon, it has moved far enough along its orbit to allow the sun to illuminate a small sliver of the side toward us, and we see a thin crescent. Night by night this crescent moon waxes (grows), until, about a week later, we see half of the side toward us illuminated by sunlight and refer to it as *first quarter.* The moon continues to wax, becoming gibbous (from the Latin for humpbacked), and then, when it is nearly opposite the sun, the side toward Earth is fully illuminated, and we see a *full moon.*

The second half of the lunar cycle reverses the first half. After reaching full, the moon wanes (shrinks) through gibbous phase to *third quarter,* then through crescent back to new moon. To distinguish between the gibbous and crescent phases of the first and second half of the cycle, we refer to *crescent waxing* and *gibbous waxing* when the moon is growing, and to *gibbous waning* and *crescent waning* when it is shrinking.

The cycle of lunar phases takes 29.53 days, the synodic period of the moon, or about 4 weeks. Thus new moon, first quarter, full moon, and third quarter occur at roughly 1-week intervals. In general, an object's **synodic period** is its orbital period relative to the sun. (*Synodic* comes from the Greek words meaning "together" and "path.") To see why the moon's synodic period is longer than its sidereal period, imagine we begin observing at new moon—that is, when the moon is near the sun in the sky. After 27.322 days, the moon's sidereal period, it has circled the sky and returned to the same place among the stars where it was last new, but the sun has moved about 27° eastward along the ecliptic. The moon needs slightly more than 2 days to catch up with the sun and reach new moon again. Thus the moon's synodic period is longer than its sidereal period.

You can make a moon-phase dial from Figure 3-8 by covering the lower half of the moon's orbit with a sheet of paper, aligning the edge of the paper to pass through the word "Full" at the left and the word "New" at the right. Push a pin through the edge of the paper at Earth's North Pole to make a pivot, and under the word "Full," write on the paper, "Eastern Horizon." Under the word "New," write "Western Horizon." The paper now represents the horizon.

You can set your moon-phase dial for a given time by rotating the diagram behind the horizon-paper. Set the dial to sunset by turning the diagram until the human figure labeled "Sunset" is standing at the top of Earth globe; the dial shows, for example, that the full moon at sunset would be at the eastern horizon.

We can now follow the cycle of the moon around the sky, and we begin at sunset. If the moon were new, it would be invisible near the sun on the western horizon. A few days after new moon, we would see the waxing crescent moon moving upward in the sunset sky; and a week after new moon, we would find the first quarter moon standing high in the sky at sunset. As the days passed, we would see the moon wax gibbous as it moved eastward, reaching full moon at the eastern horizon about two weeks after new moon.

Lunar phases after full moon are not visible in the sunset sky, so we must shift to the sunrise sky to continue. At sunrise, the full moon would be at the western horizon, and it would wane through gibbous each day until it reached third quarter high in the sunrise sky about 3 weeks after new moon. Each morning we would see the moon farther eastward as it waned through crescent phase until it reached new moon again at the eastern horizon in the sunrise sky. New moon to new moon takes just over 4 weeks.

Almost everyone is familiar with the changing phases of the moon, but those who live near the seashore are probably familiar with another phenomenon related to the lunar cycle—the periodic advance and retreat of the ocean tides.

Tides

The moon's gravity has dramatic effects on Earth. The side of Earth facing the moon is about 4000 miles closer to the moon than the center of Earth is, and the moon's gravity pulls on the near side of Earth more strongly than on Earth's center. Though we think of our planet as solid, it is not perfectly rigid, so the moon's gravity draws the rocky surface of the near side up into a bulge a few centimeters high.

The oceans respond to the force of the moon's gravity by flowing into a bulge of water on the side of Earth facing the moon. There is also a bulge on the side away from the moon, which develops because the moon pulls more strongly on Earth's center than on the far side. Thus the moon pulls Earth away from the oceans, which flow into a bulge on the far side.

We can see dramatic evidence of this effect if we watch the ocean shore for a few hours. Though Earth rotates on its axis, the tidal bulges remain fixed with respect to the moon. As the turning Earth carries us into a tidal bulge, the ocean water deepens, and the tide crawls up the beach. Later, when Earth carries us out of the bulge, the water becomes shallower, and the tide falls. Because there are two bulges on opposite sides of Earth, the tides should rise and fall twice a day on an ideal coast.

In reality, the tide cycle at any given location can be quite complex because of the latitude of the site, shape of the shore, winds, and so on. Tides in the Bay of Fundy (New Brunswick, Canada), for example, occur twice a day and can exceed 40 feet, while the northern coast of the Gulf of Mexico has only one tidal cycle, of roughly 1 foot, each day.

The sun, too, produces tidal bulges on Earth. At new moon and at full moon, the moon and sun produce tidal bulges that add together (Figure 3-9a) and produce extreme tidal changes; high tide is very high, and low tide is very low. Such tides are called **spring tides,** even though they occur at every new and full moon, and not just in the spring. **Neap tides** occur at first- and third-quarter moon, when the moon and sun pull at right angles to each other (Figure 3-9b). Then the tides caused by the sun reduce the tides caused by the moon, and the rise and fall of the ocean is less extreme than usual.

Tidal forces can have surprising effects. The friction of the ocean waters with the seabeds slows Earth's rotation and makes the length of a day grow by 0.0023 second per century. Fossils of marine animals and tidal sediments confirm that only 900 million years ago Earth's day was 18 hours long. In addition, Earth's gravitational field exerts tidal forces on the moon, and, although there are no bodies of water

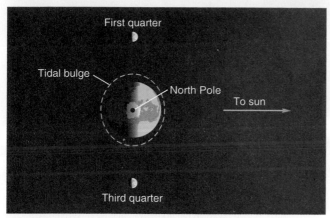

b

Figure 3-9
Looking down on Earth from above the North Pole, we can draw the shape of the tidal bulges (exaggerated here for clarity). (a) When the moon and sun pull along the same line, their tidal forces combine, and the tidal bulges are larger. (b) When the moon and sun pull at right angles, their tidal bulges do not combine, and the tidal bulges are smaller.

on the moon, friction within the flexing rock has slowed the moon's rotation to the point that it now keeps the same face toward Earth.

Tidal forces can also affect orbital motion. Friction with the ocean beds drags the tidal bulges eastward out of a direct Earth–moon line. These tidal bulges contain a large amount of mass, and their gravitational field pulls the moon forward in its orbit (Figure 3-10). As a result, the moon's orbit is growing larger by about 4 cm per year, an effect that astronomers can measure by bouncing laser beams off reflectors left on the lunar surface by the Apollo astronauts.

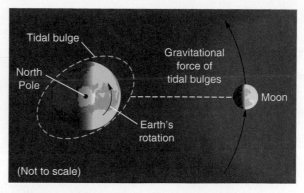

Figure 3-10
The rotation of Earth drags the tidal bulges ahead of the Earth–moon line (exaggerated here). The gravitational attraction of these masses of water does not pull directly toward Earth's center but rather pulls the moon slightly forward in its orbit (green arrow), forcing its orbit to grow larger.

These and other tidal effects are important in many areas of astronomy. In later chapters, we will see how tidal forces can pull gas away from stars, rip galaxies apart, and melt the interiors of satellites orbiting near massive planets. For now, however, we must consider other aspects of the lunar cycle. The stately progression of the lunar phases and the ebb and flow of the ocean tides are commonplace, but occasionally something peculiar happens: The moon darkens and turns copper-red in a lunar eclipse.

Lunar Eclipses

A **lunar eclipse** occurs at full moon when the moon moves through the shadow of Earth. Because the moon shines only by reflected sunlight, we see the moon gradually darken as it enters the shadow.

Earth's shadow consists of two parts (Figure 3-11). The **umbra** is the region of total shadow. If we were in

Figure 3-11
The shadows cast by a map tack resemble those of Earth and the moon. The umbra is the region of total shadow; the penumbra is the region of partial shadow.

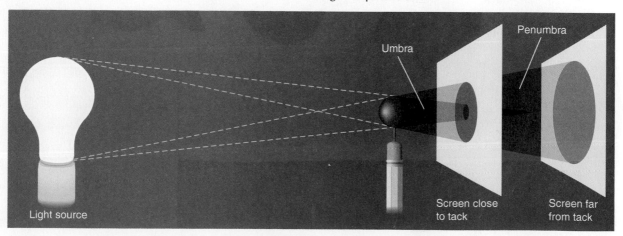

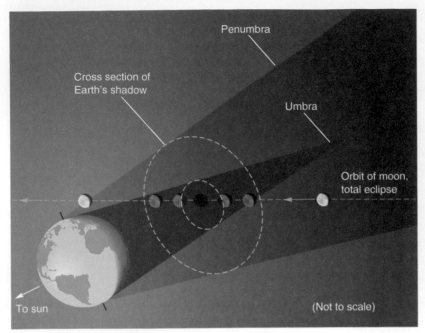

Penumbra

Cross section of
Earth's shadow

Umbra

Orbit of moon,
total eclipse

To sun

(Not to scale)

Figure 3-12
During a total lunar eclipse, the moon's orbit carries it through the
penumbra and completely into the umbra. Compare the cross
section of Earth's shadow with Figure 3-13.

the umbra of Earth's shadow, we would see no portion
of the sun. If we moved into the **penumbra,** however,
we would be in partial shadow and would see part of
the sun peeking around the edge of Earth. Thus in the
penumbra the sunlight is dimmed but not extinguished.

If the orbit of the moon carries it
through the umbra, we see a total lunar
eclipse (Figure 3-12). As we watch the
moon in the sky, it first moves into the
penumbra and dims slightly; the deeper it
moves into the penumbra, the more it
dims. In about an hour, the moon reaches
the umbra, and we see the umbral shadow
darken part of the moon. It takes about an
hour for the moon to enter the umbra completely and become totally eclipsed. Totality, the period of total eclipse, may last as
long as 1 hour 45 minutes, though the timing of the eclipse depends on where the
moon crosses the shadow.

When the moon is totally eclipsed, it
does not disappear completely (Figure 3-
13). Although it receives no direct sunlight, the moon in the umbra does receive
some sunlight refracted (bent) through
Earth's atmosphere. If we were on the
moon during totality, we would not see any part of the
sun because it would be entirely hidden behind Earth.
However, we would be able to see Earth's atmosphere
illuminated from behind by the sun. The red glow
from this "sunset" illuminates the moon during totality and makes it glow coppery red.

If the moon does not move completely into the
umbra, we see a partial lunar eclipse. The part of the
moon that remains outside the umbra receives some
direct sunlight, and the glare usually prevents our

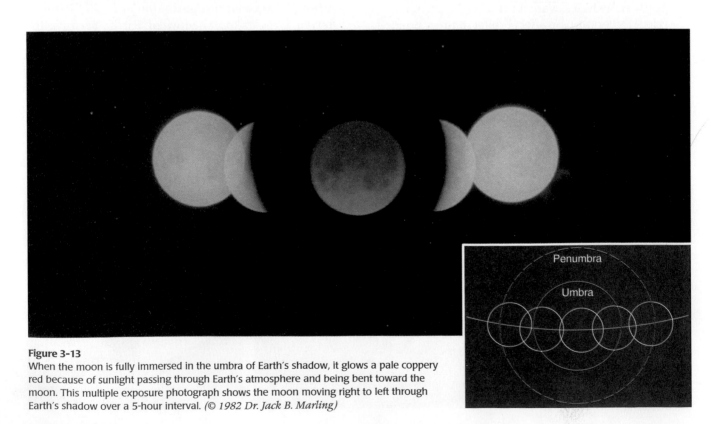

Penumbra

Umbra

Figure 3-13
When the moon is fully immersed in the umbra of Earth's shadow, it glows a pale coppery
red because of sunlight passing through Earth's atmosphere and being bent toward the
moon. This multiple exposure photograph shows the moon moving right to left through
Earth's shadow over a 5-hour interval. *(© 1982 Dr. Jack B. Marling)*

seeing the faint coppery glow of the part of the moon in the umbra.

A penumbral lunar eclipse occurs when the moon passes through the penumbra but misses the umbra entirely. Because the penumbra is a region of partial shadow, the moon is only partially dimmed. A penumbral eclipse is not very impressive.

While there are usually no more than one or two lunar eclipses each year, it is not difficult to see one. We need only be on the dark side of Earth when the moon passes through Earth's shadow. That is, the eclipse must occur between sunset and sunrise at our location. Table 3-1 will allow you to determine which upcoming total and partial eclipses will be visible from your location.

Solar Eclipses

We who live on planet Earth can see a phenomenon that is not visible from most planets. It happens that the sun is 400 times larger than our moon and, on the average, 390 times farther away, so the sun and moon have nearly equal angular diameters—0.5°. (See By the Numbers 3-1.) Thus, the moon is just the right size to cover the bright disk of the sun and cause a **solar eclipse.** If the moon covers the entire disk of the sun,

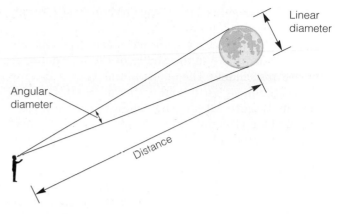

Figure 3-14
The three quantities related by the small-angle formula. Angular diameter is given in seconds of arc in the formula. Distance and linear diameter must be expressed in the same units—both in meters, both in light-years, etc.

Table 3-1 Total and Partial Eclipses of the Moon, 1999–2010

Date	Time* of Mideclipse (GMT)	Length of Totality (Min)	Length of Eclipse (Hr:Min)
1999 July 28	11:34	Partial	2:22
2000 Jan. 21	4:45	76	3:22
2000 July 16	13:57	106	3:56
2001 Jan. 9	20:22	60	3:16
2001 July 5	14:57	Partial	2:38
2003 May 16	3:41	52	3:14
2003 Nov. 9	1:20	22	3:30
2004 May 4	20:32	76	3:22
2004 Oct. 28	3:05	80	3:38
2005 Oct. 17	12:04	Partial	0:56
2006 Sept. 7	18:52	Partial	1:30
2007 Mar. 3	23:22	74	3:40
2007 Aug. 28	10:38	90	3:32
2008 Feb. 21	3:27	50	3:24
2008 Aug. 16	21:11	Partial	3:08
2009 Dec. 31	19:24	Partial	1:00
2010 June 26	11:40	Partial	2:42
2010 Dec. 21	8:18	72	3:28

*Times are Greenwich Mean Time. Subtract 5 hours for Eastern Standard Time, 6 hours for Central Standard Time, 7 hours for Mountain Standard Time, and 8 hours for Pacific Standard Time. From your time zone, lunar eclipses that occur between sunset and sunrise will be visible, and those at midnight will be best placed.

By the Numbers 3-1

The Small-Angle Formula

Figure 3-14 shows the angular diameter of an object, its linear diameter, and its distance. *Linear diameter* is the distance between an object's opposite sides. The linear diameter of the moon, for instance, is 3476 km. The *angular diameter* of an object is the angle formed by two lines extending from opposite sides of the object and meeting at our eye. Clearly, the farther away an object is, the smaller its angular diameter. We can use this formula to find any of these three quantities if we know the other two.

In the small-angle formula, we always express angular diameter in seconds of arc, and we always use the same units for distance and linear diameter:

$$\frac{\text{angular diameter}}{206,265^*} = \frac{\text{linear diameter}}{\text{distance}}$$

Example: The moon has a linear diameter of 3476 km and is about 384,000 km away. What is its angular diameter? *Solution:* We can leave linear diameter and distance in kilometers, and find the angular diameter in seconds of arc:

$$\frac{\text{angular diameter}}{206,265} = \frac{3476 \text{ km}}{384,000 \text{ km}}$$

The angular diameter is 1870 seconds, or 31 minutes, of arc—about 0.5°.

* The number 206,265 is the number of seconds of arc in a radian. When we divide by 206,265, we convert the angle from seconds of arc into radians.

we see a total eclipse. If it covers only part of the sun, we see a partial eclipse.

Whether we see a total or partial eclipse depends on whether we are in the umbra or the penumbra of the moon's shadow. The umbra of the moon's shadow barely reaches Earth and casts a small circular shadow (Figure 3.15a). The shadow is never larger than 270 km (168 miles) in diameter. If we are standing in that umbral spot, we are in total shadow, unable to see any part of the sun's surface, and the eclipse is total (Figure 3.15b). But if we are located outside the umbra, in the penumbra, we see part of the sun peeking around the edge of the moon and the eclipse is partial. Of course, if we are outside the penumbra, we see no eclipse at all.

Because of the orbital motion of the moon and the rotation of Earth, the moon's shadow sweeps rapidly

across Earth in a long, narrow path of totality. If we want to see a total solar eclipse, we must be in the path of totality. When the umbra of the moon's shadow sweeps over us, we see one of the most dramatic sights in the sky, the totally eclipsed sun.

The eclipse begins as the moon slowly crosses in front of the sun. It takes about an hour for the moon to cover the solar disk, but as the last sliver of sun disappears, dark falls in a few seconds. Automatic street lights come on, drivers of cars turn on their headlights, and birds go to roost. The sky becomes so dark we can even see the brighter stars.

The darkness lasts only a few minutes because the umbra is never more than 270 km (168 miles) in diameter and sweeps across Earth's surface at over 1600 km/hr (1000 mph). The sun cannot remain totally eclipsed for more than 7.5 minutes, and the average period of totality lasts only 2 or 3 minutes.

When the moon covers the bright surface of the sun, called the **photosphere,** we can see the bright gases of the **chromosphere** just above the photosphere, and the **corona,** the sun's faint outer atmosphere (Figure 3-16). The corona is a low-density, hot gas that glows with a pale white color. Streamers caused by the solar magnetic field streak the corona as may be seen in Figure 3-16. The chromosphere is often marked by eruptions on the solar surface called **prominences** and is bright pink. The corona, chromosphere, and prominences are visible only when the brilliant photosphere is covered. As soon as part of the photosphere reappears, the fainter corona, chromosphere, and prominences vanish in the glare, and totality is over. The moon moves on in its orbit, and in an hour the sun is completely visible again.

Just as totality begins or ends, a small part of the photosphere can peek out from behind the moon through a valley at the edge of the lunar disk. Although it is intensely bright, such a small part of the photosphere does not completely drown out the fainter corona, which forms a silvery ring of light with the brilliant spot of photosphere gleaming like a diamond (Figure 3-17). This **diamond ring effect** is one of the most spectacular of astronomical sights, but it is not visible during every solar eclipse. Its occurrence depends on the exact orientation and motion of the moon.

Sometimes when the moon crosses in front of the sun, it is too small to fully cover the sun, and we see an **annular eclipse,** a solar eclipse in which a ring (or annulus) of the photosphere is visible around the disk of the moon (Figure 3-18). With a portion of the photosphere visible, the eclipse never becomes total; it never quite gets dark; and we can't see the prominences, chromosphere, and corona. Annular eclipses occur because the moon follows a slightly elliptical orbit around Earth, and thus its angular diameter can vary (Figure 3-19). When it is at **perigee,** its point of

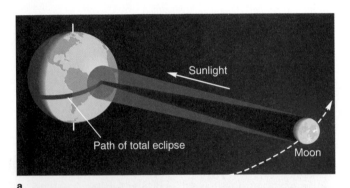

a

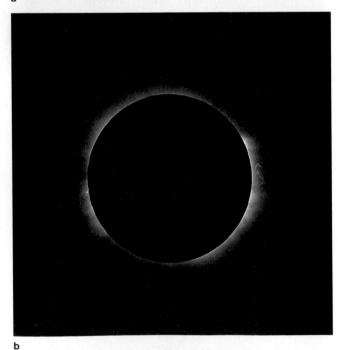

b

Figure 3-15
(a) The umbral shadow of the moon sweeps over a narrow strip of Earth. (b) From a location inside the umbral shadow, we would see the moon cover the bright surface of the sun in a total solar eclipse. *(Daniel Good)*

Figure 3-16
The sequence shows the first half of a total solar eclipse. The brilliant surface of the sun is gradually covered by the moon moving from the upper right. Once all of the bright photosphere is covered, we can see the pink prominences of the chromosphere (d). A much longer exposure is needed to photograph the fainter corona (e). *(Daniel Good)*

Figure 3-17
The diamond ring effect can sometimes occur momentarily at the beginning or end of totality if a small segment of the photosphere peeks out through a valley at the edge of the lunar disk. *(NOAO)*

a

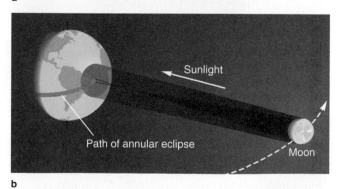

b

Figure 3-18
(a) The annular eclipse of 1994. A bright ring of photosphere remains visible around the moon, and the corona and prominences are not visible. *(Daniel Good)* (b) An annular eclipse occurs when the moon is in the farther part of its orbit and its umbral shadow does not reach Earth. From Earth, we see an annular eclipse because the moon's angular diameter is smaller than the angular diameter of the sun.

closest approach to Earth, it looks significantly larger than when it is at **apogee,** the most distant point in its orbit. Furthermore, Earth's orbit is slightly elliptical, so the Earth–sun distance varies slightly, and thus the diameter of the solar disk varies slightly. If the moon is in the farther part of its orbit during totality, its angular diameter will be less than the angular diameter of the sun, and thus we see an annular eclipse.

A list of future total and annular eclipses of the sun is given in Table 3-2. If you plan to observe a solar eclipse, remember that the sun is bright enough to burn your eyes and cause permanent damage if you look at it directly. This is true whether there is an eclipse or not. Solar eclipses can tempt us to look at the sun in spite of its brilliance and thus can tempt us to risk our eyesight. During totality, the brilliant photosphere is hidden, and it is safe to look at the eclipse, but the partial phases can be dangerous. See Figure 3-20 for a safe way to observe the partially eclipsed sun.

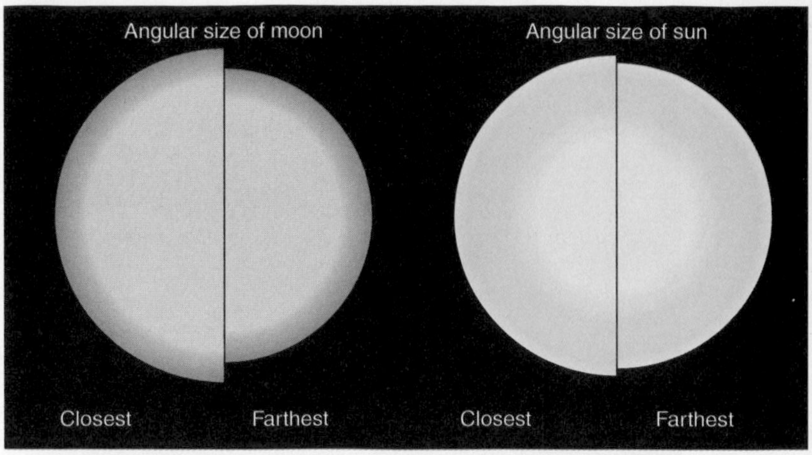

Figure 3-19
The angular diameter of the moon (left) varies dramatically because its orbit is elliptical and its distance varies from perigee (closest) to apogee (farthest). The angular diameter of the sun (right) varies by a smaller amount because Earth's orbit is nearly circular.

Predicting Eclipses

Predicting lunar or solar eclipses seems quite complex, and precise predictions require sophisticated calculations. But we can make general eclipse predictions by thinking about the geometry of an eclipse and the cyclic motions of the sun and moon.

We see a solar eclipse when the moon passes between Earth and the sun, that is, when the lunar phase is new moon. We see a lunar eclipse at full moon. However, we don't see solar eclipses at every new moon or lunar eclipses at every full moon. Why not?

Figure 3-21 is a drawing to scale of the umbral shadows of Earth and the moon. Notice that they are extremely long and narrow. Earth, moon, and sun must line up almost exactly, or the shadows miss their mark and there is no eclipse.

To be eclipsed, the moon must enter Earth's shadow. However, because the moon's orbit is tipped relative to that of Earth, the moon often misses the shadow, passing north or south of it, and there is no lunar eclipse.

To produce a solar eclipse, the moon's shadow must sweep over Earth. The inclination of the

TABLE 3-2 Total and Annular Eclipses of the Sun, 1999–2010

Date	Total/ Annular (T/A)	Time of Mideclipse[*] (GMT)	Maximum Length of Total or Annular Phase (Min:Sec)	Area of Visibility
1999 Feb. 16	A	7^h	1:19	Indian Ocean, Australia
1999 Aug. 11	T	11^h	2:23	Atlantic, Europe, S.E. and S. Asia
2001 June 21	T	12^h	4:56	Atlantic, S. Africa, Madagascar
2001 Dec. 14	A	21^h	3:54	Pacific, Central America
2002 June 10	A	24^h	1:13	Pacific
2002 Dec. 4	T	8^h	2:04	S. Africa, Indian Ocean, Australia
2003 May 31	A	4^h	3:37	Iceland, Arctic
2003 Nov. 23	T	23^h	1:57	Antarctica
2005 Apr. 8	AT	21^h	0:42	Pacific, N. of S. America
2005 Oct. 3	A	11^h	4:32	Atlantic, Spain, Africa
2006 Mar. 29	T	10^h	4:07	Atlantic, Africa, Turkey
2006 Sept. 22	A	12^h	7:09	N.E. of S. America, Atlantic
2008 Feb. 7	A	4^h	2:14	S. Pacific, Antarctica
2008 Aug. 1	T	10^h	2:28	Canada, Arctic, Siberia
2009 Jan. 26	A	8^h	7:56	S. Atlantic, Indian Ocean
2009 July 22	T	3^h	6:40	Asia, Pacific
2010 Jan. 15	A	7^h	11:10	Africa, Indian Ocean
2010 July 11	T	20^h	5:20	Pacific, S. America

The next major total solar eclipse visible from the United States will occur on August 21, 2017.

[*]Times are Greenwich Mean Time. Subtract 5 hours for Eastern Standard Time, 6 hours for Central Standard Time, 7 hours for Mountain Standard Time, and 8 hours for Pacific Standard Time.

^{h}hours.

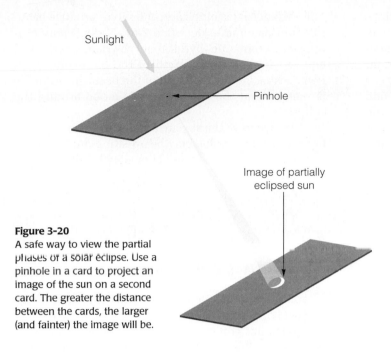

Figure 3-20
A safe way to view the partial phases of a solar eclipse. Use a pinhole in a card to project an image of the sun on a second card. The greater the distance between the cards, the larger (and fainter) the image will be.

moon's orbit, however, is such that often when the moon is new, its shadow passes north or south of Earth, and there is no solar eclipse.

For an eclipse to occur, the moon must reach full or new moon just as it passes through the plane of Earth's orbit; otherwise the shadows miss. The points where it passes through the plane of Earth's orbit are called the **nodes** of the moon's orbit, and the line connecting these is called the line of nodes. In other words, the planes of the two orbits intersect along the line of nodes. Twice a year, this line of nodes points toward the sun, and for a few weeks eclipses are possible at new moons and full moons (Figure 3-22). These intervals when eclipses are possible are called eclipse seasons, and they occur about six months apart.

Figure 3-21
Umbral shadows of Earth and the moon. Notice how easy it is for the shadows to miss their mark at full moon and at new moon and fail to produce eclipses. That is, it is easy for the moon to reach full phase and not enter Earth's shadow. It is also easy for the moon to reach new phase and not cast its shadow on Earth. (The diameters of Earth and the moon are exaggerated by a factor of 2 for clarity.)

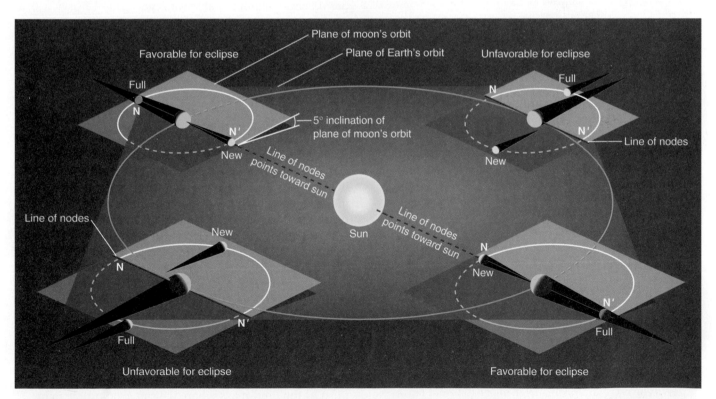

Figure 3-22
The moon's orbit is tipped about 5° to Earth's orbit. The nodes N and N′ are the points where the moon passes through the plane of the Earth's orbit. If the line of nodes does not point at the sun, the shadows miss, and there are no eclipses at new moon and full moon. At those parts of Earth's orbit where the line of nodes points toward the sun, eclipses are possible at new moon and full moon.

If the moon's orbit were fixed in space, the eclipse seasons would always occur at the same time each year. The moon's orbit precesses slowly, however, because of the gravitational pull of the sun on the moon, and the precession slowly changes the direction of the line of nodes. The line turns westward, making one complete rotation in 18.61 years. As a result, the eclipse seasons occur about 3 weeks earlier each year. The motion of the line of nodes, combined with the periodicity of the lunar phases, means that every 6585.3 days the eclipse seasons start over and the same pattern of eclipses repeats (6585.3 days equals 18 years 11.3 days or 18 years 10.3 days depending on the number of leap days during the interval). Because this cycle, termed the **Saros cycle**, contains about one-third of a day more than an integer number of days, an eclipse visible in North America will recur after 18 years 11.3 days about one-third of the way around the world in the eastern Pacific. Many ancient peoples recognized the Saros cycle from their records of previous eclipses and were able to predict when eclipses would occur, even though they did not understand what the sun and moon were or what alignments produced eclipses.

The elegant cycles of the sun and moon produce beautiful phenomena to decorate Earth's sky. Many cultures have romantic myths about the dance of the sun and moon. The Greek myth of the handsome Hippomenes racing for the hand of the beautiful Atalante is really about the sun racing the moon around the ecliptic.

The cycles in the sky are a rich part of our culture, but those same cycles may be affecting our planet in much more dramatic ways. They may be affecting the ice ages.

3-3 Astronomical Influences on Earth's Climate

Weather is what happens today; climate is the average of what happens over decades and centuries. We know that Earth has gone through past episodes, called ice ages, when the worldwide climate was cooler and dryer and thick layers of ice covered northern latitudes. The earliest known ice age occurred about 570 million years ago, and the next about 280 million years ago. The most recent ice age began only about 3 million years ago and is still going on. We are living in one of the periodic episodes when the glaciers melt and Earth grows slightly warmer. The current warm period began about 20,000 years ago.

Ice ages seem to occur with a period of roughly 250 million years, and cycles of glaciation within ice ages occur with a period of about 40,000 years. Many scientists now believe that these cyclic changes have an astronomical origin.

The Hypothesis

Sometimes a theory or hypothesis is proposed long before scientists can find the critical evidence to support it. That situation happened in 1920 when Yugoslavian meteorologist Milutin Milankovitch proposed what became known as the **Milankovitch hypothesis**—that small changes in Earth's orbit, in precession, and in inclination affect Earth's climate and trigger ice ages. We will examine each of these three motions in turn.

First, astronomers know that the shape of Earth's orbit varies slightly over a period of about 100,000 years. At present, Earth's orbit carries it about 2 percent closer than average to the sun during northern hemisphere winters and about 2 percent farther away in northern hemisphere summers. This makes the northern climate warmer, and that is critical—most of the land mass where ice can accumulate is in the northern hemisphere. If Earth's orbit became more elliptical, Milankovitch suggested, northern winters might be warm enough to prevent the accumulation

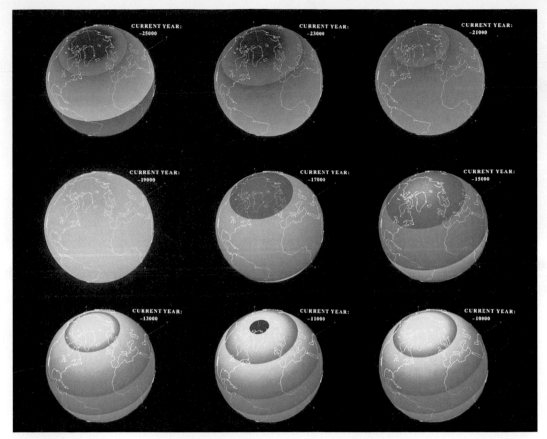

Figure 3-23

Color-coded globes show the importance of the Milankovitch effect. These globes show Earth at the summer solstice from 25,000 years ago (upper left), during the last glaciation, to only 10,000 years ago, after the end of the glaciation. Red, yellow, and green regions receive more solar energy than average; blue and violet regions receive less. Changes in Earth's rotational axis and orbital shape affect the total amount of solar energy received and thus alter the climate. *(Courtesy Arizona State University, Computer Science and Geography Departments)*

of snow and ice from forming glaciers, and Earth's climate would warm.

A second factor is also at work. Precession causes Earth's axis to sweep around a cone with a period of about 26,000 years, and that changes the location of the seasons around Earth's orbit. Northern winters now occur when Earth is 2 percent closer to the sun, but in 13,000 years northern winters will occur on the other side of Earth's orbit where Earth is farther from the sun. Northern winters will be colder, and glaciers may grow.

The third factor is the inclination of Earth's equator to its orbit. Currently at 23.5°, this angle varies from 22° to 24° with a period of roughly 41,000 years. When the inclination is greater, seasons are more severe.

In 1920, Milankovitch proposed that these three factors cycle against each other to produce complex periodic variations in Earth's climate and the advance and retreat of glaciers (Figure 3.23). But no evidence was available to test the theory in 1920, and scientists treated it with skepticism. Many thought it was laughable.

The Evidence

By the middle 1970s, Earth scientists could collect the data that Milankovitch needed. Oceanographers could drill deep into the seafloor and collect samples, and geologists could determine the age of the samples from the natural radioactive atoms they contained. From all this, scientists constructed a history of ocean temperatures that convincingly matched the predictions of the Milankovitch hypothesis (Figure 3-24).

The evidence seemed very strong, and by the 1980s, the Milankovitch hypothesis was widely discussed as the leading hypothesis. But science follows a mostly unstated set of rules that holds that a hypothesis must be tested over and over against all available evidence (Window on Science 3-2). In 1988, scientists discovered contradictory evidence.

While scuba diving in a water-filled crack in Nevada called Devil's Hole, scientists drilled out samples of calcite, a mineral that contains oxygen atoms. For 500,000 years, layers of calcite have built up in

Science is based on evidence. Every theory and conclusion must be supported by evidence obtained from experiments or from observation. If a theory is supported by many pieces of evidence but is clearly contradicted by a single experiment or observation, scientists quickly abandon it. For a theory to be true, there can be no contradictory evidence.

Of course, scientists argue about the significance of particular evidence and often disagree on the interpretation of evidence. Some observations may seem significant at first glance, but upon closer examination we may find that the procedure was flawed, and so the piece of evidence is not important. Or we might conclude that the observational fact is correct but is being misinterpreted. The observation may not mean what it seems. Some of the most famous disagreements in science, such as those surrounding Galileo and Darwin, have arisen over the interpretation of well-established factual evidence.

Furthermore, scientists are not allowed to be selective in considering evidence. A lawyer in court can call a certain witness and intentionally fail to ask a critical question that would reveal evidence harmful to the lawyer's case. But a scientist may not ignore any known evidence. The difference in their methods is revealing. The lawyer is attempting to prove only one side of the case and rightly may ignore contradictory evidence. The scientist, however, is searching for the truth and so must test any theory against all available evidence. In a sense, the scientist, in dealing with evidence, must act as both the prosecution and the defense.

As you read about any science, look for the evidence in the form of measurements or observations. Every theory or conclusion should have supporting evidence. If you can find and understand the evidence, the science will make sense. All scientists, from astronomers to zoologists, demand evidence. You should, too.

Devil's Hole, recording in their oxygen atoms the temperature of the atmosphere when rain fell there. Finding the ages of the mineral samples was difficult, but the results seemed to show that the previous ice age ended thousands of years too early to have been caused by Earth's motions.

These contradictory findings are irritating because we naturally prefer certainty, but such circumstances are common in science. The disagreement between ocean floor samples and Devil's Hole samples triggered a scramble to understand the problem. Were the ages of one or the other set of samples wrong? Were the ancient temperatures wrong? Or were scientists misunderstanding the significance of the evidence?

In 1997, a new study of the ages of the samples confirmed that those from the ocean floor are correctly dated. This seems to give scientists renewed confidence in the Milankovitch hypothesis. But the same study found that the ages of the Devil's Hole samples are also correct. Many now believe the temperatures at Devil's Hole tell us about local climate changes in the region that became the southwestern United States. The ocean floor samples, which agree with the Milankovitch hypothesis, seem to tell us about global climate. Thus Milutin Milankovitch's hypothesis, first proposed in 1920, is still being tested as we try to understand the world we live on.

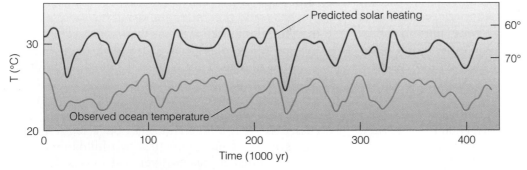

Figure 3-24

The Milankovitch theory predicts periodic changes in solar heating (shown here as the equivalent summer latitude of the sun). Over the last 400,000 years, these changes seem to have varied in step with ocean temperatures measured from fossils in sediment layers from the seabed. *(Adapted from Cesare Emiliani)*

REVIEW Critical Inquiry

How do precession and the shape of Earth's orbit interact to affect Earth's climate?

One technique that is useful in critical analysis of an idea is exaggeration. If we exaggerate the variation in the shape of Earth's orbit, we can see dramatically the influence of precession. At present, Earth reaches perihelion during winter in the northern hemisphere and aphelion during summer. The variation in distance is only about 2 percent, and that difference doesn't cause much change in the severity of the seasons. But if Earth's orbit were much more elliptical, then winter in the northern hemisphere would be much warmer, and summer would be much cooler.

Now we can see the importance of precession. As Earth's axis precesses, it points gradually in different directions, and the seasons occur at different places in Earth's orbit. In 13,000 years, northern winter will occur at aphelion, and if Earth's orbit were highly elliptical, northern winter would be terrible. Similarly, summer would occur at perihelion, and the heat would be awful. Such extremes might deposit large amounts of ice in the winter but then melt it away in the hot summer, thus preventing the accumulation of glaciers.

Continue this analysis by exaggeration. What effect would precession have if Earth's orbit were more circular?

In this chapter, we have studied the cycles of the sun and moon and in passing have seen Earth spinning on its axis and whirling around the sun. Knowledge of such movements is only a few centuries old. In the next chapter, we will tell the story of how the people of Earth discovered that they live on a moving planet.

Summary

Because Earth orbits the sun, the sun appears to move eastward along the ecliptic through the constellations. Because the ecliptic is tipped 23.5° to the celestial equator, the sun spends half the year in the northern celestial hemisphere and half the year in the southern celestial hemisphere, producing the seasons. The seasons are reversed south of Earth's equator.

Because we see the moon by reflected sunlight, its shape appears to change as it orbits Earth. The lunar phases wax from new moon to first quarter to full moon and wane from full moon to third quarter to new moon. A complete cycle of lunar phases takes 29.53 days.

The moon's gravitational field exerts tidal forces on Earth that pull the ocean waters up into two bulges, one on the side of Earth facing the moon and the other on the side away from the moon. As the rotating Earth carries the continents through these bulges of deeper water, the tides ebb and flow. Friction with the seabeds slows Earth's rotation, and the gravitational force the bulges exert on the moon force its orbit to grow larger.

When the moon passes through Earth's shadow, sunlight is cut off and the moon darkens in a lunar eclipse. If the moon only grazes the shadow, the eclipse is partial or penumbral and not total.

If the moon passes directly between the sun and Earth, it produces a total solar eclipse. During such an eclipse, the bright photosphere of the sun is covered, and the fainter corona, chromosphere, and prominences become visible. An observer outside the path of totality sees a partial eclipse. If the moon is in the farther part of its orbit, it does not cover the photosphere completely, resulting in an annular eclipse.

The motion of Earth changes in ways that can affect the climate. Changes in orbital shape, in precession, and in axial tilt can alter the planet's heat balance and may be responsible for the ice ages and glacial periods.

New Terms

rotation	lunar eclipse
revolution	umbra
ecliptic	penumbra
vernal equinox	solar eclipse
autumnal equinox	photosphere
summer solstice	chromosphere
winter solstice	corona
perihelion	prominence
aphelion	diamond ring effect
evening star	annular eclipse
morning star	perigee
sidereal period	apogee
synodic period	node
spring tides	Saros cycle
neap tides	Milankovitch hypothesis

Review Questions

1. What is the difference between the daily and annual motions of the sun?

2. If Earth did not rotate, could we still define the ecliptic? Why or why not?

3. What would our seasons be like if Earth were tipped 35° instead of 23.5°? What would they be like if Earth's axis were perpendicular to its orbit?

4. Why are the seasons reversed in the southern hemisphere?

5. Why isn't the summer solstice the hottest day of the year?

6. Why don't the planets move exactly along the ecliptic?

7. Do the phases of the moon look the same from every place on Earth, or is the moon full at different times in different locations?

8. What phase would Earth be in, if you were on the moon when the moon was full? at first quarter? at waning crescent?

9. How does the moon slow Earth's rotation, and how does Earth slow the moon's rotation?

10. Why have most people seen a total lunar eclipse, while few have seen a total solar eclipse?

11. Why isn't there an eclipse at every new moon and at every full moon?

12. Why is the moon red during a total lunar eclipse?

13. If a solar eclipse covers up part of the sun's bright disk, why do so many people feel that solar eclipses are dangerous?

14. Why should the eccentricity of Earth's orbit make winter in the northern hemisphere different from winter in the southern hemisphere?

15. How could small changes in the inclination of Earth's axis affect world climate?

Discussion Questions

1. Do planets orbiting other stars have ecliptics? Could they have seasons?

2. Why would it be difficult to see prominences if you were on the moon during a total lunar eclipse?

Problems

1. Draw a diagram like Figure 3-5, and show the path of the sun across the sky at the vernal equinox.

2. If Earth is about 5 billion (5×10^9) years old, how many precessional cycles have occurred?

3. Identify the phases of the moon if on March 21 the moon were located at (a) the vernal equinox, (b) the autumnal equinox, (c) the summer solstice, (d) the winter solstice.

4. Identify the phases of the moon if at sunset the moon were (a) near the eastern horizon, (b) high in the south, (c) in the southeast, (d) in the southwest.

5. About how many days must elapse between first-quarter moon and third-quarter moon?

6. Draw a diagram showing Earth, the moon, and shadows during (a) a total solar eclipse, (b) a total lunar eclipse, (c) a partial lunar eclipse, (d) an annular eclipse.

7. Phobos, one of the moons of Mars, is 20 km in diameter and orbits 5982 km above the surface of the planet. What is the angular diameter of Phobos as seen from Mars? (*Hint:* See By the Numbers 3-1.)

8. A total eclipse of the sun was visible from Canada on July 10, 1972. When did this eclipse occur next? From what part of Earth was it total?

9. When will the eclipse described in Problem 8 next be total as seen from Canada?

Critical Inquiries for the Web

1. Is astrology a science or a pseudoscience? The internet is full of astrological Web sites. Read the Windows on Science essays for this chapter, and examine several astrology-related Web pages. Look for evidence that a scientific approach is being applied. What points can you make to indicate that the information presented has not been generated through scientific processes?

2. Can you see the figure of a hunter in the constellation Orion? Orion's central location on the celestial sphere means that most cultures have noticed this familiar pattern of stars. Search the Internet for information on the mythologies associated with this star pattern. What stories do you find? Do the stories have common threads?

 Go to the Brooks/Cole Astronomy Resource Center (www.brookscole.com/astronomy) for critical thinking exercises, articles, and additional readings from InfoTrac College Edition, Brooks/Cole's online student library.

CHAPTER 4

The Origin of Modern Astronomy

The passions of astronomy

are no less profound because

they are not noisy.

John Steinbeck
The Short Reign of Pippin IV

Guidepost

In the 16th and 17th centuries, astronomers tried to understand the motions they saw in the sky, motions we have discussed in the last two chapters. In doing so, they invented a new way of understanding nature, what we now call science. This chapter, then, is really about the birth of science.

To understand the 16 chapters that follow, we must understand how scientists use evidence to test hypotheses and build theories to explain nature. Although science has become a powerful force in shaping our society, it remains now, as it was four centuries ago, nothing more than a logical way of thinking about nature—a way of understanding what we are and where we are.

41

The story of modern astronomy begins with an ending—the death of the great Polish astronomer Nicolaus Copernicus in May 1543 and the almost simultaneous publication of his model of the universe. That model revolutionized not only astronomy but all science, and it inspired a new consideration of our place in nature. We will trace this story over the 99 years from 1543 and Copernicus's death to 1642, the year that saw both the death of another great astronomer, Galileo, and the birth of one of the greatest scientists in history, Isaac Newton.

Two themes run through our story: Earth's place in the cosmos and the character of planetary motion. The debate over Earth's place involved theological questions and eventually brought Galileo Galilei before the Inquisition. The understanding of planetary motion involved the most basic physics of motion and eventually led Isaac Newton to understand gravitation and motion. Thus the Copernican model of the universe led to the birth of modern science.

To understand why the Copernican model was so important, we must backtrack to ancient Greece and meet the two great authorities of ancient astronomy, Aristotle and Ptolemy. These Greek astronomers struggled to understand Earth's place and the nature of planetary motion.

4-1 Pre-Copernican Astronomy

Aristotle

The history of astronomy before Copernicus extends back over 3000 years to ancient Babylon and beyond, but it is dominated by a single man, Aristotle (Figure 4-1). He lived from 384 to 322 B.C. in Greece and taught and wrote on the entire range of human knowledge. Because of his insight, he became the great authority of antiquity.

Much of what Aristotle wrote about scientific subjects was wrong, but he was not a scientist. He was trying to understand the universe by creating a system of formal philosophy. Modern science depends on evidence and hypotheses, but scientific methods had not been invented at the time of Aristotle. Rather, he attempted to use basic observations combined with first principles (ideas that he believed were obviously true) to understand the world and the heavens. His work on astronomy was accepted for almost 2000 years.

Aristotle adopted ideas from his teacher, the great philosopher Plato. Earth was imperfect and changeable, and the heavens were perfect and immutable. The most perfect geometrical figure, said Plato, was the sphere, and the natural motion of a sphere is rotation, so the heavens must be made up of rotating spheres. Thus Aristotle's description of

Figure 4-1
Aristotle (384–322 B.C.), honored on this Greek stamp, wrote on such a wide variety of subjects and with such deep insight that he became the great authority on all matters of learning. His opinions on the nature of Earth and the sky were widely accepted for almost two millennia.

the universe consisted of a perfect celestial realm rotating around Earth, which was motionless at the center (Figure 4-2).

According to Aristotle, the seven planets—Mercury, Venus, Mars, Jupiter, Saturn, the sun, and the moon—were carried by rotating, perfect, crystalline spheres centered on Earth. With 55 concentric spheres connected to each other by axes at different angles, Aristotle's system explained the general motion of the planets. Above the spheres of the planets was the sphere of fixed stars, and the entire heavenly vault rotated westward around Earth once a day and thus made the sun, moon, and stars rise in the east and set in the west. Thus, Aristotle and the classical astronomers who accepted his philosophy in the centuries following his life thought of the sky as a perfect, heavenly machine not many times larger than Earth itself.

Classical astronomers felt it was obvious that Earth did not move. They could feel no motion, of course, but more important they saw no parallax in the position of the stars. **Parallax** is the change in the apparent position of an object due to a change in the location of the observer. It is actually an old friend, though you may not have recognized its name, for you use parallax to judge the distance to things. To see how this works, close your right eye and use your thumb, held at arm's length, to cover some distant object, a building perhaps. Now look with your right eye. Your thumb seems to move to the left, uncovering the building (Figure 4-3). This apparent shift is parallax, and your brain uses it to estimate distances to objects around you.

Figure 4-2
According to Aristotle, Earth is motionless (*Terra immobilis*) at the center of the universe. Earth is surrounded by spheres of water, air, and fire (*ignus*), above which lie spheres carrying the celestial bodies beginning with the moon (*lune*) in the lowest celestial sphere. This woodcut is from C. Cornipolitanus's book *Chronographia* of 1537. *(From the Granger Collection, New York)*

For centuries astronomers reasoned that if Earth moved around the sun, we would view the stars from different places in Earth's orbit at different times of the year. Then parallax would distort the shapes of the constellations; they would look largest when Earth was nearest that part of the sky. Because they did not see this—that is, because they saw no parallax—they concluded that Earth did not move. Actually, they saw no parallax because the stars are much farther away than they supposed, and the parallax is much too small to be visible to the naked eye.

Ptolemy

One reason for the long survival of Aristotle's views was the work of the mathematician Claudius Ptolemy (Figure 4-4). His nationality and birth date are unknown, but he lived and worked in the Greek settlement at Alexandria in about A.D. 140, roughly five centuries after Aristotle. There he studied mathematics and astronomy and developed a mathematical model of the universe based on the teachings of Aristotle. Working from Aristotle's general description of the universe based on first principles, Ptolemy tried

Figure 4-4
The muse of astronomy guides Ptolemy in his study of the heavens. *(Courtesy Owen Gingerich; by permission of the Houghton Library, Harvard University)*

Seen by left eye   Seen by right eye

Figure 4-3
To demonstrate parallax, close one eye, and cover a distant object with your thumb held at arm's length. Look with the other eye, and your thumb appears to have shifted position.

to devise a mathematical model that would precisely describe the motions of the planets.

The Ptolemaic model, following Aristotle's philosophy, was a **geocentric** (Earth-centered) **universe.** Ptolemy also included the Greek belief that the heavenly bodies, being perfect, moved in **uniform circular motion,** an idea we trace back to Plato. But simple, circular paths centered on Earth do not account for the motions of the planets in the sky. The planets sometimes move faster and sometimes slower, and occasionally they appear to slow to a stop and move backward (westward) for a time in **retrograde motion** (Figure 4-5).

To describe the complicated planetary motions and yet preserve a geocentric model with uniform circular motion, Ptolemy adopted a system of wheels within wheels. Each planet moved in a small circle called an **epicycle,** and the center of the epicycle moved along a larger circle around Earth called a **deferent.** By adjusting the size of the circles and the rate of their motion, Ptolemy could account for most planetary movement. But as a final adjustment, he placed Earth off center in the deferent circle and specified that the center of the epicycle would appear to move at constant speed only if viewed from a point called the **equant,** located to the other side of the deferent's center (Figure 4-6). Thus, by using a few dozen circles of various sizes rotating at various rates, Ptolemy's system reproduced the motions of the planets (Figure 4-7).

Whereas Aristotle had attempted to describe the motions of the heavens consistent with the first principles of his system of philosophy, Ptolemy was not as concerned with first principles. He wanted accurate mathematical predictions of the positions of the planets. Motion around an epicycle or centered on an equant was not truly Earth-centered, nor was it truly uniform circular motion, but it improved the mathematical predictions. Thus, Ptolemy slightly weakened the principles of geocentrism and uniform circular motion in order to make his model more accurate.

About A.D. 140, Ptolemy included this work in a book that he called *Mathematical Syntaxis* but that is now known as *Almagest.* Ptolemy was the last great classical astronomer. In the centuries that followed, the breath of civilization passed to invading Arabs, who dominated most of the known world for centuries. Arabian astronomers translated, studied, and preserved many classical manuscripts; and in Arabic, Ptolemy's book became *Al Magisti (Greatest).* Beginning in the 1200s, European Christians drove the Arabs out of Spain and recovered their classical her-

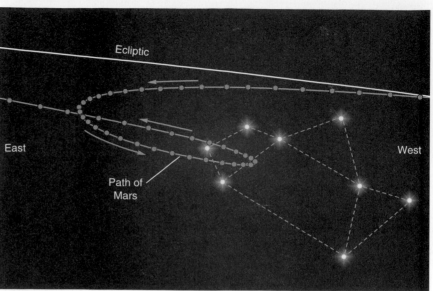

Figure 4-5
The word *planet* comes from the Greek for "wanderer," referring to the motion of the planets against the background of fixed stars. Here the motion of Mars along the ecliptic near the teapot shape of Sagittarius is shown at 4-day intervals. Though the planet usually moves eastward, it sometimes slows to a stop and moves westward in retrograde motion.

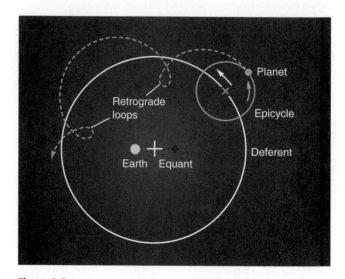

Figure 4-6
To explain the periodic retrograde motion of a planet, Ptolemy proposed that the planet traveled around a small circle called the epicycle, which revolved around a bigger circle called the deferent. To slightly adjust the speed of the planet, he supposed that Earth was slightly off center and that the center of the epicycle moved such that it appeared to move at a constant rate as seen from the equant point. A planet moving in this way would periodically pass through a retrograde loop and appear to move backward (westward) against the background of stars.

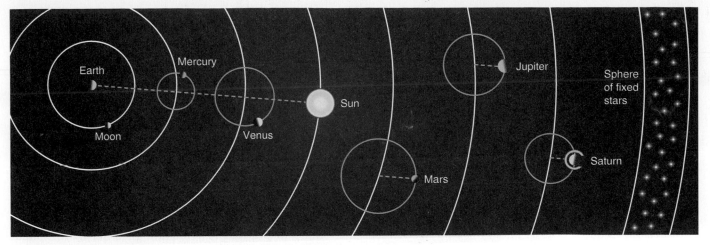

Figure 4-7
The Ptolemaic model of the universe is geocentric and based on uniform circular motion of epicycles moving around deferents. The centers of the epicycles of Mercury and Venus must always lie on the Earth–sun line, but Mars, Jupiter, and Saturn are free of this restriction. Notice that this modern illustration shows rings around Saturn and sunlight illuminating the globes of the planets, features that could not be known before the invention of the telescope.

itage through Arabic translations. In Latin, Ptolemy's book became *Almagestum* and thus our modern *Almagest.*

At first the Ptolemaic system predicted the positions of the planets with fair accuracy, but as centuries passed, errors accumulated, and Arabian and later European astronomers had to update the system, computing new constants and adjusting the epicycles. In the middle of the 13th century, a team of astronomers supported by King Alfonso X of Castile worked for 10 years revising the Ptolemaic system and publishing the result as the *Alfonsine Tables.* It was the last great adjustment of the Ptolemaic system.

REVIEW **Critical Inquiry**

Why did classical astronomers conclude the heavens were made up of spheres?

Classical astronomers did not use evidence and hypotheses the way modern scientists do. Rather they argued from first principles, and Plato argued that the perfect geometrical figure was a sphere. Then the heavens, which everyone agreed were perfect, must be made up of spheres. The natural motion of a sphere is rotation, and the only perfect motion is uniform motion, so the heavenly spheres were thought to move in uniform circular motion. Thus, classical philosophers argued that the daily motion of the heavens around Earth and the motions of the seven planets (the sun and moon were counted as planets) against the background of the stars had to be produced by the combination of uniformly rotating spheres.

Although ancient astronomers didn't use evidence as modern scientists do, they did observe the world around them. What observations led them to conclude that Earth didn't move?

Pre-Copernican astronomy consists of the elaboration of Aristotle's philosophy of the universe as expressed by Ptolemy in a mathematical model. For nearly 1500 years, astronomers believed in this geocentric universe. Earth was finally set in motion by a Polish church official with a talent for mathematics.

4-2 Copernicus

Nicolaus Copernicus was born in 1473 (Figure 4-8) in what is now Poland. At the time of his birth—and throughout his life—astronomy was based on the Ptolemaic system, even though, in spite of many revisions, it was still a poor predictor of planetary positions. Yet because of the authority of Aristotle, it was the officially accepted theory of the universe.

The Copernican Model

According to the Aristotelian universe, the most perfect region was in the heavens, and the most imperfect at Earth's center. Thus the classical geocentric universe matched the commonly held Christian geometry of heaven and hell, and anyone who criticized the geometry of the Aristotelian universe was challenging belief in heaven and hell and thus risking at least criticism and perhaps a charge of heresy.

Throughout his life, Copernicus was associated with the Church. His uncle, by whom he was raised and educated, was an important bishop in Poland, and after studying canon law and medicine in some of the major universities in Europe, Copernicus became a canon of the Church at the age of 24. He served as secretary and personal physician to his powerful uncle for 15 years. When his uncle died, Copernicus went to live in quarters adjoining the cathedral in Frauenburg. Because of this long association with the Church and his fear of persecution, he hesitated to publish his revolutionary ideas.

What was this astonishing idea? Copernicus believed that the sun and not Earth was the center of the universe. That idea was shocking because it challenged the traditional belief in Earth's place at the center of all creation. Notice that the U.S. stamp in Figure 4-8 shows Copernicus holding a sun centered, or **heliocentric,** model.

In fact, his idea was already being discussed long before he published it. His interest in astronomy had begun during his college days, and he apparently doubted the Ptolemaic system even then. About 1507, at the age of 34, he wrote a short pamphlet that discussed the motion of the sky and outlined his hypothesis that the sun, not Earth, was the center of the universe and that Earth rotated on its axis and revolved around the sun. He distributed his summary in handwritten form without a title and, in some cases, anonymously. Perhaps his caution grew from a sense of modesty, but he may also have feared criticism for challenging the Ptolemaic system and the place of Earth in the universe.

De Revolutionibus

Copernicus worked on his book *De Revolutionibus Orbium Coelestium* over a period of many years and was essentially finished by about 1530, yet he hesitated to publish it even though other astronomers knew of his theories. Church officials concerned about the reform of the calendar sought his advice and looked forward to the publication of his book.

One reason he hesitated was that the idea of a heliocentric universe was highly controversial. This was a time of rebellion in the Church—Martin Luther was speaking harshly about fundamental Church teachings, and others, both scholars and scoundrels, were questioning the authority of the Church. Even matters as abstract as astronomy could stir controversy. Remember, too, that Earth's place in astronomical theory was linked to the geometry of heaven and hell, so moving Earth from its central place was a controversial and perhaps heretical idea. Another reason Copernicus may have hesitated to publish was that his work was incomplete. His model could not accurately predict planetary positions. Thus, he did not give permission for the publication of his book until he realized he was dying.

The most important idea in the book was placing the sun at the center of the universe. That single innovation had an astonishing consequence—the retrograde motion of the planets was immediately explained in a straightforward way without the large epicycles that Ptolemy used. In the Copernican system, Earth moves faster along its orbit than the planets that lie farther from the sun. Consequently, Earth periodically overtakes and passes these planets. Imagine that you are in a race car, driving rapidly along the inside lane of a circular race track. As you pass slower cars driving in the outer lanes, they fall behind, and if you did not know you were moving, it would seem that the cars in the outer lanes occasionally slowed to a stop and then backed up for a short interval. The same thing happens as Earth

Figure 4-8
Nicolaus Copernicus (1473–1593) pursued a lifetime career in the Church, but he was also a talented mathematician and astronomer. His work triggered a revolution in human thought. These stamps were issued in 1973 to mark the 500th anniversary of his birth.

passes a planet such as Mars. Although Mars moves steadily along its orbit, as seen from Earth it appears to slow to a stop and move westward (retrograde) as Earth passes it (Figure 4-9). Because the planetary orbits do not lie in precisely the same plane, a planet does not resume its eastward motion in precisely the same path it followed earlier. Consequently, it describes a loop whose shape depends on the angle between the orbital planes.

Copernicus could explain retrograde motion without epicycles, and that was impressive. The Copernican system was elegant and simple compared to the whirling epicycles and off-center equants of the Ptolemaic system. However, *De Revolutionibus* failed to disprove the geocentric model for one critical reason—the Copernican theory could not predict the positions of the planets any more accurately than the Ptolemaic system could. To understand why it failed

this critical test, we must understand Copernicus and his world.

Although Copernicus proposed a revolutionary idea in making the planetary system heliocentric, he was a classical astronomer with tremendous respect for the old concept of uniform circular motion. In fact, Copernicus objected strongly to Ptolemy's use of the equant. It seemed arbitrary to Copernicus, a direct violation of the elegance of Aristotle's philosophy of the heavens. Copernicus called equants "monstrous" in that they violated both geocentrism and uniform circular motion. In devising his model, Copernicus returned to a strong belief in uniform circular motion. Although he did not need epicycles to explain retrograde motion, Copernicus discovered that the sun, moon, and planets suffered other smaller variations in their motions that he could not explain with uniform circular motion centered on the sun. Today we recognize those variations as typical of objects following elliptical orbits, but Copernicus held firmly with uniform circular motion, so he had to introduce small epicycles to reproduce these minor variations in the motions of the sun, moon, and planets.

Because Copernicus imposed uniform circular motion on his model, it could not accurately predict the motions of the planets. The *Prutenic Tables* (1551) were based on the Copernican model, and they were not significantly more accurate than the 13th century *Alfonsine Tables* that were based on Ptolemy's model. Both could be in error by as much as 2°, which is four times the angular diameter of the full moon.

The Copernican *model* is inaccurate. It includes small epicycles and thus does not precisely describe the motions of the planets. But the Copernican *hypothesis* that the universe is heliocentric is correct. Why that hypothesis gradually won acceptance in spite of the inaccuracy of the epicycles and deferents is a question historians still debate. There are probably a number of reasons, including the revolutionary temper of the times, but the most important factor may be the elegance of the idea. Placing the sun at the center of the universe produced a symmetry among the motions of the planets that is pleasing to the eye as well as to the intellect (Figure 4-10). No longer do Venus and Mercury revolve around empty points located between Earth and the sun. Now they, like the rest of the planets, move in orbits around the sun.

The most astonishing consequence of the Copernican hypothesis was not what it said about the sun, but what it said about Earth. By placing the sun at the center, Copernicus made Earth move along an orbit like the other planets. By making Earth a planet, Copernicus revolutionized humanity's view of its place in the universe and triggered a controversy that would eventually bring the astronomer Galileo Galilei before the Inquisition, a controversy that continues even today.

Figure 4-9
The Copernican explanation of retrograde motion. As Earth overtakes Mars (a–c), Mars appears to slow its eastward motion. As Earth passes Mars (d), Mars appears to move westward. As Earth draws ahead of Mars (e–g), Mars resumes its eastward motion against the background stars. Compare with Figure 4-5. The positions of Earth and Mars are shown at equal intervals of 1 month.

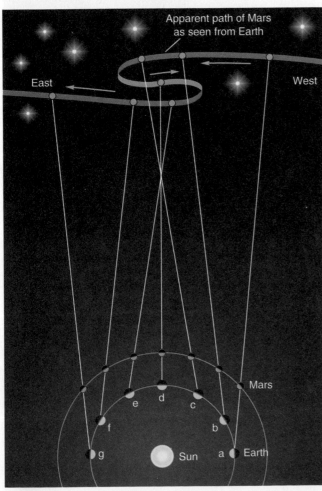

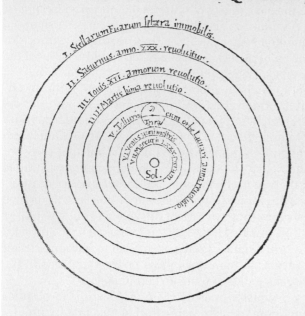

a

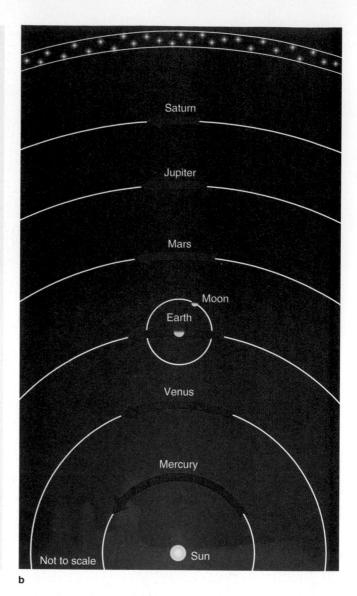

b

Figure 4-10

(a) The Copernican universe as reproduced in *De Revolutionibus*. Earth and all the known planets revolve in separate circular orbits about the sun (Sol) at the center. The outermost sphere carries the immobile stars of the celestial sphere. Notice the orbit of the moon around Earth (Terra). *(Courtesy Yerkes Observatory)* (b) The model is elegant not only in the arrangement of the planets but also in their motions. Orbital velocities (red arrows) decrease from that of Mercury, the fastest, to that of Saturn, the slowest. Compare the elegance of this model with the complexity of the Ptolemaic model in Figure 4-7.

Although astronomers throughout Europe read and admired *De Revolutionibus,* they did not usually accept the Copernican hypothesis. The mathematics was elegant, and the astronomical observations and calculations were of tremendous value; but few astronomers believed, at first, that the sun actually was the center of the planetary system and that Earth moved. How the Copernican hypothesis was gradually recognized as correct has been called the Copernican revolution because it was not just the adoption of a new idea but a total change in the way astronomers think about Earth's place (Window on Science 4-1).

REVIEW Critical Inquiry

Why do we say the Copernican hypothesis was correct but the model was wrong?

The Copernican hypothesis was that the sun and not Earth was the center of the universe. Given the limited knowledge of the Renaissance astronomers about distant stars and galaxies, that hypothesis was correct. The sun is at the center of our planetary system. That hypothesis produces an elegant model of the

The Copernican revolution is often cited as the perfect example of a scientific revolution. Over a few decades, astronomers abandoned a way of thinking about the universe that was almost 2000 years old and adopted a new set of ideas and assumptions. The American philosopher of science Thomas Kuhn has referred to a commonly accepted set of scientific ideas and assumptions as a scientific **paradigm.** Thus, the pre-Copernican astronomers had a geocentric paradigm that included uniform circular motion and the perfection of the heavens. That paradigm survived for many centuries until a new generation of astronomers was able to overthrow the old paradigm and establish a new paradigm that included heliocentrism and the motion of the earth.

A scientific paradigm is powerful because it shapes our perceptions. It determines what we judge to be important questions and what we judge to be significant evidence. Thus, it is often difficult for us to recognize how our paradigms limit what we can understand. For example, the geocentric paradigm contained problems that seem obvious to us, but because astronomers before Copernicus lived and worked inside that paradigm, they had difficulty seeing the problems. Overthrowing an outdated paradigm is not easy, because we must learn to see nature in an entirely new way. Galileo, Kepler, and Newton saw nature from a new paradigm that would have been almost incomprehensible to astronomers of earlier centuries.

We can find examples of scientific revolutions in many fields, including biology, geology, genetics, and psychology. These scientific revolutions have been difficult and controversial because they have involved the overthrow of accepted paradigms. But that is why scientific revolutions are exciting. They give us not just a new idea or a new theory, but an entirely new insight into how nature works—a new way of seeing the world.

---

universe without large epicycles and equants and explains retrograde motion in a simple way. The Copernican model, however, included not only the heliocentric hypothesis but also uniform circular motion. The model is wrong because the planets don't really follow circular orbits. Thus, the Copernican model could not make accurate predictions of planetary position even though it placed the sun at the center of the universe.

The Copernican hypothesis won converts because it is elegant and can explain retrograde motion. How does its explanation of retrograde motion work, and how is it more elegant than the Ptolemaic explanation?

The central issue in the Copernican hypothesis is the place of Earth, but that was linked to a second issue, planetary motion. Thus, our story of the Copernican revolution shifts from Copernicus, a mathematician and theorist, to one of the first great observational astronomers, a man who measured the daily positions of the sun, moon, and planets.

4-3 Tycho Brahe

The great observational astronomer of our 99-year story was a Danish nobleman, Tycho Brahe (Figure 4-11), born December 14, 1546, only 3 years after the publication of *De Revolutionibus.* In histories of astronomy, the great astronomers are often referred to by their last name, but Tycho Brahe is usually called Tycho. Were he alive today, he would no doubt object to such familiarity from his obvious inferiors — he was well known for his vanity and lordly manners.

The New Star

During his university days, Tycho was officially studying law, but he made it clear to his family that his real passion was mathematics and astronomy. Yet his life was shaped by two events outside his studies — a duel and a new star in the sky.

While at the university, Tycho fought a duel over some supposed insult and received a wound that disfigured his nose. For the rest of his life, he wore false noses made of gold and silver and stuck on with wax (see Figure 4-11). He was known as a proud and haughty nobleman, and the injury did little to improve his disposition.

Figure 4-11
Tycho Brahe (1546–1601), a Danish nobleman, established an observatory at Hveen and measured planetary positions with high accuracy. His artificial nose is suggested in this engraving.

Early in his university days, Tycho showed an interest in astronomy and began measuring the positions of the planets in the sky. In 1563, Jupiter and Saturn passed very near each other in the sky, nearly merging into a single point on the night of August 24. Tycho found that the *Alphonsine Tables* were a full month in error and that the *Prutenic Tables* were in error by a number of days.

Then in 1572 a "new star" (now called Tycho's supernova) appeared in the sky, shining more brightly than Venus. Tycho carefully measured its position and concluded that it displayed no parallax. To understand the significance of this observation, we must note that the rotation of Earth carries us eastward and that the positions of celestial objects near Earth, such as the moon, change slightly through the night because the observer is in different positions. Tycho believed Earth was stationary and that the heavens were rotating westward, but that makes no difference. Objects moving with the heavens near a stationary Earth should still have measurable parallax. That he detected no change in the position of the new star through the night proved it was farther away than the moon. Therefore, he concluded, it was a new star in the heavens. Because Aristotle and Ptolemy held that the heavens were perfect and unchanging, the new star led Tycho to question the Ptolemaic system. He summarized his results in a small book, *De Stella Nova* (*The New Star*), published in 1573.

The book attracted the attention of astronomers throughout Europe, and soon Tycho was summoned to the court of the Danish King Frederik II and offered funds to build an observatory on the island of Hveen just off the Danish coast. Tycho also received a steady source of income as landlord of a coastal district from which he collected rents. (He was not a popular landlord.) On Hveen, Tycho constructed a luxurious home with four towers especially equipped for astronomy and populated it with servants, assistants, and a dwarf to act as jester. Soon Hveen was an international center of astronomical study.

Tycho Brahe's Legacy

Tycho made no direct contribution to astronomical theory. Because he could measure no parallax for the stars, he concluded that Earth had to be stationary, thus rejecting the Copernican hypothesis. However, he also rejected the Ptolemaic model because of its inaccurate predictions. Instead, he devised a complex model in which Earth was the immobile center of the universe around which the sun and moon moved. The other planets circled the sun. This model thus incorporated part of the Copernican model, but Earth—rather than the sun—was stationary. In this way, Tycho preserved the central, immobile Earth described in scripture (Figure 4-12). Although Tycho's model was very popular at first, the Copernican model replaced it within a century.

The true value of Tycho's work was observational. Because he was able to devise new and better instruments, he was able to make highly accurate observations of the positions of the stars, sun, moon, and

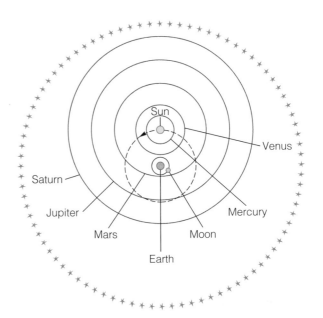

Figure 4-12
Tycho Brahe's model of the universe retained the most important of the classical principles. The stars were in a starry sphere. Earth was fixed and immobile at the center with the sun and moon revolving around Earth. The rest of the planets circled the sun.

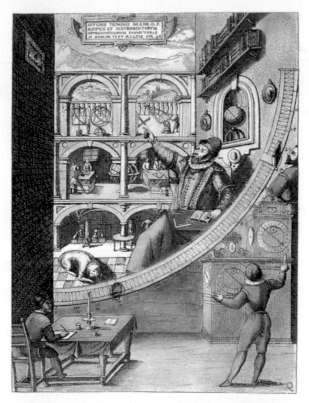

Figure 4-13
Much of Tycho's success was due to his skill in designing large, accurate instruments such as this one known as the mural quadrant. In this engraving, the figure of Tycho at the center, his dog, and the scene in the background are a mural painted on the wall within the arc of the quadrant. The observer at the extreme right (also Tycho) peers through a sight out the loophole in the wall at the upper left and thus measures a celestial body's angular distance above the horizon. *(The Granger Collection, New York)*

planets. Tycho had no telescopes—they were not invented until the next century—so his observations were made by the naked eye peering along sights. All of his instruments were designed to measure angles in the sky. For example, his quadrant could measure angles up to 90° above the horizon (Figure 4-13). By designing and building large instruments with great care, he was able to measure angles to high precision. He measured the positions of 777 stars to better than 4 minutes of arc and measured the positions of the sun, moon, and planets almost daily for the 20 years he stayed on Hveen.

Unhappily for Tycho, King Frederik II died in 1588, and his young son took the throne. Suddenly Tycho's temper, vanity, and noble presumptions threw him out of favor. In 1596, taking most of his instruments and books of observations, he went to Prague, the capital of Bohemia, and became imperial mathematician to the Holy Roman Emperor Rudolph II. His goal was to revise the *Alphonsine Tables* and publish the revision as a monument to his new patron. It would be called the *Rudolphine Tables*.

Tycho did not intend to base the *Rudolphine Tables* on the Ptolemaic system but rather on his own Tychonic system, proving once and for all the validity of his hypothesis. To assist him, he hired a few mathematicians and astronomers, including one Johannes Kepler. Then in November 1601, Tycho collapsed at a nobleman's home. Before he died, 9 days later, he asked Rudolph II to make Kepler imperial mathematician. Thus the newcomer, Kepler, became Tycho's replacement (though at one-sixth Tycho's salary).

R E V I E W Critical Inquiry

Why did Brahe conclude that the new star of 1572 contradicted the Ptolemaic universe?

According to Aristotle, the heavens were perfect and therefore unchanging. Any change in the heavens had to occur below the moon, which was attached to the lowest of the crystalline spheres. When Tycho saw the new star of 1572 in the sky, he could detect no parallax. If the star lay below the moon's sphere, then it should appear to be east of its average position when it was rising in the east, and it should appear west of its average position when it was setting in the west. That is, it should show parallax. Tycho could measure no parallax at all, which meant the new star had to be at least above the moon's sphere and was probably part of the starry sphere itself. Thus, it was a change in the supposedly unchanging heavenly realm.

Tycho didn't believe in the Ptolemaic universe, and he produced his own model to explain planetary motion. In what ways did his model incorporate ideas from classical astronomy and from the Copernican model?

Tycho Brahe tried to resolve the problem of Earth's place by assuming the universe was geocentric, but his attempts to understand planetary motion were cut short by his death. The newly appointed imperial mathematician, Johannes Kepler, used Tycho's observations to finally solve the second great problem of the age, the motion of the planets.

4-4 Johannes Kepler
An Astronomer of Humble Origins

No one could have been more different from Tycho Brahe than Johannes Kepler (Figure 4-14). He was born on December 27, 1571, to a poor family in a region now

Figure 4-14
Johannes Kepler (1571–1630) derived three laws of planetary motion from Tycho Brahe's observations of the positions of the planets. This Romanian stamp commemorates the 400th anniversary of Kepler's birth. Ironically, it contains an astronomical error—the horns of the crescent moon should be inclined upward from the horizon.

included in southwest Germany. His father was unreliable and shiftless, principally employed as a mercenary soldier fighting for whoever paid enough. He finally failed to return from a military expedition, either because he was killed or because he found circumstances more to his liking elsewhere. Kepler's mother was apparently an unpleasant and unpopular woman. She was accused of witchcraft in later years, and Kepler had to defend her in a trial that dragged on for 3 years. She was finally acquitted but died the following year.

Kepler was the oldest of six children, and his childhood was no doubt unhappy. The family was not only poor and often lacking a father, but it was also Protestant in a predominantly Catholic region. In addition, Kepler was never healthy, even as a child, so it is surprising that he did well in the pauper's school he attended, eventually winning a scholarship to the university at Tübingen, where he studied to become a Lutheran pastor.

During his last year of study, Kepler accepted a job in Graz teaching mathematics and astronomy, a job he resented because he knew little about the subjects. Evidently he was not a good teacher either—he had few students his first year, and none at all his second. His superiors put him to work teaching a few introductory courses and preparing an annual almanac that contained astronomical, astrological, and weather predictions. Through good luck, in 1595 some of his weather predictions were fulfilled, and he gained a reputation as an astrologer and seer. Even in later life he earned money from his almanacs.

While still a college student, Kepler had become a believer in the Copernican hypothesis, and at Graz he used his extensive spare time to study astronomy. By 1596, the same year Tycho left Hveen, Kepler was ready to solve the mystery of the universe. That year

he published a book called *The Forerunner of Dissertations on the Universe, Containing the Mystery of the Universe*. The book, like nearly all scientific works of that age, was written in Latin, and is now known as *Mysterium Cosmographicum.*

By modern standards, the book contains almost nothing of value. It begins with a long appreciation of Copernicanism and then goes on to speculate on the reasons for the spacing of the planetary orbits. Kepler felt he had found the underlying architecture of the universe in the five regular solids*—the cube, tetrahedron, dodecahedron, icosahedron, and octahedron. Because these five solids were the only regular solids, he supposed they were the "spacers" between the planetary orbits (Figure 4-15). In fact, Kepler concluded that there could be only six planets (Mercury, Venus, Earth, Mars, Jupiter, and Saturn) because there were only five regular solids to act as spacers between their spheres. He advanced astrological, numerological, and even musical arguments for his theory.

The second half of the book is no better than the first, but it has one virtue—as Kepler tried to fit the five solids to the planetary orbits, he demonstrated that he was a talented mathematician and that he was well versed in astronomy. He sent copies to Tycho and to Galileo in Rome, and both recognized his talent in spite of the mystical content of the book.

Joining Tycho

Life was unsettled for Kepler because of the persecution of Protestants in the region, so when Tycho Brahe invited him to Prague in 1600, Kepler went readily, eager to work with the famous Danish astronomer. Tycho's sudden death in 1601 left Kepler in a position to use the observations from Hveen to analyze the motions of the planets and complete the *Rudolphine Tables.* Tycho's family, recognizing that Kepler was a Copernican and guessing that he would not follow the Tychonic system in completing the *Rudolphine Tables,* sued to recover the instruments and books of observations. The legal wrangle went on for years. Tycho's family did get back the instruments Tycho had brought to Prague, but Kepler had the books and he kept them.

Whether Kepler had any legal right to Tycho's records is debatable, but he put them to good use. He began by studying the motion of Mars, trying to deduce from the observations how the planet moved. By 1606, he had solved the mystery, this time correctly. The orbit of Mars is an ellipse and not a circle. Thus he abandoned the 2000-year-old belief in the circular motion of the planets. But the mystery was even more complex. The planets do not move at a uniform speed along their elliptical orbits. Kepler recognized that

---

*A regular solid is a three-dimensional body, each of whose faces is the same. A cube is a regular solid, each of whose faces is square.

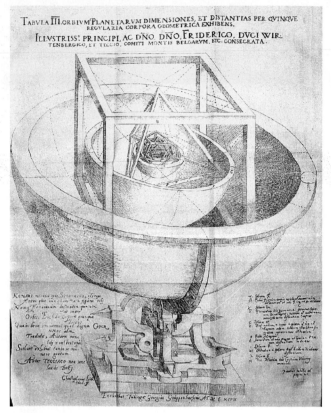

Figure 4-15
Kepler believed the five regular solids were the spacers between the spheres containing the planetary orbits. His book *Mysterium Cosmographicum* contained this fold-out illustration of the spheres and spacers. *(Courtesy Owen Gingerich; by permission of the Houghton Library, Harvard University)*

they move faster when close to the sun and slower when farther away. Thus Kepler abandoned both uniform motion and circular motion and finally solved the problem of planetary motion.

Kepler published his results in 1609 in a book called *Astronomia Nova* (*New Astronomy*). Like Copernicus's book, *Astronomia Nova* was written in Latin for other scientists and was highly mathemati-

cal. In some ways the book was surprisingly advanced. For instance, Kepler discussed the force that holds the planets in their orbits and came within a paragraph of stating the principle of mutual gravitation.

In spite of the abdication of Rudolph II in 1611, Kepler continued his astronomical studies. He wrote about a supernova that had appeared in 1604 (now known as Kepler's supernova) and about comets, and he wrote a textbook about Copernican astronomy. In 1619, he published *Harmonice Mundi* (*The Harmony of the World*) in which he returned to the cosmic mysteries of *Mysterium Cosmographicum.* The only thing of note in *Harmonice Mundi* is his discovery that the radii of the planetary orbits are related to the planet's orbital period. That and his two previous discoveries have become known as the most fundamental rules of orbital motion.

Kepler's Three Laws of Planetary Motion

Although Kepler dabbled in the philosophical arguments of the day, he was a mathematician, and his triumph was the solution of the problem of the motion of the planets. The key to his solution was the ellipse.

An **ellipse** is a figure drawn around two points, called the *foci,* in such a way that the distance from one focus to any point on the ellipse and back to the other focus equals a constant. This makes it easy to draw ellipses with two thumbtacks and a loop of string. Press the thumbtacks into a board, loop the string about the tacks, and place a pencil in the loop. If you keep the string taut as you move the pencil, it traces out an ellipse (Figure 4-16a).

The geometry of an ellipse is described by two simple numbers. The **semimajor axis, *a*,** is half of the longest diameter (Figure 4-16b). The **eccentricity** of an ellipse, *e,* is the distance between the foci divided by the longest diameter. To draw a circle with the string and tacks shown in Figure 4-16a, we would move the two thumbtacks together, which shows that a circle is really just an ellipse with eccentricity equal

Figure 4-16
The geometry of elliptical orbits. (a) Drawing an ellipse with two tacks and a loop of string. (b) The semi-major axis, *a,* is half of the longest diameter. (c) Kepler's second law is demonstrated by a planet that moves from *A* to *B* in 1 month and from *A'* to *B'* in the same amount of time. The two blue segments have the same area.

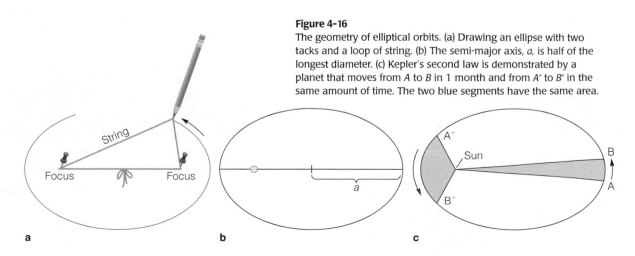

Even scientists misuse the words *hypothesis, theory,* and *law.* We must try to distinguish these terms from one another because they are key elements in science.

A **hypothesis** is a single assertion or conjecture that must be tested. It could be true or false. "All Texans love chili" is a hypothesis. To know whether it is true or false, we need to test it against reality by making observations or performing experiments. Copernicus asserted that the universe was heliocentric; his assertion was a hypothesis subject to testing.

In Chapter 2, we saw that a model (Window on Science 2-1) is a description of some natural phenomenon; it can't be right or wrong. A model is not a conjecture of truth but merely a convenient way to think about a natural phenomenon.

Thus, a model such as the celestial sphere cannot be a hypothesis. Copernicus used his hypothesis to build a model of the universe, and, to the extent that the model described the motion of the planets, it was useful.

A **theory** is a system of rules and principles that can be applied to a wide variety of circumstances. A theory may have begun as one or more hypotheses, but it has been tested, expanded, and generalized. Many textbooks refer to the "Copernican theory," but some experts argue that it was not complete in that it lacked a precise description of orbital motion and gravitation, features added by Kepler and Newton. Thus, it is probably more accurate to call it the Copernican hypothesis.

A **natural law** is a theory that almost everyone accepts as true. Such a theory is usually very specific and has been tested many times and applied to many situations. Thus, natural laws are the most fundamental principles of scientific knowledge. Kepler's laws of planetary motion are good examples.

Science is not always consistent about these words. For example, Einstein's theory of relativity is much more precise than Newton's laws of gravity and motion, but tradition dictates the words we use. Darwin's theory of evolution is well tested and widely accepted, but no one refers to it as Darwin's law. The distinctions are subtle and sometimes depend more on custom than on the definition of the words.

to zero. As we move the thumbtacks farther apart, the ellipse becomes flatter and the eccentricity moves closer to 1.

Kepler used ellipses to describe the motion of the planets in three fundamental rules that have been tested and confirmed so many times that astronomers now refer to them as natural laws (Window on Science 4-2). They are commonly called Kepler's laws of Planetary Motion (Table 4-1).

Kepler's first law states that the orbits of the planets around the sun are ellipses with the sun at one focus. Thanks to the precision of Tycho's observations and the sophistication of Kepler's mathematics, Kepler was able to recognize the elliptical shape of the orbits even though they are nearly circular. Of the planets known to Kepler, Mercury has the most elliptical orbit, but even it deviates only slightly from a circle (Figure 4-17).

Kepler's second law states that a line from the planet to the sun sweeps over equal areas in equal intervals of time. This means that when the planet is closer to the sun and the line connecting it to the sun is shorter, the planet moves more rapidly to sweep over the same area that is swept over when the planet is farther from the sun. Thus the planet in Figure 4-16b would move from point A' to point B' in one month, sweeping over the area shown. But when the planet is farther from the sun, one month's motion would be shorter, from A to B.

The time that a planet takes to travel around the sun once is its orbital period, P, and its average dis-

Table 4-1 Kepler's Laws of Planetary Motion

I. The orbits of the planets are ellipses with the sun at one focus.

II. A line from a planet to the sun sweeps over equal areas in equal intervals of time.

III. A planet's orbital period squared is proportional to its average distance from the sun cubed:

$$P_y^2 = a_{AU}^3$$

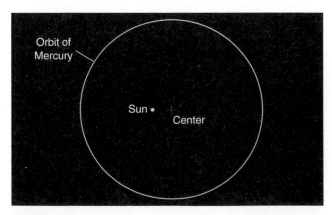

Figure 4-17
The orbits of the planets are nearly circular. Of the planets known to Kepler, Mercury has the most elliptical orbit.

tance from the sun equals the semimajor axis of its orbit, *a*. Kepler's third law tells us that these two quantities are related; the orbital period squared is proportional to the semimajor axis cubed. If we measure *P* in years and *a* in astronomical units, we can summarize the third law as:

$$P_y^3 = a_{AU}^3$$

For example, Jupiter's average distance from the sun is roughly 5.2 AU. The semimajor axis cubed would be about 140.6, so the period must be the square root of 140.6, about 11.8 years.

It is important to notice that Kepler's three laws are empirical. That is, they describe a phenomenon without explaining why it occurs. Kepler derived them from Tycho's extensive observations, not from any fundamental assumption or theory. In fact, Kepler never knew what held the planets in their orbits or why they continued to move around the sun. His books are a fascinating blend of careful observation, mathematical analysis, and mystical theory.

The *Rudolphine Tables*

In spite of Kepler's recurrent involvement with astrology and numerology, he continued to work on the *Rudolphine Tables*. At last, in 1627, they were ready, and he financed their printing himself, dedicating them to the memory of Tycho Brahe. In fact, Tycho's name appears in larger type on the title page than Kepler's own. This is especially surprising when we recall that the tables were based on the heliocentric model of Copernicus and the elliptical orbits of Kepler, and not on the Tychonic system. The reason for Kepler's evident deference was Tycho's family, still powerful and still intent on protecting the memory of Tycho. They even demanded a share of the profits and the right to censor the book before publication, though they changed nothing but a few words on the title page.

The *Rudolphine Tables* was Kepler's masterpiece, the precise model of planetary motion that Copernicus had sought but failed to find. The accuracy of the *Rudolphine Tables* was strong evidence that both Kepler's model for planetary motion and the Copernican hypothesis for the place of Earth were correct. Copernicus would have been pleased.

Kepler died on November 15, 1630. He had solved the problem of planetary motion, and his *Rudolphine Tables* demonstrated his solution. Although he did not understand why the planets moved or why they followed ellipses, insights that had to wait half a century for Isaac Newton, Kepler's three rules worked. In science the only test of a theory is, "Does it describe reality?" Kepler's laws have been used for almost four centuries as a true description of orbital motion.

REVIEW Critical Inquiry

What principles of classical astronomy made it difficult for Kepler to understand planetary motion?

Critical analysis is more than careful thinking. We can think carefully about what Kepler did, but until we analyze the paradigm (the network of assumptions and principles) within which he worked, we cannot understand how impressive his achievement was. Kepler lived at a time when every astronomer believed in uniform circular motion. The heavens were assumed perfect, and, according to classical philosophers such as Plato and Aristotle, circular motion at a constant rate is the only perfect motion. Before Kepler could adopt elliptical orbits and variable planetary speeds, he had to step outside that classical paradigm and see nature in a new way. Only then could he recognize how the planets really moved.

This makes us wonder what false paradigms surround us and prevent us from understanding nature better. Kepler was never confused by believing in geocentrism. From his days in college he accepted that the sun was the center of the universe. Of course, his first law partially violates a strict rule of heliocentrism. How?

Kepler was a Copernican, and he wrote a number of books that proclaimed his belief, but he was never seriously persecuted because he lived in northern Europe, beyond the sway of the Inquisition. Others were not so lucky. The next astronomer in our story was born in southern Europe under the rule of the Inquisition.

4-5 Galileo Galilei

Most people know two facts about Galileo, and both are wrong. Galileo did not invent the telescope, and he was not condemned by the Inquisition for believing that Earth moved around the sun. Then why is Galileo so important that in 1979, almost 400 years after his trial, the Vatican reopened his case? As we discuss Galileo, we will discover that his trial concerned not just the place of Earth and the motion of the planets, but also a new method of understanding nature, a method we call *science*.

Telescopic Observations

Galileo Galilei (Figure 4-18) was born in Pisa (in what is now Italy) in 1564, and he studied medicine at the university there. His true love, however, was mathematics, and although he had to leave school early for financial reasons, he returned only 4 years later as a

Figure 4-18
Galileo Galilei (1564–1642), remembered
as the great defender of Copernicanism,
also made important discoveries in the
physics of motion. He is honored here on
an Italian 2000-lira note. The reverse side
shows one of his telescopes at lower right
and a modern observatory above it.

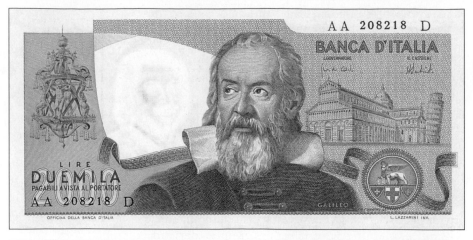

professor of mathematics. Three years after that he became professor of mathematics at the university at
Padua, where he remained for 18 years.

During this time, Galileo seems to have adopted
the Copernican model, although he admitted in a
1597 letter to Kepler that he did not support Copernicanism publicly because of the criticism such a declaration would bring. It was the telescope that drove
Galileo to publicly defend the heliocentric model.

Galileo did not invent the telescope. It was apparently invented around 1608 by lens makers in Hol-
land. Galileo, hearing descriptions in the fall of 1609,
was able to build working telescopes in his workshop.
In fact, Galileo was not the first person to look at the
sky through a telescope, but he was the first person to
observe the sky systematically and apply his observations to the theoretical problem of the day—the place
of Earth (Figure 4-19).

What Galileo saw through his telescopes was so
amazing he rushed a small book into print. *Sidereus
Nuncius* (*The Sidereal Messenger*) reported three
major discoveries. First, the moon was not perfect. It
had mountains and valleys on its surface, and Galileo
used the shadows to calculate the height of the mountains. Aristotle's philosophy held that the moon was

Figure 4-19
Galileo's telescopic discoveries generated intense interest and
controversy. Some critics refused to look through a telescope lest
it deceive them. *(Courtesy Yerkes Observatory)*

perfect, but Galileo showed that it was not only imperfect but was also a world like Earth.

The second discovery reported in the book was that the Milky Way was made up of a myriad of stars too faint to see with the unaided eye. While intriguing, this discovery could not match the third discovery. Galileo's telescope revealed four new "planets" circling Jupiter, planets that we know today as the Galilean moons of Jupiter.

The moons of Jupiter supported the Copernican model over the Ptolemaic model. Critics of Copernicus had said Earth could not move because the moon would be left behind. But Jupiter moved yet kept its satellites, so Galileo's discovery proved that Earth, too, could move and keep its moon. Also, Aristotle's philosophy included the belief that all heavenly motion was centered on Earth. Galileo showed that Jupiter's moons revolve around Jupiter, so there could be other centers of motion.

Soon after *Sidereus Nuncius* was published, Galileo made two additional discoveries. When he observed the sun, he discovered sunspots, raising the suspicion that the sun was less than perfect. Further, by noting the movement of the spots, he concluded that the sun was a sphere and that it rotated on its axis. When he observed Venus, Galileo saw that it was going through phases like those of the moon. In the Ptolemaic model, Venus moves around an epicycle centered on a line between Earth and the sun. Thus, it would always be seen as a crescent (Figure 4-20a). But Galileo saw Venus go through a complete set of phases, which proved that it did indeed revolve around the sun (Figure 4-20b).

Sidereus Nuncius was very popular and made Galileo famous. He became chief mathematician and philosopher to the Grand Duke of Tuscany in Florence. In 1611, Galileo visited Rome and was treated with great respect. He had long, friendly discussions with the powerful Cardinal Barberini, but he also made enemies. Personally, Galileo was outspoken, forceful, and sometimes tactless. He enjoyed debate, but most of all he enjoyed being right. Thus, in lectures, debates, and letters he offended important people who questioned his telescopic discoveries.

By 1616, Galileo was the center of a storm of controversy. Some critics said he was wrong, and others said he was lying. Some refused to look through a telescope lest it mislead them, and others looked and claimed to see nothing (hardly surprising given the awkwardness of those first telescopes). Pope Paul V decided to end the disruption, so when Galileo visited Rome in 1616 Cardinal Bellarmine interviewed him privately and ordered him to cease debate. There is some controversy today about the nature of Galileo's instructions, but he did not pursue astronomy for some years after the interview. Books relevant to Copernicanism were banned, although *De Revolutionibus,* recognized as an important and useful book in astronomy, was only suspended pending revision. Everyone who owned a copy of the book was required to cross out certain statements and add handwritten corrections stating that Earth's motion and the central location of the sun were only theories and not facts.

Dialogo and Trial

In 1621 Pope Paul V died, and his successor, Pope Gregory XV, died in 1623. The next pope was Galileo's friend Cardinal Barberini, who took the name Urban VIII. Galileo rushed to Rome hoping to have the prohibition of 1616 lifted, and although the new pope did not revoke the orders, he did encourage Galileo. Thus, Galileo

Figure 4-20

(a) If Venus moved in an epicycle centered on the Earth–sun line (see Figure 4-7), it would always appear as a crescent. (b) Galileo's telescope showed that Venus goes through a full set of phases, proving that it must orbit the sun.

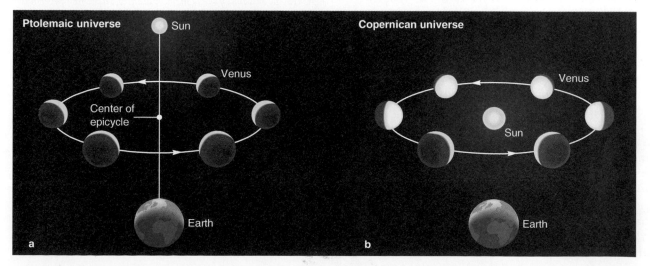

began to write his great defense of the Copernican model, finally completing it on December 24, 1629. After some delay, the book was approved by both the local censor in Florence and the head censor of the Vatican in Rome. It was printed in February 1632.

Called *Dialogo Dei Due Massimi Sistemi* (*Dialogue Concerning the Two Chief World Systems*), it confronts the ancient astronomy of Aristotle and Ptolemy with the Copernican model (Figure 4-21). Galileo wrote the book as a debate among three friends. Salviati, a swift-tongued defender of Copernicus, dominates the book; Sagredo is intelligent but largely uninformed. Simplicio is the dismal defender of Ptolemy. In fact, he does not seem very bright.

The publication of *Dialogo* created a storm of controversy, and it was sold out by August 1632, when the Inquisition ordered sales stopped. The book was a clear defense of Copernicus, and, either intentionally or unintentionally, Galileo exposed the pope's authority to ridicule. Urban VIII was fond of arguing that, as God was omnipotent, He could construct the universe in any form while making it appear to us to have a different form, and thus we could not deduce its true nature by mere observation. Galileo placed the pope's argument in the mouth of Simplicio, and Galileo's enemies showed the passage to the pope as an example of Galileo's disrespect. The pope thereupon ordered Galileo to face the Inquisition.

Galileo was interrogated by the Inquisition four times and was threatened with torture. He must have thought often of Giordano Bruno, tried, condemned, and burned at the stake in Rome in 1600. One of Bruno's offenses had been Copernicanism. But the trial did not center on Galileo's belief in Copernicanism. *Dialogo* had been approved by two censors. Rather, the trial centered on the instructions given Galileo in 1616. From his file in the Vatican, his accusers produced a record of the meeting between Galileo and Cardinal Bellarmine that included the statement that Galileo was "not to hold, teach, or defend in any way" the principles of Copernicus. Many historians believe that this document, which was signed neither by Galileo nor by Bellarmine nor by a legal secretary, was a forgery and that Galileo's true instructions were much less restrictive. But Bellarmine was dead, and Galileo had no defense.

The Inquisition condemned him not for heresy but for disobeying the orders given him in 1616. On June 22, 1633, at the age of 70, kneeling before the Inquisition, Galileo read a recantation admitting his er-

Figure 4-21
Aristotle and Ptolemy discuss astronomy with Copernicus (right) in this frontispiece from Galileo's book *Dialogue Concerning the Two Chief World Systems*. Notice the geocentric model in Ptolemy's hands and the heliocentric model held by Copernicus. (*Courtesy Owen Gingerich; by permission of the Houghton Library, Harvard University*)

rors. Tradition has it that as he rose he whispered *"E pur si muove"* ("Still it moves"), referring to Earth. Although he was sentenced to life imprisonment, he was actually confined at his villa for the next 10 years, perhaps through the intervention of the pope. He died there on January 8, 1642, 99 years after the death of Copernicus.

Galileo was not condemned for heresy, nor was the Inquisition interested when he tried to defend Copernicanism. He was tried and condemned on a charge we might call a technicality. Then why is his trial so important that historians have studied it for almost four centuries? Why have some of the world's greatest authors, including Bertolt Brecht, written about Galileo's trial? Why in 1979 did Pope John Paul II create a commission to reexamine the case against Galileo?

To understand the trial, we must recognize that it was the result of a conflict between two ways of understanding our universe. Since the Middle Ages, scholars had taught that the only path to true understanding was through religious faith. St. Augustine (A.D. 354–430) wrote *Credo ut intelligame,* which can be translated as "Believe in order to understand." But Galileo and other scientists of the Renaissance used their own observations to try to understand the universe, and when their observations contradicted Scripture, they assumed their observations were correct. Galileo paraphrased Cardinal Baronius in saying, "The Bible tells us how to go to heaven, not how the heavens go."

The trial of Galileo was not about Earth's place in the universe. It was not about Copernicanism. It wasn't really about the instructions Galileo received in 1616. It was about the birth of modern science as a rational way to understand our universe. The commission appointed by John Paul II in 1979, reporting its conclusions in October 1992, said of Galileo's inquisitors, "This subjective error of judgment, so clear to us today, led them to a disciplinary measure from which Galileo 'had much to suffer.'" Galileo was not found innocent in 1992 so much as the Inquisition was forgiven for having charged him in the first place.

REVIEW Critical Inquiry

How were Galileo's observations of the moons of Jupiter evidence against the Ptolemaic model?

Galileo presented his arguments in the form of evidence and conclusions, and the moons of Jupiter were key evidence. Ptolemaic astronomers argued from first principles given by Aristotle that the heavens were made up of perfect spheres rotating uniformly around Earth at the center. Then Earth could be the only center of motion. Although undermined by Ptolemy's epicycles and equants, this belief was still critical to classical astronomers. When Galileo saw four moons revolving around Jupiter, he knew that Earth was not the only center of motion in the universe, and thus at least that one aspect of classical astronomy had to be wrong.

Furthermore, Ptolemaic astronomers had argued that Earth could not move or it would leave the moon behind. The fact that the moon continued to circle Earth seemed to prove that Earth was immobile. But Galileo saw moons orbit Jupiter, and everyone agreed that Jupiter moved. Thus, if Jupiter could move and keep its moons, then Earth might move and keep its moon.

Of all of Galileo's telescopic observations, the moons of Jupiter caused the most debate. But the craters on the moon and the phases of Venus were also critical evidence. How did they argue against the Ptolemaic model?

Galileo faced the Inquisition because of a conflict between two ways of knowing and understanding the world. The Church taught faith and understanding through revelation, but the scientists of the age were inventing a new way to understand nature that relied on the evidence of observation and measurement. Thus, the problems of the place of Earth and the motion of the planets triggered the birth of modern science. The final episode in that adventure has at its center a man who wondered what makes an apple fall.

4-6 Isaac Newton and Orbital Motion

We date the birth of modern astronomy and of modern science from the 99 years between the deaths of Copernicus and Galileo. The Renaissance is commonly taken to be the period between 1350 and 1600, and thus the 99 years of this story lie at the culmination of the reawakening of learning in all fields (Figure 4-22). Not only did the world adopt a new model of the universe, but it also adopted a new way of understanding our place in nature.

The problem of the place Earth was resolved by the Copernican Revolution, but the problem of planetary motion was only partly solved by Kepler's laws. For the last 10 years of his life, Galileo studied the nature of motion, especially the accelerated motion of falling bodies. Although he made some important progress, he was not able to relate his discoveries about motion to the heavens. That final step fell to Isaac Newton.

Isaac Newton

Galileo died in January 1642. Some 11 months later, on Christmas day 1642,* a baby was born in the English village of Woolsthorpe. His name was Isaac Newton (Figure 4-23), and his life represented the first flower of the seeds planted by the four astronomers of our story.

*Because England had not yet reformed its calendar, December 25, 1642, in England was January 4, 1643, in Europe. It is only a small deception to use the English date and thus include Newton's birth in our 99-year history.

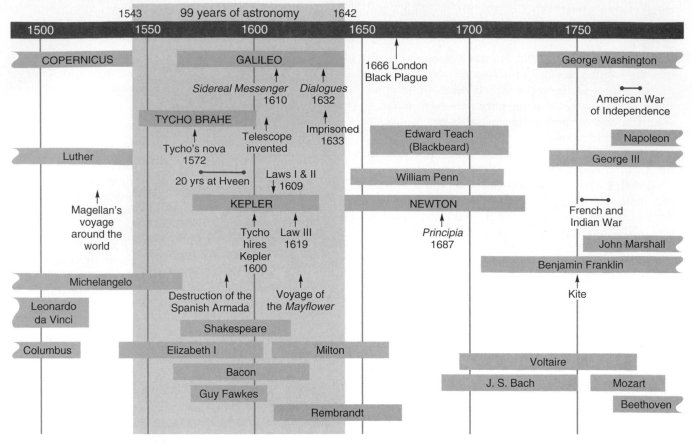

Figure 4-22

The 99 years between the death of Copernicus in 1543 and the birth of Newton in 1642 marked the transition from the ancient astronomy of Ptolemy and Aristotle to the revolutionary theory of Copernicus. This period saw the birth of modern scientific astronomy.

Figure 4-23

Isaac Newton (1642–1727) worked from the discoveries of Galileo and Kepler to study motion and gravitation. He and some of his discoveries are honored on this English 1-pound note.

Newton was a quiet child from a farming family, but his work at school was so impressive that his uncle financed his education at Trinity College, where he studied mathematics and physics. In 1665, the Black Plague swept through England, and the colleges were closed. During 1665 and 1666, Newton spent his time in Woolsthorpe, thinking and studying. It was during these years that he made most of his discoveries in optics, mechanics, and mathematics. Among other things, he studied optics, developed three laws of motion, divined the nature of gravity, and invented differential calculus. The publication of his work in his book *Principia* in 1687 placed science on a firm analytical base.

One of the most often used and least often stated principles of science is cause and effect, and we could argue that Newton's second law of motion was the first clear statement of the principle. Ancient philosophers such as Aristotle argued that objects moved because of tendencies. They said that earth and water, and objects made of earth and water, had a natural tendency to move toward the center of the universe. This natural motion had no cause but was inherent in the nature of the objects. But Newton's second law says $F = ma$. If an object of mass m changes its motion by an acceleration a, then it must be acted on by a force F.

The principle of cause and effect goes far beyond motion. The principle of cause and effect gives scientists confidence that every effect has a cause. Hearing loss in certain laboratory rats, color changes in certain chemical dyes, and explosions on certain stars are all effects that must have causes. All of science is focused on understanding the causes of the effects we see. If the universe were not rational, then we could never expect to discover causes. Newton's second law of motion was arguably the first statement that the behavior of the universe depends rationally on causes.

It is beyond the scope of this book to analyze all of Newton's work, but his laws of motion and gravity had an important impact on the future of astronomy. From his study of the work of Galileo, Kepler, and others, Newton extracted three laws that relate the motion of a body to the forces acting on it (Table 4-2). These laws made it possible to predict exactly how a body would move if the forces were known (Window on Science 4-3).

When Newton thought carefully about motion, he realized that some force must pull the moon toward Earth's center. If there were no such force altering the moon's motion, it would continue moving in a straight line and leave Earth forever. It can circle Earth only if Earth attracts it. Newton's insight was to recognize that the force that holds the moon in its orbit is the same as the force that makes apples fall from trees. The force of gravity is universal—that is, all objects attract all other objects with a force that depends only on their masses and on the distance between their centers.

The **mass** of an object is a measure of the amount of matter in the object, usually expressed in kilograms. Mass is not the same as weight. An object's *weight* is the force that Earth's gravity exerts on the object. Thus an object in space far from Earth might have no weight, but it would contain the same amount of matter and would thus have the same mass that it has on Earth.

Newton also realized that the distance between the objects is important. He recognized that the force of gravity decreases as the square of the distance between the objects increases. Specifically, if the distance from, say, Earth to the moon were doubled, the gravitational force between them would decrease by a factor of 2^2, or 4. If the distance were tripled, the force would decrease by a factor of 3^2, or 9. This relationship is known as the **inverse square relation.** (We will discuss it in more detail in Chapter 8 where we apply it to the intensity of light.)

With these definitions of mass and the inverse square relation, we can describe Newton's law of gravity in a simple equation:

$$F = -G\frac{Mm}{r^2}$$

Here M and m are the masses of two objects, r is the distance between their centers, G is the gravitational constant, and F is the force of gravity acting between the two objects. That is, the force of gravity attracting two objects to each other equals the gravitational constant times the product of their masses divided by the square of the distance between the objects.

Orbital Motion

Newton's laws of motion and gravity solved a problem that had perplexed Kepler—why the planets move along their orbits. Kepler speculated that magnetic fields might pull the planets along their orbits, and he also wondered if angels beating their wings

---

Table 4-2 Newton's Three Laws of Motion

I. A body continues at rest or in uniform motion in a straight line unless acted upon by some force.

II. A body's change of motion is proportional to the force acting on it and is in the direction of the force.

III. When one body exerts a force on a second body, the second body exerts an equal and opposite force back on the first body.

pushed the planets around the sun. Newton's work gives us a much deeper insight into planetary motion.

To illustrate the principle of orbital motion, imagine that we place a large cannon at the top of a mountain, point the cannon horizontally, and fire it. The cannonball falls to Earth some distance from the foot of the mountain. The more gunpowder we use, the faster the ball travels, and the farther from the foot of the mountain it falls. If we use enough powder, the ball travels so fast it never strikes the ground. Our planet's gravity pulls it toward Earth's center, but Earth's surface curves away from it at the same rate at which it falls. Thus we say it is in orbit (Figure 4-24). If the cannonball is high above the atmosphere, where there is no friction, it will fall around Earth forever.

Simple physics says that an object in motion tends to stay in motion in a straight line unless acted upon by some force. Thus the cannonball in this example travels in a curve around Earth only because Earth's gravity acts to pull it away from its straight-line motion.

How fast does an object have to go to remain in orbit? Newton could answer that question because he could describe the force of gravity mathematically. The **circular velocity** is the velocity needed to maintain a circular orbit, and it is given by the formula in By the Numbers 4-1. A satellite such as the space shuttle just above Earth's atmosphere must travel about 7800 m/sec, which equals a bit over 17,000 mph. The moon, having a lower mass, has a lower circular velocity. A satellite orbiting the moon just above its surface need only travel 1700 m/sec (3800 mph).

Figure 4-24
A cannon on a high mountain could put its projectile into orbit if it could achieve a high enough velocity. Newton published a similar figure in *Principia*.

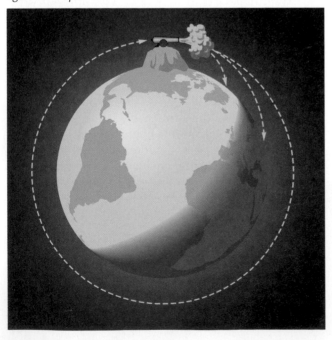

Newton's explanation of orbital motion is only one example of the power of his conception of nature. His laws of motion and gravity were *general* laws that described the motions of all bodies under the action of external forces. In addition, the laws were *productive* because they made possible specific calculations that could be tested by observation. For example, Newton's laws of motion could be used to derive Kepler's third law from the law of gravity.

Newton's discoveries remade astronomy into an analytical science in which astronomers could measure the positions and motions of celestial bodies, calculate the gravitational forces acting on them, and predict their future motion (Window on Science 4-4).

Were we to trace the history of astronomy after Newton, we would find scientists predicting the motion of comets, the gravitational interaction of the planets, the motions of double stars, and so on. Astronomers built on the discoveries of Newton, just as he had built on the discoveries of Copernicus, Tycho, Kepler, and Galileo. It is the nature of science to build

By the Numbers 4-1
Circular Velocity

Circular velocity is the velocity a satellite must have to remain in a circular orbit around a larger body. If we assume the mass of the satellite is small compared to the central body, then the circular velocity is given by

$$V_c = \sqrt{\frac{GM}{r}}$$

In this formula, M is the mass of the central body in kilograms, r is the radius of the orbit in meters, and G is the gravitational constant, 6.67×10^{-11} N m²/kg², where N stands for a newton, a measure of force. This formula is all we need to calculate how fast an object must travel to stay in a circular orbit.

For example, how fast does the moon travel in its orbit? The mass of Earth is 5.98×10^{24} kg, and the radius of the moon's orbit is 3.84×10^{8} m. The the moon's velocity is:

$$V_c = \sqrt{\frac{6.67 \times 102aa \times 5.98 \times 10^{24}}{3.84 \times 10^{8}}}$$

$$= \sqrt{\frac{39.9 \times 10^{13}}{3.84 \times 10^{8}}}$$

$$= \sqrt{1.04 \times 10^{6}} = 1020 \text{ m/sec}$$

This calculation shows that the moon travels 1.02 km along its orbit each second. That is the circular velocity of the moon.

When you read about any science, you should notice that scientific theories face in two directions. They look back into the past and explain phenomena we have previously observed. For example, Newton's laws of motion and gravity explained how the planets moved. But theories also face forward in that they enable us to make predictions about what we should find as we explore further. Thus, Newton's laws allowed astronomers to calculate the orbits of comets, predict their return, and eventually understand their origin.

Scientific predictions are important in two ways. First, if a theory leads to a prediction and scientists later discover the prediction was true, the theory is confirmed, and scientists gain confidence that it is a true description of nature. But predictions are important in science in a second way. Using an existing theory to make a prediction may lead us into an unexplored avenue of knowledge. Thus, the first theories of genetics made predictions that confirmed the genetic theory of inheritance, but those predictions also created a new understanding of how living creatures evolve.

As you read about any scientific theory, think about both what it can explain and what it can predict.

on the discoveries of the past, and Newton was thinking of that when he wrote, "If I have seen farther than other men, it is because I stood upon the shoulders of giants."

REVIEW Critical Inquiry

What do the words *universal* and *mutual* mean in the term *universal mutual gravitation*?

Newton realized that gravity was a force that drew the moon toward Earth. But his third law of motion says that forces always act in pairs, so if Earth attracts the moon, then the moon must attract Earth. That is, gravitation is *mutual* between any two objects.

When Newton looked beyond the Earth–moon system, he realized that Earth's gravity must attract the sun as well as the moon, and if Earth attracts the sun, then the sun must attract Earth. Furthermore, if the sun has gravity that acts on Earth, then the same gravity acts on all the planets and even on the distant stars. Step by step, Newton's third law of motion led him to conclude that gravitation applies to all masses in the universe. That is, it is *universal*.

Aristotle explained gravity in a totally different way. How did Aristotle account for a falling apple? Could that explanation account for a hammer falling on the surface of the moon?

Newton said he stood on the shoulders of giants, and we too have that vantage point. We understand what we see in the sky because we understand the work of Copernicus, Kepler, and Newton—not just their discoveries, but also their creation of a new way to understand nature. Modern astronomers gaze at the sky with powerful, modern tools, the subject of the next chapter.

Summary

Classical astronomy was based on the writings of the Greek philosopher Aristotle. He taught that Earth was the immobile center of the universe and that the stars and planets were carried around Earth by great crystalline spheres. This model of the universe was given mathematical form about A.D. 140 in the *Almagest,* the great work of Ptolemy. Ptolemy preserved the classical belief in geocentrism, but he replaced the concentric, Earth-centered heavenly spheres with a system of epicycles, deferents, and equants and thus tried to create a mathematical model that could accurately predict the positions of the sun, moon, and planets.

In contrast to the geocentric universe of classical astronomy, the universe devised by Copernicus was heliocentric, or sun-centered. One advantage of a heliocentric universe is that retrograde motion, the occasional westward motion of the planets, is easily explained. Copernicus did not publish his book *De Revolutionibus* until 1543, the year he died. The teachings of Aristotle had become part of Church teachings. As a critic of the classical view that Earth is at the center of the universe, Copernicus was exploring controversial ideas, ideas that some would claim were heretical.

The Danish astronomer Tycho Brahe did not accept the Ptolemaic or the Copernican model but rather developed his own, in which the sun and moon circled Earth, and the planets circled the sun. Although his hypothesis was not correct, Tycho made precise observations of planetary positions that later led to a true understanding of planetary motion.

Johannes Kepler, Tycho Brahe's assistant, inherited the Danish astronomer's records in 1601 and used his observations to uncover three laws of planetary motion. Kepler discovered that the planets follow ellipses with the sun at one focus, that they move faster when near the sun, and that a planet's period squared is proportional to its orbital radius cubed.

Galileo Galilei was a great defender of the Copernican hypothesis. Galileo was the first person to use a telescope to observe the heavens and to recognize the significance of what he saw. His discoveries of the phases of Venus, the satellites of Jupiter, the mountains of the moon, and other phenomena helped undermine the Ptolemaic universe. In 1633, Galileo was finally condemned before the Inquisition for refusing to halt his defense of Copernicanism.

Born in 1642, the same year that Galileo died, Isaac Newton used the work of Kepler and Galileo to discover three laws of motion and the law of gravity. These laws made it possible to understand the orbital motion of the planets as a consequence of the sun's gravity. In addition, Newton's work made it possible to analyze the motion of any celestial body and predict its path in the future.

The 99 years from the death of Copernicus to the birth of Newton marked the birth of modern science. From that time on, science depended on evidence to support theories and relied on the analytic methods first demonstrated by Newton.

New Terms

parallax	ellipse
geocentric universe	semimajor axis, *a*
uniform circular motion	eccentricity, *e*
retrograde motion	hypothesis
epicycle	theory
deferent	natural law
equant	mass
heliocentric	inverse square relation
paradigm	circular velocity

Review Questions

1. Why did Greek astronomers conclude that the heavens were made up of perfect crystalline spheres moving at constant speeds?

2. Why did classical astronomers conclude that Earth had to be motionless?

3. How did the Ptolemaic model explain retrograde motion?

4. In what ways were the models of Ptolemy and Copernicus similar?

5. Why did the Copernican hypothesis win gradual acceptance?

6. Why is it difficult for scientists to replace an old paradigm with a new paradigm?

7. Why did Tycho Brahe expect the new star of 1572 to show parallax? Why was the lack of parallax evidence against the Ptolemaic model?

8. How was Brahe's model of the universe similar to the Ptolemaic model? How did it resemble the Copernican model?

9. Explain how Kepler's laws contradict uniform circular motion.

10. What is the difference between a hypothesis, a theory, and a law?

11. How did the *Alphonsine Tables,* the *Prutenic Tables,* and the *Rudolphine Tables* differ?

12. Review Galileo's telescope discoveries, and explain why they supported the Copernican model and contradicted the Ptolemaic model.

13. Galileo was condemned by the Inquisition, but Kepler, also a Copernican, was not. Why not?

14. Why did Newton conclude that gravitation had to be universal?

15. Explain why we might describe the orbital motion of the moon with the statement, "The moon is falling."

Discussion Questions

1. Why might Copernicus have delayed publication of his book? Discuss as many reasons as possible.

2. Why might Tycho Brahe have hesitated to hire Kepler? Why do you suppose he appointed Kepler his scientific heir?

Problems

1. If you lived on Mars, which planets would describe retrograde loops? Which would never be visible as crescent phases?

2. Galileo's telescope showed him that Venus has a large angular diameter (61 seconds of arc) when it is a crescent and a small angular diameter (10 seconds of arc) when it is nearly full. Use the small-angle formula to find the ratio of its maximum distance to its minimum distance. Is this ratio compatible with the Ptolemaic universe shown in Figure 4-7?

3. Galileo's telescopes were not of high quality by modern standards. He was able to see the rings of Saturn, but he never reported seeing features on Mars. Use the small-angle formula to find the angular diameter of Mars when it is closest to Earth. How does that compare with the maximum diameter of Saturn's rings?

4. If a planet had an average distance from the sun of 10 AU, what would its orbital period be?

5. If a space probe were sent into an orbit around the sun that brought it as close as 0.5 AU to the sun and as far away as 5.5 AU, what would its orbital period be?

6. Pluto orbits the sun with a period of 247.7 years. What is its average distance from the sun?

7. Calculate the circular velocity of Venus and Saturn

around the sun. (*Hint:* The mass of the sun is 2×10^{30} kg.)

8. The circular velocity of Earth around the sun is about 30 km/sec. Are the arrows for Venus and Saturn correct in Figure 4-10b? (*Hint:* See Problem 7.)

9. What is the orbital velocity of an Earth-satellite 42,000 km from Earth? How long does it take to circle its orbit once?

Critical Inquiries for the Web

1. The trial of Galileo is an important event in the history of science. We now know, and the Church now recognizes, that Galileo's view was correct, but what were the arguments on both sides of the issue as it was un-folding? Research the Internet for documents chronicling the trial, Galileo's observations and publications, and the position of the Church. Use this information to outline a case for and against Galileo in the context of the times in which the trial occurred.

2. It's hard to imagine that an observatory could exist before the invention of the telescope, but Tycho Brahe's observatory at Hveen was a great astronomical center of its day. Search the Web sites on Tycho and his instruments, and describe what an observing session at Hveen might have involved.

 Go to the Brooks/Cole Astronomy Resource Center (www.brookscole.com/astronomy) for critical thinking exercises, articles, and additional readings from InfoTrac College Edition, Brooks/Cole's online student library.

CHAPTER 5

Astronomical Tools

He burned his house down for the

fire insurance and spent the proceeds

on a telescope.

Robert Frost
The Star-Splitter

Guidepost

Previous chapters have described the sky as it appears to our unaided eyes, but modern astronomers turn powerful telescopes on the sky. Chapter 5 introduces us to the modern astronomical telescope and its delicate instruments.

The study of the universe is so challenging that astronomers cannot ignore any source of information, and so they use the entire spectrum, from gamma rays to radio waves. This chapter shows how critical it is for astronomers to understand the nature of light.

In each of the 15 chapters that follow, we will study the universe using information gathered by the telescopes and instruments described in this chapter.

Starlight is going to waste. Every night, light from the stars falls on trees, oceans, roofs, and empty parking lots, and it is all wasted. To an astronomer, nothing is so precious as starlight. It is our only link to the sky, and the astronomer's quest is to gather as much starlight as possible and extract from it the secrets of the stars.

The telescope is the emblematic tool of the astronomer because its purpose is to gather and concentrate light for analysis. Nearly all of the interesting objects in the sky are faint sources of light, so modern astronomers are driven to build the largest possible telescopes to gather the maximum amount of light. Thus, our discussion of astronomical tools is concentrated on large telescopes and the specialized tools used to analyze light.

If we wish to gather visible light, a normal telescope will do; but, as we will see in this chapter, visible light is only one kind of radiation. We can extract information from other forms of radiation by using specialized telescopes. Radio telescopes, for example, give us an entirely different view of the sky. Some of these specialized telescopes can be used from Earth's surface, but some must go into orbit above Earth's atmosphere. Telescopes that observe X rays, for instance, must be placed in orbit.

In addition to using telescopes to gather light, astronomers need instruments to analyze light. Such instruments measure brightness, record images, or form spectra.

Astronomers no longer study the sky by mapping constellations, charting the phases of the moon, or following the retrograde motion of the planets. Modern astronomers use the most sophisticated telescopes and instruments to analyze starlight. Thus, we begin this chapter with a discussion of the nature of light.

5-1 Radiation: Information from Space

Just as a book on bread baking might begin with a discussion of flour, our chapter on telescopes begins with a discussion of light—not just visible light, but the entire range of radiation from the sky.

Light as a Wave and a Particle

If you have admired the colors in a soap bubble, you have seen light behave as a wave. But when that light enters the light meter on a camera, it behaves as a particle. How it behaves depends on how we observe it— it is both wave and particle.

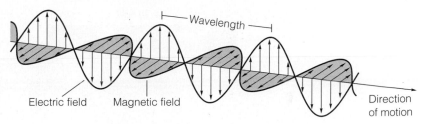

Figure 5-1
Electric and magnetic fields travel together through space as electromagnetic radiation. The wavelength, commonly represented by the Greek letter lambda (λ), is the distance between successive peaks in the wave.

We experience waves whenever we hear sound. Sound waves are a mechanical disturbance that travels through the air from source to ear. Sound requires a medium; so on the moon, where there is no air, there can be no sound. In contrast, light is made up of electric and magnetic waves that can travel through empty space. Unlike sound, light does not require a medium and thus can travel through a perfect vacuum.

Because light is made up of both electric and magnetic fields, we refer to it as **electromagnetic radiation.** As we are about to see, visible light is only one form of electromagnetic radiation. These oscillating electric and magnetic fields travel through space at about 300,000 km/sec (186,000 m/sec). This is commonly referred to as the speed of light c, but it is in fact the speed of all such radiation.

Electromagnetic radiation is a wave phenomenon—that is, it is associated with a periodically repeating disturbance, a wave. We are familiar with waves in water: If we disturb a pool of water, waves spread across the surface. Imagine that we use a meter stick to measure the distance between the successive peaks of a wave. This distance is the **wavelength,** usually represented by the Greek letter lambda (λ) (Figure 5-1).

While light does behave as a wave, it also behaves as a particle. We call a particle of light a **photon,** and we can think of a photon as a bundle of waves.

The amount of energy a photon carries depends inversely on its wavelength. That is, shorter wavelength photons carry more energy, and longer wavelength photons carry less. We can express this relationship in a simple formula:

$$E = \frac{hc}{\lambda}$$

Here h is Planck's constant (6.6262×10^{-34} joule sec), and c is the speed of light (3×10^8 m/sec). The important point in this formula is not the value of the constants but the inverse relationship between the energy E and the wavelength λ. As λ gets smaller, E gets larger. Thus a photon of visible light carries a very small

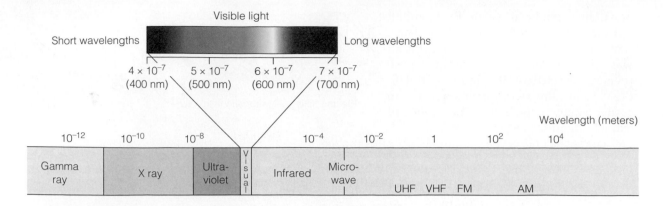

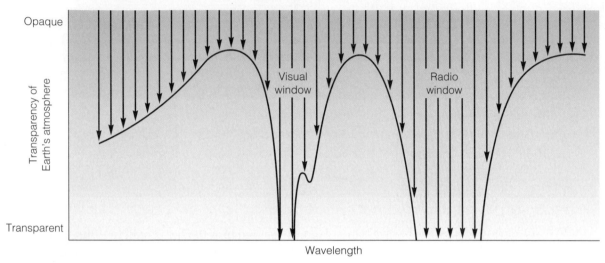

Figure 5-2
The spectrum of visible light, extending from red to violet, is only part of the electromagnetic spectrum. Most radiation is absorbed in Earth's atmosphere, and only radiation in the visual window and the radio window can reach Earth's surface.

amount of energy, but a photon with a wavelength much shorter than that of visible light can carry much more energy.

The Electromagnetic Spectrum

A spectrum is an array of electromagnetic radiation in order of wavelength. We are most familiar with the spectrum of visible light, which we see in rainbows. The colors of the spectrum differ in wavelength with red having the longest wavelength and violet the shortest, as shown in the visible spectrum at the top of Figure 5-2.

The average wavelength of visible light is about 0.0005 mm. We could put 50 light waves end to end across the thickness of a sheet of household plastic wrap. It is too awkward to measure such short distances in millimeters, so we will measure the wavelength of light using the **nanometer (nm),** one billionth of a meter (10^{-9} m). Another unit that astronomers commonly use is called the **angstrom (Å)** (named after the Swedish astronomer Anders Jonas Ångström). One angstrom is 10^{-10} m. The wavelength of visible light

ranges between 400 nm and 700 nm or between 4000 Å and 7000 Å. Radio astronomers often refer to wavelength using meters, centimeters, or millimeters.

Figure 5-2 shows how the visible spectrum makes up only a small part of the electromagnetic spectrum. Beyond the red end of the visible spectrum lies infrared radiation, where wavelengths range from 700 nm to about 1 mm. Our eyes are not sensitive to this radiation, but our skin senses it as heat. A heat lamp is nothing more than a bulb that gives off large amounts of infrared radiation.

Beyond the infrared part of the electromagnetic spectrum lie radio waves. The radio radiation used for AM radio transmissions has wavelengths of a few kilometers down to a few hundred meters, while FM, television, military, governmental, and ham radio transmissions have wavelengths that range down to a few tens of centimeters. Microwave transmissions, used for radar and long-distance telephone communications, for instance, have wavelengths from a few centimeters down to about 1 mm.

You may not think of radio waves in terms of wavelength because radio dials are marked in units of frequency, the number of waves that pass a stationary point in one second. To calculate the wavelength of a radio wave, divide the speed of light by the frequency. Thus, when you tune in your favorite FM station at 89.5 MHz (million cycles per second), you are adjusting your radio to detect radio photons with a wavelength of 335 cm.

The distinction between the wavelength ranges is not sharp. Long-wavelength infrared radiation and the shortest microwave radio waves are the same. Similarly, there is no clear division between the short-wavelength infrared and the long-wavelength part of the visible spectrum. It is all electromagnetic radiation.

Look once again at the electromagnetic spectrum in Figure 5-2, and notice that electromagnetic waves shorter than violet are called *ultraviolet.* Electromagnetic waves even shorter are called *X rays,* and the shortest are gamma rays. Again, the boundaries between these wavelength ranges are not clearly defined.

X rays and gamma rays can be dangerous, and even ultraviolet photons have enough energy to do us harm. Small doses produce a suntan; larger doses can cause sunburn, and extreme doses might produce skin cancers. Contrast this to the lower-energy infrared photons. Individually, they have too little energy to affect skin pigment, a fact that explains why you can't get a tan from a heat lamp. Only by concentrating many low-energy photons in a small area, as in a microwave oven, can we transfer significant amounts of energy.

We are interested in electromagnetic radiation because it brings us clues to the nature of stars, planets, and other celestial objects. Earth's atmosphere is opaque to most electromagnetic radiation, as shown by the graph at the bottom of Figure 5-2. Gamma rays, X rays, and some radio waves are absorbed high in Earth's atmosphere, and a layer of ozone (O_3) at an altitude of about 30 km absorbs ultraviolet radiation. Water vapor in the lower atmosphere absorbs the longer-wavelength infrared radiation. Only visible light, some shorter-wavelength infrared, and some radio waves reach Earth's surface through two wavelength regions called **atmospheric windows.** Obviously, if we wish to study the sky from Earth's surface, we must look out through one of these windows.

REVIEW Critical Inquiry

What could we see if our eyes were sensitive only to X rays?

Sometimes the critical analysis of an idea is easier if we try to imagine a totally new situation. In this case, we might at first expect to be able to see through walls, but remember that our eyes detect only light that already exists. There are almost no X rays bouncing around at Earth's surface, so if we had X-ray eyes, we would be in the dark and would be unable to see anything. Even when we looked up at the sky, we would see nothing, because Earth's atmosphere is not transparent to X rays. If Superman can see through walls, it is not because his eyes can detect X rays.

But suppose our eyes were sensitive only to radio waves. Would we be in the dark?

Now that we know something about electromagnetic radiation, we can study the tools astronomers use to gather and analyze that radiation.

5-2 Optical Telescopes

Astronomers build optical telescopes to gather light and focus it into sharp images. This requires sophisticated optical and mechanical designs, and it leads astronomers to build gigantic telescopes on the tops of high mountains. To begin, we need to understand the terminology of telescopes, but it is more important to understand how different kinds of telescopes work and why some are better than others.

Two Kinds of Telescopes

Astronomical telescopes focus light into an image in one of two ways, as shown in Figure 5-3. A lens bends (refracts) the light as it passes through the glass and brings it to a focus to form a small inverted image. A mirror — a concave piece of glass with a reflective surface — forms an image by reflecting the light. In either case, the **focal length** is the distance from the lens or mirror to the image formed of a distant light source, such as a star (Figure 5-4). Short-focal-length lenses and mirrors must be strongly curved, and long-focal-length lenses and mirrors are less strongly curved. Grinding the proper shape on a lens or mirror is a delicate, time-consuming, and expensive process.

The main lens in a refracting telescope is called the **primary lens,** and the main mirror in a reflecting telescope is called the **primary mirror.** These are also called the **objective lens** and **mirror.** Both kinds of telescopes form a very small, inverted image that is difficult to observe directly, so astronomers use a small lens called the **eyepiece** to magnify the image and make it convenient to view.

Because there are two ways to focus light, there are two kinds of astronomical telescopes. **Refracting telescopes** use a large lens to gather and focus the light, while **reflecting telescopes** use a concave mirror

Figure 5-3

(a) To see how a lens can focus light, we trace three light rays from the flame and base of a candle through a lens, where they are refracted to form an inverted image. (b) A mirror forms an image by reflection from a concave surface. Notice that the light is reflected from the aluminized front surface of the mirror and does not enter the glass. Thus, mirrors do not produce chromatic aberration.

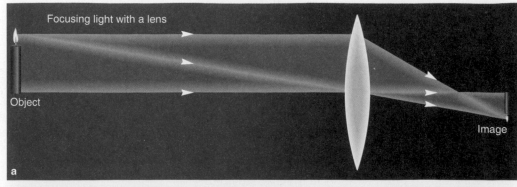

Focusing light with a lens

Object

Image

a

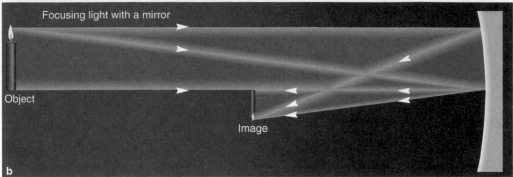

Focusing light with a mirror

Object

Image

b

Figure 5-4

The focal length of a lens is the distance from the lens to the point where parallel rays of light come to a focus. The lens at the top has a longer focal length than does the lens at the bottom.

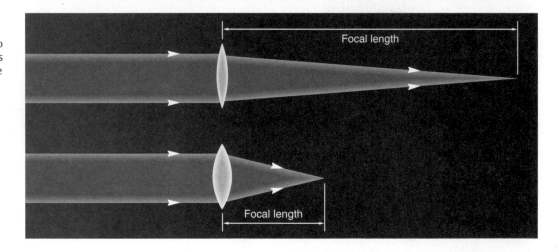

Focal length

Focal length

(Figure 5-5). The advantages of the reflecting telescope have made it the preferred design for modern observatories.

Refracting telescopes suffer from a serious optical distortion that limits their usefulness. When light is refracted through glass, shorter wavelengths bend more than longer wavelengths, and blue light comes to a focus closer to the lens than does red light (Figure 5-6a). If we focus the eyepiece on the blue image, the red light is out of focus, and we see a red blur around the image. If we focus on the red image, the blue light blurs. The color separation is called **chromatic aberration.** Telescope designers can grind a telescope lens of two components made of different kinds of glass and so bring two different wavelengths to the same focus (Figure 5-6b). This does improve the image, but these **achromatic lenses** are not totally free of chromatic aberration, because other wavelengths still blur. Telescopes made with such lenses were popular until the end of the 19th century.

The primary lens of a refracting telescope is very expensive to make because it must be achromatic, and the glass must be pure and flawless because the light passes through the lens. The four surfaces must be ground precisely, and the lens can be supported only along its edge. The largest refracting telescope in the world is at Yerkes Observatory in Wisconsin. Its lens, 1 m (40 inches) in diameter, weighs half a ton, and the glass sags under its own weight. Larger refracting telescopes are prohibitively expensive.

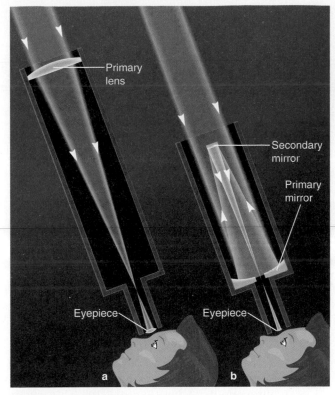

Figure 5-5
(a) A refracting telescope uses a primary lens to focus starlight into an image that is magnified by a lens called an eyepiece. The primary lens has a long focal length, and the eyepiece has a short focal length. (b) A reflecting telescope uses a primary mirror to focus the light by reflection. A small secondary mirror reflects the starlight back down through a hole in the middle of the primary mirror to the eyepiece.

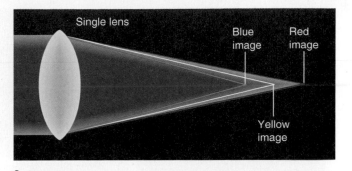

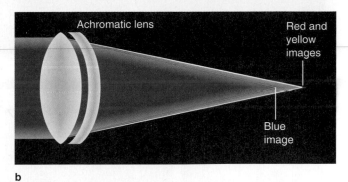

Figure 5-6
(a) A normal lens suffers from chromatic aberration because short wavelengths bend more than long wavelengths. (b) An achromatic lens, made in two parts, can bring any two colors to the same focus, but other colors remain slightly out of focus.

Reflecting telescopes are much less expensive because only the front surface of the mirror must be ground to precise shape. The glass of the mirror need not be perfectly transparent, and the mirror can be supported over its back surface to reduce sagging. Most important, reflecting telescopes do not suffer from chromatic aberration, because the light is reflected toward the focus before it can enter the glass. Thus, every large astronomical telescope built in this century has been a reflecting telescope.

The only difficulty with a reflecting telescope is looking through it. The mirror forms an image in the middle of the telescope tube where it is inconvenient to place your eye. Isaac Newton solved this problem by using a small mirror to direct the light out the side of the telescope tube (Figure 5-7), and these **Newtonian telescopes** have been popular for smaller instruments. In the largest telescopes, an astronomer can ride in a cage inside the tube to make observations at the **prime focus.** In the **Cassegrain telescope,** the light is reflected back through a hole in the middle of the primary mirror to a convenient observing location. Many large telescopes are Cassegrains (Figure 5-8). **Schmidt-Cassegrain telescopes,** using a thin lens

added to a Cassegrain telescope, are popular with amateur astronomers. Many other focal arrangements are used for special purposes, such as photography.

Whatever their focal arrangement, nearly all professional telescopes are operated from a separate control room. This is more comfortable for the astronomers, but it also makes more efficient use of the telescope. Some telescopes can now be operated from control rooms halfway around the world.

In addition to gathering and focusing light, astronomical telescopes must move easily and smoothly to point anywhere in the sky, and that places special restrictions on the machinery that supports the telescope.

Telescope Mountings

A telescope mounting can be more expensive than the telescope itself. The mounting must support the optics, protect against vibration, move accurately to designated objects, and then compensate for Earth's rotation. Because Earth rotates eastward, the telescope mounting must contain a **sidereal drive** (literally, a star drive) that can move the telescope smoothly westward to follow the celestial object being studied. To simplify this motion, most telescope mountings are **equatorial mountings** with one of its two axes of rotation, the **polar axis,** parallel to Earth's axis. Rotation

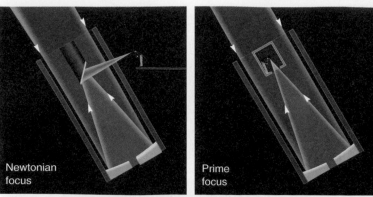

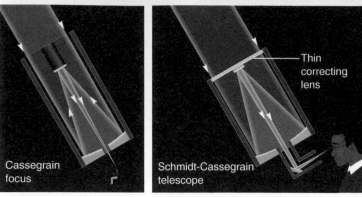

Figure 5-7
The mirror of a reflecting telescope forms the image in the middle of the telescope tube, so astronomers have devised various focal arrangements to bring the light to a convenient place for observing. Observing in a prime focus cage is possible only in the largest telescopes. Most large telescopes use the Cassegrain focus, but many amateur astronomers use the smaller Schmidt-Cassegrain telescopes.

around the polar axis moves the telescope parallel to the celestial equator (Figure 5-9).

Improvements in computer technology are making telescope mountings and observatories much less expensive. Instead of building a large, awkward equatorial mounting inside a large, heavy dome, astronomers now take advantage of the short focal lengths typical of reflecting telescopes. Such telescopes are so short they can be mounted like a cannon; that is, the mounting can move the telescope both in altitude (perpendicular to the horizon) and in azimuth (parallel to the horizon). Such **alt-azimuth mountings** are compact, but they must move simultaneously about both axes at varying rates in order to track stars in different parts of the sky. This complicated motion is made possible by computers that guide the telescope.

Figure 5-8
The 4-meter Mayall telescope at Kitt Peak National Observatory can be used as a prime focus telescope or, with the secondary mirror in place, as a Cassegrain telescope. Note the human figure at the lower right for scale. *(© Association of Universities for Research in Astronomy, Inc., Kitt Peak National Observatory)*

Such computer-controlled alt-azimuth mountings are stronger and smaller and do not require buildings as large as those required by equatorial mountings. In fact, some of the newest telescopes are designed so that the telescope is part of the observatory building. Like the turret of a tank, the entire building rotates as the telescope moves.

Astronomical telescopes are obviously large and expensive. We might wonder what advantages they give us.

The Powers of a Telescope

A telescope can aid our eyes in only three ways. These are called light-gathering power, resolving power, and magnifying power.

Most interesting celestial objects are faint sources of light, so we need a telescope that can gather large amounts of light to produce a bright image. **Light-gathering power** refers to the ability of a telescope to collect light. Catching light in a telescope is like catching rain in a bucket — the bigger the bucket, the more rain it catches (Figure 5-10). This is why astronomers use large telescopes and why they refer to telescopes by diameter.

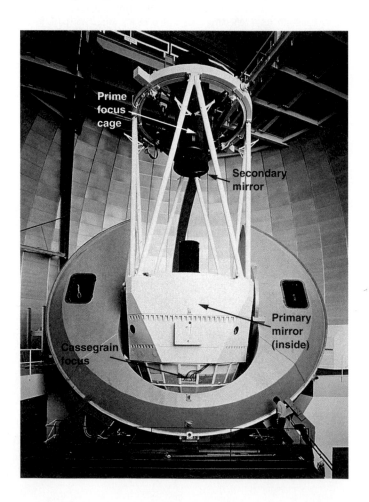

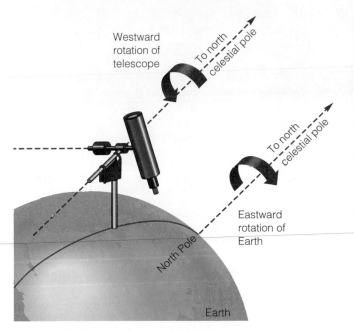

Figure 5-9
Westward motion around the polar axis of an equatorial mounting counters Earth's eastward rotation and keeps the telescope pointed at a given star.

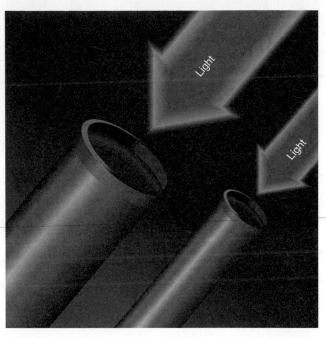

Figure 5-10
Gathering light is like catching rain in a bucket. A large-diameter telescope gathers more light and has a brighter image than does a smaller telescope of the same focal length.

The second power, **resolving power,** refers to the ability of the telescope to reveal fine detail. Because light acts as a wave, it produces a small **diffraction fringe** around every point of light in the image, and we cannot see any detail smaller than the fringe (Figure 5-11). Astronomers can't eliminate diffraction fringes, but the larger a telescope is in diameter, the smaller the diffraction fringes are. Thus the larger the telescope, the better its resolving power.

Two other factors—optical quality and atmospheric conditions—limit resolving power. A telescope must contain high-quality optics to achieve its full potential resolving power. Even a large telescope

shows us little detail if its optics are marred with imperfections. In addition, when we look through a telescope, we look through miles of turbulent air in Earth's atmosphere, which makes the image dance and blur, a condition called **seeing** (Figure 5-12). A related phenomenon is the twinkling of a star. The twinkles are caused by turbulence in Earth's atmosphere, and a star near the horizon, where we look through more air, will twinkle more than a star overhead.

On a night when the atmosphere is unsteady, the stars twinkle, the images are blurred, and the seeing is bad. Even under good seeing conditions, the detail visible through a large telescope is limited, not by its

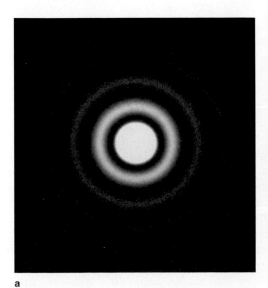

a

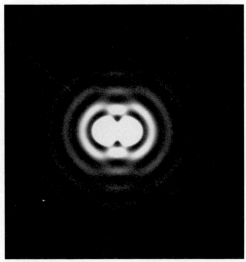

b

Figure 5-11
(a) Stars are so far away that their images are points, but the wave nature of light surrounds each star image with diffraction fringes (much magnified in this computer model). (b) Two stars close to each other have overlapping diffraction fringes and become impossible to detect separately. *(Computer model by M. A. Seeds)*

Have you ever seen a movie in which the hero magnifies a newspaper photo and reads some tiny detail? It isn't really possible, because newspaper photos are made up of tiny dots of ink, and no detail smaller than a single dot will be visible no matter how much you magnify the photo. In fact, all images are made up of elements of some sort, and that means there is a limit to the amount of detail you can see in an image. In an astronomical image, the resolution is often set by seeing. It is foolish to attempt to see a detail in the image that is smaller than the resolution.

This limitation is true of all measurements in science. A zoologist might be trying to measure the length of a live snake, or a sociologist might be trying to measure the attitudes of people toward drunk driving, but both face limits to the resolution of their measurements. The zoologist might specify that the snake was 43.28932 cm long, and the sociologist might say that 98.2491 percent of people oppose drunk driving, but a critic might point out that it isn't possible to make these measurements that accurately. The resolution of the techniques does not justify the accuracy implied.

Science is based on measurement, and whenever we make a measurement we should ask ourselves how accurate that measurement can be. The accuracy of the measurement is limited by the resolution of the measurement technique, just as the amount of detail in a photograph is limited by the resolution of the photo.

diffraction fringes, but by the air through which the observer must look. A telescope performs best on a high mountaintop, where the air is thin and steady, but even there Earth's atmosphere limits the detail the telescope can reveal.

This limitation on the amount of information in an image is related to the limitation on the accuracy of a measurement. All measurements have some built-in uncertainty (Window on Science 5-1), and scientists must learn to work within those limitations.

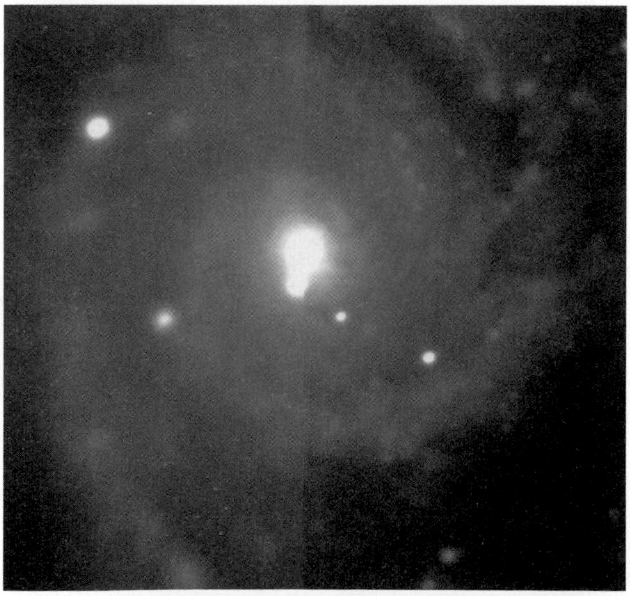

The third and least important power of a telescope is **magnifying power,** the ability to make the image bigger. Since the amount of detail we can see is limited by the seeing conditions and the resolving power, very high magnification does not necessarily show us more detail. Also, we can change the magnification by changing the eyepiece, but we cannot alter the telescope's light-gathering or resolving power.

Compare an astronomer's telescope with a biologist's microscope. A microscope is designed primarily to magnify and thus show us things too small to see. An astronomer's telescope solves a different problem. Its primary function is to gather light and show us things too faint to see.

If you visit a department store to shop for a telescope, you will probably find telescopes described according to magnification. One may be labeled an "80-power telescope" and another a "40-power telescope." However, the magnifying power really tells us little about the telescopes. Astronomers identify telescopes by diameter because that determines both light-gathering power and resolving power.

Nearly all major observatories are located far from major cities and usually on high mountains. Astronomers avoid cities because **light pollution,** the brightening of the night sky by light scattered from artificial outdoor lighting, can make it impossible to see faint objects (Figure 5-13). In fact, many residents of cities are unfamiliar with the beauty of the night sky because they can see only the brightest stars. Astronomers prefer to place their telescopes on carefully selected high mountains. The air there is thin and more transparent, but, most important, astronomers select mountains where the air flows smoothly and is not tur-

Figure 5-12
The left half of this photograph of a galaxy is from an image recorded on a night of poor seeing. Small details are blurred. The right half of the figure is from an image recorded on a night when Earth's atmosphere above the telescope was steady and the seeing was better. Much more detail is visible under good seeing conditions. *(Courtesy William Keel)*

Figure 5-13
This satellite view of the continental United States at night shows the light pollution produced by outdoor lighting. Not only does the glare drown out the fainter stars and interfere with astronomy, but it wastes electrical power. Many astronomers work with city governments to enact laws that improve lighting on the ground and reduce light scattered into the night sky. *(National Geophysical Data Center)*

bulent. This produces the best seeing. Building an observatory on top of a high mountain far from civilization is difficult and expensive, but the dark sky and steady seeing make it worth the effort (Figure 5-14).

When we compare telescopes, we should consider their powers. By the Numbers 5-1 shows how to calculate the powers of a telescope. This will be useful if you decide to buy a telescope of your own.

Buying a Telescope

Thinking about how we should shop for a new telescope will not only help us if we decide to buy one but will also illustrate some important points about astronomical telescopes.

Assuming we have a fixed budget, we should buy the highest-quality optics and the largest-diameter telescope we can afford. Of the two things that limit what we see, optical quality is under our control. We can't make the atmosphere less turbulent, but we should buy good optics. If we buy a telescope from a toy store and it has plastic lenses, we shouldn't expect to see very much. Also, we want to maximize the light-gathering power of our telescope, so we want to purchase the

Figure 5-14
The southern Gemini telescope, 8 meters in diameter, is being built on top of an 8895-ft-high mountain in the Chilean Andes. A twin telescope is being built at 13,882 ft atop the extinct volcano Mauna Kea in Hawaii. The difficulty of building giant telescopes on remote mountaintops is justified by the dark skies, clear air, and the outstanding seeing. *(Gemini 8-m Telescopes Project)*

The Powers of a Telescope

Light-gathering power is proportional to the area of the telescope objective. A lens or mirror with a large area gathers a large amount of light. The area of a circular lens or mirror of diameter D is $\pi(D/2)^2$. To compare the relative light-gathering powers (LGP) of two telescopes A and B, we can calculate the ratio of the areas of their objectives, which reduces to the ratio of their diameters (D) squared.

$$\frac{LGP_A}{LGP_B} = \left(\frac{D_A}{D_B}\right)^2$$

Example A: Suppose we compare a 4-cm telescope with a 24-cm telescope. How much more light will the large telescope gather? *Solution:*

$$\frac{LGP_{24}}{LGP_4} = \left(\frac{24}{4}\right)^2 = 6^2 = 36 \text{ times more light}$$

Example B: Our eye acts like a telescope with a diameter of about 0.8 cm, the diameter of the pupil. How much more light can we gather if we use a 24-cm telescope? *Solution:*

$$\frac{LGP_{24}}{LGP_{eye}} = \left(\frac{24}{0.8}\right)^2 = (30)^2 = 900 \text{ times more light}$$

The resolving power of a telescope is the angular distance between two stars that are just barely visible through the telescope as two separate images. The resolving power α, in seconds of arc, equals 11.6 divided by the diameter of the telescope in centimeters:

$$\alpha = \frac{11.6}{D}$$

Example C: What is the resolving power of a 25-cm telescope? *Solution:*

$$\alpha = \frac{11.6}{25} = 0.46 \text{ seconds of arc}$$

If the lenses are of good quality and if the seeing is good, we should be able to distinguish as separate points of light any pair of stars farther apart than 0.46 seconds of arc. If the stars are any closer together, diffraction fringes blur the stars together into a single image (see Figure 5-11).

The magnification M of a telescope is the ratio of the focal length of the objective lens or mirror F_o divided by the focal length of the eyepiece F_e:

$$M = \frac{F_o}{F_e}$$

Example D: What is the magnification of a telescope whose focal length is 80 cm used with an eyepiece whose focal length is 0.5 cm? *Solution:* The magnification is 80 divided by 0.5, or 160 times.

largest-diameter telescope we can afford. Given a fixed budget, that means we should buy a reflecting telescope (which includes Schmidt-Cassegrains) rather than a refracting telescope. Not only will we get more diameter per dollar, but our telescope will not suffer from chromatic aberration.

We can safely ignore magnification. Department stores and camera stores may advertise telescopes by quoting their magnification, but it is not an important number. What we can see is fixed by light-gathering power, optical quality, and Earth's atmosphere. Besides, we can change the magnification by changing eyepieces.

Other things being equal, we should choose a telescope with a solid mounting that will hold the telescope steady and allow us to point at objects easily. A sidereal drive would be very useful even on a small telescope, and computer-controlled pointing systems are available for a price on many small telescopes. A good telescope on a poor mounting is almost useless.

We might be buying a telescope to put in our backyard, but we must think about the same issues astronomers consider when they design giant telescopes to go on mountaintops. In fact, some of the newest telescopes solve these tradiational problems in new ways.

New-Generation Telescopes

For most of this century, astronomers faced a serious limitation on the size of telescopes. If a large telescope mirror were to sag under its own weight, it would not be able to focus the light precisely. To avoid this sag, astronomers made telescope mirrors very thick, but that produced two further problems—weight and cost.

Supporting a heavy mirror requires a massive telescope. The two largest traditional telescopes in the world are the 6-m (236-inch) reflector in the former Soviet Union and the 5-m (200-inch) Hale Telescope on Mount Palomar. The 5-m mirror weighs 14.5 tons, and the mounting needed to support this mirror weighs about 530 tons. Grinding one of these mirrors to shape is expensive, because it requires the slow removal of large amounts of glass. The 5-m mirror, for instance, was begun in 1934 and finished in 1948, and 5 tons of glass were ground away.

Astronomers have been developing new techniques to make large mirrors that weigh and cost less. For example, the Steward Observatory Mirror Laboratory has built a revolving oven under the football stands at the University of Arizona that can produce preshaped mirrors. The oven turns like a merry-go-round, and the molten glass flows outward in the mold to form a concave upper surface. After they slowly cool, the preshaped mirrors can be quickly ground to final shape. The oven is currently able to cast mirrors as large as 8.4 m in diameter.

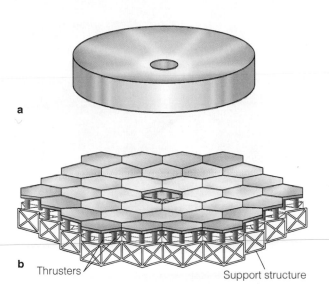

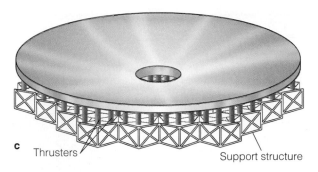

Figure 5-15
The problem and two solutions: Conventional telescope mirrors (a) had to be made very thick so they could not sag out of shape. To reduce the weight and increase the diameter, astronomers have begun making mirrors of segments (b) or as relatively thin disks (c). To hold these mirrors in shape, computer-controlled thrusters (red) are positioned under the mirrors.

a

b
Thrusters
Support structure

c
Thrusters
Support structure

A mirror of conventional thickness with diameter of 8 m would be too heavy to support in a telescope, so astronomers have devised ways to make the mirrors thinner. One technique is to make the mirror in segments (Figure 5-15a and b). Small segments are less expensive and sag less under their own weight. The largest general-purpose optical telescopes in the world are the twin 10-m Keck I and Keck II Telescopes in Hawaii. Each of these giant telescopes uses 36 hexagonal mirror segments held in alignment by computer-controlled thrusters to form a single mirror (Figure 5-16a). The special purpose Hobby–Eberly Telescope in Texas uses 91 hexagonal segments to make up a mirror 11 m in diameter; it is especially designed to focus starlight into a sophisticated spectrograph (Figure 5-16b).

Another way to reduce the weight of telescope mirrors is to make them thin. Thin mirrors sag so easily under their own weight that they are called floppy mirrors, but a computer can control their shape in what astronomers call **active optics.** The New Technology Telescope at the European Southern Observatory in Chile contains a 3.58-m (141-inch) mirror only 24 cm (10 inches) thick. With its active optics, the telescope produces very high-quality images.

Segmented or floppy, thin telescope mirrors have yet another advantage—they cool quickly. After sunset the air cools quckly, but a massive telescope mirror cannot cool rapidly and uniformly. As some parts of the mirror cool faster than others, the material expands and contracts, producing strains in the mirror that distort the image. Thin telescope mirrors cool more quickly and uniformly as night falls and thus produce better images.

Some astronomers used to hesitate to build big Earth-based telescopes because seeing limits the detail

Figure 5-16
Segmented mirrors can be assembled to create very-large-diameter telescopes. (a) The hexagonal mirror segments in the Keck I Telescope surround a worker crouching in the Cassegrain opening. Each segment is 1.8 m (70 inches) in diameter and is supported and aligned by computer-controlled actuators *(© Russ Underwood/W. M. Keck Observatory)* (b) The 91 hexagonal mirrors of the Hobby–Eberly Telescope in Texas are visible here reflecting the supporting frame on the interior of the dome. Computer controlled, the mirror segments make up a mirror 11 m in diameter. *(Photo courtesy Dr. Thomas G. Barnes III, Mc-Donald Observatory, University of Texas at Austin)*

a

b

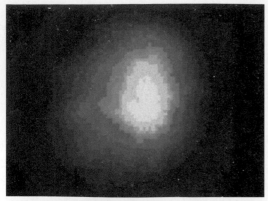

a Adaptive optics off

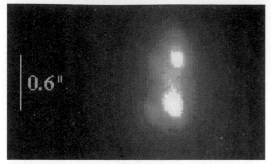

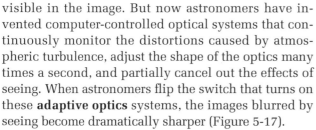

b Adaptive optics on

Figure 5-17
(a) With the adaptive optics system turned off, the telescope images a star as a broad blur distorted by seeing. (b) With the adaptive optics turned on, seeing is partially eliminated, and the object is revealed to be a double star. Even the best seeing limits the detail we can see to about 1 second of arc. This adaptive optics system reveals much smaller details. *(Paul Kalas)*

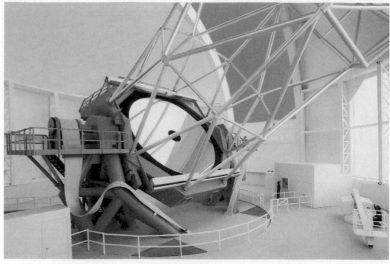

a

b

Figure 5-18
(a) This model of one of the two Gemini Telescopes shows how the 8-m mirror will be carried in an alt-azimuth mount that can rotate horizontally and tip up and down. Shutters in the sides of the dome (right) will open at night to help the telescope cool and to reduce air turbulence inside the dome. (See Figure 5-14.) *(The Gemini 8-m Telescope Project)* (b) The four 8.2-m telescopes of the Very Large Telescope in the Andes of Chile are housed in separate domes, but they can combine their light as one telescope. *(ESO)*

visible in the image. But now astronomers have invented computer-controlled optical systems that continuously monitor the distortions caused by atmospheric turbulence, adjust the shape of the optics many times a second, and partially cancel out the effects of seeing. When astronomers flip the switch that turns on these **adaptive optics** systems, the images blurred by seeing become dramatically sharper (Figure 5-17).

The technical advances discussed here have stimulated the construction of a number of giant telescopes. A consortium of nations is building the Gemini Telescopes, a pair of 8-m thin-mirror telescopes, one in Hawaii and one in Chile (Figure 5-18a). Japanese astronomers have built the Subaru (Pleiades) Telescope with an 8.3-m mirror, and Arizona astronomers are building the Large Binocular Telescope, which will carry two 8.4-m mirrors on a single mounting. The European Southern Observatory has built the first of four 8.2-m telescopes that will eventually observe as a single telescope, the Very Large Telescope, from the Andes in Chile (Figure 5-18b). Even more giant telescopes are being planned.

REVIEW Critical Inquiry

Why do astronomers build observatories at the tops of mountains?

Astronomers have joked that the hardest part of building a new observatory is constructing the road to the top of the mountain. It certainly isn't easy to build a large, delicate telescope at the top of a high mountain, but it is worth the effort. A telescope on top of a high mountain is above the thickest part of Earth's atmosphere. There is less air to dim the light, and there is less water vapor to absorb infrared radiation. Even more important, the thin air on a mountaintop causes less disturbance to the image, and thus the seeing is better. A large telescope with modern, high-quality optics is capable of detecting very small details. It is the seeing that limits the detail that astronomers can detect. It really is worth the trouble to build telescopes atop high mountains where the air is thin, dry, and steady.

Astronomers not only build telescopes on mountaintops; they also build gigantic telescopes many meters in diameter. What are the problems and advantages in building such giant telescopes?

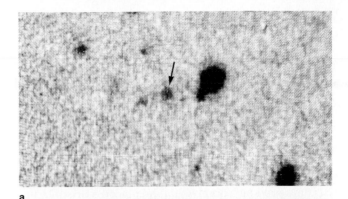

a

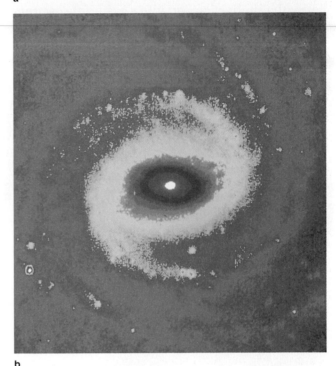

b

Figure 5-19
(a) The arrow points to an electronic image of a faint galaxy so distant that light took 10 billion years to reach us. This image is reproduced as a negative. The sky is light, and the stars and galaxy are black. *(Hyron Spinrad)* (b) This false-color image of the galaxy NGC 1232 displays different levels of brightness as different colors. *(CARA)*

Astronomers sometimes refer to a telescope that produces distorted images as a "light bucket." In a sense, all astronomical telescopes are just light buckets, because the light they focus into images tells us very little until it is recorded and analyzed by special instruments attached to the telescopes.

5-3 Special Instruments

Just looking through a telescope doesn't tell us much. To use an astronomical telescope to learn about stars, we must be able to analyze the light the telescope gathers. Special instruments attached to the telescope make that possible.

Imaging Systems

The original imaging device in astronomy was the photographic plate. It could record faint objects in long exposures and could be stored for later analysis, but photographic plates have been almost entirely replaced in astronomy by electronic imaging systems.

Most modern astronomers use **charge-coupled devices (CCDs)** to record images. A CCD is a specialized computer chip containing roughly a million microscopic light detectors arranged in an array about the size of a postage stamp. These devices can be used like a small photographic plate, but they have dramatic advantages. They can detect both bright and faint objects in a single exposure, are much more sensitive than a photographic plate, and can be read directly into computer memory for later analysis. Although CCDs for astronomy are extremely sensitive and therefore expensive, less sophisticated CCDs are used in most video cameras and digital electronic cameras.

The image from a CCD is stored in computer memory as numbers, so it is easy to manipulate the image to bring out details that would not otherwise be visible. For example, astronomical images are often reproduced as negatives with the sky white and the stars dark. This makes the faint parts of the image easier to see (Figure 5-19a). Astronomers also manipulate images to produce **false-color images** in which the colors represent different levels of intensity and are not related to the true colors of the object (Figure 5-19b).

Measurements of intensity and color were made in the past using a photometer, a highly sensitive light meter attached to a telescope. Today, however, most such measurements are made on CCD images. Because the CCD image is easily digitized, brightness and color can be measured to high precision.

The Spectrograph

To analyze light in detail, we need to spread the light out according to wavelength into a spectrum, a task performed by a **spectrograph.** We can understand how this works if we reproduce an experiment performed by Isaac Newton in 1666. Boring a hole in his window shutter, Newton admitted a thin beam of sunlight into his darkened bedroom. When he placed a prism in the beam, the sunlight spread into a beautiful spectrum on the far wall. From this Newton concluded that white light was made of a mixture of all the colors.

Newton didn't think in terms of wavelength, but we can use that modern concept to see that the light passing through the prism is bent at an angle that depends on the wavelength. Violet (short wavelength) bends most, and red (long wavelength) least. Thus, the white light entering the prism is spread into a spectrum (Figure 5-20a). A typical prism spectrograph contains more than one prism to spread the light farther and lenses to guide the light into the prism and to focus the light onto a photographic plate.

Nearly all modern spectrographs use a grating in place of a prism. A **grating** is a piece of glass with thousands of microscopic parallel lines scribed onto its surface. Different wavelengths of light reflect from the grating at slightly different angles, so white light is spread into a spectrum and can be recorded, often by a CCD camera.

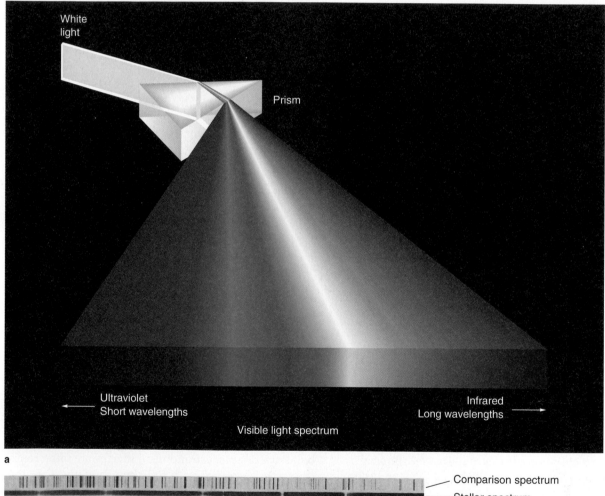

White light

Prism

Ultraviolet
Short wavelengths

Infrared
Long wavelengths

Visible light spectrum

a

Comparison spectrum
Stellar spectrum
Comparison spectrum

b

Figure 5-20
(a) A prism bends light by an angle that depends on the wavelength of the light. Short wavelengths bend most and long wavelengths least. Thus, white light passing through a prism is spread into a spectrum. (b) In this negative image, the bright spectrum of the star is the dark streak running from left to right in the middle of the image. The gaps in the star's spectrum are produced by different gases in the star. Above and below the stellar spectrum are the lines of a comparison spectrum, which astronomers use to calibrate the spectrum of the star and measure wavelengths to high precision. *(Palomar Observatory)*

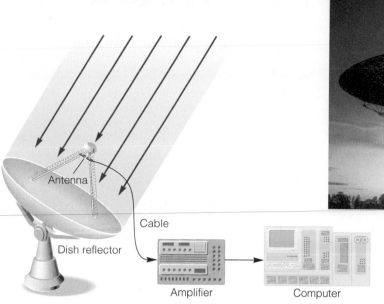

Because astronomers need to measure the precise location of features in a spectrum, most spectrographs can add a **comparison spectrum** above and below the spectrum of the observed object (Figure 5-20b). The features in this comparison spectrum have known wavelengths carefully measured in the laboratory, and thus the astronomer can use the comparison spectrum as a precise calibration of the wavelengths in the spectrum being studied.

Because astronomers understand how light interacts with matter, a spectrum carries a tremendous amount of information (as we will see in the next chapter), and that makes a spectrograph the astronomer's most powerful instrument. An astronomer recently remarked, "We don't know anything about an object 'til we get a spectrum," and that is only a slight exaggeration.

REVIEW Critical Inquiry

What is the difference between light going through a lens and light passing through a prism?

A refracting telescope producing chromatic aberration and a prism dispersing light into a spectrum are two examples of the same thing, but one is bad and one is good. When light passes through the curved surfaces of a lens, different wavelengths are bent by slightly different amounts, and the different colors of light come to focus at different focal lengths. This produces the color fringes in an image called chromatic aberration, and that's bad. But the surfaces of a prism are made to be precisely flat, so all of the light enters the prism at the same angle, and any given wavelength is

bent by the same amount wherever it meets the prism. Thus, white light is dispersed into a spectrum. We could call the dispersion of light by a prism "controlled chromatic aberration," and that's good.

CCDs have been very good for astronomy, and they are now widely used. Explain why they are more useful than photographic plates.

So far, our discussion has been limited to visual wavelengths. Now it is time to consider the rest of the electromagnetic spectrum.

5-4 Radio Telescopes
Operation of a Radio Telescope

A radio telescope usually consists of four parts: a dish reflector, an antenna, an amplifier, and a recorder (Figure 5-21). The components, working together, make it possible for astronomers to detect radio radiation from celestial objects.

The dish reflector of a radio telescope, like the mirror of a reflecting telescope, collects and focuses radiation. Because radio waves are much longer than light waves, the dish need not be as smooth as a mirror. In some radio telescopes, the reflector may not even be dish-shaped, or the telescope may contain no reflector at all.

Though a radio telescope's dish may be many meters in diameter, the antenna may be as small as your hand. Like the antenna on a TV set, its only function is to absorb the radio energy and direct it along a cable to an amplifier. After amplification, the signal goes to some kind of recording instrument. Most radio observatories record data on magnetic tape or feed it directly to a computer. However it is recorded, an observation with a radio telescope measures the amount of radio energy coming from a specific point on the sky.

Because humans can't see radio waves, astronomers must convert them into something perceptible. One way is to measure the strength of the radio signal at various places in the sky and draw a map in which contours mark areas of uniform radio intensity. We might compare such a map to a seating diagram for a baseball stadium in which the contours mark areas in which the seats have the same price (Figure 5-22a). Contour maps are very common in radio astronomy and are often reproduced using false colors (Figure 5-22b).

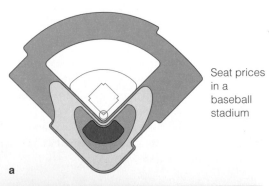

Seat prices in a baseball stadium

a

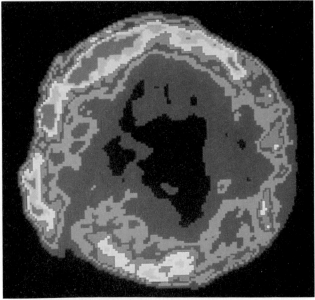

b

Figure 5-22
(a) A contour map of a baseball stadium shows regions of similar admission prices. The most expensive seats are those behind home plate. (b) A false-color-image radio map of Tycho's supernova remnant, the expanding shell of gas produced by the explosion of a star in 1572. The radio contour map has been color-coded to show intensity. Red is the strongest radio intensity, and violet the weakest. *(Courtesy NRAO)*

Limitations of the Radio Telescope

A radio astronomer works under three handicaps: poor resolution, low intensity, and interference. We saw that the resolving power of an optical telescope depends on the diameter of the objective lens or mirror. It also depends on the wavelength of the radiation. At very long wavelengths, like those of radio waves, images become fuzzy because of the large diffraction fringes. As with an optical telescope, the only way to improve the resolving power is to build a bigger telescope. Consequently, radio telescopes must be quite large.

Even so, the resolving power of a radio telescope is not good. A dish 30 m in diameter receiving radiation with a wavelength of 21 cm has a resolving power of about 0.5°. Such a radio telescope would be unable to show us any details in the sky smaller than the moon. Fortunately, radio astronomers can combine two or more radio telescopes to improve the resolving power. Such a linkup of radio telescopes is called a **radio interferometer** and has the resolving power of a radio telescope whose diameter equals the separation of the radio telescopes (Figure 5-23a). For example, the Very Large Array (VLA) radio interferometer shown in Figure 5-23b uses multiple radio dishes spread across the New Mexico–Arizona desert to simulate a single radio telescope with a diameter of 40 km (25 miles). It can produce radio maps with a resolution better than 1 second of arc. The Very Long Baseline Array (VLBA) consists of matching radio dishes spread from Hawaii to the Virgin Islands and has an effective diameter of about 8000 km. HALCA, the Highly Advanced Laboratory for Communications and Astronomy, consists of an 8-m radio dish in orbit around Earth. It can be used with ground-based radio telescopes to produce maps as good as a radio telescope 32,000 km in diameter. In these ways, radio astronomers use interferometers to compensate for the low resolving power of a single radio telescope.

The second handicap radio astronomers face is the low intensity of the radio signals. We saw earlier that the energy of a photon depends on its wavelength. Photons of radio energy have such long wavelengths that their individual energies are quite low. In order to get strong signals focused on the antenna, the radio astronomer must build large collecting dishes.

The largest radio dish in the world is 300 m (1000 ft) in diameter. So large a dish can't be supported in the usual way, so it is built into a mountain valley in Arecibo, Puerto Rico. The reflecting dish is a thin metallic surface supported above the valley floor by cables attached near the rim, and the antenna hangs above the dish on cables from three towers built on three mountain peaks that surround the valley (Figure 5-24). Although this telescope can look only overhead, the operators can change its aim slightly by

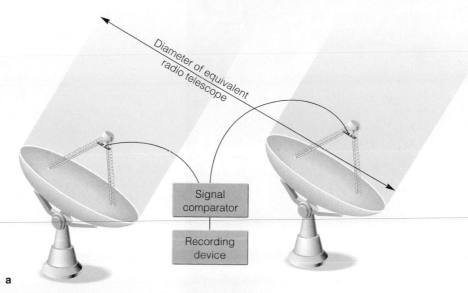

Figure 5-23

(a) A radio interferometer consists of two or more radio telescopes whose signals are combined to give the resolving power of a much larger instrument. (b) The Very Large Array radio interferometer uses up to 27 dishes along the 20-km-long arms of a Y-shaped layout to simulate a radio telescope 40 km in diameter. *(NRAO)*

Diameter of equivalent radio telescope

Signal comparator

Recording device

a

b

moving the antenna and by waiting for Earth's rotation to point the telescope in the proper direction. This may sound clumsy, but the telescope's ability to detect weak radio sources, together with its good resolution, makes it one of the most important radio observatories in the world. Chinese astronomers are now designing a similar 500-m radio telescope they hope to build in a mountain valley in China.

The third handicap the radio astronomer faces is interference. A radio telescope is an extremely sensitive radio receiver listening to radio signals thousands of times weaker than artificial radio and TV transmissions. Such weak signals are easily drowned out by interference. Sources of such interference include everything from poorly designed transmitters in Earth satellites to automobiles with faulty ignition systems.

To avoid this kind of interference, radio astronomers locate their telescopes as far from civilization as possible. Hidden deep in mountain valleys, they are able to listen to the sky protected from human-made radio noise.

Advantages of Radio Telescopes

Building large radio telescopes in isolated locations is expensive, but three factors make it all worthwhile. First, and most important, a radio telescope can show us where clouds of cool hydrogen are located between the stars. Because 90 percent of the atoms in the universe are hydrogen, that is important information. Large clouds of cool hydrogen are completely invisible to normal telescopes, because they produce no visible light of their own and reflect too little to be detected on photographs. However, cool hydrogen emits a radio signal at the specific wavelength of 21 cm. (We will see how the hydrogen produces this radiation when we discuss our galaxy in Chapter 12.) The only way we can detect these clouds of gas is with a radio telescope that receives the 21-cm radiation. These hydrogen clouds are the places where stars are born, and that is one reason that radio telescopes are so important.

Nevertheless, there is a second reason. Because radio signals have relatively long wavelengths, they can penetrate the vast clouds of dust that obscure our view at visual wavelengths. Light waves are short, and they interact with tiny dust grains floating in space; thus, the light is scattered and never penetrates the

Figure 5-24
The largest radio telescope in the world is the 300-m (1000-ft) dish suspended in a valley in Arecibo, Puerto Rico. The antenna hangs above the dish on cables stretching from towers. The Arecibo Observatory is part of the National Astronomy and Ionosphere Center, which is operated by Cornell University under contract with the National Science Foundation. *(David Parker/Science Photo Library)*

dust to reach optical telescopes on Earth. However, radio signals from far across the galaxy pass unhindered through the dust, giving us an unobscured view.

Finally, a radio telescope can detect objects that are more luminous at radio wavelengths than at visible wavelengths. This includes everything from the coldest clouds of gas to the hottest stars. Some of the most distant objects in the universe, for instance, are detectable only at radio wavelengths.

REVIEW Critical Inquiry

Why do optical astronomers build big telescopes, while radio astronomers build groups of widely separated smaller telescopes?

Comparison of seemingly similar concepts is a powerful tool in critical analysis because it often reveals subtle differences. In this case, optical astronomers are trying to maximize light-gathering power, but radio astronomers try to maximize resolving power. Because radio waves are so much longer than light waves, a single radio telescope can't see details in the sky much smaller than the moon. By linking radio telescopes miles apart, radio astronomers build a radio interferometer that can simulate a radio tele-

scope miles in diameter and thus increase the resolving power.

This difference between the wavelengths of light and radio waves makes a big difference in building the best telescopes. But why don't radio astronomers build their telescopes on mountaintops as optical astronomers do?

Our atmosphere causes trouble for Earth's astronomers in two ways. It distorts images, and it absorbs many wavelengths. The only way to avoid the limitations completely is to send telescopes above the atmosphere, into space.

5-5 Space Astronomy

Ground-based telescopes can operate only at wavelengths in the visual and radio windows of the atmosphere. Most of the rest of the electromagnetic radiation—infrared, ultraviolet, X ray, and gamma ray—never reaches Earth's surface. To observe at these wavelengths, telescopes must fly above the atmosphere in high-flying aircraft, rockets, balloons, and satellites. The only exception is some observations of the shorter infrared wavelengths that can be made from high mountains.

Infrared Astronomy

Some infrared radiation does leak through our atmosphere. This radiation enters narrow, partially open atmospheric windows scattered from 1200 nm to about 40,000 nm. Infrared astronomers usually measure wavelength in micrometers (10^{-6} meters), so they refer to this wavelength range as 1.2 to 40 micrometers. In this range, called the near infrared, much of the radiation is absorbed by water vapor, carbon dioxide, and oxygen molecules in Earth's atmosphere, so it is an advantage to place telescopes on mountains where the air is thin and dry. A number of important infrared telescopes, for example, observe from the 4150-m (13,600-ft) summit of Mauna Kea in Hawaii. At this altitude, they are above much of the water vapor, which is the main absorber of infrared.

The far-infrared range, which includes wavelengths longer than 40 micrometers, can tell us about planets, comets, forming stars, and other cool objects, but these wavelengths are absorbed high in the atmosphere. To observe in the far infrared, telescopes must venture to high altitudes. Remotely operated infrared telescopes suspended under balloons have reached altitudes as high as 41 km (25 miles). For many years, a NASA jet transport carried a 91-cm infrared tele-

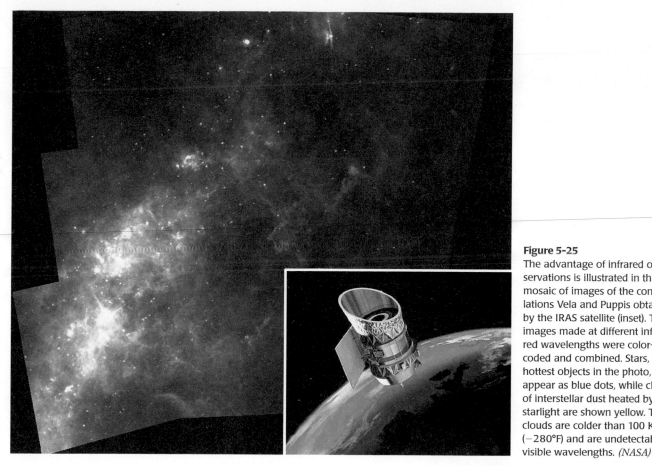

Figure 5-25
The advantage of infrared observations is illustrated in this mosaic of images of the constellations Vela and Puppis obtained by the IRAS satellite (inset). Three images made at different infrared wavelengths were color-coded and combined. Stars, the hottest objects in the photo, appear as blue dots, while clouds of interstellar dust heated by starlight are shown yellow. These clouds are colder than 100 K (−280°F) and are undetectable at visible wavelengths. *(NASA)*

scope and a crew of astronomers to altitudes of 12,000 m (40,000 ft) to get above 99 percent of the water vapor in Earth's atmosphere. Now retired from service, that airborne observatory will soon be replaced with the Stratospheric Observatory for Infrared Astronomy (SOFIA), a Boeing 747 that will carry a 2.5-m telescope to the fringes of the atmosphere.

The ultimate solution is to place infrared telescopes in space above the atmosphere. In the early 1980s, the Infrared Astronomy Satellite (IRAS) mapped the sky at infrared wavelengths (Figure 5-25). In the middle 1990s, European astronomers launched the Infrared Space Observatory, which carried detectors more sensitive than IRAS. NASA now plans for the Space Infrared Telescope Facility (SIRTF) to be a major infrared observatory in space.

If a telescope observes at far-infrared wavelengths, then it must be cooled. Infrared radiation is heat, and if the telescope is warm it will emit many times more infrared radiation than that coming from a distant object. Imagine trying to look at a dim, moonlit scene through binoculars that are glowing brightly. In a telescope observing near-infrared wavelengths, only the detector, the element on which the infrared radiation is focused, must be cooled. To observe in the far infrared, however, as IRAS did, the entire telescope must be cooled.

Ultraviolet Astronomy

Ultraviolet radiation with wavelengths shorter than about 290 nm, the far ultraviolet, is completely absorbed by the ozone layer in our atmosphere. The ozone layer extends from 20 km to about 40 km above Earth's surface. Telescopes to observe in the far ultraviolet must get above the ozone layer, and that means they must go into space.

One of the most successful ultraviolet observatories was the International Ultraviolet Explorer (IUE). Launched in 1978, it carried a 45-cm (18-inch) telescope and used TV systems to record spectra from 320 nm to 115 nm. Although IUE was expected to last only a year or two, it survived and was widely used by astronomers from all around the world until, partially shut down because of budget cutbacks, it finally failed in 1996.

The Extreme Ultraviolet Explorer (EUVE) telescope was launched in 1992 to observe at wavelengths from 100 nm to 10 nm. With four telescopes and more modern detectors than IUE, EUVE surveyed the entire sky and studied specific targets at its shorter wavelength range.

Observations in the infrared reveal cool material, but observations in the ultraviolet tend to show hot, excited regions. Hot stars and hot gas are mapped by such telescopes.

X-Ray Astronomy

Beyond the UV, at wavelengths from 10 nm to 0.01 nm, lie the X rays. These photons can be produced only by high-energy processes, so we see them coming from very hot regions in stars and from violent events such as matter smashing onto a neutron star—a subject we will discuss in Chapter 11. X-ray images can give us information about the heavens that we can get in no other way.

Although early X-ray observations of the sky were made from balloons and small rockets in the 1960s, the age of X-ray astronomy did not really begin until 1970, when an X-ray telescope named Uhuru (Swahili for "freedom") was put into orbit. Uhuru detected nearly 170 separate sources of celestial X rays. In the late 1970s, the three High Energy Astronomy Observatories (HEAO) satellites, carrying more sensitive and more sophisticated equipment, pushed the total to many hundreds. The second HEAO satellite, named the Einstein Observatory, used special optics to produce X-ray images. A number of space telescopes now observe the sky at X-ray wavelengths, including satellites that image the sun (Figure 5-26).

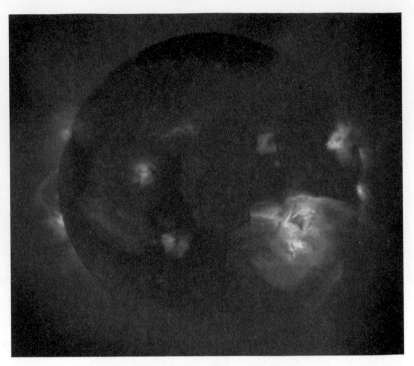

Figure 5-26
The power of X-ray astronomy is illustrated by this X-ray image of the sun recorded by the Yohkoh orbiting telescope. Bright yellow regions are million-degree swirls of gas high in the sun's upper atmosphere. The tangled filaments reveal the presence of twisted magnetic fields that confine and heat the gas. Such features are undetectable at visual wavelengths. *(Courtesy Japan/US Yohkoh team)*

The Hubble Space Telescope

Astronomers have dreamed of having a giant telescope in orbit since the 1920s. Such a telescope would not only allow observation at nonvisible wavelengths but would also avoid the turbulence in Earth's atmosphere. That dream was fulfilled on April 24, 1990, when the Hubble Space Telescope (HST) was released into orbit (Figure 5-27a). Named after the man who discovered the expansion of the universe (Chapter 15), it is the largest orbiting telescope ever built. As big as a large bus, the HST carries a 2.4-m (96-inch) mirror. The light can be directed to any of its spectrographs and cameras.

Soon after the telescope was launched, NASA engineers and astronomers discovered a defect in the shape of the mirror that prevented it from focusing sharp images. While some astronomers devised computer programs to partially remove the distortion from images, others developed corrective optics. Although the press emphasized the problem, astronomers used the telescope with great success even before astronauts visited the telescope in December 1993 and installed the corrective optics package in a series of five space walks. With its optics corrected, the telescope has exceeded its original design criteria. It can detect objects so faint it can look 50 times farther into space than can Earth-based telescopes. With its precise optics and its location above Earth's dis-

torting atmosphere, it can resolve details 10 times smaller than can be resolved by any conventional telescope on Earth. In early 1997, astronauts again visited the telescope in orbit and installed new instruments to make it even more sensitive and productive.

HST is controlled from the Space Telescope Science Institute located at Johns Hopkins University in Baltimore, Maryland. Astronomers there plan the telescope's observing schedule to maximize the scientific return while engineers study the operation of the telescope. HST transmits its data to the Institute, where it is analyzed. Because the telescope is a national facility, anyone may propose research projects for the telescope. Competition is fierce, and only the most worthy projects win approval. Nevertheless, even some amateur astronomers have projects approved for the Space Telescope.

The Hubble Space Telescope has been a phenomenal success, exploring such mysteries as the age of the universe, the birth and death of stars, and weather on nearby planets (Figure 5-27b). Visits by astronauts will keep it in operation for years to come, but astronomers are already planning an even larger space telescope that will someday orbit above Earth's atmosphere.

a   b

Figure 5-27

(a) The Hubble Space Telescope is the largest orbiting telescope ever launched. It was carried into orbit by the space shuttle in 1990. (b) This image of Mars recorded by the Hubble Space Telescope reveals thin clouds drifting around high volcanoes at the left and details of the polar caps at the top. *(NASA)*

R E V I E W   Critical Inquiry

Why can infrared astronomers observe from high mountaintops, while X-ray astronomers must observe from space?

In this analysis, we find similar consequences springing from different causes. Although both infrared and X-ray telescopes have been put into orbit, they differ dramatically. Infrared radiation is absorbed by water vapor in Earth's atmosphere, which is confined to the lower atmospheric layers. If we built an infrared telescope on top of a high mountain, we would be above most of the water vapor, and we could collect some infrared radiation from the stars. The longer-wavelength infrared is absorbed much higher in the atmosphere, so telescopes to observe in the far infrared must go into space. X rays are absorbed in the uppermost layers of the atmosphere, and no mountain is high enough to extend above those layers. To observe the stars at X-ray wavelengths, a telescope must go into space.

X-ray and far-infrared telescopes must observe from space, but the Hubble Space Telescope observes in the visual wavelength range. Then why must it observe from orbit?

The tools of the astronomer are designed to gather radiation from the sky and extract information. Perhaps no tool is as important as the spectrograph, because no form of observation is as loaded with information as a spectrum. In the next chapter, we will see how we can harvest the information in a star's spectrum.

Summary

Electromagnetic radiation is an electric and magnetic disturbance that transports energy at the speed of light. The electromagnetic spectrum includes gamma rays, X rays, ultraviolet radiation, visible light, infrared radiation, and radio waves.

We can think of a particle of light, a photon, as a bundle of waves that sometimes acts as a particle and sometimes as a wave. The energy a photon carries depends on its wavelength. The wavelength of visible light, usually measured in nanometers (10^{-9} m), ranges from 400 nm to 700 nm. Infrared and radio photons have longer wavelengths and carry less energy. Ultraviolet, X-ray, and gamma-ray photons have shorter wavelengths and carry more energy.

To obtain data, astronomers use telescopes to gather light, see fine detail, and magnify the image. The first two of these three powers of the telescope depend on the telescope's diameter; thus, astronomical telescopes often have large diameters.

Astronomical telescopes are of two types, refracting and reflecting. A refracting telescope uses a lens to bend the light and focus it into an image. Because of chromatic aberration, refracting telescopes cannot bring all colors to the same focus, resulting in color fringes around the images. An achromatic lens partially corrects for this, but such lenses are expensive and cannot be made larger than about 1 m in diameter.

Reflecting telescopes use a mirror to focus the light and are less expensive than refracting telescopes of the same diameter. In addition, reflecting telescopes do not suffer from chromatic aberration. Thus most recently built large telescopes are reflectors.

Astronomers build observatories atop high mountains for two reasons. Turbulence in Earth's atmosphere blurs the image in an astronomical telescope, a phenomenon that astronomers refer to as seeing. Atop a mountain, the air is steady, and the seeing is better. The air at a mountaintop is also thin and dry, and thus it is more transparent, especially in the infrared.

The light gathered by an astronomical telescope can be recorded and analyzed by special instruments attached to the telescope. For many decades, astronomers have used photographic plates to record images at the telescope, but modern electronic systems such as CCDs have now replaced photographic plates in most applications. Spectrographs spread starlight out according to wavelength to form a spectrum.

To observe radio signals from celestial objects, we need a radio telescope, which usually consists of a dish reflector, an antenna, an amplifier, and a recorder. Such an instrument can measure the intensity of radio signals over the sky and construct radio maps. The poor resolution of the radio telescope can be improved by combining it with another radio telescope to make a radio interferometer. Radio telescopes have three important features—they can detect cool hydrogen, they can see through dust clouds in space, and they can detect certain objects invisible at other wavelengths.

Earth's atmosphere admits radiation primarily through two wavelength intervals, or windows—the visual window and the radio window. At other wavelengths, our atmosphere absorbs radiation. To observe in the far infrared, astronomers must fly telescopes high in balloons or aircraft, though they can work at some wavelengths in the near infrared from high mountaintops. To observe in the ultraviolet and X-ray range and some parts of the infrared, they must send their telescopes into space to get above our atmosphere.

New Terms

electromagnetic radiation	objective lens, mirror
wavelength	eyepiece
photon	refracting telescope
nanometer (nm)	reflecting telescope
angstrom (Å)	chromatic aberration
atmospheric window	achromatic lens
focal length	Newtonian telescope
primary lens, mirror	prime focus

Cassegrain telescope	magnifying power
Schmidt-Cassegrain telescope	light pollution
sidereal drive	active optics
equatorial mounting	adaptive optics
polar axis	charge-coupled device (CCD)
alt-azimuth mounting	false-color image
light-gathering power	spectrograph
resolving power	grating
diffraction fringe	comparison spectrum
seeing	radio interferometer

Review Questions

1. Why would you not plot sound waves in the electromagnetic spectrum?

2. If you had unlimited funds to build a large telescope, which type would you choose, a refractor or a reflector? Why?

3. Why do nocturnal animals usually have large pupils in their eyes? How is that related to astronomical telescopes?

4. Why do optical astronomers sometimes put their telescopes at the tops of mountains, while radio astronomers sometimes put their telescopes in deep valleys?

5. Optical and radio astronomers both try to build large telescopes but for different reasons. How do these goals differ?

6. What are the advantages of making a telescope mirror thin? What problems does this cause?

7. Small telescopes are often advertised as "200 power" or "magnifies 200 times." As someone knowledgeable about astronomical telescopes, how would you improve such advertisements?

8. An astronomer recently said, "Some people think I should give up photographic plates." Why might she change to something else?

9. What purpose do the colors in a false-color image or false-color radio map serve?

10. How is chromatic aberration related to a prism spectrograph?

11. Why would radio astronomers build identical radio telescopes in many different places around the world?

12. Why do radio telescopes have poor resolving power?

13. Why must telescopes observing in the far infrared be cooled to low temperature?

14. What might we detect with an X-ray telescope that we could not detect with an infrared telescope?

15. If the Hubble Space Telescope observes at visual wavelengths, why must it observe from space?

Discussion Questions

1. Why does the wavelength response of the human eye match so well the visual window of Earth's atmosphere?

2. Basic research in chemistry, physics, biology, and similar sciences is supported in part by industry. How is astronomy different? Who funds the major observatories?

Problems

1. The thickness of the plastic in plastic bags is about 0.001 mm. How many wavelengths of red light is this?

2. Measure the actual wavelength of the wave in Figure 5-1. In what portion of the electromagnetic spectrum would it belong?

3. Compare the light-gathering powers of a 5-m telescope and a 0.5-m telescope.

4. How does the light-gathering power of the largest telescope in the world compare with that of the human eye? (*Hint:* Assume that the pupil of your eye can open to about 0.8 cm.)

5. What is the resolving power of a 25-cm telescope? What do two stars 1.5 seconds of arc apart look like through this telescope?

6. Most of Galileo's telescopes were only about 2 cm in diameter. Should he have been able to resolve the two stars mentioned in Problem 5?

7. How does the resolving power of the 5-m telescope compare with that of the Hubble Space Telescope? Why does the Hubble Space Telescope outperform the 5-m telescope?

8. If we build a telescope with a focal length of 1.3 m, what focal length should the eyepiece have to give a magnification of 100 times?

9. Astronauts observing from a space station need a telescope with a light-gathering power 15,000 times that of the human eye, capable of resolving detail as small as 0.1 second of arc and having a magnifying power of 250. Design a telescope to meet their needs. Could you test your design by observing stars from Earth?

10. A spy satellite orbiting 400 km above Earth is supposedly capable of counting individual people in a crowd. What minimum-diameter telescope must the satellite carry? (*Hint:* Use the small-angle formula.)

Critical Inquiries for the Web

1. How do professional astronomers go about making observations at major astronomical facilities? Visit several observatory Web sites to determine the process an astronomer would go through to secure observing time and make observations at the facility.

2. NASA is in the process of completing a fleet of four space-based "Great Observatories." (The Hubble Space Telescope is one; what are the others?) Examine the current state of these missions by visiting their home pages on the Internet. What advantages would these facilities have over ground-based observatories?

Go to the Brooks/Cole Astronomy Resource Center (www.brookscole.com/astronomy) for critical thinking exercises, articles, and additional readings from InfoTrac College Edition, Brooks/Cole's online student library.

CHAPTER 6

Atoms and Starlight

Awake! for Morning in the Bowl

of Night

Has flung the stone that put the

Stars to Flight!

And lo! the Hunter of the East

has caught

The Sultan's Turret in a Noose

of Light.

Edward FitzGerald, translator
The Rubaiyat of Omar Khayyám

Guidepost

Unlike the chapters in traditional textbooks, the chapters in this book do more than present facts: They attempt to present organized knowledge. But this chapter is special. It presents a tool: the understanding of how atoms interact with light. The way light and matter interact gives astronomers clues about the nature of the heavens, but the clues are meaningless unless astronomers understand how atoms leave their traces on starlight. Thus, we dedicate an entire chapter to this critically important tool.

This chapter marks a transition in the way we look at nature. Up to this point, the chapters have described what we see with our eyes and explained our perceptions using models and theories. With this chapter, we turn to modern astrophysics, the application of physics to the study of the sky. Now we can search out secrets of the stars that lie beyond what we can see.

If this chapter presents us with a tool, then we should use it immediately. In the next chapter, we will use this tool to study the sun.

Figure 6-1
So vast that light takes 20 years to cross its diameter, the cloud of gas called the Eagle Nebula glows deep red. Nothing in the appearance of the swirling gas and the cluster of bright stars just above and to the right of the nebula's center helps us make sense of what we see. But by analyzing the light from the gas and stars, astronomers have learned that the nebula is mostly hydrogen gas and that the star cluster is a young group of stars that have formed recently from the nebula. *(NOAO)*

No laboratory jar on Earth holds a sample labeled "star stuff," and no instrument has ever probed inside a star. The stars are far beyond our reach, and the only information we can obtain about them comes to us hidden in light (Figure 6-1). Whatever we want to know about the stars we must catch in a noose of light.

We earthbound humans knew almost nothing about stars until the early 19th century, when the Munich optician Joseph von Fraunhofer studied the solar spectrum and found it interrupted by some 600 dark lines. As scientists realized that the lines were related to the various atoms in the sun and found that stellar spectra had similar patterns of lines, the door to an understanding of stars finally opened.

In this chapter, we will go through that door by considering how atoms interact with light to produce spectral lines. We begin with the hydrogen atom because it is the most common atom in the universe, as

well as the simplest. Other atoms are larger and more complicated, but in many ways their properties resemble those of hydrogen. Another reason for studying hydrogen is to consider how the structure of its atoms can give rise to the 21-cm-wavelength radiation that is so important to radio astronomers (see Chapter 5).

Once we understand how an atom's structure can interact with light to produce spectral lines, we will recognize certain patterns in stellar spectra. By classifying the spectra according to these patterns, we can arrange the stars in a sequence according to temperature. One of the most important pieces of information revealed in a star's spectrum is its temperature.

But, properly analyzed, a stellar spectrum can tell us much more. The spectrum gives us information about the chemical composition of the star and the star's motion relative to Earth.

6-1 Atoms

A Model Atom

To think about atoms and how they can interact with light, we must create a working model of an atom. In Chapter 2, we created a working model of the sky, the celestial sphere. We identified and named the important parts and described how they were located and how they interacted. We begin our study of atoms by creating a model of an atom.

In our model atom, we identify a positively charged **nucleus** at the center; this nucleus consists of two kinds of particles. **Protons** carry a positive electrical charge, and **neutrons** have no charge. Thus, the nucleus has a net positive charge.

The nucleus in our model atom is surrounded by a whirling cloud of orbiting **electrons,** low-mass particles with a negative charge. In a normal atom, the number of electrons equals the number of protons, and the positive and negative charges balance to produce a neutral atom. Because protons and neutrons each have a mass 1836 times greater than that of an electron, most the mass of the atom lies in the nucleus. The hydrogen atom is the simplest of all atoms. The nucleus is a single proton orbited by a single electron, with a total mass of only 1.67×10^{-27} kg, about a trillionth of a trillionth of a gram.

An atom is mostly empty space. To see this we can construct a simple scale model. The nucleus of a hydrogen atom is a proton with a diameter of about 0.0000016 nm, or 1.6×10^{-15} m. If we multiply this by one trillion (10^{12}) we can represent the nucleus of our model atom with a grape seed, which is about 0.16 cm in diameter. The region of a hydrogen atom containing the whirling electron has a diameter of about 0.4 nm or 4×10^{-10} m. Multiplying by a trillion magnifies the diameter to about 400 m, or about 4.5 football fields laid end to end (Figure 6-2). When you imagine a grape seed in the midst of a sphere 4.5 football fields in diameter, you can see that an atom is mostly empty space.

Different Kinds of Atoms

There are over a hundred kinds of atoms, called chemical elements. Which element an atom is depends only on the number of protons in the nucleus. For example, carbon has six protons in its nucleus. An atom with one more proton than this is nitrogen, and an atom with one fewer proton is boron.

Although the number of protons in an atom of an element is fixed, we can change the number of neutrons in an atom's nucleus without changing the atom significantly. For instance, if we add a neutron to the carbon nucleus, we still have carbon, but it is slightly heavier than normal carbon. Atoms that have the

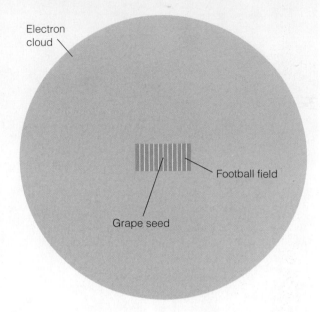

Figure 6-2
Magnifying a hydrogen atom by 10^{12} makes the nucleus the size of a grape seed and the diameter of the electron clouds about 4.5 times longer than a football field. The electron itself is still too small to see.

same number of protons but a different number of neutrons are **isotopes.** Carbon has two stable isotopes. One form contains six protons and six neutrons for a total of 12 particles and is thus called carbon-12. Carbon-13 has six protons and seven neutrons in its nucleus.

Protons and neutrons are bound tightly into the nucleus, but the electrons are held loosely in the electron cloud. Running a comb through your hair creates a static charge by removing a few electrons from their atoms. This process is called **ionization,** and the atom that has lost one or more electrons is an **ion.** A carbon atom is neutral if it has six electrons to balance the positive charge of the six protons in its nucleus. If we ionize the atom by removing one or more electrons, the atom is left with a net positive charge. Under some circumstances, an atom may capture one or more extra electrons, giving it more negative charges than positive. Such a negatively charged atom is also considered an ion.

Atoms that collide may form bonds with each other by exchanging or sharing electrons. Two or more atoms bonded together form a **molecule.** Atoms do collide in stars, but the high temperatures cause violent collisions that are unfavorable for chemical bonding. Only in the coolest stars are the collisions gentle enough to permit the formation of chemical bonds. We will see later that the presence of molecules such as titanium oxide (TiO) in a star is a clue that the star is cool. In later chapters, we will see that molecules can form in cool gas clouds in space and in the atmospheres of planets.

Quantum mechanics is the set of rules that describe how atoms and subatomic particles behave. When we think about large objects, such as stars, planets, aircraft carriers, and hummingbirds, we don't have to think about quantum mechanics; but on the atomic scale, particles behave in ways that seem unfamiliar to us.

One of the principles of quantum mechanics specifies that we cannot know simultaneously the exact location and motion of a particle. This is why physicists refer to the electrons in an atom as if they were a cloud of negative charge surrounding the nucleus. Because we can't know the position and motion of the electron, we can't really describe it as a small particle following an orbit. We can use that image as a model to help our imaginations, but the reality is much more interesting, and describing the electrons as a charge cloud gives us a better and more sophisticated model of an atom.

This raises some serious questions about reality. Is an electron really a particle at all? Quantum mechanics describes particles as waves, and waves as particles. If we can't know simultaneously the position and motion of a specific particle, how can we know how it will react to a collision with a photon or another particle? The answer is that we can't know, and that seems to violate the principle of cause and effect (Window on Science 4-3).

Needless to say, we can't explore these discrepancies here. Scientists and philosophers of science continue to struggle with the meaning of reality on the quantum-mechanical level. Here we should note that the reality we see on the scale of stars and hummingbirds is only part of nature. We have constructed some models to help us think about nature on the scale of atoms, but the truth is much more interesting and much more exciting than anything we see on larger scales. Although we use models of atoms to study stars, there is still much to learn about the atoms themselves.

Electron Shells

We mentioned the whirling cloud of orbiting electrons in a general way, but the specific way electrons behave within the cloud is very important in astronomy.

The electrons are bound to the atom by the attraction between their negative charge and the positive charge on the nucleus. This attraction is known as the **Coulomb force,** after the French physicist Charles-Augustin de Coulomb (1736–1806). If we wish to ionize an atom, we need a certain amount of energy to pull an electron away from its nucleus. This energy is the electron's **binding energy,** the energy that holds it to the atom.

An electron may orbit the nucleus at various distances. If the orbit is small, the electron is close to the nucleus, and a large amount of energy is needed to pull it away. Therefore, its binding energy is large. An electron orbiting farther from the nucleus is held more loosely, and less energy will pull it away. It therefore has less binding energy. The size of an electron's orbit is related to the energy that binds it to the atom.

Nature permits atoms only certain amounts (quanta) of binding energy, and the laws that describe how atoms behave are called the laws of **quantum mechanics** (Window on Science 6-1). Much of our discussion of atoms is based on the laws of quantum mechanics.

Because atoms can have only certain amounts of binding energy, our model atoms can have orbits of only certain sizes, called **permitted orbits.** These are like steps in a staircase: You can stand on the number-one step or the number-two step, but not on the number-one-and-one-quarter step. The electron can occupy any permitted orbit but not orbits in between.

The arrangement of permitted orbits depends primarily on the charge of the nucleus, which in turn depends on the number of protons. Thus, each kind of element has its own pattern of permitted orbits (Figure 6-3). Isotopes of the same elements have nearly

Figure 6-3
The electron in atoms may occupy only certain, permitted orbits. Since different elements have different charges on their nuclei, the elements have different patterns of permitted orbits.

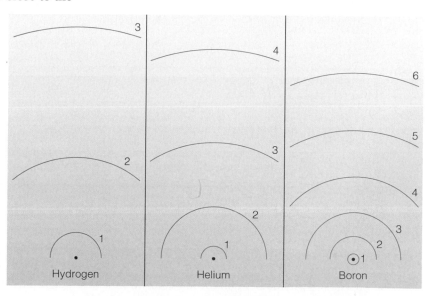

the same pattern because they have the same number of protons. However, ionized atoms have orbital patterns that differ from their un-ionized forms. Thus the arrangement of permitted orbits differs for every kind of atom and ion.

REVIEW

Critical Inquiry

How many hydrogen atoms would it take to cross the head of a pin?

This is not a frivolous question. In answering it, we will discover how small atoms really are, and we will see how powerful physics can be as a way to understand nature. First, we will assume that the head of a pin is about 1 mm in diameter. That is 0.001 m. The size of a hydrogen atom is represented by the diameter of the electron cloud, and we will assume that the electron in our atom is in the second orbit. Then the diameter of the electron cloud is about 0.4 nm. Since 1 nm equals 10^{-9} m, we multiply and discover that 0.4 nm equals 4×10^{-10} m. To find out how many atoms would stretch 0.001 m, we divide the diameter of the pinhead by the diameter of an atom. That is, we divide 0.001 m by 4×10^{-10} m, and we get 2.5×10^6. It would thus take 2.5 million hydrogen atoms lined up side by side to cross the head of a pin.

This shows how tiny an atom is and also how powerful basic physics is. A bit of arithmetic gives us a view of nature beyond the capability of our eyes. Now use another bit of arithmetic to calculate how many hydrogen atoms you would need to equal the mass of a paper clip (1g).

The energy levels in atoms are familiar territory to astronomers, because the electrons in those levels can interact with light. Such interactions fill starlight with clues to the nature of the stars.

6-2 The Interaction of Light and Matter

We begin our study of light and matter by considering the hydrogen atom. As we noted earlier, hydrogen is both simple and common. Roughly 90 percent of all atoms in the universe are hydrogen.

The Excitation of Atoms

Each orbit in an atom represents a specific amount of binding energy, so physicists commonly refer to the orbits as **energy levels.** Using this terminology, we can say that an electron in its smallest and most tightly bound orbit is in its lowest permitted energy level. We can move the electron from one energy level to another by supplying enough energy to make up the difference between the two energy levels. It is like moving a flowerpot from a low shelf to a high shelf; the greater the distance between the shelves, the more energy we need to raise the pot. The amount of energy needed to move the electron is the energy difference between the two energy levels.

If we move the electron from a low energy level to a higher energy level, we say the atom is an **excited atom.** That is, we have added energy to the atom in moving its electron. If the electron falls back to the lower energy level, that energy is released.

An atom can become excited by collision. If two atoms collide, one or both may have electrons knocked into a higher energy level. This happens very commonly in hot gas, where the atoms move rapidly and collide often.

Another way an atom can get the energy that moves an electron to a higher energy level is to absorb a photon. Only a photon with exactly the right amount of energy can move the electron from one level to another. If the photon has too much or too little energy, the atom cannot absorb it. Because the energy of a photon depends on its wavelength, only photons of certain wavelengths can be absorbed by a given kind of atom. Figure 6-4 shows the lowest four energy levels of the hydrogen atom along with three photons the atom could absorb. The longest-wavelength photon only has enough energy to excite the electron to the second energy level, but the shorter-wavelength photons can excite the electron to higher levels. A photon with too much or too little energy cannot be absorbed. Because the hydrogen atom has many more energy levels than shown in Figure 6-4, it can absorb photons of many different wavelengths.

Atoms, like humans, cannot exist in an excited state forever. The excited atom is unstable and must eventually (usually within 10^{-6} to 10^{-9} sec) give up the energy it has absorbed and return its electron to the lowest energy level. Because the electrons eventually tumble down to this bottom level, physicists call it the **ground state.**

When the electron drops from a higher to a lower energy level, it moves from a loosely bound level to one more tightly bound. The atom then has a surplus of energy—the energy difference between the levels—that it can emit as a photon. Study the sequence of events in Figure 6-5 to see how an atom can absorb and emit photons. Because each type of atom or ion has its unique set of energy levels, each type absorbs and emits photons with a unique set of wavelengths. Thus we can identify the elements in a gas by studying the characteristic wavelengths of light absorbed or emitted.

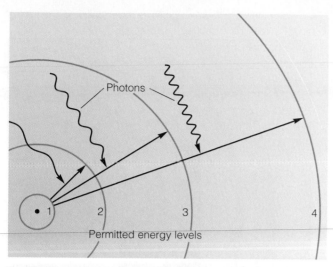

Figure 6-4
A hydrogen atom can absorb only those photons that move the atom's electron to one of the higher-energy orbits. Here three different photons are shown along with the change they would produce if they were absorbed.

The process of excitation and emission is a common sight in urban areas at night. A neon sign glows when atoms of the neon gas in the tube are excited by electricity flowing through the tube. As the electrons in the electric current flow through the gas they collide with the neon atoms and excite them. As we have seen, immediately after an atom is excited, its electron drops back to a lower energy level, emitting the surplus energy as a photon of a certain wavelength. The photons emitted by excited neon produce a reddish-orange glow. Signs of other colors, erroneously called "neon," contain other gases or mixtures of gases instead of pure neon.

Radiation from a Heated Object

To begin our discussion of the interaction of light and matter, we must consider a simple but very important phenomenon. When we heat an object up, it begins to glow. Think of a blacksmith heating a horseshoe in a forge. The hot iron glows bright yellow-orange. What produces this light?

Recall that light is a changing electric and magnetic field. Whenever we produce a changing electric field, we produce electromagnetic radiation. If we run a comb through our hair, we disturb electrons in both hair and comb, producing static electricity. Because each electron is surrounded by an electric field, any sudden change in the electron's motion gives rise to electromagnetic radiation. Running a comb through your hair while standing near an AM radio produces radio static. This illustrates an important principle: Whenever we change the motion of an electron, we generate electromagnetic waves.

To see what this has to do with a heated object, think of what we mean by heat. **Heat** refers to the amount of energy stored in a body as agitation among its particles. The molecules and atoms in an object are in constant motion, and, in a heated object, they are more agitated than in a cooler object. **Temperature,** on the other hand, refers to the speed with which the particles move. Hot cheese contains a great deal of heat and can burn our tongue, but green beans at the same temperature contain less heat and are less likely to burn us. *Heat* refers to the quantity of thermal energy, while *temperature* refers to its intensity.

When we refer to the temperature of astronomical objects such as stars, we will use the **Kelvin temperature scale.** On this scale, zero degrees Kelvin (written 0 K) is **absolute zero** (-459.7 °F), the temperature at which an object contains no heat energy that can be extracted. Water freezes at 273 K and boils at 373 K. The Kelvin temperature scale is useful in astronomy, because it is based on absolute zero and thus is related to the motion of the particles in an object.

Now we can understand why a hot object glows. The hotter an object is, the more motion among its particles. The agitated particles collide with electrons, and when electrons are accelerated, part of the energy is carried away as a photon. The radiation emitted by a heated object is called **black body radiation,** a name that refers to the way a perfect emitter of radiation would behave. A perfect emitter would also be a perfect absorber and at room temperature would look black. Thus we often use the term black body radiation to refer to objects that glow brightly.

Black body radiation is quite common. In fact, it is responsible for the light emitted by an incandescent lightbulb. Electricity flowing through the filament of the lightbulb heats it to high temperature, and it glows. We also recognize the light emitted by a heated horseshoe in the blacksmith's forge as black body radiation.

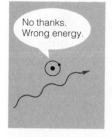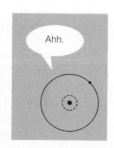

Figure 6-5
An atom can absorb a photon only if the photon has the correct amount of energy. The excited atom is unstable and within a fraction of a second returns to a lower energy level, reradiating the photon in a random direction.

Many objects in astronomy, including stars, emit radiation as if they were nearly perfect black bodies.

Hot objects emit black body radiation, but so do cold objects. Ice cubes are cold, but their temperature is higher than absolute zero, so they contain some heat and must emit some black body radiation. The coldest gas drifting in space has a temperature only a few degrees above absolute zero, but it too emits black body radiation.

We must discuss two important features of black body radiation. First, the hotter an object is, the more black body radiation it emits. Hot objects emit more radiation because their agitated particles collide more often and more violently with electrons. Thus we expect a glowing coal from a fire to emit more total energy than an ice cube of the same size.

The second feature we must discuss is the relationship between the temperature of the object and the wavelengths of the photons it emits. The wavelength of a photon emitted when a particle collides with an electron depends on the violence of the collision. Only a violent collision can produce a short-wavelength (high-energy) photon. Because extremely violent collisions don't occur very often, short-wavelength photons are rare. Similarly, most collisions are not extremely gentle, so long-wavelength (low-energy) photons are also rare. Consequently, black body radiation is made up of photons with a distribution of wavelengths, and very short and very long wavelengths are rare. The **wavelength of maximum intensity** (λ_{max}), the wavelength at which the object emits the most radiation, occurs at some intermediate wavelength.

Figure 6-6 shows the intensity of radiation versus wavelength for three objects of different temperatures. As we expect, the objects emit most of their radiation at intermediate wavelengths, and the hotter object (top) emits more total radiation than the cooler object (bottom). We can also see that the wavelength of maximum intensity depends on temperature. The hotter the object, the shorter the wavelength of maximum. Notice in Figure 6-6 how the temperature determines the color of a glowing black body. The hotter object emits more blue light than red and thus looks blue, and the cooler object emits more red than blue and consequently looks red. We will see this among stars. Hot stars look blue and cool stars red. The properties of black body radiation are described in By the Numbers 6-1.

Notice that cool objects may emit little visible radiation but are still producing black body radiation. For example, the human body has a temperature of 310 K and emits black body radiation mostly in the infrared part of the spectrum. Infrared security cameras can detect burglars by the radiation they emit. We humans emit very few visible wavelength photons and almost never emit X-ray or gamma-ray photons. Our wavelength of maximum intensity lies in the infrared part of the spectrum.

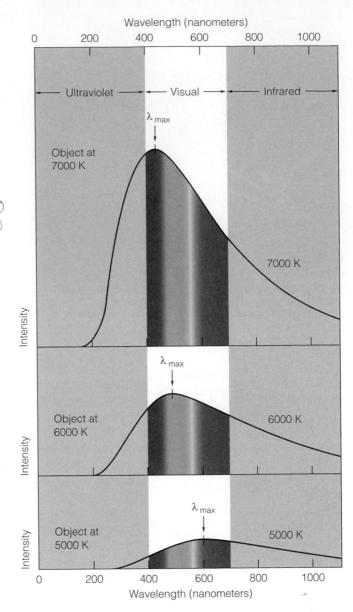

Figure 6-6
Black body radiation from three bodies at different temperatures demonstrates that a hot body radiates more total energy and that the wavelength of maximum intensity is shorter for hotter objects. The hotter object here will look blue to our eyes, while the cooler object will look red.

Now that we understand black body radiation, we can understand how a star emits light. More important, we can understand how to get information from a star's spectrum.

The Formation of a Spectrum

To see how an astronomical object like a star can produce a spectrum, we can construct an imaginary experiment with an incandescent lightbulb and a cloud of hydrogen gas in space. We will discover there are three kinds of spectra and that our single experiment is capable of producing all three.

Our incandescent lightbulb contains a hot filament and emits black body radiation. As we saw in the previous section, a black body radiates at all wavelengths, and if we spread the light from the bulb out to form a spectrum we will find all wavelengths represented. This is called a **continuous spectrum,** as illustrated in the top spectrum in Figure 6-7.

In our imaginary experiment, the light from the bulb must pass through the hydrogen gas before it can reach our telescope (Figure 6-8). Most of the photons will pass through the gas unaffected because they have wavelengths the hydrogen atoms cannot absorb, but a few photons will have the right wavelengths. These photons cannot pass through the gas because they are absorbed by the first atom they meet. The atom is excited for a fraction of a second, and the electron then drops back to a lower energy level and a new photon is emitted. The original photon was traveling through the gas toward our telescope, but the new photon is emitted in some random direction. Very few of these new photons leave the cloud in the direction of our telescope, so the light that finally enters the telescope

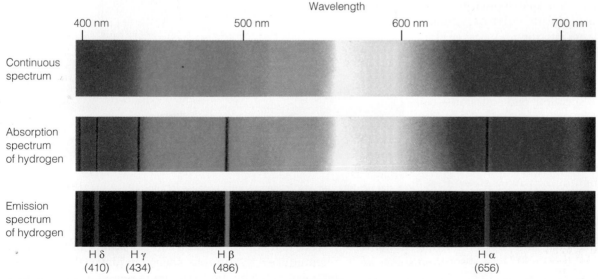

Figure 6-7
The three types of spectra. A continuous spectrum (top) contains no bright or dark lines, but an absorption spectrum (middle) is interrupted by dark absorption lines. An emission spectrum (bottom) is dark except at certain wavelengths where there are emission lines. Note that the dark lines in the absorption spectrum of hydrogen have the same wavelength as the bright lines in the emission spectrum of hydrogen.

has very few photons at the wavelengths the atoms can absorb. When we form a spectrum from this light, photons of these wavelengths are missing, and the spectrum has dark lines at the positions these photons would have occupied. These dark lines, like those the Munich optician Fraunhofer saw in the solar spectrum in 1814 and 1815, are called **absorption lines** because the atoms absorbed the photons. A spectrum containing absorption lines is an **absorption spectrum** (also called a **dark-line spectrum**). The middle spectrum in Figure 6-7 illustrates the absorption spectrum of hydrogen.

What happens to the photons that were absorbed? They bounce from atom to atom, being absorbed and emitted over and over until they escape from the cloud. If, instead of aiming our telescope at the bulb, we swing it to one side so that no light from the bulb enters the telescope, we can photograph a spectrum of the light emitted by the gas atoms (Figure 6-9). In this case, the only photons entering the telescope are photons that were absorbed and re-emitted. A spectrum of this light is almost entirely dark except for the wavelengths corresponding to the photons the gas can absorb and re-emit. Thus we will see a spectrum containing only bright lines on a dark background. These bright lines are called **emission lines,** and a spectrum with emission lines is an **emission spectrum** (also called a **bright-line spectrum**). The lower spectrum in Figure 6-7 illustrates the emission spectrum of hydrogen. The spectrum of a neon sign is an emission spectrum. Also, the bluish-purple color of mercury-vapor streetlights and the pink-orange color of sodium-vapor streetlights are produced by the emission lines of these elements.

The properties of the three spectra — continuous, absorption, and emission — can be described by the three rules known as **Kirchhoff's laws.** Law I states that a heated solid, a liquid, or a dense gas will produce a continuous spectrum. We see this in the filament of a lightbulb. Law II states that a low-density gas excited to emit light will produce an emission spectrum. We see this in neon signs. And Law III states that if light comprising a continuous spectrum passes through a cool, low-density gas, the result will be an absorption spectrum. We see this in stars.

Stellar spectra are absorption spectra formed as light from the bright surface, the photosphere, travels outward through the stellar atmosphere (Figure 6-10). The bright surface of the star is sufficiently dense and

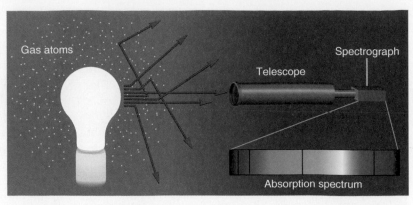

Figure 6-8
Photons of the proper wavelengths can be absorbed by the gas atoms and re-emitted in random directions. Because most of these particular photons do not reach the telescope, the spectrum is dark at the wavelengths of the missing photons.

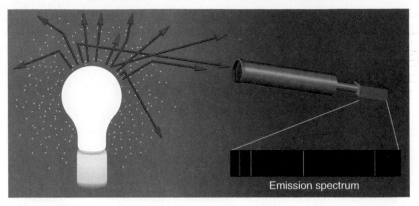

Figure 6-9
Pointing the telescope away from the bulb, we can receive only those photons the atoms can absorb and re-emit, producing emission lines in the spectrum.

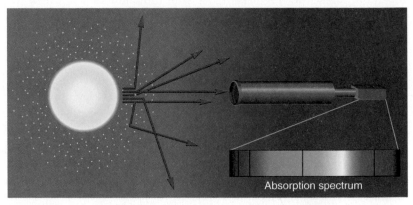

Figure 6-10
A star produces an absorption spectrum because its atmosphere absorbs certain wavelength photons in the spectrum.

hot that collisions between gas atoms and electrons produce a continuous spectrum just as does the filament of the lightbulb in Figure 6-8. The lower-density gases in the stellar atmosphere absorb certain wavelength photons, and thus the light that finally reaches our telescope is missing those specific wavelengths, and we see an absorption spectrum.

The absorption lines in stellar spectra provide a windfall of data about the star's atmosphere. By studying the spectral lines we can identify the elements in the stellar atmosphere and find the temperature of the atoms. To see how to get all this information, we need to look carefully at the way the hydrogen atom produces lines in a star's spectrum.

The Hydrogen Spectrum

As you must have gathered by now, each element has its own spectrum, unique as a human fingerprint; each element can be recognized by its spectrum across trillions of miles. To see how hydrogen produces its unique spectrum, we need a detailed diagram of its permitted orbits drawn so that the size of each orbit is proportional to its energy. That is, we need an accurate diagram of its energy levels such as that in Figure 6-11. An energy level diagram will allow us to study the way hydrogen atoms interact with light.

A **transition** occurs in an atom when an electron changes energy levels. In our diagram of a hydrogen atom, arrows pointing from one level to another represent transitions. If the transition results in the absorption or emission of a photon, the length of the arrow tells us its energy. Long arrows represent large amounts of energy and thus short-wavelength photons. Short arrows represent smaller amounts of energy and longer-wavelength photons.

We can divide the possible transitions in a hydrogen atom into groups called series, according to their lowest energy level. Those arrows whose lower ends rest on the ground state represent the **Lyman series;** those resting on the second energy level, the **Balmer series;** and those resting on the third, the **Paschen series.** In principle, each series contains an infinite number of transitions, and there are an infinite number of series. Figure 6-11 shows the first few transitions in the first few series.

The Lyman series transitions are large changes in energy, as shown by the long arrows in Figure 6-11. These energetic transitions produce lines in the ultraviolet part of the spectrum, where they are invisible to the human eye. The Paschen series also lies

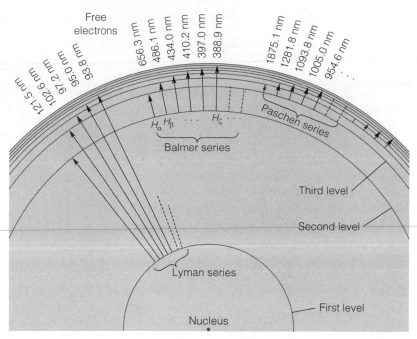

Figure 6-11
The electron orbits in this diagram of the hydrogen atom are spaced to represent the energy the electron would have in each orbit. Thus we can refer to the orbits as energy levels. The permitted energy-level changes, drawn as arrows, can be grouped into series according to their lowest energy level. This drawing shows only a few of the infinite number of transitions and series possible.

outside the visible part of the spectrum. These transitions are small changes in energy and thus produce spectral lines in the infrared.

Balmer series transitions produce the only spectral lines of hydrogen in the visible part of the spectrum. Figure 6-11 shows that the first few Balmer series transitions are intermediate between the energetic Lyman transitions and the low-energy Paschen transitions. These Balmer lines are labeled by Greek letters for easy identification. H_α is a red line, H_β is blue, and

Figure 6-12
The Lagoon Nebula in Sagittarius is a cloud of gas and dust about 60 ly in diameter. Its gases are excited by the ultraviolet radiation of the hot, young stars within, and it glows in the pink color produced by the mixture of the red, blue, and violet Balmer lines. (NOAO)

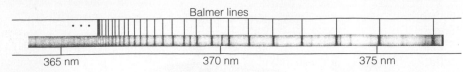

Figure 6-13
The Balmer lines photographed in the near ultraviolet. *(The Observatories of the Carnegie Institution of Washington)*

H_γ and H_δ are violet. These four lines create the purple-pink color characteristic of glowing clouds of hydrogen in space (Figure 6-12). The remaining Balmer lines have wavelengths too short to see, though they can be photographed easily (Figure 6-13).

R E V I E W   Critical Inquiry

What kind of spectrum would we see if we observed molten iron in a steel mill?

Molten iron is a dense liquid; the atoms and molecules collide so often that they emit all wavelengths, and we would see a continuous spectrum. That is Kirchhoff's first law. But in order to see the molten iron, we would have to look through the hot vapors rising from it. Photons on their way to our spectrograph would pass through these gases, and atoms in the gases would absorb certain wavelengths. So what we would really see would be the continuous spectrum of the molten iron with weak absorption lines caused by the gases above the iron. This is what Kirchhoff's third law describes.

But suppose we had a very sensitive spectrograph that could look at the hot gases above the molten iron from the side so as to avoid looking directly at the molten iron. What kind of spectrum would the hot gases emit? Why?

Whatever kind of spectrum astronomers look at, the most common spectral lines are the Balmer lines of hydrogen, the only hydrogen lines we can study from Earth's surface. In the next section, we will see how Balmer lines can tell us a star's temperature.

6-3 Stellar Spectra

Science is a way of understanding nature, and the spectrum of a star tells us a great deal about such things as temperature, motion, and composition. In later chapters, we will use spectra to study galaxies and planets, but we begin by studying the spectra of stars, including the sun. Stellar spectra are the easiest to understand, and the nature of stars is central to our study of all celestial objects.

The Balmer Thermometer

We can use the Balmer lines as a thermometer to find the temperatures of stellar surfaces. From our discussion of black body radiation, we know that we can estimate stellar temperatures from color—red stars are cool, and blue stars are hot. But the Balmer lines give us much greater accuracy. We will use the Kelvin temperature scale to refer to stellar temperatures.

Note that when we discuss the temperature of a star, we mean the surface temperature—typically 40,000 to 2000 K. The centers of stars are much hotter—many millions of degrees—but the spectra tell us about the surface layers from which the light originates.

The Balmer thermometer works because the Balmer absorption lines are produced by hydrogen atoms whose electrons are in the second energy level (see Figure 6-11). If the surface of a star is as cool as the sun or cooler, there are few violent collisions between atoms to excite the electrons, and most atoms have their electrons in the ground state. These atoms can't absorb photons in the Balmer series. As a result, we should expect to find weak Balmer lines in the spectra of cool stars.

In the surface layers of stars hotter than about 20,000 K, however, there are many violent collisions between atoms, exciting electrons to high energy levels or knocking the electrons completely out of most atoms. That is, most atoms become ionized. Therefore, few atoms have electrons in the second energy level to form Balmer absorption lines, and we should expect hot stars, like cool stars, to have weak Balmer absorption lines.

At an intermediate temperature, roughly 10,000 K, the collisions have the correct amount of energy to excite large numbers of electrons into the second energy level. With many atoms excited to the second level, the gas absorbs Balmer wavelength photons well and thus produces strong Balmer lines.

To summarize, the strength of the Balmer lines depends on the temperature of the star's surface layers. Both hot and cool stars have weak Balmer lines, but medium-temperature stars have strong Balmer lines.

Theoretical calculations can predict just how strong the Balmer lines should be for stars of various temperatures. Such calculations are the key to finding temperatures from stellar spectra. The curve in Figure 6-14a shows the calculated strength of the Balmer lines for various stellar temperatures. We could use this as a temperature indicator, except that the curve gives us two answers. A star with Balmer lines of a certain strength might have either of two temperatures, one high and one low. How do we know the true temperature of the star? We must examine other spectral lines to arrive at the correct temperature.

We have seen how the strength of the Balmer lines depends on temperature. Temperature has a similar effect on the spectral lines of other elements, such as ionized calcium (Figure 6-14b). That is, their lines are weak at high and low temperatures and strong at some intermediate temperature. But the temperature at which their lines reach maximum strength is different for each element. If we add a number of these elements to our graph, we get a powerful tool for finding the temperature of stars (Figure 6-14c).

How do we use this tool? We determine a star's temperature by comparing the strengths of its spectral lines with our graph. For instance, if we photographed a spectrum of a star and found medium-strength Balmer lines and strong helium lines, we could conclude it had a temperature of about 20,000 K. But if the star had weak hydrogen lines and strong lines of ionized iron, we would assign it a temperature of about 5800 K, similar to that of the sun.

The spectra of stars cooler than about 3000 K contain dark bands produced by molecules such as titanium oxide (TiO). Because of their structure, molecules can absorb photons at many wavelengths, producing numerous, closely spaced spectral lines that blend together to form bands. These molecular bands appear only in the spectra of the coolest stars because, as mentioned before, molecules in cool stars are not subject to the violent collisions that would break up molecules in hotter stars. Thus the presence of dark bands in a star's spectrum indicates that the star is very cool.

From stellar spectra, astronomers have found that the hottest stars have surface temperatures above 40,000 K and the coolest about 2000 K. Compare these with the surface temperature of the sun, which is about 5800 K.

Spectral Classification

We have seen that the strengths of spectral lines depend on the surface temperature of the star. From this we can predict that all stars of a given temperature should have similar spectra. If we learn to recognize the pattern of spectral lines produced by a 6000-K star, for instance, we need not use Figure 6-14c every time we see that kind of spectrum. We can save time by classifying stellar spectra rather than analyzing each one individually.

The first widely used classification system was devised by astronomers at Harvard during the 1890s and 1900s. One of them, Annie J. Cannon, personally inspected and classified the spectra of over 250,000 stars. The spectra were first classified in groups labeled A through Q, but some groups were later dropped, merged with others, or reordered. The final classification includes the seven **spectral classes,** or **types,** still used today: O, B, A, F, G, K, M.*

---

*Generations of astronomy students have remembered the spectral sequence using the mnemonic "Oh, Be A Fine Girl (Guy), Kiss Me."

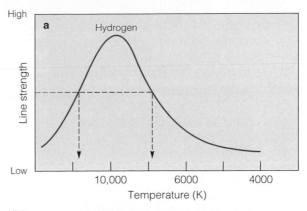

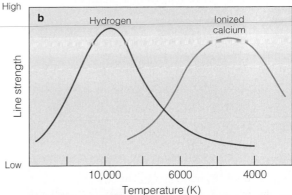

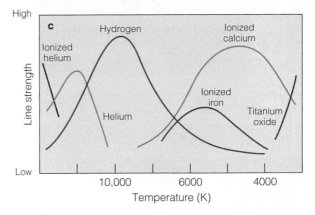

Figure 6-14
(a) The strength of the Balmer lines in a stellar spectrum depends on the temperature of the star. The dashed lines indicate that a star with medium-strength Balmer lines could have one of two possible temperatures. (b) The curves for lines of hydrogen and ionized calcium reach maximum strength at different temperatures. (c) The strengths of the spectral lines produced by each atom, ion, and molecule depend on the temperature of the star. Astronomers can compare the strengths of lines in a stellar spectrum with a diagram such as this to find the temperature of the star.

This sequence of spectral types, called the **spectral sequence,** is important because it is a temperature sequence. The O stars are the hottest, and the temperature continues to decrease down to the M stars, the coolest. For maximum precision, astronomers divide each spectral class into 10 subclasses. For example, spectral class A consists of the subclasses A0, A1, A2, . . . A8, A9. Next comes F0, F1, F2, and so on. This finer division gives us a star's temperature to an accuracy

within about 5 percent. The sun, for example, is not just a G star, but a G2 star,
with a temperature of 5800 K.

We classify a star by the lines and
bands in its spectrum as shown in Table 6-
1. For example, if it has weak Balmer lines
and lines of ionized helium, it must be an
O star. This table is based on the same information we used in Figure 6-14c.

Figure 6-15 shows spectra for stars of
different spectral types. Notice that the
spectra extend from the visible into the
near ultraviolet to reveal the calcium H
and K lines; the human eye cannot see to
wavelengths shorter than about 420 nm.

Table 6-1 Spectral Classes

Spectral Class	Approximate Temperature (K)	Hydrogen Balmer Lines	Other Spectral Features
O	40,000	Weak	Ionized helium
B	20,000	Medium	Neutral helium
A	10,000	Strong	Ionized calcium weak
F	7500	Medium	Ionized calcium weak
G	5500	Weak	Ionized calcium medium
K	4500	Very weak	Ionized calcium strong
M	3000	Very weak	Titanium oxide strong

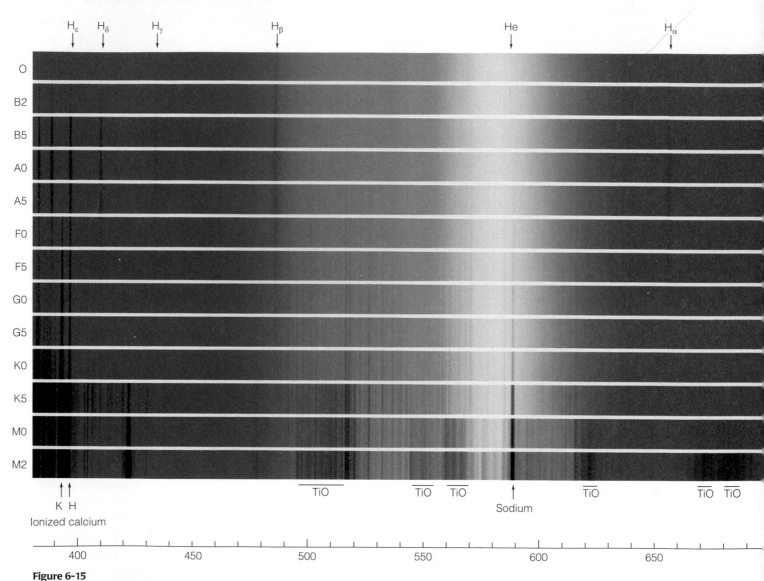

Figure 6-15
These stellar spectra were produced by computer simulation of real spectra. The spectra run from the hot
O star at the top to the cool M2 star at the bottom. In the spectra of the hot stars we see weak Balmer
lines and barely visible helium lines. The Balmer lines are strongest about A0 but are very weak in cool
stars. The two ultraviolet lines of ionized calcium are strong in cooler stars, while sodium lines and Titanium oxide (TiO) bands are strong in the spectra of the coolest stars. These spectra have been extended
past the short-wavelength limit for the human eye, about 420 nm, to show the H and K lines of calcium.
(Courtesy Roger Bell and Michael Briley)

With the rise of computers, astronomers began displaying spectra as graphs of intensity versus wavelength. Such graphs show more detail than photographic plates, and they can be measured directly. Figure 6-16 shows graphical spectra for stars of different spectral types. In these spectra, absorption lines are sharp dips in the curve. Compare Figures 6-15 and 6-16 to find similar lines and bands in the spectra, and you will see how precise the graphical spectra are.

When we look at spectra for stars from O to M we see how spectral lines change from class to class. Note that the Balmer lines are strongest in A stars, where the temperature is moderate but still high enough to excite the electrons in hydrogen atoms to the second energy level, where they can absorb Balmer wavelength photons. In the hotter stars (O and B), the Balmer lines are weak because the higher temperature excites the electrons to energy levels

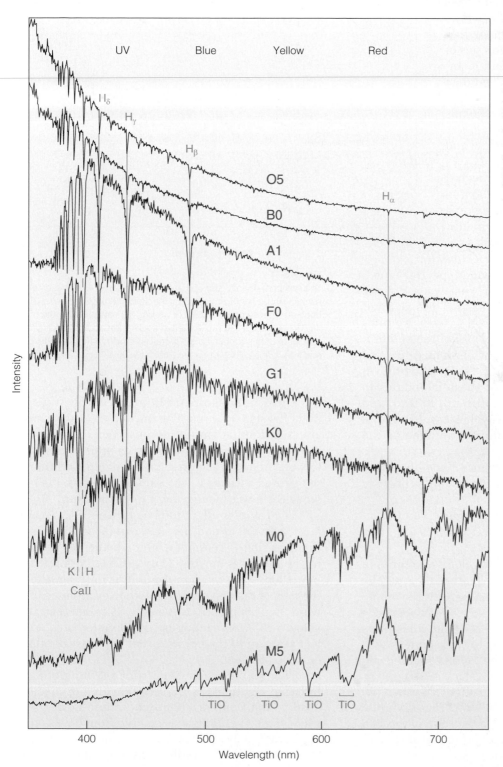

Figure 6-16
Modern digital spectra are often represented by graphs of intensity versus wavelength. Dark absorption lines are dips in intensity. The hottest stars are at the top and the coolest at the bottom. Hydrogen Balmer lines are strongest at about A0, while lines of ionized calcium (CaII) are strong in K stars. Titanium oxide (TiO) bands are strongest in the coolest stars. Compare these spectra with Figure 6-14c and 6-15. *(Courtesy NOAO, G. Jacoby, D. Hunter, and C. Christian)*

above the second or ionizes the atoms. The Balmer lines in cooler stars (F through M) are also weak but for a different reason. The lower temperature cannot excite many electrons to the second energy level, so few hydrogen atoms are capable of absorbing Balmer wavelength photons.

The spectral lines of other atoms also change from class to class. Helium is visible only in the spectra of the hottest classes, and titanium oxide bands only in the coolest. Two lines of ionized calcium, labeled H and K, increase in strength from A to K and then decrease from K to M. Because the strength of these spectral lines depends on temperature, it requires only a few minutes to study a star's spectrum and determine its temperature.

The century-old study of spectral types might seem a closed book, but in 1998 astronomers surveying the sky in infrared wavelengths found a new class of star. The extremely cool, faint stars have been added to the spectral sequence after M and are known as spectral type L stars or **L dwarfs.** They are clearly a new class of star and not an extension of the M stars. The spectra of M stars contain bands produced by metal oxides such as titanium oxide, but spectra of the cooler L stars contain bands produced by chromium hydride and iron hydride (Figure 6-17). Their temperatures must be less than 2000 K for these molecules to form. The L dwarfs are faint, difficult to find, and difficult to study at visual wavelengths. Nevertheless, they are helping astronomers understand the coolest stars.

Not only can a star's spectrum tell us the surface temperature of the star, but it can also tell us how the star is moving through space.

The Doppler Effect

Astronomers can measure the wavelengths of lines in a star's spectrum and find the velocity of the star. The **Doppler effect** is the apparent change in the wavelength of radiation caused by the motion of the source.

If a star is moving toward Earth, the lines in its spectrum will be shifted slightly toward shorter wavelengths. That is, they are shifted toward the blue end of the spectrum—a **blue shift.** If a star is moving away from Earth, the lines are shifted slightly toward the red end of the spectrum—a **red shift.** These Doppler shifts are small and don't change the color of the star. But it is easy to detect these changes in wavelength in a star's spectrum.

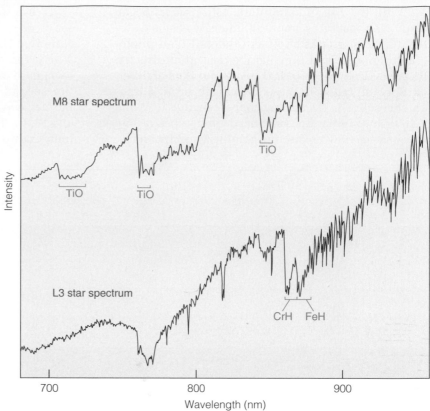

Figure 6-17

The spectrum of an M star and an L star compared: M-star spectra contain strong bands produced by the titanium oxide molecule (TiO), but L-star spectra contain weak or absent TiO bands. Notice the missing TiO band near 710 nm in the L-star spectrum above. Instead, the spectra of L stars contain bands produced by chromium hydride (CrH) and iron hydride (FeH). *(Adapted from data by J. Davy Kirkpatrick)*

Figure 6-18a shows the Doppler effect in two spectra of the star Arcturus. The lines in the top spectrum are slightly blue shifted because the spectrum was recorded when Earth, following its orbit, was moving toward Arcturus. The lines in the bottom spectrum are red shifted because it was recorded six months later, when Earth was moving away from Arcturus.

The Doppler effect tells us how rapidly the distance between us and the source of light is increasing or decreasing. It does not matter whether we are moving or the star is moving. Only the relative velocity is important. Also, the Doppler shift is only sensitive to the part of the velocity directed away from us or toward us. This part of the velocity is called the **radial velocity (V_r).** We cannot use the Doppler effect to detect any part of the velocity that is perpendicular to our line of sight.

The Doppler effect occurs for any kind of radiation; not just light. Radio astronomers, for instance, observe the Doppler effect when they measure the wavelength of radio waves coming from objects moving toward or away from Earth. Whatever wavelength range we consider—radio, X rays, gamma rays—we

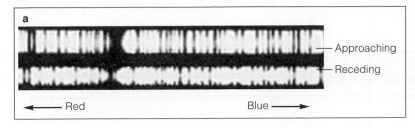

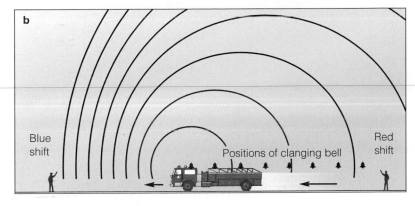

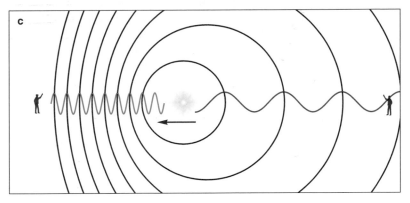

Figure 6-18

The Doppler effect. (a) A red shift appears in the spectrum of a star approaching Earth (top spectrum). A blue shift appears in the spectrum of a star moving away from Earth (bottom spectrum). *(The Observatories of the Carnegie Institution of Washington)* (b) The clanging bell on a moving fire truck produces sounds that move outward (black circles). An observer ahead of the truck hears the clangs closer together, while an observer behind the truck hears them farther apart. (c) A moving source of light emits waves that move outward (black circles). An observer in front of the light source observes a shorter wavelength (a blue shift), and an observer behind the light source observes a longer wavelength (a red shift).

We can understand how the Doppler shift works by thinking about a fire truck approaching us with a bell clanging once a second. When the bell clangs, the sound travels ahead of the truck to reach our ears. One second later, the bell clangs again, but not at the same place. During that one second, the fire truck moved closer to us, so the bell is closer when it clangs. Now the sound has a shorter distance to travel and reaches our ears sooner than it would have if the truck were not approaching. The third time the bell clangs, it is even closer. By timing the bell, we could observe that the clangs are slightly less than 1 second apart, all because the fire truck is approaching. If the fire truck were moving away from us, we would hear the clangs sounding more than one second apart, because each successive clang of the bell occurs farther from us.

Figure 6-18b shows a fire truck moving toward one observer and away from another observer. The position of the bell at each clang is shown by the small black bells with the sound spreading outward as blue circles. We can see that the clangs are squeezed together ahead of the fire truck and stretched apart behind.

Now we can substitute a source of light waves for the clanging of the bell. If the source is approaching, then each time the source emits the peak of a wave it will be slightly closer to us, and we will observe a shorter wavelength. If it is moving away, we will observe a longer wavelength. This is shown in Figure 6-18c, where the peaks of the waves appear compressed in front of the moving light source and stretched out behind the source.

Of course, how great the change in wavelength is depends on the velocity. Just as a slowly moving car has a smaller Doppler effect in sound than does a high-speed airplane, a slowly moving star has a smaller Doppler effect in light than does a high-speed star. Thus, we can measure the velocity by measuring the amount by which the spectral lines are shifted in wavelength. A large Doppler shift means a high radial velocity. This is a common tool in law enforcement. Police transmit radio signals toward passing autos and measure speed by measuring the wavelength shifts in the reflected signal. By the Numbers 6-2 shows how we could do a Doppler effect calculation.

Stellar spectra are filled to bursting with information about stars and their motion, but the astronomer must know some physics to interpret a spectrum. For example, the physics we know is enough to allow us

will still refer to a lengthening of the observed wavelength as a red shift and a shortening of the observed wavelength as a blue shift.

We can even detect the Doppler shift in sound. Sounds with long wavelengths have a lower pitch, and sounds with short wavelengths have a higher pitch. When we hear a truck roar past us on the highway, we hear the sound of the truck drop in pitch as it passes. Its sound is blue shifted while it is approaching and red shifted after it passes.

Table 6-2 The Most Abundant Elements in the Sun

Element	Percentage by Number of Atoms	Percentage by Mass
Hydrogen	91.0	70.9
Helium	8.9	27.4
Carbon	0.03	0.3
Nitrogen	0.008	0.1
Oxygen	0.07	0.8
Neon	0.01	0.2
Magnesium	0.003	0.06
Silicon	0.003	0.07
Sulfur	0.002	0.04
Iron	0.003	0.01

To derive accurate chemical abundances, we must use the physics that describes the interaction of light and matter to analyze a star's spectrum, take into account the star's temperature, and calculate the amounts of the elements present in the star. Such results show that nearly all stars have compositions similar to the sun's—about 92 percent of the atoms are hydrogen, and 7.8 percent are helium, with small traces of heavier elements (Table 6-2). We will use these results in later chapters when we analyze the life stories of the stars, the history of our galaxy, and the origin of the universe.

to estimate the chemical composition of stars from their spectra.

Chemical Composition

Identifying the elements in a star by identifying the lines in the star's spectrum is a relatively straightforward procedure. For example, two dark absorption lines appear in the yellow region of the solar spectrum at the wavelengths 589 nm and 589.6 nm. The only atom that can produce this pair of lines is sodium, so we must conclude that the sun contains sodium. Over 90 elements in the sun have been identified this way.

However, just because the spectral lines characteristic of an element are missing, we cannot conclude that the element itself is absent. For example, the hydrogen Balmer lines are weak in the sun's spectrum, yet 90 percent of the atoms in the sun are hydrogen. The reason for this apparent paradox is that the sun is too cool to produce strong Balmer lines. Similarly, an element's spectral lines may be absent from a star's spectrum because the star is too hot or too cool to excite those atoms to the energy levels that produce visible spectral lines.

REVIEW Critical Inquiry

Why are helium lines weak and calcium lines strong in the visible spectrum of the sun?

To analyze this problem, we must recall that the ability of an atom or ion to absorb light depends on temperature. Helium is quite abundant in the sun, but the surface of the sun is too cool to excite helium atoms and enable them to easily absorb visible-wavelength photons. On the other hand, calcium is easily ionized, and calcium atoms that have lost an electron are very good absorbers of photons at temperatures like those at the sun's surface. Thus, calcium lines are strong in the solar spectrum even though calcium ions are rare, and helium lines are weak even though helium atoms are common.

If we are to analyze a star's spectrum and consider the strength of spectral lines, we must take into account the temperature of the star. Why don't we need to take temperature into account when we use the Doppler effect to measure a star's radial velocity?

The spectra of the stars are filled with clues about the gas emitting the light. Much of astronomy is based on unraveling these clues. We will begin in the following chapter by studying the nearest star—the sun.

Summary

An atom consists of a nucleus surrounded by a cloud of electrons. The nucleus is made up of two kinds of particles: positively charged protons and uncharged neutrons. The number of protons in an atom determines which element it is. Atoms of the same element (that is, having the same number of protons) with different numbers of neutrons are called *isotopes*.

The motion among the atoms in a solid, a liquid, or a dense gas causes the emission of black body radiation. The hotter the object, the more radiation it emits. This produces a continuous spectrum, and the energy is most intense at the wavelength of maximum intensity, λ_{max}, which depends on the object's temperature. Hot objects emit mostly short-wavelength radiation, while cool objects emit mostly long-wavelength radiation. This effect gives us clues to the temperatures of stars—hot stars are blue, and cool stars are red.

The negatively charged electrons surrounding an atomic nucleus may occupy various permitted orbits. An electron may be excited to a higher orbit during a collision between atoms, or it may move from one orbit to another by absorbing or emitting a photon of the proper energy.

Because only certain orbits are permitted, only photons of certain wavelengths can be absorbed or emitted. Each kind of atom has its own characteristic set of spectral lines. The hydrogen atom has the Lyman series in the ultraviolet, the Balmer series in the visible, and the Paschen series (and others) in the infrared.

If light passes through a low-density gas on its way to our telescope, the gas can absorb photons of certain wavelengths, and we will see dark lines in the spectrum at those positions. Such a spectrum is called an *absorption spectrum*. If we look at a low-density gas that is excited to emit photons, we see bright lines in the spectrum at those positions. Such a spectrum is called an *emission spectrum*.

Most stellar spectra are absorption spectra, and the hydrogen lines we see there are Balmer lines. In cool stars, the Balmer lines are weak because atoms are not excited out of the ground state. In hot stars, the Balmer lines are weak because atoms are excited to higher orbits or are ionized. Only at medium temperatures are the Balmer lines strong. We can use this effect as a thermometer for determining the temperature of a star. In its simplest form, this amounts to classifying the star's spectra in the spectral sequence: O, B, A, F, G, K, M.

When a source of radiation is approaching us, we observe shorter wavelengths, and when it is receding, we observe longer wavelengths. This Doppler effect makes it possible for the astronomer to measure a star's radial velocity, that part of its velocity directed toward or away from Earth.

We can also find the composition of a gas from its spectrum. Such studies of the sun show that it is mostly hydrogen, contains only about 7.8 percent helium atoms, and includes only traces of heavier elements.

New Terms

nucleus	joule
proton	continuous spectrum
neutron	absorption line
electron	absorption spectrum (dark-line spectrum)
isotope	
ionization	emission line
ion	emission spectrum (bright-line spectrum)
molecule	
Coulomb force	Kirchhoff's laws
binding energy	transition
quantum mechanics	Lyman series
permitted orbit	Balmer series
energy level	Paschen series
excited atom	spectral class or type
ground state	spectral sequence
heat	L dwarf
temperature	Doppler effect
Kelvin temperature scale	blue shift
absolute zero	red shift
black body radiation	radial velocity (V_r)
wavelength of maximum intensity (λ_{max})	

Review Questions

Same Prot *diff neut* [handwritten]

1. Why might we say that atoms are mostly empty space?
2. What is the difference between an isotope and an ion?
3. Why is the binding energy of an electron related to the size of its orbit? *clos more binding Energy* [handwritten] *Loss e/e.* [handwritten]
4. Explain why ionized calcium can form absorption lines but ionized hydrogen cannot.
5. Describe two ways an atom can become excited.
6. Why do different atoms have different lines in their spectra?
7. Why does the amount of black body radiation emitted depend on the temperature of the object?
8. Why do hot stars look bluer than cool stars?
9. What kind of spectrum does a neon sign produce? *Emission Spectrum Page 100* [handwritten]
10. Why are Balmer lines strong in the spectra of medium-temperature stars and weak in the spectra of hot and cool stars?
11. *102 page →* [handwritten] Why are titanium oxide features visible in the spectra of only the coolest stars? *Bi its a Classified as an M 3000* [handwritten]
12. Explain the similarities among Table 6-1, Figure 6-14c, Figure 6-15, and Figure 6-16.
13. Why does the Doppler effect detect only radial velocity?

14. How can the Doppler effect explain shifts in both light and sound?

15. Explain why the presence of spectral lines of a given element in the solar spectrum tells us that element is present in the sun, but the absence of the lines would not mean the element was absent from the sun?

Discussion Questions

1. In what ways is our model of an atom a scientific model? In what ways is it incorrect?

2. Can you think of classification systems we use to simplify what would otherwise be complex measurements? Consider foods, movies, cars, grades, and clothes.

Problems

1. Human body temperature is about 310 K (98.6°F). At what wavelength do humans radiate the most energy? What kind of radiation do we emit?

2. If a star has a surface temperature of 20,000 K, at what wavelength will it radiate the most energy?

3. Infrared observations of a star show that it is most intense at a wavelength of 2000 nm. What is the temperature of the star's surface?

4. If we double the temperature of a black body, by what factor will the total energy radiated per second per square meter increase?

5. If one star has a temperature of 6000 K and another star has a temperature of 7000 K, how much more energy per second will the hotter star radiate from each square meter of its surface?

6. Transition *A* produces light with a wavelength of 500 nm. Transition *B* involves twice as much energy as *A*. What wavelength light does it produce?

7. Determine the temperatures of the following stars based on their spectra. Use Figures 6-15 and 6-16.
   a. medium-strength Balmer lines, strong helium lines
   b. medium-strength Balmer lines, weak ionized-calcium lines
   c. strong TiO bands
   d. very weak Balmer lines, strong ionized-calcium lines

8. To which spectral classes do the stars in Problem 7 belong?

9. In a laboratory, the Balmer beta line has a wavelength of 486.1 nm. If the line appears in a star's spectrum at 486.3 nm, what is the star's radial velocity? Is it approaching or receding?

10. The highest-velocity stars an astronomer might observe have velocities of about 400 km/sec. What change in wavelength would this cause in the Balmer gamma line? (*Hint:* Wavelengths are given in Figure 6-11.)

Critical Inquiries for the Web

1. The name for the element helium has astronomical roots. Search the Internet for information on the discovery of helium. How and when was it discovered, and how did it get its name? Why do you suppose it took so long for helium to be recognized?

2. How was the model of the atom presented in the text you read developed? Search the Web for information on historical models of the atom, and compile a time line of important developments leading to our current understanding. What evidence exists that supports our model?

Go to the Brooks/Cole Astronomy Resource Center (www.brookscole.com/astronomy) for critical thinking exercises, articles, and additional readings from InfoTrac College Edition, Brooks/Cole's online student library.

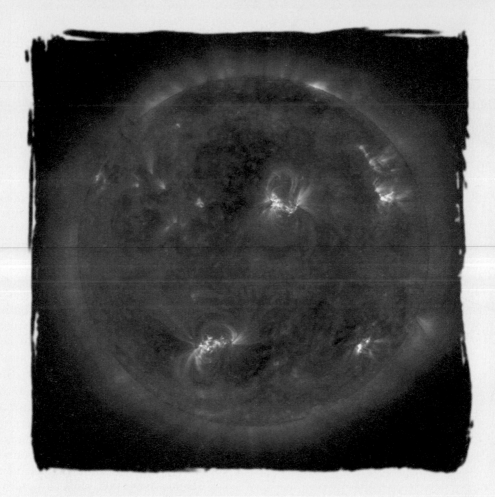

CHAPTER 7

The Sun —
Our Star

Guidepost

The preceding chapter described how atoms interact with light to produce spectra. This knowledge is a powerful tool, and in this chapter we will see what we can learn about the sun by applying our new tool to the analysis of sunlight.

This chapter gives us our first close look at scientists at work, and we discover that much of science consists of confirmation of previous hypotheses and the gradual consolidation of our understanding.

Most of all, in this chapter we get our first look at a real star. The chapters that follow will concentrate on the billions of stars in the heavens, but this chapter tells us that each of them is both complex and beautiful. Each is a sun.

The Sun Data File One

An image of the sun in visible light shows a few sunspots. The Earth–moon system is added for scale. *(Daniel Good)*

Average distance from Earth	1.00 AU (1.495979×10^8 km)
Maximum distance from Earth	1.0167 AU (1.5210×10^8 km)
Minimum distance from Earth	0.9833 AU (1.4710×10^8 km)
Average angular diameter seen from Earth	0.53° (32 minutes of arc)
Period of rotation	25 days at equator
Radius	6.9599×10^5 km
Mass	1.989×10^{30} kg
Average density	1.409 g/cm³
Escape velocity at surface	617.7 km/sec
Luminosity	3.826×10^{26} J/sec
Surface temperature	5800 K
Central temperature	15×10^6 K
Spectral type	G2 V
Apparent visual magnitude	−26.74
Absolute visual magnitude	4.83

A wit once remarked that solar astronomers would know a lot more about the sun if it were farther away. This comment contains a grain of truth; the sun is only a humdrum star, and there are billions like it in the sky, but the sun is the only one close enough to show surface detail. Solar astronomers can see so much detail in the swirling currents of gas and arching bridges of magnetic force that present theories seem inadequate to describe it. Yet the sun is not a complicated object. It is just a star.

In their general properties, stars are very simple. They are great balls of hot gas held together by their own gravity. Their gravity would make them collapse into small, dense bodies were they not so hot. The tremendously hot gas inside stars has such a high pressure that the stars would surely explode were it not for their own confirming gravity. Like soap bubbles, stars are simple structures balanced between opposing forces that individually would destroy them. Thus, we study the sun as a close-up example of a star.

Another reason to study the sun is that life on Earth depends critically on the sun. Should the sun's energy output vary by even a small amount, life on Earth might vanish. In addition, we get nearly all our energy from the sun—oil and coal are merely stored sunlight—and our pleasant climate is maintained by energy from the sun. In fact, the sun's atmosphere of very thin gas reaches out past Earth's orbit, and thus any change in the sun, such as an eruption or a magnetic storm, can have a direct effect on Earth.

Finally, we study the sun because it is beautiful. Our analysis of sunlight will reveal that the sun is both powerful and delicate. Thus, we study the sun not only because it is *a star*, not only because it is *our star*, but because it is the sun.

7-1 The Solar Atmosphere

The sun is hot gas from its highest layers right down to its center, and we can see this from its density (Data File One). It is a bit more dense than water but not nearly as dense as rock. When we look at the sun, we see only the highest layers of this gas, the solar atmosphere. As we have seen in our discussion of solar eclipses, the atmosphere of the sun is divided into three layers, the photosphere, chromosphere, and corona. We are now ready to explore these layers in detail.

Below the Surface

The nature of the sun's atmosphere is determined by processes that go on deep below the visible surface. Near the center of the sun nuclear fusion processes generate energy, and that energy flows outward toward the surface where it is radiated into space as heat and light and other forms of electromagnetic radiation. We will discuss the process by which the sun generates its energy and the way that energy flows outward in later chapters, when we consider the structure, evolution, and death of stars. Here, it is sufficient to know that heat from the solar interior flows out through the surface.

As we observe the solar surface in this chapter, we will be alert for evidence of energy flow from the interior. We will find such evidence in the form of hot and cool regions, gas motions, magnetic fields, and so on. All of the beautiful and violent activity we will discover on the sun is driven by energy flowing outward. All stars make their energy near their centers, so all stars are dominated by the outward flow of energy. The features we see on the surface of the sun are characteristic of the features on the surfaces of other stars.

The Photosphere

The visible surface of the sun, the photosphere, is not a solid surface. In fact, the sun is gaseous from its outer atmosphere right down to its center. The photosphere is the thin layer of gas from which we receive most of the sun's light. It is less than 500 km deep and has an average temperature of about 6000 K. If the sun magically shrank to the size of a bowling ball, the photosphere would be no thicker than a layer of tissue paper wrapped around the ball (Figure 7-1a). For comparison, the chromosphere lies above the photosphere and is only a few times deeper, but the corona, beginning above the chromosphere, extends far from the sun (Figure 7-1b).

Below the photosphere, the gas is denser and hotter and therefore radiates plenty of light, but that light cannot escape from the sun because of the outer layers of gas. Thus, we cannot detect light from these deeper layers. Above the photosphere, the gas is less dense and so is unable to radiate much light. The photosphere is the layer in the sun's atmosphere that is dense enough to emit plenty of light but not so dense that the light can't escape.

Although the photosphere appears to be substantial, it is really a very-low-density gas. Even in the deepest and densest layers visible, the photosphere is 3400 times less dense than the air we breathe. To find gases as dense as the air we breathe, we would have to descend about 7×10^4 km below the photosphere, about 10 percent of the way to the sun's center. With a fantastically efficient insulation system, we could fly a spaceship right through the photosphere.

The spectrum of the sun is an absorption spectrum, and that can tell us a great deal about the photosphere. We know from Kirchhoff's third law that an absorption spectrum is produced when a source of a

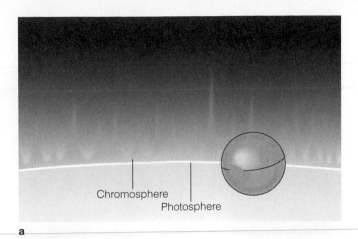

Chromosphere
Photosphere

a

b

Figure 7-1
(a) A cross section at the edge of the sun shows the relative thickness of the photosphere and chromosphere. Earth is shown for scale. On this scale, the disk of the sun would be more than 1.5 m (5 feet) in diameter. (b) The corona extends from the top of the chromosphere to great height above the photosphere. This photograph, made during a total solar eclipse, shows only the inner part of the corona. *(Daniel Good)*

continuous spectrum is viewed through a gas. In the case of the photosphere, the deeper layers are dense enough to produce a continuous spectrum, but atoms in the photosphere absorb photons of specific wavelengths, producing the absorption lines we see.

In good photographs, the photosphere has a mottled appearance because it is made up of dark-edged regions. The regions are called granules, and the visual pattern is called **granulation** (Figure 7-2a). Each granule is about the size of Texas and lasts for only 10 to 20 minutes before fading away. Faded granules are continuously replaced by new granules. Spectra of these granules show that the centers are a few hundred

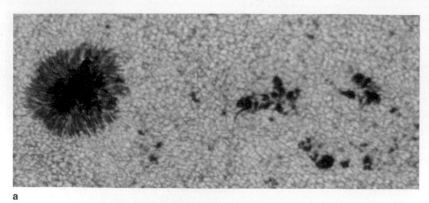

a

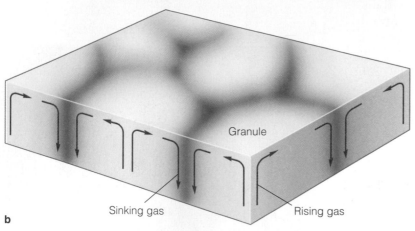

Granule

Sinking gas   Rising gas

b

Figure 7-2

(a) A visible-light photo of the sun's surface shows granulation. *(NOAO)* (b) This model explains granulation as the tops of rising convection currents just below the photosphere. Heat flows upward as rising currents of hot gas and downward as sinking currents of cool gas. The rising currents heat the solar surface in small regions that we see as granules.

degrees hotter than the edges, and Doppler shifts reveal that the centers are rising and the edges are sinking at speeds of about 1 km/sec.

From this evidence, astronomers recognize granulation as the surface effects of convection just below the photosphere. **Convection** occurs when hot fluid rises and cool fluid sinks, as when, for example, a convection current of hot gas rises above a candle flame. You can create convection in a liquid by adding a bit of cool nondairy creamer to an unstirred cup of hot coffee. The cool creamer sinks, warms, rises, cools, sinks again, and so on, creating small regions on the surface of the coffee that mark the tops of convection currents. Viewed from above, these regions look much like solar granules. Rising currents of hot gas heat small regions of the photosphere, which, being slightly hotter, emit more black body radiation and look brighter. The cool sinking gas of the edges emits less light and thus looks darker (Figure 7-2b). Thus the granulation reveals that energy flows upward from below by convection and heats the photosphere.

The Chromosphere

The chromosphere is roughly 1000 times fainter than the photosphere, so without specialized telescopes we can see it only during a total solar eclipse. When the moon covers the brilliant photosphere, we see the chromosphere flash into view as a narrow layer of pink gas just above the photosphere. The term *chromosphere* comes from the Greek word *chroma,* meaning "color."

We may turn to a spectrum to learn about conditions in the chromosphere. The spectrum of the chromosphere is an emission spectrum, and Kirchoff's laws tell us that means the chromosphere is an excited, low-density gas. The pink color results from the blending of the red, blue, and violet Balmer emission lines produced by ionized hydrogen, plus emission lines of other elements. Analysis of the spectrum shows that the temperature of the gas in the chromosphere ranges from 10,000 K up to 1,000,000 K or more (Figure 7-3a).

Solar astronomers can take advantage of some simple physics to study details in the chromosphere. The gases of the chromosphere are mostly transparent to visible light, but atoms such as hydrogen are very good at absorbing specific wavelengths. For instance, few photons of Balmer wavelengths can escape from the photosphere, so the photons of these wavelengths we see leaving the sun must escape from the chromosphere. This makes the chromosphere visible. Photographs called **filtergrams** are recorded through filters that admit only photons at the wavelengths of a certain strong absorption line such as H_α or H_β. In this way filtergrams reveal detail in the uppermost layers of the chromosphere.

Filtergrams such as the one shown in Figure 7-3b reveal **spicules,** flamelike structures 100 to 1000 km in diameter. Although the chromosphere is only a few thousand kilometers thick, the spicules can extend up to 10,000 km above the photosphere. Spicules, which last from 5 to 15 minutes, are extensions of relatively cool chromospheric gases up into the much hotter corona. Seen at the edge of the solar disk, these spicules blend together and look like flames covering a burning prairie, but filtergrams of spicules located near the center of the solar disk show that they spring up around the edges of regions called **supergranules.** Over twice the diameter of Earth, supergranules appear to be caused by heat carried upward by larger convection currents deeper below the photosphere—further evidence of the flow of energy out of the sun's interior.

We will see evidence later in this chapter that the solar surface is dominated by a strong magnetic field,

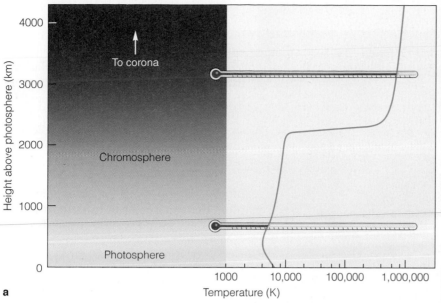

Figure 7-3

The chromosphere. (a) If we could place thermometers in the sun's atmosphere, we would discover that the temperature increases from 6000 K at the photosphere to 10^6 K at the top of the chromosphere. (b) An H_α filtergram shows that the chromosphere is not uniform. Spicules appear like black grass blades at the edges of supergranules that are twice the diameter of Earth. *(© 1971 NOAO/NSO)*

and much of that field emerges from below the photosphere at the edges of the supergranules. Thus, the spicules are clearly governed by magnetic fields that are rooted far below the photosphere.

The Corona

The sun's atmosphere above the chromosphere is called the corona, after the Latin word for "crown." During a total solar eclipse, the moon covers the brilliant photosphere, and the corona shines with a milky glow not even as bright as the full moon (see Figure 3-15b). Photographs taken from the ground can trace the corona out to a distance of a few solar radii, but images taken from high-flying balloons or aircraft can trace the corona out to 12 solar radii or more (Figure 7-4). Ground-based telescopes of special design can record the lower corona of the uneclipsed sun, but the best studies of the corona have been made from spacecraft above Earth's atmosphere.

When we analyze the spectrum of the corona we discover the light comes from a number of different sources. Some of the coronal light we see during total solar eclipses is just sunlight reflected from interplanetary dust near the sun. Some of the light is scattered when it encounters free electrons in the coronal gas, and we might expect this light to produce a solar absorption spectrum. This doesn't happen, however, because the gas is very hot (about 1 million K), and the electrons travel very fast in the corona. When photons scatter off of these electrons, they suffer very large

Doppler shifts that smear out any absorption lines in the reflected solar spectrum, and thus we see a continuous spectrum in the corona.

Superimposed on the corona's continuous spectrum are emission lines of highly ionized atoms. That is, the atoms have lost more than just one or two electrons. This high ionization tells us that the gas is very hot. In the lower corona, the atoms are not as highly ionized as they at higher altitudes, and this tells us that the temperature of the corona rises with altitude. Just above the chromosphere, the temperature of the corona can be as low as 50,000 K; but the outer corona can be as hot as 3,000,000 K. The density of the gas making up the outer corona must be very low indeed, or it would emit a great deal of black body radiation and would be easily visible. In fact, the density of the outer corona is only about 1 to 10 atoms/cm³—fewer atoms per cubic centimeter than the best vacuum on Earth.

The high temperature of the corona has been a long-standing mystery. We have seen that heat flows upward from inside the sun to the surface, but how can heat flow from the relatively cool photosphere out into the hot corona? Heat flows from hot to cool, not from cool to hot. Far ultraviolet observations made with a satellite called SOHO (Solar and Heliospheric Observatory) may have solved the mystery. Using SOHO, astronomers were able to map small loops of magnetic field scattered all over the solar surface. Looking like the loops of yarn in a shag carpet, the loops in this **magnetic carpet** constantly thrash

Figure 7-4
Streamers in the solar corona. An image from the Solar Maximum Mission satellite (inset) has been computer-enhanced to produce a false-color image that reveals subtle variations in brightness. *(NASA/JPL)*

Helioseismology

It seems that the light we receive from the sun can tell us nothing about the layers below the photosphere, but solar astronomers have found a way to explore the sun's interior by **helioscismology,** the study of the modes of vibration of the sun. Just as geologists can study Earth's interior by observing how sound waves produced by earthquakes are reflected and transmitted by the layers of Earth's interior, so too can solar astronomers explore the sun's interior by studying how vibrations (sound waves) travel through the sun.

As these vibrations rise from the interior and reach the surface of the sun, we see the photosphere move up and down in a complicated pattern that depends on the frequency of the sound waves (Figure 7-5). The Doppler shifts caused by the moving surface, although very small, can tell solar astronomers which frequency vibrations are present. Although 10 million different frequencies are possible, some penetrate deeper than others, and conditions in the sun's interior layers can weaken or strengthen the waves. From the frequencies that are actually observed, solar astronomers can determine the temperature, density, pressure, composition, and motion of the sun's internal layers.

Helioseismology sounds almost magical, but we can understand it better if we think of a duck pond. If we stood at the shore of a duck pond and looked down at the water, we would see ripples arriving from all parts of the pond. Because every duck on the pond contributes to the ripples, we could, in principle, study the ripples near the shore and draw a map showing the position and velocity of every duck on the pond. Of course, it would be difficult to untangle the different ripples, but all of the information would be there, lapping at the rocks at our feet.

Just as we can map the ducks on a pond, helioseismologists can study the vibrations in the surface of the sun and deduce the characteristics of the sun's interior. Because there are so many possible modes of vibration, they need large masses of data in order to separate the frequencies. The Global Oscillation Network Group (GONG) has set up small telescopes around the world to observe the sun nonstop for up to 3 years. Needless to say, supercomputers are needed to analyze the data and map the sun's interior. This exciting form of solar astronomy is just beginning to produce results.

around, break, and reconnect. This agitates the atoms above the photosphere and deposits heat in the upper chromosphere and corona. Thus the corona is heated by energy flowing from the sun's interior not as heat but as magnetic energy.

The outer corona is so hot the sun is unable to hold it. The high-velocity gas atoms stream away from the sun in a continuous breeze called the **solar wind.** It contains mostly protons and electrons (ionized hydrogen) but also carries heavier particles. With an average density of a few atoms per cubic centimeter, this solar wind blows past Earth at 300 to 800 km/sec with irregular gusts that can reach 1000 km/sec. Continuing out past the planets, the solar wind eventually mixes with the gases between the stars. Thus Earth is bathed in the corona's hot breath.

Do other stars have chromospheres and coronas like the sun's? Ultraviolet spectra taken by orbiting space telescopes such as the IUE (see Chapter 5) suggest that the answer is yes. The spectra of many stars contain emission lines in the far ultraviolet that could have been formed only in low-density, high-temperature gases like those of the upper chromosphere and lower corona. Thus, the sun, for all its complexity, seems to be a normal star.

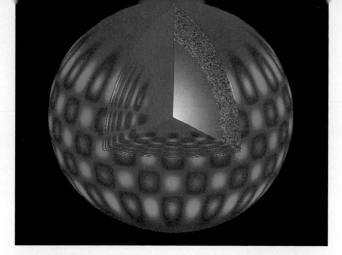

Figure 7-5
Helioseismology is the study of the modes of vibration of the sun. This computer image shows one of nearly 10 million possible modes of oscillation. Red regions are receding and blue approaching. Helioseismological studies can reveal details about the sun's interior. *(NOAO)*

REVIEW Critical Inquiry

How deeply into the sun can we see?

This is a simple question, but it has a very interesting answer. To analyze this question we must avoid being misled by the words we use. When we look into the layers of the sun, our sight does not really penetrate *into* the sun. Rather, our eyes record photons that have escaped *from* the sun and traveled outward through the layers of the sun's atmosphere. If we observe at a wavelength at the center of a dark absorption line, then the photosphere and lower chromosphere are opaque, photons can't escape to our eyes, and the only photons we can see come from the upper chromosphere. What we see are the details of the upper chromosphere—a filtergram. On the other hand, if we observe at a wavelength that is not easily absorbed (a wavelength between spectral lines), the atmosphere is more transparent, and photons from deeper inside the photosphere can escape to our eyes.

By choosing the proper wavelength, solar astronomers can observe different depths in the sun's photosphere and chromosphere. But the corona is so thin and the gas below the photosphere so dense that this method doesn't work in these regions. How can we observe the corona and the deeper layers of the sun?

So far we have thought of the sun as a static, unchanging ball of gas with energy flowing outward from the interior and through the atmospheric layers. In fact, the sun is a highly variable body whose appearance is constantly changing. It is now time to think of the active sun.

7-2 Solar Activity

Solar activity refers to features on the sun that change over minutes, days, or years. We will use spectroscopic analysis to explore this aspect of solar astronomy, and we will discover that solar activity is shaped by magnetic fields and driven by the powerful flow of energy rising from the sun's interior.

Sunspots

Solar activity is often visible with even a small telescope, but we should exercise great caution in observing the sun. Not only is sunlight intense, but it also includes infrared radiation (heat), which can burn our eyes. It is not safe to look at the sun directly, and it is even more dangerous to look at the sun through a telescope, which concentrates the sunlight. Figure 7-6a illustrates a safe way to observe the sun with a small telescope.

When we observe the sun at visual wavelengths, we see the gases of the photosphere. Dark spots on the photosphere are called **sunspots,** and observations over a period of days will show the sun rotating and the sunspots growing larger or shrinking away (Figure 7-6b). The dark center of a sunspot, called the *umbra,* is surrounded by a lighter region called the *penumbra* (Figure 7-7a). A typical sunspot is about twice the diameter of Earth, but there is a wide range of sizes. Our visual observations would show that sunspots tend to occur in groups or pairs (Figure 7-7b) and that a typical large group can last as long as two months.

Sunspots look dark because they are cooler than the photosphere. By noting which atoms are excited in the spectra of sunspots, astronomers can tell that the umbra is cool, with a temperature of about 4240 K. The photosphere is about 6000 K. Because the amount of black body radiation emitted depends on the temperature to the fourth power, this small difference in temperature makes a big difference in brightness, and the spot looks quite dark compared to the photosphere. In fact, a sunspot emits quite a bit of radiation. If the sun were removed and only an average-size sunspot were left behind, it would glow a brilliant orange-red and would be brighter than the full moon.

The total number of sunspots visible on the sun is not constant. In 1843, the German amateur astronomer Heinrich Schwabe noticed that the number of sunspots varies with a period of about 11 years. This is now known as the sunspot cycle (Figure 7-8a). At sunspot maximum there are often as many as 100 spots visible at any one time, but at sunspot minimum there are only a few small spots. A sunspot maximum occurred about 1990, with the next expected about 2001.

At the beginning of each sunspot cycle, the spots begin to appear in the sun's middle latitudes about

Figure 7-6
Looking through a telescope at the sun is dangerous, but you can always view the sun safely with a small telescope by projecting its image on a white screen (a). If you sketch the location and structure of sunspots on successive days (b), you will see the rotation of the sun and gradual changes in the size and structure of sunspots.

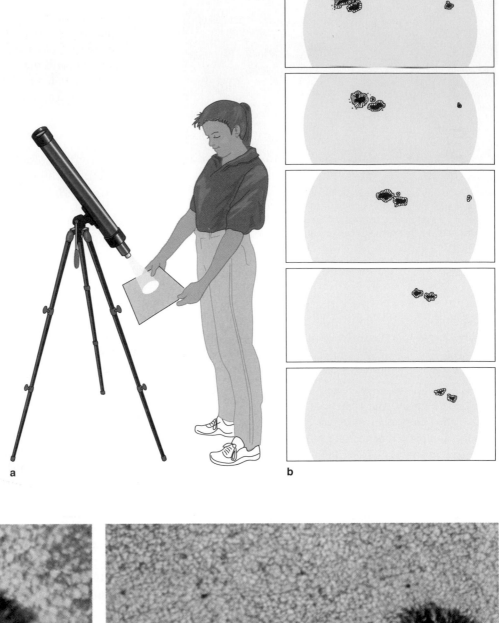

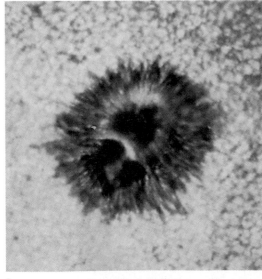 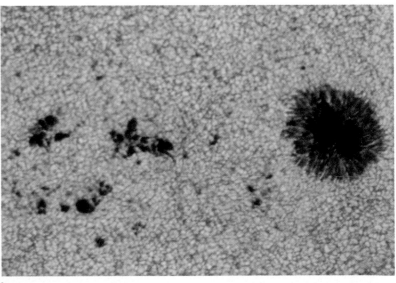

Figure 7-7
(a) This visible-light photograph of a sunspot is shown digitally enhanced to reveal radial structure in the penumbra and in the umbra. These markings are caused by the magnetic field in the sunspot. *(Courtesy William Livingston)* (b) A dark sunspot slightly larger than Earth (right) is part of a group of spots. Notice the granulation of the solar surface throughout the photo. *(NOAO)*

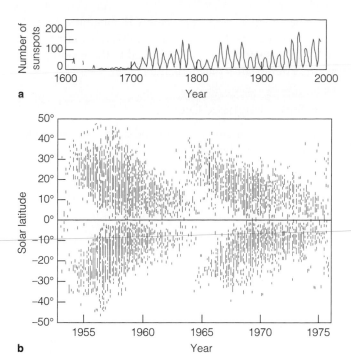

a

b

Figure 7-8

(a) The number of sunspots varies over an 11-year period. (b) The Maunder butterfly diagram shows that the spots first occur at higher latitudes and later nearer the solar equator.

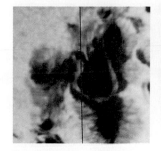

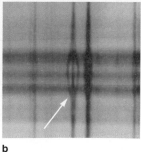

a   b

Figure 7-9

(a) The slit of a spectrograph has been placed across a complex sunspot. Thus, light from the sunspot enters the spectrograph. (b) The resulting spectrum shows a typical absorption line split into three components (arrow) by the magnetic field in the sunspot. *(NOAO)*

35° above and below the sun's equator. As the cycle proceeds, the spots appear at lower latitudes, until, near the end of the cycle, they are appearing within 5° of the sun's equator. If we plot the latitude of the appearance of sunspots over time, the diagram takes on the appearance of butterfly wings (Figure 7-8b). Such diagrams are now known as **Maunder butterfly diagrams,** named after E. Walter Maunder of the Greenwich Observatory.

Maunder also noticed that records show very few sunspots from 1645 to 1715. Modern studies show that this **Maunder minimum** (Figure 7-8a) coincided with a period of reduced solar activity and may be linked to the "little ice age," a period of unusually cool weather in Europe and North America from about 1430 to 1850.

Active Regions and Sunspots

Just what are sunspots? Strong evidence convinces astronomers that sunspots are the visible traces of great magnetic storms on the solar surface.

Critical evidence linking sunspots with magnetic fields appears when we examine the **Zeeman effect,** the splitting of single spectral lines into multiple components through the influence of a magnetic field (Figure 7-9). When an atom is in a magnetic field, the electron orbits are al-

tered, and the atom is able to absorb a number of different wavelength photons even though it was originally limited to a single wavelength. In the spectrum, we see single lines split into multiple components, with the separation between components proportional to the strength of the magnetic field.

Using the Zeeman effect, astronomers find that the field in a typical sunspot is about 1000 times stronger than the sun's average field (Figure 7-10a). Apparently this powerful magnetic field inhibits gas motion below the photosphere, and rising gas cannot deliver its heat to the surface. Thus, the area cools, and we see a sunspot. Infrared observations give us further evidence. Infrared images of sunspots shows that the photosphere around the sunspots is slightly brighter than the rest of the surface. Evidently the heat rising from the interior that did not emerge through the sunspot is deflected and emerges around the sunspot (Figure 7-10b).

Figure 7-10

(a) A magnetogram, a magnetic map of the sun constructed using the Zeeman effect, confirms that the regions around sunspots (white) contain powerful magnetic fields. *(NOAO)* (b) This infrared image of a sunspot shows a brightening of the photosphere around the sunspot, apparently caused by the escape of energy that was unable to emerge through the strongly magnetic sunspot. *(Dan Gezari, NASA/Goddard)*

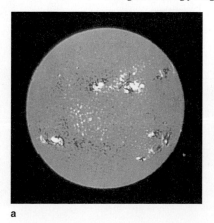

a

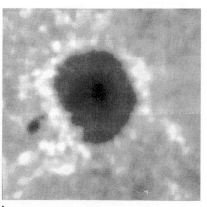

b

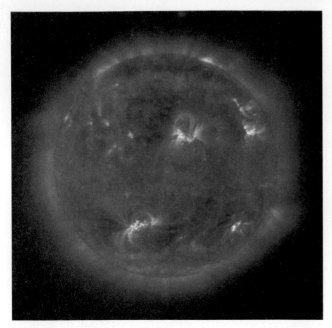

Figure 7-11

This far ultraviolet image of the sun was recorded by the SOHO space-craft. It shows arched magnetic structures in the upper chromosphere and lower corona. At visual wavelengths we would see a sunspot group at the location of each of these active regions. *(SOHO/EIT, ESA and NASA)*

Observations at nonvisible wavelengths reveal that the chromosphere and corona above sunspots are violently disturbed. These **active regions** are highly complex with ionized gas trapped in twisted magnetic fields arching above sunspot groups (Figure 7-11). The shape alone suggests magnetic fields, and the Zeeman effect confirms that these arches are magnetic.

The sunspot groups are merely the visible traces of the active regions. But what causes this magnetic activity? The answer appears to be linked to the waxing and waning of the sun's magnetic field.

The Sun's Magnetic Cycle

Sunspots are magnetic phenomena, so the 11-year cycle of sunspots must be caused by cyclical changes in the sun's magnetic field. To explore that idea, we must begin with the sun's rotation.

The sun does not rotate as a rigid body. It is a gas from its outermost layers down to its center, so some parts of the sun rotate faster than other parts. The equatorial region of the photosphere rotates faster than do regions at higher latitudes (Figure 7-12a). At the equator, the photosphere rotates once every 25 days, but at a latitude of 45° one rotation takes 27.8 days. Helioseismology shows that deeper layers of gas rotate slower than the surface and that gases near the poles rotate the slowest of all (Figure 7-12b). This phenomenon is called **differential rotation,** and it is clearly linked with the magnetic cycle.

Astronomers believe that the sun's overall magnetic field is produced by its rotation and the outward flow of energy. The gases of the sun are highly ionized, so they are very good conductors of electricity. When an electrical conductor rotates rapidly and is stirred by convection, it can convert some of the energy flowing outward as convection into a magnetic field. This process is called the **dynamo effect,** and it is believed to produce Earth's magnetic field as well. Helioseismologists have found evidence that the sun's magnetic field is generated in convection currents deep under the pho-

Figure 7-12
(a) At the photosphere, the sun rotates faster at the equator than at higher latitudes. If we started five sunspots in a row, they could not stay aligned as the sun rotates. (b) Helioseismology shows how the interior of the sun rotates relative to the surface. In this cutaway diagram, the photosphere at the equator rotates the fastest and the poles the slowest. *(Courtesy Kenneth Libbrecht, Caltech)* (c) The magnetic polarity of sunspot groups in the sun's southern hemisphere is reverse of that in the northern hemisphere.

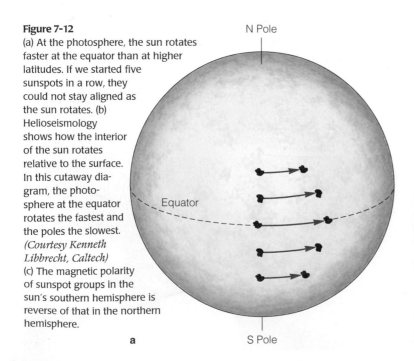

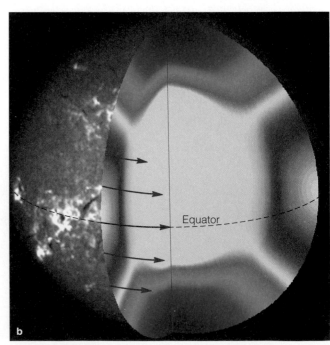

tosphere. Once the magnetic field is created, the convection and differential rotation stretch, twist, and tangle the field to produce a magnetic cycle.

The magnetic behavior of sunspots gives us an insight into how the magnetic cycle works. Sunspots tend to occur in pairs, and the magnetic field around the pair resembles that around a bar magnet with one end magnetic north and the other end magnetic south. At any one time, sunspot pairs south of the sun's equator have reversed polarity compared to those north of the sun's equator. Figure 7-12c illustrates this by showing sunspot pairs south of the sun's equator with magnetic south poles leading and sunspots north of the sun's equator with magnetic north poles leading. At the end of an 11-year sunspot cycle, the new spots appear with reversed magnetic polarity.

This magnetic cycle is not fully understood, but the **Babcock model** (named for its inventor) explains the magnetic cycle as a progressive tangling of the solar magnetic field. Because the electrons in an ionized gas are free to move, the gas is a very good conductor of electricity, and any magnetic field in the gas is "frozen" into the gas. If the gas moves, the magnetic field must move with it. Thus the sun's magnetic field is frozen into its gases, and the differential rotation wraps this field around the sun like a long string caught on a hubcap. Rising and sinking gas currents twist the field into ropelike tubes, which tend to float upward. Where these magnetic tubes burst through the sun's surface, sunspot pairs occur (Figure 7-13).

The Babcock model explains the reversal of the sun's magnetic field from cycle to cycle. As the magnetic field becomes tangled, adjacent regions of the sun's surface are dominated by magnetic fields that

Figure 7-13
The Babcock model suggests that the differential rotation of the sun winds up the magnetic field. When the field becomes tangled and bursts through the surface, it forms sunspot pairs. For the sake of clarity, only one line in the sun's magnetic field is shown here.

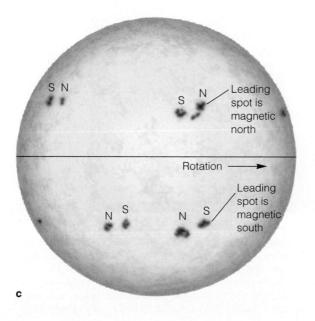

While many textbooks describe science as the process of testing hypotheses by observation and experiment, you should not think that every astronomer approaches the telescope expecting to make an observation that will disprove long-held beliefs and trigger a revolution in science. Then what is the daily grind of science really about?

First, many observations and experiments merely confirm already-tested hypotheses. The biologist knows that all worker bees in a hive are sisters, but a careful study of the DNA from different workers further confirms that hypothesis. By repeatedly confirming a hypothesis, scientists build confidence in the hypothesis and may be able to extend it to a wider application. Of course, there is always the chance that a new observation or experiment will disprove the hypothesis, but that is usually very unlikely. Much of the daily grind of science is confirmation.

Another aspect of routine science is consolidation, the linking of a hypothesis to other well-studied phenomena. Chemists may understand certain kinds of carbon molecules shaped like rings, but by repeated study they find a carbon molecule shaped like a hollow sphere. To consolidate their findings, they must show that the chemical bonding in the two molecules follows the same rules and that the molecules have certain properties in common. No hypothesis is overthrown, but the chemists consolidate their knowledge and understand carbon molecules better.

The Babcock model of the solar magnetic cycle is an astronomical example of the scientific process. Solar astronomers know that the model explains some solar features but has shortcomings. Although most astronomers don't expect to discard the entire model, they work through confirmation and consolidation to better understand how the solar magnetic cycle works and how it is related to cycles in other stars.

---

point in different directions. After about 11 years of tangling, the field becomes so complex that adjacent regions of the solar surface begin changing their magnetic field to agree with neighboring regions. Quickly the entire field rearranges itself into a simpler pattern, and differential rotation begins winding it up to start a new cycle. But the newly organized field is reversed, and the next sunspot cycle begins with magnetic north replaced by magnetic south. Thus the complete magnetic cycle is 22 years long, and the sunspot cycle is 11 years long.

Notice the power of a scientific model. The Babcock model may in fact be incorrect in some or all details, but it gives us a framework on which to organize all of the complex solar activity. Even though our models of the sky (see Chapter 2) and the atom (see Chapter 6) were only partially correct, they served as organizing themes to guide our thinking. Similarly, although the precise details of the solar magnetic cycle are not yet understood, the Babcock model gives us a general picture of the behavior of the sun's magnetic field (Window on Science 7-1).

If the sun is truly a representative star, we might expect to find similar magnetic cycles on other stars, but stars other than the sun are too distant for spots to be directly visible. Some stars, however, vary in brightness over a period of days in a way that reveals they are marked with dark spots believed to resemble sunspots (Figure 7-14). Other stars have spectral features that vary over periods of years, suggesting that they are subject to magnetic cycles much like the sun's. In fact, some stars display sudden flares that may resemble solar eruptions.

Prominences and Flares

A **prominence** is composed of hot gas trapped in magnetic fields extending above active regions into the lower corona. We can tell that the gas in a prominence is hot because it is ionized and produces an emission spectrum. At visible wavelengths, the Balmer lines give prominences a bright pink color, and they are visible during a total solar eclipse protruding beyond the

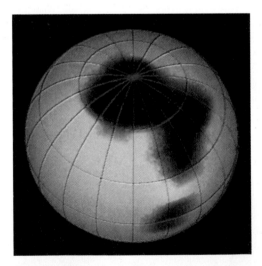

Figure 7-14
This computer-generated map shows the distribution of dark spots over the surface of the star HR1099. The map was constructed by analyzing the Doppler shifts in the spectrum of the rotating star. Although more extensive than sunspots, these dark regions are believed to be related to magnetic activity similar to that responsible for sunspots. *(Courtesy Steven Vogt, Artie Hatzes, and Don Penrod)*

a

b

Figure 7-15

(a) A loop prominence near the edge of the sun shows by its shape the magnetic fields that control the gas. *(Sacramento Peak Observatory)* (b) This far ultraviolet photograph of the sun shows a very large arched prominence extending out into the corona. Note the complexity of the magnetic field. *(SOHO/EIT, ESA and NASA)*

edge of the moon. (See Figure 3-15b.) We can tell that prominences are dominated by magnetic fields because of their arched shapes (Figure 7-15).

We can identify two kinds of prominences. Eruptive prominences burst out of active regions and in a few hours extend into the lower corona only to fall back. Quiescent prominences sometimes hang in the lower corona for many days, sometimes with gas streaming downward into the active region along magnetic fields.

Although we think of the gas trapped in a prominence as hot and excited, it is not as hot as the gas in the lower corona. Thus prominences are complex objects that originate in the hot gas of an active region and extend up into the even hotter regions of the corona.

A solar **flare** is a violent outburst on the solar surface that grows to a maximum for a few minutes and decays in an hour or less. During that time it emits vast amounts of X-ray, ultraviolet, and visible radiation, plus streams of high-energy protons and electrons. Figure 7-16 shows a **shock wave** (a naturally occurring sonic boom) rushing away from a solar flare. A large flare can release energy equivalent to 10 billion megatons of TNT. (Large hydrogen bombs on Earth typically release about 10 megatons of energy.)

Solar flares are linked to the sun's magnetic field. Flares almost always occur near active regions, where the magnetic field is strong and twisted into great arches. Small flares can recur over and over at the same place. A large active region may experience 100 small flares a day. Flares are believed to occur when arches of magnetic force encounter each other and reconnect in such a way that they cancel each other out. These **reconnections** release the tremendous energy

Figure 7-16

This sequence of images recorded about 5 minutes apart shows the result of a solar flare. A shock wave spreads outward across the solar disk as a ripple up to 3 km high, traveling 50 km/sec, and equivalent to an earthquake 40,000 times stronger than the quake that destroyed San Francisco in 1906. *(SOHO/MDI, ESA and NASA)*

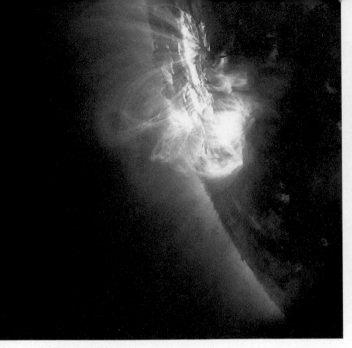

a   b

Figure 7-17
(a) This ultraviolet image of the edge of the sun shows an outburst of hot gas extending up into the corona produced when two magnetic fields just above the photosphere recombined, canceled each other out, and released their energy to the surrounding gas. *(NASA)* (b) Eruptions on the sun can bathe Earth in high-energy particles that produce auroras commonly seen at high latitudes. Notice the two vertical curtains of light within the glowing cloud in this photograph of an aurora seen in southern Australia. *(David Miller)*

stored in the magnetic fields as X rays and as high-speed electrons that heat the gas nearby (Figure 7-17a). During a major solar flare, the gas can be heated to over 5,000,000 K. X-ray observations detect photons of exactly the energies produced in certain nuclear reactions, and that assures us that the gas in a solar flare is very hot indeed. Although flares are common on the sun, most are detected only in filtergrams. So-called white-light flares, which are detectable in visible-light photographs, occur only about once a year.

Solar flares can have important effects on Earth. The X-ray and ultraviolet radiation reaches Earth in only 8 minutes and increases the ionization in Earth's upper atmosphere. This alters the reflection of short-wave radio signals and can absorb them completely, thereby interfering with communications. Flares can eject high-energy particles at a third the speed of light, but most particles ejected from flares have lower velocities and reach Earth hours or days after the flare as gusts in the solar wind. Such gusts interact with Earth's magnetic field and generate tremendous electrical currents (as much as a million megawatts), which flow down into Earth's atmosphere near the magnetic poles. There they can excite the atoms of the upper atmosphere at altitudes of 100 to 400 km to glow in displays called **auroras** (Figure 7-17b). These currents can also disturb Earth's magnetic field and cause magnetic storms in which compasses behave erratically.

Other effects of solar flares include surges in high-voltage power lines and radiation hazards to passengers in supersonic transports and in spacecraft. The U.S. Air Force watches for flares from observatories around the world, and solar astronomers have developed ways of predicting which flares will affect Earth.

Coronal Activity

Even the corona of the sun takes part in the magnetic cycle. At sunspot minimum, when there are few active regions and the magnetic field is not strongly tangled, eclipse observers see a small, slightly flattened corona. But at sunspot maximum, when the solar magnetic field produces many active regions, eclipse observers are treated to a blazing corona that is nearly circular. Figure 7-1b shows the corona near sunspot maximum.

We don't need to depend on total solar eclipses to study the corona. Specialized telescopes on Earth and ultraviolet and X-ray telescopes in space can image the corona every day. The results show that the corona is composed of streamers shaped by the solar magnetic field. Some parts of the corona are filled with hot gas trapped by magnetic fields that loop back down to the solar surface. In other regions the magnetic field does not loop back, and the hot gas streams away from the sun. These regions are called **coronal holes** because they appear as dark regions in ultraviolet and X-ray

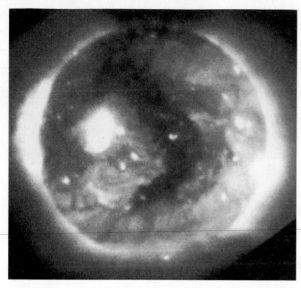

Figure 7-18
This X-ray image of the sun shows intense X-ray emission from the corona and a dark coronal hole snaking down the center of the disk. *(NASA Skylab)*

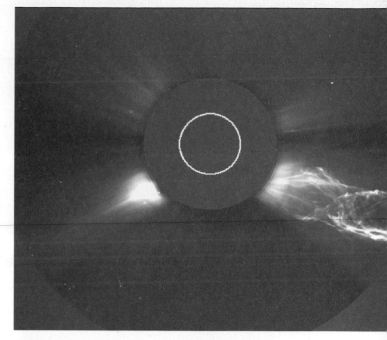

Figure 7-19
In this ultraviolet image of the corona, the disk of the sun (white circle) is hidden behind a circular screen. A violent eruption near the surface of the sun has ejected hot gases out into the corona toward the lower right. Such matter usually escapes from the sun completely. *(SOHO/LASCO, ESA and NASA.)*

images of the sun (Figure 7-18). There are permanent coronal holes at the sun's north and south poles, but coronal holes in other regions come and go as they are shaped by the changing magnetic field.

The corona is strongly affected by matter ejected from active regions on the sun's surface. When arcs of magnetic field reconnect, tremendous energy can be released, and great bursts of hot gas are expelled into the corona, as shown in Figure 7-19. If one of these eruptions strikes Earth, it can have dramatic effects on communications and electrical power networks.

It should not surprise us that Earth is affected by disturbances in the corona. The solar wind is derived from the coronal holes, so it is not an exaggeration to say that Earth orbits within the expanding corona.

REVIEW Critical Inquiry

What kind of activity would the sun have if it didn't rotate differentially?

This is a really difficult question, because we can see only one star close up and thus have no other examples. Nevertheless, we can make an educated guess by thinking about the Babcock model. If the sun didn't rotate differentially, with its equator traveling faster than higher latitudes, then the magnetic field would not get twisted and rise through the surface to form sunspot pairs and active regions. Of course, the magnetic field would not get progressively more wound up, so there would also be no magnetic cycle. Loops of magnetic field might not rise above the sur-

face to trap ionized gases and form prominences, and twists and kinks might never form in the field; and thus there would be no flares. In fact, with no differential rotation to stir up the magnetic field, the chromosphere and corona might not be heated to such high temperatures.

This is very speculative, but sometimes in the critical analysis of ideas it helps to imagine a change in a single important factor and try to understand what might happen. For example, what do you think the sun would be like if it had no convection inside?

The sun is beautiful and complex, with great eruptions and spots sweeping across its surface like magnetic weather. All of this activity is driven by the steady flow of energy outward from the sun's interior through its photosphere and into space. Other stars generate energy and radiate it into space, so we can conclude that many of those stars must have surface activity similar to that on our sun. But other stars are so far away we can't see their surfaces. Through the largest telescopes, they look like nothing more than points of light. How can we learn about other stars? We begin to answer that question in the next chapter.

Summary

The atmosphere of the sun consists of three layers: the photosphere, chromosphere, and corona. The photosphere, or visible surface, is a thin layer of low-density gas that is the level in the sun from which visible photons most easily escape. It is marked by granulation, a pattern produced by gas currents rising from below the photosphere.

The chromosphere is most easily visible during total solar eclipses, when it flashes into view for a few seconds. It is a thin, hot layer of gas just above the photosphere, and its brilliant pink color is caused by the emission lines in its spectrum. Filtergrams of the chromosphere reveal large jets called spicules extending up into the corona.

The corona is the sun's outermost atmospheric layer. It is composed of a very-low-density, very hot gas extending many solar radii from the sun. Its high temperature—up to 3×10^6 K—is believed to be maintained by the magnetic field and the rotation of the sun. Parts of the corona give rise to the solar wind, a breeze of low-density ionized gas streaming away from the sun. Thus Earth, which is bathed in the solar wind, is orbiting within the solar atmosphere.

Solar astronomers can study the motion, density, and temperature of gases below the solar surface by analyzing the way the solar surface oscillates. Known as helioseismology, this process requires large amounts of data and extensive computer analysis. It is beginning to reveal the nature of the layers below the solar surface.

Sunspots are the most prominent example of solar activity. A sunspot seems dark because it is slightly cooler than the rest of the photosphere. The average sunspot is about twice the size of Earth and contains magnetic fields about 1000 times stronger than the sun's average field. Sunspots are thought to form because the magnetic field inhibits rising currents of hot gas and allows the surface to cool.

The average number of sunspots varies over a period of about 11 years and appears to be related to a magnetic cycle. The sunspot cycle does not repeat exactly each cycle, and the decades 1645 to 1715, known as the Maunder minimum, seem to have been a time when solar activity was very low.

Alternate sunspot cycles have reversed magnetic polarity, and this has been explained by the Babcock model of the magnetic cycle. In this theory the differential rotation of the sun winds up the magnetic field. Tangles in the field rise to the surface and cause active regions visible to our eyes as sunspot pairs. When the field becomes strongly tangled, it reorders itself into a simpler but reversed field, and the cycle starts over.

Prominences and flares are other examples of solar activity. Prominences occur in the chromosphere; their arched shapes show that they are formed of ionized gas trapped in the magnetic field. Flares, too, seem to be related to the magnetic field. They are sudden eruptions of X-ray, ultraviolet, and visible radiation, and high-energy particles from the twisted magnetic fields around sunspot groups. Flares are important because they can have dramatic effects on Earth such as communications blackouts and auroras.

Activity in the corona is also guided by the magnetic field. The corona seems to be composed of streamers of thin, hot gas escaping from the magnetic field. In some regions of the corona, the magnetic field does not loop back to the sun, and the gas escapes unimpeded. These regions are called coronal holes and are believed to be the source of the solar wind.

New Terms

granulation	Zeeman effect
convection	active region
filtergram	differential rotation
spicule	dynamo effect
supergranule	Babcock model
magnetic carpet	prominence
solar wind	flare
helioseismology	shock wave
sunspot	reconnection
Maunder butterfly diagram	aurora
Maunder minimum	coronal hole

Review Questions

1. Why would an astronomer joke that solar astronomers would know more about the sun if it were farther away?

2. Why can't we see deeper than the photosphere?

3. What evidence do we have that granulation is caused by convection?

4. How are granules and supergranules related? How do they differ?

5. Why is the chromosphere pink?

6. How can a filtergram reveal structure in the chromosphere?

7. What observations would you make to test the hypothesis that spicules are confined by magnetic fields?

8. What evidence do we have that the corona has a very high temperature?

9. What heats the chromosphere and corona to high temperature?

10. How are astronomers able to explore the layers of the sun below the photosphere?

11. What evidence do we have that sunspots are magnetic phenomena?

12. How does the Babcock model explain the sunspot cycle?

13. What does the spectrum of a prominence tell us? What does the shape tell us?

14. How can flares on the sun affect Earth?

15. What evidence do we have that other stars have surface activity like the sun's?

Discussion Questions

1. What energy sources on Earth cannot be thought of as stored sunlight?

2. What would the spectrum of an auroral display look like? Why?

3. What observations would you make if you were ordered to set up a system that could warn astronauts in orbit of dangerous solar flares? Such a system exists.

Problems

1. The radius of the sun is 0.7 million km. What percentage of the radius is taken up by the chromosphere?

2. The smallest detail visible with ground-based solar telescopes is about 1 second of arc. How large a region does this represent on the sun? (*Hint:* Use the small-angle formula.)

3. What is the angular diameter of a star like the sun located 5 ly from Earth? Is the Hubble Space Telescope able to detect detail on the surface of such a star?

4. If a sunspot has a temperature of 4200 K and the solar surface has a temperature of 5800 K, how many times brighter is the surface compared to the sunspot? (*Hint:* Use the Stefan-Boltzmann law, Chapter 7.)

5. At what wavelength will a sunspot radiate most intensely? What color would the umbra of a sunspot be if we could see it alone?

6. A solar flare can release 10^{25} J. How many megatons of TNT would be equivalent? (*Hint:* A 1-megaton bomb produces about 4×10^{15} J.)

7. The United States consumes about 2.5×10^{19} J of energy in all forms in a year. How many years could we run the United States on the energy released by the solar flare in Problem 6?

8. Neglecting energy absorbed or reflected by our atmosphere, the solar energy hitting 1 square meter of Earth's surface is 1360 J/sec (the solar constant). How long does it take a baseball diamond (90 ft on a side) to receive 1 megaton of solar energy? (*Hint:* See Problem 6.)

Critical Inquiries for the Web

1. Do disturbances in one layer of the solar atmosphere produce effects in other layers? We have seen that filtergrams are useful in identifying the layers of the solar atmosphere and the structures within them. Visit a Web site that provides daily solar images, choose today's date (or one near it), and examine the sun in several wavelengths to explore the relation between disturbances in various layers.

2. Figure 7-17b shows the auroral displays caused by interaction of particles from the sun with our magnetosphere and ultimately our atmosphere. Explore the Web to find out how auroral activity is affected as solar activity rises and falls through the solar cycle. What changes in auroral visibility occur during this cycle? In what other ways can the increased activity associated with a solar maximum affect Earth?

 Go to the Brooks/Cole Astronomy Resource Center (www.brookscole.com/astronomy) for critical thinking exercises, articles, and additional readings from InfoTrac College Edition, Brooks/Cole's online student library.

If it can't be expressed in figures, it

is not science; it is opinion.

Robert Heinlein
The Notebooks of Lazarus Long

CHAPTER 8

The Properties of Stars

Guidepost

Science is based on measurement, but measurement in astronomy is very difficult. Even with the powerful modern telescopes described in Chapter 5, it is impossible to look at a star and easily find such simple properties as its diameter. This chapter shows how we can use the simple observations that can be made, combined with the basic laws of physics, to discover the properties of stars.

With this chapter we leave our sun behind and begin our study of the billions of stars that dot the sky. In a sense, the star is the basic building block of the universe. If we hope to understand what the universe is, what our sun is, what our Earth is, and what we are, we must understand stars.

Once we know how to discover the basic properties of stars, we will be ready to trace the history of stars from birth to death, a story we will begin in the next chapter.

The stars are unimaginably remote. The nearest star is the sun, only 150 million km (93 million miles) away, so close that light takes only about 8 minutes to reach Earth. The next nearest star is nearly 300,000 times farther away, a distance so great that the starlight takes over 4 years to reach Earth. The distances from Earth to the stars are so large we must express them in light-years. A light-year is the distance light travels in 1 year—about 9.5 trillion kilometers or about 5.9 trillion miles.

In spite of their great distance from us, the stars are the key to the secrets of the universe. The universe is filled with stars, and if we are to understand the universe, we must discover how stars are born, live, and die. We begin our study in this chapter by gathering data about stars (Figure 8-1). How much energy do they make? How big are stars? How much matter do the contain? These data will help us understand stars, and we will use these data in later chapters to deduce the life stories of different kinds of stars.

Unfortunately, finding out what a star is like is quite difficult. When we look at a star through a telescope, we see only a point of light. Just looking tells us almost nothing about the star's energy production, diameter, or mass. Rather than just looking at stars, we must analyze starlight with great care. We will concentrate in this chapter on finding out three things about stars—how much energy they emit, how big they are, and how much mass they contain. These three properties, combined with stellar temperatures—already discussed in Chapter 6—will give us an overview of stars.

Although we begin with three things to learn about stars, we immediately meet a short detour. To find out how much energy a star emits, we must know how far away it is. If at night we see lights approaching on the highway, we cannot tell whether we are looking at the bright headlights of a distant truck or the faint lights of a pair of nearby bicycles. Only when we know the distance to the lights can we judge their intrinsic brightness. In the same way, to find the intrinsic brightness of a star—or, more precisely, the amount of energy it emits—we must know its distance from us. Our short detour will provide us with a method of measuring stellar distances.

After we reach our three goals, we will put our data together to find out what the average star is like. How dense are stars? Which types of stars are most common? Which are rare? By the time we finish this chapter, we will know what stars are like.

Figure 8-1
The modern quest to understand the universe is manifest in this photo of the cloud of gas and dust called the Trifid Nebula. Near the center of the cloud, new hot stars have been born. The hundreds of stars scattered in the background of this photo assure us that this is a common process. To understand how stars are born, evolve, and die, we must discover their properties. *(NOAO and Nigel Sharp)*

8-1 Measuring the Distances to Stars

Distance is the most important and the most difficult measurement in astronomy, and astronomers have found many different ways to estimate the distance to stars. Yet each of those ways depends on a direct geometrical method that is much like the method surveyors would use to measure the distance across a river they cannot cross. We will begin by reviewing this method and then apply it to stars.

The Surveyor's Method

To measure the distance across a river, a team of surveyors begins by driving two stakes into the ground. The distance between the stakes is the baseline of the measurement. The surveyors then choose a landmark on the opposite side of the river, a tree perhaps, thus establishing a large triangle marked by the two stakes

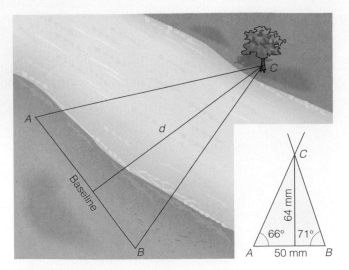

Figure 8-2
You can find the distance *d* across a river by measuring the baseline and the angles *A* and *B* and then constructing a scale drawing of the triangle.

and the tree. Using their surveyor's instruments, they sight the tree from the two ends of the baseline and measure the two angles on their side of the river.

Knowing two angles of this large triangle and the length of the side between them, the surveyors can then find the distance across the river by simple trigonometry. Another way to find the distance is to construct a scale drawing. For example, if the baseline is 50 m and the angles are 66° and 71°, we can draw a line 50 mm long to represent the baseline. Using a protractor, we can construct angles of 66° and 71° at each end of the baseline, and then, as shown in Figure 8-2, extend the two sides until they meet at *C*. Point *C* on our drawing is the location of the tree. Measuring the height of our triangle, we find it to be 64 mm and thus conclude that the distance from the baseline to the tree is 64 m.

Modern surveyors use computers for these problems and not simple drawings. The point is not how the problem is solved, but that it can be solved. Surveyors use simple triangulation to find the distance across a river.

The Astronomer's Method

To find the distance to a star, we must use a very long baseline, the diameter of Earth's orbit. If we take a photograph of a nearby star and then wait 6 months, Earth will have moved halfway around its orbit. We can then take another photograph of the star. This second photograph is taken at a point in space 2 AU (astronomical units) from the

point where the first photograph was taken. Thus our baseline equals the diameter of Earth's orbit, or 2 AU.

We then have two photographs of the same part of the sky taken from slightly different locations in space. When we examine the photographs, we will discover that the star is not in exactly the same place in the two photographs. This apparent shift in the position of the star is called *parallax*.

Parallax is the apparent change in the position of an object due to a change in the location of the observer. We saw in Chapter 4 that parallax is an everyday experience. Our thumb, held at arm's length, appears to shift position against a distant background when we look with first one eye and then with the other (see Figure 4-3). In this case, the baseline is the distance between our eyes, and the parallax is the angle through which our thumb appears to move when we change eyes. The farther away we hold our thumb, the smaller the parallax. If we know the length of the baseline and measure the parallax, we can calculate the distance from our eyes to our thumb.

Because the stars are so distant, their parallaxes are very small angles, usually expressed in seconds of arc. The quantity that astronomers call **stellar parallax (*p*)** is half the total shift of the star, as shown in Figure 8-3. Astronomers measure the parallax and surveyors measure the angles at the ends of the baseline, but both measurements tell us the same thing—the shape of the triangle and thus the distance to the object in question.

Figure 8-3
We can measure the parallax of a nearby star by photographing it from two points along Earth's orbit. For example, we might photograph it now and again in 6 months. Half of the star's total change in position from one photograph to the other is its stellar parallax *p*.

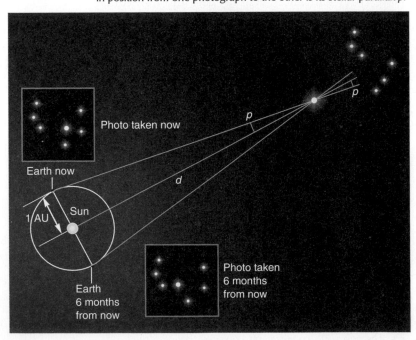

Measuring the parallax p is very difficult because it is such a small angle. The nearest star, α Centauri, has a parallax of only 0.76 second of arc, and the more distant stars have even smaller parallaxes. To see how small these angles are, hold a piece of paper edgewise at arm's length. The thickness of the paper covers an angle of about 30 seconds of arc.

We cannot use scale drawings to find the distances to stars because the angles are so small and the distances are so large. Even for the nearest star, the triangle would have to be 300,000 times longer than it was wide. If the baseline in our drawing were 1 cm, the triangle would have to be about 3 km long. By the Num-

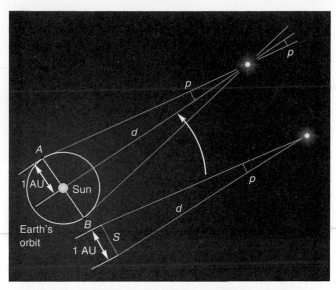

Figure 8-4
To measure the parallax of a star we must examine the long, thin triangle formed by Earth, the sun, and the star. The short side of the triangle, S, is the baseline of our measurement, 1 AU. If we were located at the star and looked back at Earth, the angular distance from Earth to the sun would equal the parallax, p. This means we can use the small-angle formula (see By the Numbers 3-1) to find the star's distance, d, using its parallax.

By the Numbers 8-1

Parallax and Distance

We wish to find the distance to a star from its measured parallax. To see how this is done, imagine that we observe Earth from the star. Figure 8-4 shows that the angular distance we observe between the sun and Earth equals the star's parallax p. To find the distance, we recall that the small-angle formula (see By the Numbers 3-1) relates an object's angular diameter, its linear diameter, and its distance. In this case, the angular diameter is p and the linear diameter is 1 AU. Then the small-angle formula, rearranged slightly, tells us that the distance to the star in AU is equal to 206,265 divided by the parallax in seconds of arc:

$$d = \frac{206{,}265}{p}$$

Because the parallaxes of even the nearest stars are less than 1 second of arc, the distances in AU are inconveniently large numbers. To keep the numbers manageable, astronomers have defined the parsec as their unit of distance in a way that simplifies the arithmetic. One parsec equals 206,265 AU, so the equation becomes:

$$d = \frac{1}{p}$$

Thus a parsec is the distance to an imaginary star whose parallax is 1 second of arc.

Example: The star Altair has a parallax of 0.20 second of arc. How far away is it? *Solution:* The distance in parsecs equals 1 divided by 0.2, or 5 pc:

$$d = \frac{1}{0.2} = 5 \text{ pc}$$

Because 1 pc equals about 3.26 ly, Altair is about 16.3 ly away.

bers 8-1 describes how we could find the distance from the parallax without drawing scale triangles.

The distances to the stars are so large that it is not convenient to use astronomical units. As By the Numbers 8-1 explains, when we measure distance via parallax, it is convenient to use the unit of distance called a **parsec (pc).** The word *parsec* was created by combining *par*allax and *sec*ond of arc. One parsec equals the distance to an imaginary star that has a parallax of 1 second of arc. A parsec is 206,265 AU, roughly 3.26 ly (light-years).[*]

The blurring of Earth's atmosphere limits the accuracy of parallax measurements made from the ground to about 0.002 seconds of arc. If we measure a parallax of 0.002 seconds of arc, the uncertainty is as big as the quantity we are trying to measure. To say that another way, our uncertainty is 100 percent. To limit our uncertainty to only 10 percent with ground-based observations, we can't measure parallaxes smaller than about 0.02 seconds of arc, which corresponds to a distance of 50 pc. Thus ground-based parallax measurements have been limited to only the closest stars. The first stellar parallax was measured in 1838, and since then ground-based astronomers have been able to measure accurate parallaxes for only a few thousand stars.

---

[*]The parsec is used throughout astronomy because it simplifies the calculation of distance. However, there are instances in which the light-year is also convenient. Consequently, the chapters that follow use either parsecs or light-years as convenience and custom dictate.

In 1989, the European Space Agency launched the satellite Hipparcos to measure stellar parallaxes from above Earth's atmosphere. After 4 years of observations and 3 years of data reduction, the data were published in 1997 as two catalogues. One contains 120,000 stars with parallaxes accurate to 0.001 seconds of arc, and the other contains over a million stars with parallaxes as accurate as ground-based observations. The Hipparcos data are giving astronomers new insights into stars.

Distance is the key to the secrets of the stars. Now that we know how to find parallaxes, we are ready to think about stellar brightness.

8-2 Intrinsic Brightness

Our eyes tell us that some stars look brighter than others, and in Chapter 2 we used the apparent magnitude scale to refer to stellar brightness. The faintest stars we can see with the naked eye are about 6th magnitude, with brighter stars having smaller magnitudes. In fact, the brightest stars we see in the sky have negative magnitudes, such as Sirius, whose apparent magnitude is −1.47.

The scale of apparent magnitudes only tells us how bright stars look, however, and we want to know their true or intrinsic brightness. "Intrinsic" means "belonging to the thing." When we refer to the intrinsic brightness of a star, we mean a measure of the total amount of light the star emits. An intrinsically very bright star might appear faint if it were far away. Thus to find out the true brightness of a star, we must correct for the influence of its distance.

Brightness and Distance

When we look at a bright light, our eyes respond to the visual-wavelength energy falling on the eye's retina, which tells us how bright the object looks. Thus, brightness is related to the flux of energy entering our eye. **Flux** is the energy in joules (J) per second falling on one square meter. (We commonly refer to 1 J/sec as 1 watt.)

If we placed a screen 1 meter square near a lightbulb, a certain amount of flux would fall on the screen. If we moved the screen twice as far from the bulb, the light that previously fell on the screen would be spread to cover an area four times larger, and the screen would receive only one-fourth as much light. If we tripled the distance to the screen, it would receive only one-ninth as much light. Thus the flux we receive from a light source is inversely proportional to the square of the distance to the source. This is known as the inverse square relation (Figure 8-5). (We first encountered the inverse square relation in Chapter 4, where it was applied to the strength of gravity.)

Absolute Visual Magnitude

If all the stars were the same distance away, we could compare one with another and decide which was emitting more light and which less. Of course, the stars are scattered at different distances, and we can't shove them around to line them up for comparison. If, however, we know the distance to a star, we can use the inverse square relation to calculate the brightness the star would have at some standard distance. Astronomers take 10 pc as the standard distance and refer to the intrinsic brightness of the star as its **absolute visual magnitude (M_v)**, the apparent visual magnitude the star would have if it were 10 pc away.

The symbol for absolute visual magnitude is a capital M with a subscript v. The subscript tells us it is a visual magnitude based only on the wavelengths of light we can see. Other magnitude systems are based on other parts of the electromagnetic spectrum, such as the infrared and ultraviolet.

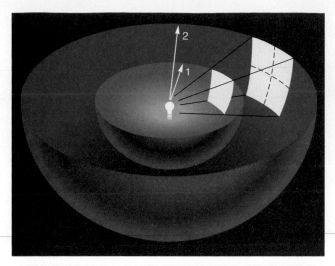

Figure 8-5
The inverse square relation. A light source is surrounded by spheres with radii of 1 unit and 2 units. The light falling on an area of 1 m² on the inner sphere spreads to illuminate an area of 4 m² on the outer sphere. Thus, the brightness of the light source is inversely proportional to the square of the distance.

It is not difficult to find the absolute magnitude of a nearby star. We begin by measuring the apparent magnitude, which is an easy task in astronomy. Then we find the distance to the star. If the star is nearby, we can measure its parallax and from that find the distance. Once we know the distance, we can use a simple formula to correct the apparent magnitude for the distance and find the absolute magnitude (By the Numbers 8-2).

We can find the sun's absolute magnitude because we know the distance to the sun and can measure its apparent magnitude. The absolute magnitude of the sun is about 4.78. If the sun were only 10 pc from Earth (not a great distance in astronomy), it would look no brighter than the faintest star in the handle of the Little Dipper.

Our detour to find the distance to stars has led us to absolute magnitude, a measure of the true brightness of the stars. We are now ready to reach the first of our three goals in this chapter.

Luminosity

The first of our three goals for this chapter was to find out how much energy the stars emit. With the absolute magnitudes of the stars in hand, we can now compare other stars with our sun.

The intrinsically brightest stars known have absolute magnitudes of about −8, which means that if such a star were 10 parsecs away from Earth, we would see it nearly as bright as the moon. Such stars are 13 magnitudes brighter than the sun, so they must be emitting over 100,000 times more light than the sun. Yet the intrinsically faintest stars have absolute

magnitudes of +15 or fainter. They are 10 magnitudes fainter than the sun, meaning they are emitting 10,000 times less light than the sun.

Our goal is to discover the **luminosities (L)** of the stars. That is, we want to know how much energy per second each star is emitting. We can use absolute magnitude to compare a star with the sun, but a small correction is necessary. Absolute visual magnitude refers to light, but we want to know the luminosity, which includes all energy. Hot stars emit a great deal of ultraviolet radiation that we can't see, and cool stars emit infrared. To add in the energy we can't see, astronomers make a small correction that depends on the temperature of the star. With that correction, the absolute magnitudes can tell us the luminosities of the stars.

Astronomers know the luminosity of the sun because they can send satellites above Earth's atmosphere and measure the amount of energy Earth receives from the sun in one second. The luminosity of the sun is about 4×10^{26} J/sec.

We can express a star's luminosity in two ways. For example, we can say that the star Capella, which is 100 times more luminous than the sun, has a luminosity of 100 solar luminosities. We can also express this in real energy units by multiplying by the luminosity of the sun. The luminosity of Capella is 4×10^{28} J/sec.

When we look at the night sky, the stars look much the same, yet some are astonishingly more luminous than others. Clearly, the stars are much more than just twinkling points in the sky.

REVIEW Critical Inquiry

Why do extremely cool stars look fainter than we might expect from their luminosities and distances?

Stars like the sun radiate most of their energy at visible wavelengths, and we can see those wavelengths, so they look about right for their luminosities and distances. But the coolest stars radiate the vast majority of their photons in the infrared, which we can't see. The luminosity includes all radiated energy. If we use that luminosity and a cool star's distance to predict how bright it should look, we might be disappointed. Most of the energy cool stars radiate is invisible to our eyes, so they will look fainter than we might expect. Of course, the same thing happens for very hot stars, which radiate most of their energy in the ultraviolet part of the spectrum.

The luminosity is a very important piece of information about any star, but we must remember all the details that go into the measurement of luminosity. Explain exactly how errors in measuring the positions of blurred star images could introduce errors into our measurements of luminosity.

Although we detoured to find the distances to the stars, we have now reached our first goal. We have found the luminosities of the stars. When we reach the next goal, we will know even more about the stars.

8-3 The Diameters of Stars

Our second goal in this chapter is to find the diameters of stars. We know little about stars until we know their diameters. Are they all the same size as our sun, or are some larger and some smaller? We certainly can't see their diameters through a telescope; the images of the stars are much too small for us to resolve their disks and measure their diameters. But there is a way to find out how big stars really are. If we know

their temperatures and luminosities, we can find their diameters. That relationship will introduce us to the most important diagram in astronomy, where we will discover the family relations among the stars.

Luminosity, Radius, and Temperature

The luminosity and temperature of a star can tell us its diameter if we understand the two factors that affect a star's luminosity: surface area and temperature. For example, you can eat dinner by candlelight because the candle flame has a small surface area, and, although it is very hot, it cannot radiate much heat; it has a low luminosity. However, if the candle flame were 12 ft tall, it would have a very large surface area from which to radiate, and, although it would be no hotter than a normal candle flame, its luminosity would drive you from the table.

In a similar way, a star's luminosity is proportional to its surface area. A hot star may not be very luminous if it has a small surface area, although it could be highly luminous if it were larger. Even a cool star could be luminous if it had a large surface area. (See By the Numbers 8-3.) Because of this dependence on both temperature and surface area, we can use stellar luminosities to determine the diameters of stars if we can separate the effects of temperature and surface area.

The **Hertzsprung-Russell (H-R) diagram,** named after its originators, Ejnar Hertzsprung and Henry Norris Russell, is a graph that separates the effects of temperature and surface area on stellar luminosities and enables us to sort the stars according to their diameters. Before we discuss the details of the H-R diagram (as it is often called), let us look at a similar diagram we might use to sort automobiles.

We can plot a diagram such as Figure 8-6 to show horsepower versus weight for various makes of cars. And in so doing, we find that in general the more a car weighs, the more horsepower it has. Most cars fall somewhere along the sequence of cars, running from heavy, high-powered cars to light, low-powered models. We might call this the main sequence of cars. But some cars have much more horsepower than normal for their weight—the sport or racing models—and the economy models have less power than normal for cars of the same weight. Just as this diagram helps us understand the different kinds of autos, the H-R diagram helps us understand the kinds of stars.

The H-R diagram is a graph with luminosity on the vertical axis and temperature on the horizontal axis. A star is represented by a point on the graph that tells us the luminosity of the star and its temperature. The schematic H-R diagram in Figure 8-7 also contains a scale of spectral type across the top. Because a star's spectral type is determined by its temperature, we could use either spectral type or temperature on the horizontal axis. Also, because absolute magnitude

Luminosity, Radius, and Temperature

The luminosity L of a star depends on two things — its size and its temperature. If the star has a large surface area from which to radiate, it can radiate a great deal. Recall from our discussion of black body radiation in By the Numbers 6-1 that the amount of energy emitted per second from each square meter of the star's surface is σT^4. Thus, the star's luminosity can be written as its surface area in square meters times the amount it radiates per square meter:

$$L = \text{area} \times \sigma T^4$$

Because a star is a sphere, we can use the formula area $= 4\pi R^2$. Then the luminosity is:

$$L = 4\pi R^2 \sigma T^4$$

This seems complicated, but if we express luminosity, radius, and temperature in terms of the sun, we get a much simpler form.*

$$\frac{L}{L_\odot} = \left(\frac{R}{R_\odot}\right)^2 \left(\frac{T}{T_\odot}\right)^4$$

Example A: Suppose we want to find the luminosity of a particular star. This star is 10 times the sun's radius but only half as hot. How luminous is it? *Solution:*

$$\frac{L}{L_\odot} = \left(\frac{10}{1}\right)^2 \left(\frac{1}{2}\right)^4 = \frac{100}{1} \times \frac{1}{16} = 6.25$$

The star has 6.25 times the sun's luminosity.

We can also use this formula to find diameters.

Example B: Suppose we found a star whose absolute magnitude is +1 and whose spectrum shows it is twice the sun's temperature. What is the diameter of the star? *Solution:* The star's absolute magnitude is 4 magnitudes brighter than the sun, and we recall from By the Numbers 2-1 that 4 magnitudes is a factor of 2.512^4, or about 40. The star's luminosity is therefore about 40 $L_\odot$. With the luminosity and temperature, we can find the radius:

$$\frac{40}{1} = \left(\frac{R}{R_\odot}\right)^2 \left(\frac{2}{1}\right)^4$$

Solving for the radius we get:

$$\left(\frac{R}{R_\odot}\right)^2 = \frac{40}{2^4} = \frac{40}{16} = 2.5$$

So the radius is:

$$\frac{R}{R_\odot} = \sqrt{2.5} = 1.58$$

The star is 58 percent larger in radius than the sun.

*In astronomy the symbols $\odot$ and $\oplus$ refer respectively to the sun and Earth. Thus $L_\odot$ refers to the luminosity of the sun, $T_\odot$ refers to the temperature of the sun, and so on.

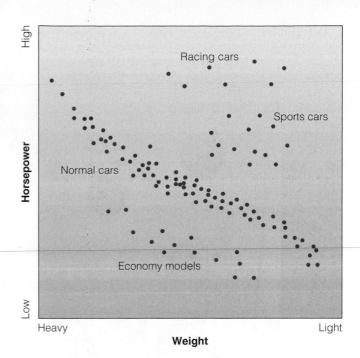

Figure 8-6
We could analyze automobiles by plotting their horsepower versus their weight and thus reveal relationships between various models. Most would lie somewhere along the main sequence of "normal" cars.

is related to luminosity, we could use absolute magnitude on the vertical axis, as shown at the right edge of Figure 8-7.

In an H-R diagram, the location of a point representing a star tells us a great deal about the star. Points near the top of the diagram represent very luminous stars, and points near the bottom represent very low-luminosity stars. Also, points near the right edge of the diagram represent very cool stars, and points near the left edge of the diagram represent very hot stars. Notice in Figure 8-7 how the artist has used the colors of the stars to represent temperature. Cool stars are red, and hot stars are blue.

Astronomers use H-R diagrams so often that they usually skip the words "the point that represents the star." Rather they will say that a star is located in a certain place in the diagram. Of course, they mean the point that represents the luminosity and temperature of the star and not the star itself.

Notice that the location of a star in the H-R diagram has nothing to do with the location of the star in space. Furthermore, a star may move in the H-R diagram as it ages and its luminosity and temperature change, but such motion in the diagram has nothing to do with the star's motion in space.

The **main sequence** is the region of the H-R diagram running from upper left to lower right. It includes roughly 90 percent of all stars. In Figure 8-7, the main sequence is represented by a curved dashed line with symbols for stars plotted along it. As we

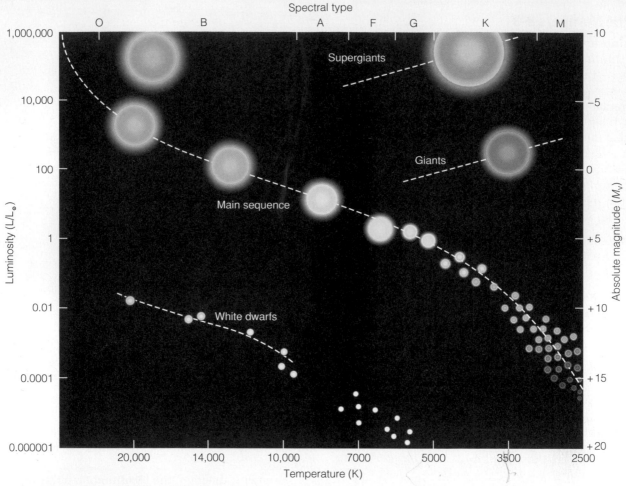

Figure 8-7
An H-R diagram is a graph of luminosity versus temperature on which stars are represented a points. Spectral types have been added at the top to show how they are related to temperature. Absolute magnitude has been added at the right to show its relationship to luminosity. Star diameters are not to scale in this schematic diagram.

might expect, the hot main-sequence stars are more luminous than the cool main-sequence stars.

We have learned, however, that both temperature and size determine a star's luminosity. In the H-R diagram, the stars at the top are larger in diameter, and stars at the bottom are smaller. The artist represented this variation in size in Figure 8-7 by drawing the symbols for stars at the top of the diagram larger than the symbols at the bottom.

The relation among luminosity, size, and temperature is very precise, as was shown in By the Numbers 8-3. In fact, the mathematical relationship can be used to draw lines of constant radius across an H-R diagram. Figure 8-8 is an H-R diagram on which slanting dashed lines show the location of stars of certain radii. For example, locate the dashed line labeled 1 $R_\odot$. That line passes through the point marked "Sun" and represents the location of any star whose radius equals that of the sun. Of course, the line slants down to the right because cooler stars are always fainter that hotter stars of the same size.

Now we can identify certain families of stars. The **giant stars** lie at the right above the main sequence. Although these stars are cool, they are luminous because they are 10 to 100 times larger than the sun. The **super giants** lie near the top of the H-R diagram and are 10 to 1000 times the size of the sun. Betelgeuse in Orion is a supergiant. If it magically replaced the sun at the center of our solar system, it would swallow up Mercury, Venus, Earth, Mars, and would just reach Jupiter. μ Cephei is believed to be over three times larger than Betelgeuse. It would reach nearly to the orbit of Uranus.

At the bottom of the H-R diagram lie the economy models, stars that are very low in luminosity because they are very small. In fact, the hottest of these stars are among the hottest stars known, so they were named **white dwarfs** because they shine blue-white. On average, these stars are about the size of Earth.

Notice the great range of sizes among stars. The largest stars are 100,000 times larger than the tiny white dwarfs. Look at the symbols plotted in Figure

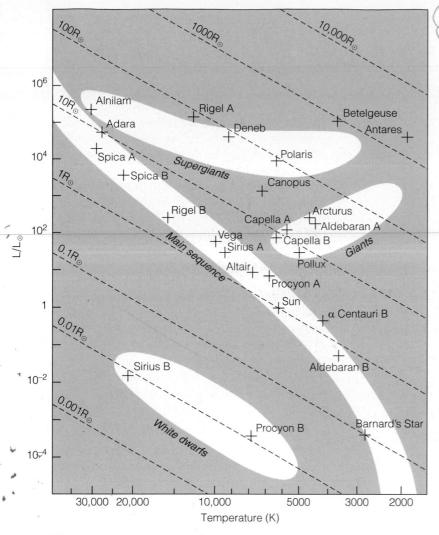

Figure 8-8
An H-R diagram showing the luminosity and temperature of many well-known stars. The dashed lines are lines of constant radius. (Individual stars that orbit each other are designated A and B, as in Spica A and Spica B.)

8-7. How big would the symbol for a supergiant have to be if it were 100,000 times larger than the symbols used for white dwarfs?

Luminosity Classification

We can tell from a star's spectrum whether it is a main sequence star, a giant, or a supergiant. The larger a star is, the less dense its atmosphere is, and the

widths of spectral lines are partially determined by the density of the gas.

Some spectral lines are particularly sensitive to density. If the atoms collide often in a dense gas, their energy levels become distorted and the spectral lines are broadened. Hydrogen Balmer lines are an example. In the spectrum of a main-sequence star, the Balmer lines are broad because the star's atmosphere is dense and the hydrogen atoms collide often. In the spectrum of a giant star, the lines are narrower (Figure 8-9), because the giant star's atmosphere is less dense and the hydrogen atoms collide less often. In the spectrum of a supergiant star, the Balmer lines are very narrow.

Thus we can look at a star's spectrum and tell roughly how big it is. These are called **luminosity classes,** because the size of the star is the dominating factor in determining luminosity. Supergiants, for example, are very luminous because they are very large.

The luminosity classes are represented by the roman numerals I through V, with supergiants further subdivided into types Ia and Ib, as follows:

Luminosity Classes

Ia	Bright supergiant
Ib	Supergiant
II	Bright giant
III	Giant
IV	Subgiant
V	Main-sequence star

We can distinguish between the bright supergiants (Ia) such as Rigel (β Orionis) and the regular supergiants (Ib) such as Polaris, the North Star. The star Adhara (ε Canis Majoris) is a bright giant (II), Capella (α Aurigae) is a giant (III), and Altair (α Aquilae) is a subgiant (IV). The sun is a main-sequence star (V). The luminosity class appears after the spectral type, as in G2 V for the sun. White dwarfs don't enter into this classification, because their spectra are peculiar.

We can plot the positions of the luminosity classes on the H-R diagram (Figure 8-10). Remember that these are rather broad classifications and that the lines on the diagram are only approximate. A star of lumi-

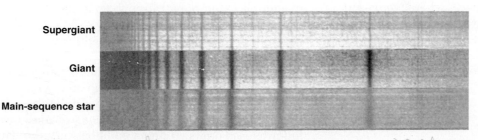

Figure 8-9
Differences in widths and strengths of spectral lines distinguish the spectra of supergiants, giants, and main-sequence stars. *(Adapted from H. A. Abt, A. B. Meinel, W. W. Morgan, and J. W. Tapscott,* An Atlas of Low-Dispersion Grating Stellar Spectra, *Kitt Peak National Observatory, 1968.)*

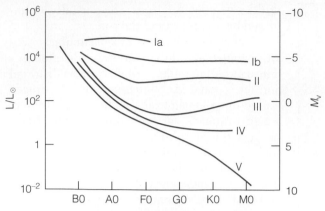

Figure 8-10
The approximate location of the luminosity classes on the H-R diagram.

nosity class III may lie slightly above or below the line labeled III.

Luminosity classification is subtle and not too accurate, but it is an important tool in modern astronomy. As we will see in the next section, luminosity classification gives us a way to find the distance to stars that are too far away to have measurable parallaxes.

Spectroscopic Parallax

We can measure the stellar parallax of nearby stars, but most stars are too distant to have measurable parallaxes. We can find the distances to these stars if we can photograph their spectra and determine their luminosity classes in a process called **spectroscopic parallax**—the estimation of the distance to a star from its spectral type, luminosity class, and apparent magnitude. Spectroscopic parallax is not an actual measure of parallax, but it does tell us the distance to the star.

The method of spectroscopic parallax depends on the H-R diagram. If we photograph the spectrum of a star, we can determine its spectral class, and that tells us its horizontal location in the H-R diagram. We can also determine its luminosity class by looking at the widths of its spectral lines, and that tells us the star's vertical location in the diagram. Once we plot the point that represents the star in the H-R diagram, we can read off its absolute magnitude. As we have seen earlier in this chapter, we find the distance to a star by comparing its apparent and absolute magnitudes.

For example, Spica is classified B1 V, and its apparent magnitude is +1. We can plot this star in an H-R diagram such as Figure 8-10, where we would find it should have an absolute magnitude of about −3. Using the apparent and absolute magnitudes, we can find the distance from the equation in By the Numbers 8-2. Spica is about 63 pc from Earth.

The first two goals of this chapter have been achieved; we know how to find the luminosities and diameters of stars. The next step is to find their masses.

8-4 The Masses of Stars

Our third goal is to find out how much matter stars contain, that is, to know their mass. Do they all contain about the same mass as our sun, or are some more massive and others less? Unfortunately, it's difficult to determine the mass of a star. Looking through a telescope at a star, we see only a point of light that tells nothing about mass. To find the masses of stars, we study **binary stars,** pairs of stars that orbit each other.

Binary Stars in General

The key to finding the mass of a binary star is understanding orbital motion, which provides clues to mass.

In Chapter 4, we illustrated orbital motion by imagining a cannonball fired from a high mountain (see Figure 4-24). If Earth's gravity didn't act on the cannonball, it would follow a straight-line path and leave Earth forever. Because Earth's gravity pulls it away from its straight-line path, the cannonball follows a curved path around Earth—an orbit. When two stars orbit each other, their mutual gravitation pulls them away from straight-line paths and makes them follow closed orbits around a point between the stars.

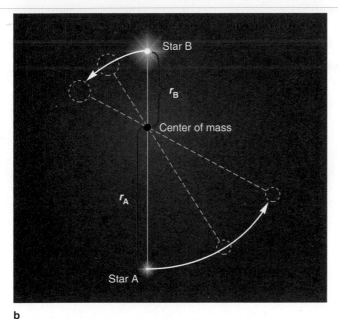

b

Figure 8-11
(a) A seesaw balances at the center of mass. If one child is more massive than the other, the balance point is closer to the more massive child. If a pair of binary stars could be connected with a massless rod, their balance point would be nearer the more massive star. (b) As stars in a binary star system revolve around each other, the line connecting them always passes through the center of mass, and the more massive star is always closer to the center of mass.

Each star in a binary system moves in its own orbit around the system's **center of mass,** the balance point of the system. If the stars were connected by a massless rod and placed in a uniform gravitational field such as that near Earth's surface, the system would balance at its center of mass like a child's seesaw (Figure 8-11a). If one star were more massive than its companion, then the massive star would be closer to the center of mass and would travel in a smaller orbit, while the lower-mass star would whip around in a larger orbit (Figure 8-11b). The ratio of the masses of the stars M_A/M_B equals r_B/r_A, the inverse of the ratio of the radii of the orbits. If one star has an orbit twice as large as the other star's orbit, then it must be half as massive. Getting the ratio of the masses is easy, but that doesn't tell us the individual masses of the stars,

which is what we really want to know. That takes further analysis.

To find the mass of a binary star system, we must know the size of the orbits and the orbital period—the length of time the stars take to complete one orbit. The smaller the orbits are and the shorter the orbital period is, the stronger the stars' gravity must be to hold each other in orbit. For example, if two stars whirl rapidly around each other in small orbits, then their gravity must be very strong to prevent their flying apart. Such stars would have to be very massive. From the size of the orbits and the orbital period, we can figure out how much mass the stars contain as explained in By the Numbers 8-4. Calculations such as those in By the Numbers 8-4 yield the total mass, which, combined with the ratio of the masses found

Scientists can rarely observe the things they really want to know, so they must construct chains of inference. We can't observe the mass of stars directly, so we must find a way to use what we can observe, orbital period and angular separation, and step by step figure out the parameters we need to reach our goal. In this case, the goal is the masses of the two stars. We follow the chain of inference from the observable parameters to the unobservable quantities we want to know.

Chains of inference can be mathematical, as when geologists use earthquakes to calculate the temperature and density of Earth's interior. There is no way to drill a hole to Earth's center and lower a thermometer or recover a sample, so the internal temperature and density are not directly observable. Nevertheless, the speed of vibrations from distant earthquakes depends on the temperature and density of the rock they pass through. Geologists can't measure the speed of the vibrations deep inside Earth; but they can measure the delay in the arrival time at different locations on the surface, and that allows them to work their way back to the speed and, finally, the temperature and density.

Chains of inference can also be nonmathematical. Biologists studying the migration of whales can't follow individual whales for years at a time, but they can observe feeding and mating in different locations; take into consideration food sources, ocean currents, and water temperatures; and construct a chain of inference that leads back to the seasonal migration pattern for whales.

This chapter contains a number of chains of inference because it describes how we know what stars are like. Almost all sciences use chains of inference. If you can link the observable parameters step by step to the final conclusions, you have gained a strong insight into the nature of science.

from the relative sizes of the orbits, can tell us the individual masses of the stars.

Actually, figuring out the mass of a binary star system is not as easy as it might seem from the examples in By the Numbers 8-4. The orbits of the two stars may be elliptical, and, although the orbits lie in the same plane, that plane can be tipped at an unknown angle to our line of sight, further distorting the shapes of the orbits. Astronomers must find ways to correct for these distortions. In addition, astronomers analyzing binary systems must find the distances to the stars so they can estimate the true size of the orbits in astronomical units. Notice that finding the masses of binary stars requires a number of steps to get from what we can observe to what we really want to know, the masses. Constructing such sequences of steps is an important part of science (Window on Science 8-1).

Although there are many different kinds of binary stars, three types are especially important for determining stellar masses. We will discuss these separately in the next sections.

Visual Binary Systems

In a **visual binary system,** the two stars are separately visible in the telescope. Only a pair of stars with large orbits can be separated visually, for if the orbits are small, the star images blend together in the telescope and we see only a single point of light. Because visual binary systems must have large orbits, they also have long orbital periods. Some take hundreds or even thousands of years to complete a single orbit.

Astronomers study visual binary systems by measuring the position of the two stars directly at the telescope or in images. In either case, the astronomers need measurements over many years to map the orbits. The first frame of Figure 8-12 shows a photograph of the bright star Sirius, which is a visual binary system made up of the bright star Sirius A and its white dwarf companion Sirius B. The photo was taken in 1960. Successive frames in Figure 8-12 show the motion of the two stars as observed since 1960 and the orbits the stars follow. The orbital period is 50 years.

Binary systems are common; more than half of all stars are members of binary star systems. Few, however, can be analyzed completely. Many are so far apart that their periods are much too long for practical mapping of their orbits. Others are so close together they are not visible as separate stars.

Spectroscopic Binary Systems

If the stars in a binary system are close together, the telescope, limited by diffraction and by seeing, shows us a single point of light. Only by looking at a spectrum, which is formed by light from both stars and contains spectral lines from both, can we tell that there are two stars present and not one. Such a system is a **spectroscopic binary system.**

Figure 8-13 shows a pair of stars orbiting each other in identical circular orbits. From the diagram we can guess that the two stars have equal masses, and we notice that the circular orbit appears elliptical

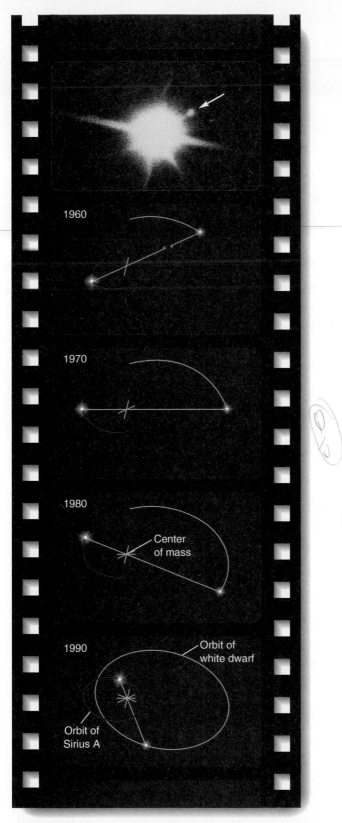

Figure 8-12
Sirius and its white dwarf companion (arrow) are shown in the top frame in a 1960 photograph. Spikes are caused by diffraction in the telescope. *(Lick Observatory photograph.)* Sequential measurements of the positions of the two stars reveal the shapes and sizes of their orbits. The intersection of the lines connecting the stars at different times defines the center of mass. The more massive star has the smaller orbit.

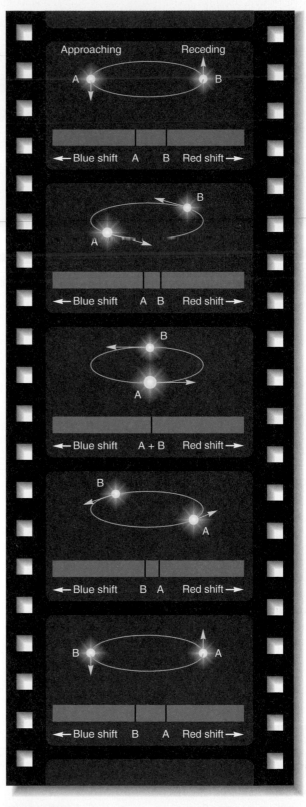

Figure 8-13
As the stars of a spectroscopic binary system revolve around their common center of mass, they alternately approach and recede from Earth. The Doppler shifts cause their spectral lines to move back and forth across each other.

because we see it nearly edge on. Of course, if this were a spectroscopic binary system, we would not see the separate stars. Nevertheless, the Doppler shift would tell us there were two stars orbiting each other. In the first frame of the figure, star A is approaching us while star B recedes from us. In the spectrum, we see a spectral line from star A blue shifted while the same spectral line from star B is red shifted. As we watch the two stars revolve around their orbits, they alternately approach and recede from us, and we see their spectral lines Doppler shifted first toward the blue and then toward the red. In a real spectroscopic binary, we can't see the individual stars, but the sight of pairs of spectral lines moving back and forth across each other would alert us that we were observing a spectroscopic binary (Figure 8-14).

We can find the orbital period of a spectroscopic binary system by waiting to see how long it takes for the spectral lines to return to their starting positions. We can measure the size of the Doppler shifts to find the orbital velocities of the two stars. If we multiply velocity times orbital period, we can find the circumference of the orbit, and from that we can find the radius of the orbit. Of course, if we know the orbital period and the size of the orbit we can calculate the mass. One important detail is missing, however. We don't know how much the orbits are inclined to our line of sight.

We can find the inclination of a visual binary system because we can see the shape of the orbits. In a spectroscopic binary system, however, we cannot see the individual stars, so we can't find the inclination or untip the orbits. Recall that the Doppler effect only tells us the radial velocity, the part of the velocity directed toward or away from us. Because we cannot find the inclination, we cannot correct these radial velocities to find the true orbital velocities. Consequently, we cannot find the true masses. All we can find from a spectroscopic binary system is a lower limit to the masses.

More than half of all stars are in binary systems, and most of those are spectroscopic binary systems. Many of the familiar stars in the sky are actually pairs of stars orbiting each other (Figure 8-15).

We might wonder what happens when the orbits of a spectroscopic binary system lie exactly edge on to Earth. The result is the most useful kind of binary system.

Eclipsing Binary Systems

As we said earlier, the orbits of the two stars in a binary system always lie in a single plane. If that plane is nearly edge-on to Earth, then the stars can cross in front of each other as seen from Earth. Imagine a model of a binary star system in which a cardboard disk represents the orbital plane and balls represent

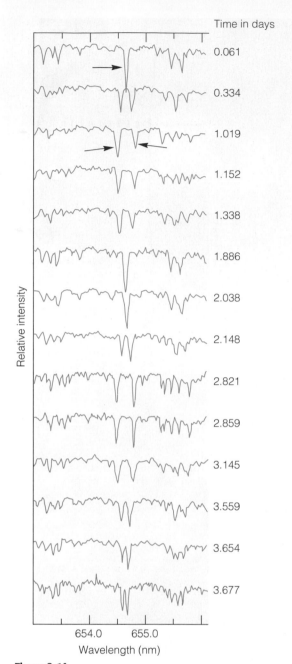

Figure 8-14
Fourteen spectra of the star HD80715 are shown here as graphs of intensity versus wavelength. A single spectral line (arrow in top spectrum) splits into a pair of spectral lines (arrows in third spectrum), which then merge and split apart again. These changing Doppler shifts reveal that HD80715 is a spectroscopic binary. *(Adapted from data courtesy of Samuel C. Barden and Harold L. Nations)*

the stars, as in Figure 8-16. If we see the model from the edge, then the balls that represent the stars can move in front of each other as they follow their orbits. The small star crosses in front of the large star and then, half an orbit later, the large star crosses in front of the small star. When one star moves in front of the other, it blocks some of the light, and we say the star is eclipsed. Such a system is called an **eclipsing binary system.**

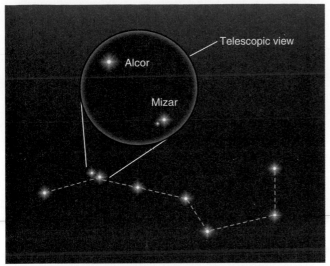

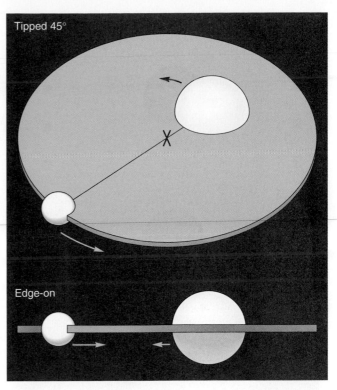

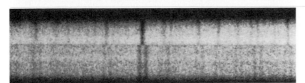

a

b

Figure 8-15
(a) At the bend of the handle of the Big Dipper lies a pair of stars, Mizar and Alcor. Through a telescope we discover that Mizar has a fainter companion and so is a member of a visual binary system. (b) Spectra of Mizar recorded at different times show that it is itself a spectroscopic binary system rather than a single star. In fact, both the faint companion to Mizar and the nearby star Alcor are also spectroscopic binary systems. *(The Observatories of the Carnegie Institution of Washington)*

Seen from Earth, the two stars are not visible separately. The system looks like a single point of light. But when one star moves in front of the other star, part of the light is blocked, and the total brightness of the point of light decreases. Figure 8-17 shows a smaller star moving in an orbit around a larger star, first eclipsing the larger star and then being eclipsed as it moves behind. The resulting variation in the brightness of the system is shown as a graph of brightness versus time, a **light curve.**

The light curves of eclipsing binary systems contain tremendous amounts of information about the stars, but the curves can be difficult to analyze. Figure 8-17 shows an idealized system. Figure 8-18 shows the light curve of a real system in which the stars have dark spots on their surfaces and are so close to each other that their shapes are distorted.

Once the light curve of an eclipsing binary system has been accurately observed, we can construct a chain of inference that leads to the masses of the two stars. We can find the orbital period easily, and we can get spectra showing the Doppler shifts of the two stars. We can find the orbital velocity because we

Figure 8-16
Imagine a model of a binary system with balls for stars and a disk of cardboard for the plane of the orbits. Only if we view the system edge-on do we see the stars cross in front of each other.

don't have to untip the orbits; we know they are nearly edge-on or there would not be eclipses. Then we can find the size of the orbits and the masses of the stars.

Earlier we used luminosity and temperature to calculate the radii of stars, but eclipsing binary systems give us a way to measure the sizes of stars directly. From the light curve we can tell how long it took for the small star to cross the large star. Multiplying this time interval by the orbital velocity of the small star gives us the diameter of the larger star. We can also determine the diameter of the small star by noting how long it took to disappear behind the edge of the large star. For example, if it took 300 seconds for the small star to disappear while traveling 500 km/sec relative to the large star, then it must be 150,000 km in diameter.

Of course, there are complications due to the inclination and eccentricity of orbits, but often these effects can be taken into account, and the system can tell us not only the masses of its stars but also their diameters.

Algol (β Persei) is one of the best-known eclipsing binaries because its eclipses are visible to the naked eye. Normally, its magnitude is about 2.15, but its brightness drops to 3.4 in eclipses that occur every 68.8 hours. Although the nature of the star was not recognized until 1783, its periodic dimming was

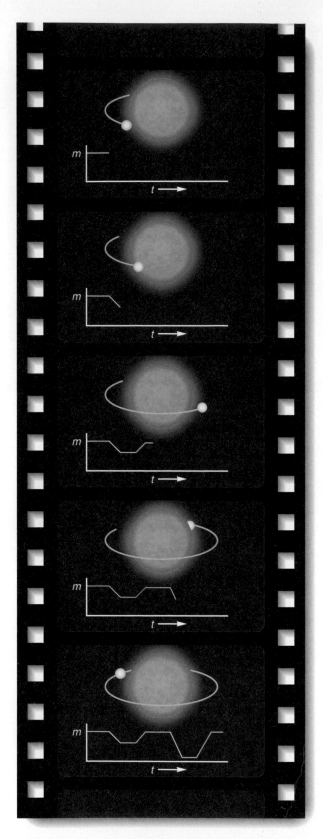

Figure 8-17
As the stars in an eclipsing binary cross in front of each other, the total brightness of the system changes as shown in the graph of magnitude (*m*) versus time (*t*). In this example, a small hot star orbits a giant cool star.

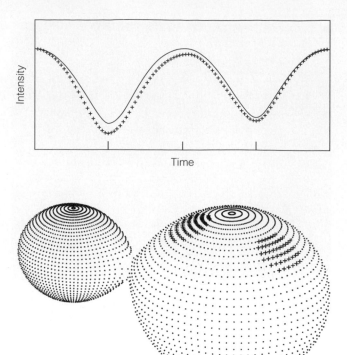

Figure 8-18
The observed light curve of the binary star VW Cephei (lower curve) shows that the two stars are so close together their gravity distorts their shapes. Slight distortions in the light curve reveal the presence of dark spots at specific places on the star's surface. The upper curve shows what the light curve would look like if there were no spots. *(Graphics created with Binary Maker 2.0)*

probably known to the ancients. *Algol* comes from the Arabic for "the demon's head," and it is associated in constellation mythology with the severed head of Medusa, the sight of whose serpentine locks turned mortals to stone (Figure 8-19). Indeed, in some accounts, Algol is the winking eye of the demon.

When we look at the light curve for an eclipsing binary system with total eclipses, how can we tell which star is hotter?

If we assume that the two stars in an eclipsing binary are not the same size, then we can refer to them as the larger star and the smaller star. When the smaller star moves behind the larger star, we lose the light coming from the total area of the small star. And when the smaller star moves in front of the larger star it blocks off light from the same amount of area on the larger star. In both cases, the same amount of area, the same number of square meters, is hidden from our sight. Then the amount of light lost during an eclipse depends only on the temperature because that is what determines how much a single square meter can radiate per second. When the surface of the hotter star is hidden, the brightness will fall dramatically, but when the surface of the cooler star is hidden, the brightness will not fall as much. So we can look at the light curve and point to the deeper of the two eclipses and say, "That is where the hotter star is behind the cooler star."

But eclipsing binaries also tell us the diameters of the stars. How could we look at the light curve of an eclipsing binary with total eclipses and find the ratio of the diameters?

Binary stars give us a powerful tool with which to discover the masses of the stars. With that information, we are now ready to assemble our data and begin trying to understand the secrets of the stars.

8-5 The Family of Stars

We have achieved the three goals we set for ourselves at the start of this chapter. We know how to find the luminosities, diameters, and masses of stars. Now we can put that data (Window on Science 8-2) together to paint a family portrait of the stars; and like most family portraits, both similarities and differences are important clues to the history of the family. As we begin trying to understand how stars are born and how they die, we ask a simple question: What is the average star like? Answering that question is both challenging and illuminating.

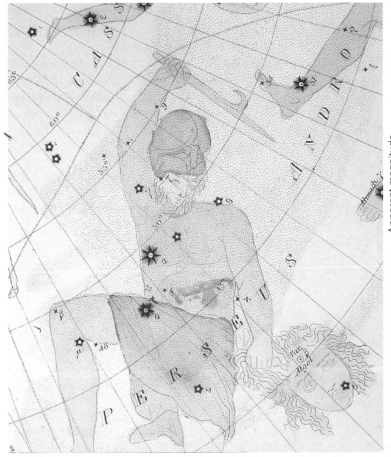

a

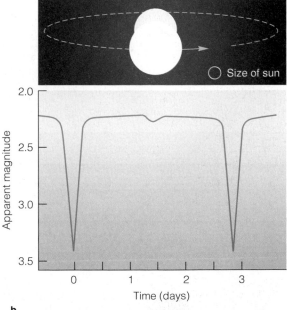

b

Figure 8-19

(a) The eclipsing binary system Algol is the star on Medusa's forehead in this engraving of Perseus. *(From Duncan Bradford,* Wonders of the Heavens, *Boston: John B. Russell, 1837.)* (b) Analysis of the light curve reveals the system contains a hot star and a cooler star orbiting in a plane that is slightly tipped such that neither star is ever totally eclipsed.

In a simple sense, science is the process by which we look at data and search for relationships that tell us how nature works. But, as a result, science sometimes requires large amounts of data. For example, astronomers need to know the masses and luminosities of many stars before they can begin looking for relationships.

Compiling basic data is one of the common forms of scientific study. This work may not seem very exciting to an outsider, but scientists often love their work not so much because they want to know nature's secrets but because they love the process of studying nature. Using a microscope or a telescope is fun. Gathering plants in the rain forest or geological samples from a cliff face can be

tremendously exciting and satisfying. It is sometimes the process of science that is most rewarding, and enjoyment of the process can lead scientists to gather significant amounts of information.

Solving a single binary star to find the masses of the stars does not tell an astronomer a great deal about nature, but solving a binary is like solving a puzzle. It is fun and it is satisfying. Over the years, many astronomers have added their results to the growing data file on stellar masses. We can now analyze that data to search for relationships between the masses of stars and other parameters such as diameter and luminosity.

The history of science is filled with stories of hardworking scientists who com-

piled large amounts of data that later scientists used to make important discoveries. For example, Tycho Brahe spent 20 years recording the positions of the stars and planets (Chapter 4). He must have loved his work to spend 20 years on his windy island, but he did not live to use his data. It was his successor, Johannes Kepler, who used Brahe's data to discover the laws of planetary motion.

Whatever science you study, you will encounter accumulations of measurements and observations that have been compiled over the years, including everything from the hardness of rocks to the attention span of infants. Determining these basic scientific data is as much a part of science as is testing a hypothesis.

Mass, Luminosity, and Density

The H-R diagram is filled with patterns that give us clues as to how stars are born, how they age, and how they die. When we add our data, we see traces of those patterns.

If we label an H-R diagram with the masses of the plotted stars, as is done in Figure 8-20, we discover that the main-sequence stars are ordered by mass. The most massive main-sequence stars are the hot stars. As we run our eye down the main sequence, we find lower-mass stars, and the lowest-mass stars are the coolest, faintest main-sequence stars.

Stars that do not lie on the main sequence are not in order according to mass. Some giants and supergiants are massive, while others are no more massive

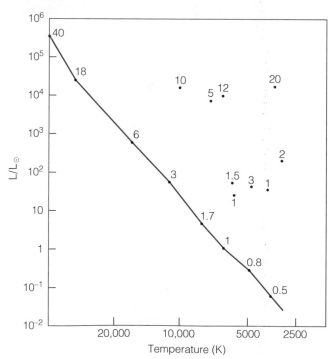

Figure 8-20
The masses of the plotted stars are labeled on this H-R diagram. Notice that the masses of main-sequence stars decrease from top to bottom but that masses of giants and supergiants are not arranged in any ordered pattern.

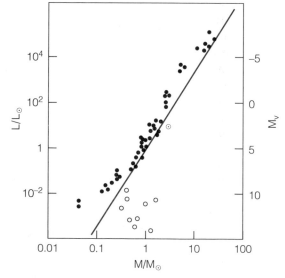

Figure 8-21
The mass–luminosity relation shows that the more massive a main-sequence star is, the more luminous it is. The open circles represent white dwarfs, which do not obey the relation. The red line represents the equation in By the Numbers 8-5.

than our sun. All white dwarfs have about the same mass, somewhere in the narrow range of 0.5 to about 1 solar mass.

Because of the systematic ordering of mass along the main sequence, these main-sequence stars obey a **mass–luminosity relation**—the more massive a star is, the more luminous it is (Figure 8-21). In fact, the mass–luminosity relation can be expressed as a simple formula. (See By the Numbers 8-5.) Giants and supergiants do not follow the mass–luminosity relation very closely, and white dwarfs not at all. In the next chapters, the mass–luminosity relation will help us understand how stars generate their energy.

Though mass alone does not reveal any pattern among giants, supergiants, and white dwarfs, density does. Once we know a star's mass and diameter, we can calculate its average density by dividing its mass by its volume. Stars are not uniform in density but are most dense at their centers and least dense near their surface. The center of the sun, for instance, is about 100 times as dense as water; its density near the visible surface is about 3400 times less dense than Earth's atmosphere at sea level. A star's average density is intermediate between its central and surface densities. The sun's average density is approximately 1 gm/cm^3—about the density of water.

Main-sequence stars have average densities similar to the sun's, but giant stars, being large, have low average densities, ranging from 0.1 to 0.01 gm/cm^3. The enormous supergiants have still lower densities, ranging from 0.001 to 0.000001 gm/cm^3. These densities are thinner than the air we breathe, and if we could insulate ourselves from the heat, we could fly an airplane through these stars. Only near the center would we be in any danger, for there the material is very dense—about 3,000,000 gm/cm^3.

The white dwarfs have masses about equal to the sun's, but are very small, only about the size of Earth. Thus, the matter is compressed to densities of 3,000,000 gm/cm^3 or more. On Earth, a teaspoonful of this material would weigh about 15 tons.

Density divides stars into three groups. Most stars are main-sequence stars with densities like the sun's. Giants and supergiants are very-low-density stars, and white dwarfs are high-density stars. We will see in later chapters that these densities reflect different stages in the evolution of stars.

Surveying the Stars

What is the most common kind of star? To answer that question we must go beyond the H-R diagram and actually take a survey of large numbers of stars. If we assume the sun is located in a typical place in the disk of our galaxy, then we can survey stars in the solar neighborhood and assume the results apply to stars in general. Unfortunately, there are three problems that make our survey difficult.

All we want to know is the number of stars of different kinds in every 1,000,000 cubic parsecs, but finding that information means we must observe the spectrum of each star to find its spectral type and measure its distance to find its absolute magnitude. A sphere containing 1,000,000 pc^3 is 62 pc in radius, but we can't measure parallax accurately at such a distance. Spectroscopic parallax isn't accurate enough for this purpose, so distances in our survey will be uncertain and that will make luminosities uncertain. That is our first problem.

One way to solve the parallax problem is to observe only stars near the sun, but that is where the second problem comes in. Some stars, such as the lower-main-sequence M stars and white dwarfs, have very low luminosities. Even if they are only a few dozen parsecs away, they are very hard to find. We could miss large numbers of these faint stars in our survey.

The third problem is that the most luminous stars are very rare. There are no O stars at all within 62 pc of the sun, and we must extend our survey to a very great distance before we find many of these stars. At such distances, parallaxes are too small to be measured, and we have to find distance in other, less accurate ways, such as through spectroscopic parallax.

The Frequency of Stellar Types

In spite of the difficulties in surveying stars, astronomers have been able to use statistical methods to study the abundance of stars in the solar neighborhood.

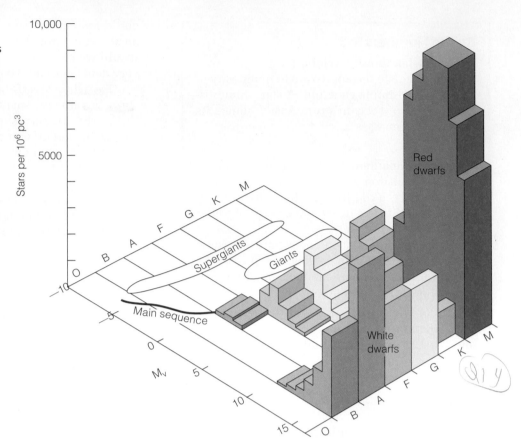

Figure 8-22
In this histogram, bars rise from an H-R diagram to represent the numbers of stars of different spectral types and absolute magnitudes found in 1 million cubic parsecs. The main-sequence M stars are the most common stars, followed closely by the white dwarfs. The upper-main-sequence stars and the giants and supergiants are so rare their bars are not visible in this diagram.

These studies help us understand what the average star is like by telling us the numbers of different kinds of stars in space. Shown in Figure 8-22, these data show that some kinds of stars are extremely rare, and others are very common. Compare Figure 8-22 with Figure 8-7 to locate the main-sequence stars, giants, white dwarfs, and supergiants.

The lowest-mass main-sequence stars, being faint and cool, are called **red dwarfs.** Notice how common these stars are; there are about 65,000 in every million cubic parsecs. White dwarfs are also very common. Fortunately for us, these stars are very faint, for if they were as luminous as supergiants, the sky would be filled with their glare, and we would see hardly anything else. As it is, the most common stars are so faint they are hard to find even with large telescopes.

Notice that the more luminous stars are rare. The stars become less common as we go up the main sequence; A stars are rare, and type O and B stars are so rare we cannot plot them on this diagram. There are only about 0.036 of the type O stars per million cubic parsecs. That is, we would have to search through about 30,000,000 pc³ to find an O star. Put yet another way, only 1 star in 4 million is an O star. Giants and supergiants are also very rare. Every million cubic parsecs contain only about 100 giants and only 0.07 supergiants. Luckily, these very rare kinds of stars are also very luminous—we can see them from great dis-

tances. If they were as faint as the main-sequence M stars, we might never know they existed.

We can see the consequences of the luminosities of the stars on any clear night. The stars we see in the sky with our unaided eyes are in fact very rare, highly luminous stars (Figure 8-23a). They are far away, and we can see them at all only because they are so very luminous. What we don't see in our sky are the common, low-luminosity stars. In a sample of the nearest stars, nearly all are red dwarfs or white dwarfs (Figure 8-23b), but we cannot see those stars with our unaided eyes because they are so faint. Thus the family relations among the stars determine what we see when we look at the night sky.

R E V I E W Critical Inquiry

Why is it difficult to be sure how common O and B stars are?

In our survey of the stars, the number of O and B stars per million cubic parsecs is difficult to determine because these stars are quite rare. We know that O and B stars are not common because there are none near us in space. If we take our survey by counting stars within 63 pc of the sun, a distance that would include a million cubic parsecs, we find no O or B stars that

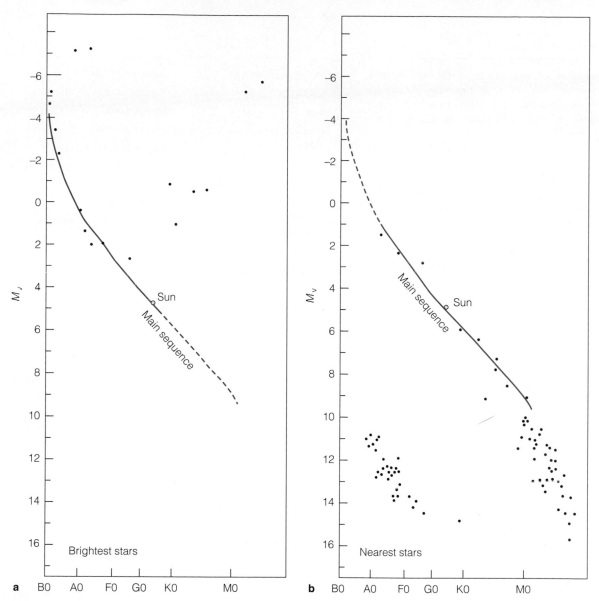

Figure 8-23
Representative H-R diagrams of (a) the brightest stars in the sky and (b) the nearest stars in space show that the most common kind of star is a lower-main-sequence M star or a white dwarf. Of the bright stars we see in the sky, many are giants and supergiants, and none are lower-main-sequence M stars or white dwarfs.

close. To find these stars we must look to great distances, and at such distances we can't measure their distance by parallax. Rather we must use the spectroscopic parallax method, which is uncertain. Thus, the chain of inference we use to find the number of sunlike stars in space does not work well for these more luminous stars.

Now consider the chain of inference for the most common stars. Why is it difficult to be sure just how common these stars are?

In this chapter we set out to measure the basic properties of stars. Once we found the distance to the stars, we were able to find their luminosities, diameters, and masses—rather mundane data. But we have now discovered a puzzling situation. The largest and most luminous stars are so rare we might joke that they hardly exist, and the average stars are such dinky low-mass things they are hard to see even if they are near us in space. Why does nature make stars in this peculiar way? To answer that question, we must explore the birth, life, and death of stars. We begin that quest in the next chapter.

Summary

Our goal in this chapter was to characterize the stars by finding their luminosities, diameters, and masses. Before we could begin, we needed to find the distances to stars. Only by first knowing the distance could we then find the other properties of the stars.

We can measure the distance to the nearer stars by observing their parallaxes. The more distant stars are so far away that their parallaxes are unmeasurably small. To find the distances to these stars, we must use spectroscopic parallax. Stellar distances are commonly expressed in parsecs. One parsec is 206,265 AU—the distance to an imaginary star whose parallax is 1 second of arc.

Once we know the distance to a star, we can find its intrinsic brightness, expressed as its absolute magnitude or its luminosity. A star's absolute magnitude is the apparent magnitude we would see if the star were only 10 pc away. The luminosity of the star is the total energy radiated in 1 second, usually expressed in terms of the luminosity of the sun.

The H-R diagram plots stars according to their brightness and their surface temperature. In the diagram, roughly 90 percent of all stars fall on the main sequence, the more massive being hotter, larger, and more luminous. The giants and supergiants, however, are much larger and lie above the main sequence. They are more luminous than main-sequence stars of the same temperature. Some of the white dwarfs are hot stars, but they fall below the main sequence because they are so small.

The large size of the giants and supergiants means their atmospheres have low densities and their spectra have sharper spectral lines than the spectra of main-sequence stars. In fact, it is possible to assign stars to luminosity classes by the widths of their spectral lines. Class V stars are main-sequence stars with broad spectral lines. Giant stars (III) have sharper lines, and supergiants (I) have extremely sharp spectral lines.

The only direct way we can find the mass of a star is by studying binary stars. When two stars orbit a common center of mass, we can find their masses by observing the period and sizes of their orbits.

Given the mass and diameter of a star, we can find its average density. On the main sequence, the stars are about as dense as the sun, but the giants and supergiants are very low-density stars. Some are much thinner than air. The white dwarfs, lying below the main sequence, are tremendously dense.

The mass–luminosity relation says that the more massive a star is, the more luminous it is. Main-sequence stars follow this rule closely, the most massive being the upper-main-sequence stars and the least massive the lower-main-sequence stars. Giants and supergiants do not follow the relation precisely, and white dwarfs not at all.

A survey in the neighborhood of the sun shows us that lower-main-sequence stars are the most common type. The hot stars of the upper main sequence are very rare. Giants and supergiants are also rare, but white dwarfs are quite common, although they are faint and hard to find.

New Terms

stellar parallax (p)

parsec (pc)

flux

absolute visual magnitude (M_v)

luminosity (L)

Hertzsprung-Russell (H-R) diagram

main sequence

giant star

supergiant star

white dwarf star

luminosity class

spectroscopic parallax

binary stars

center of mass

visual binary system

spectroscopic binary system

eclipsing binary system

light curve

mass–luminosity relation

red dwarf

Review Questions

1. Why are parallax measurements limited to the nearest stars?

2. Why is the Hipparcos satellite able to make more accurate parallax measurements than are ground-based telescopes?

3. What do the words *absolute* and *visual* mean in the definition of absolute visual magnitude?

4. What does luminosity measure that is different from what absolute visual magnitude measures?

5. Why does the luminosity of a star depend on both its radius and its temperature?

6. How can we be sure that giant stars really are larger than main-sequence stars?

7. Why do we conclude that white dwarfs must be very small?

8. What observations would we make to classify a star according to its luminosity? Why does that method work?

9. Why does the orbital period of a binary star depend on its mass?

10. What observations would you make to study an eclipsing binary star?

11. Why don't we know the inclination of a spectroscopic binary? How do we know the inclination of an eclipsing binary?

12. How do the masses of stars along the main sequence illustrate the mass–luminosity relation?

13. Why is it difficult to find out how common the most luminous stars are? The least luminous stars?

14. What is the most common kind of star?

15. If you look only at the brightest stars in the night sky, what kind of star are you likely to be observing? Why?

Discussion Questions

1. If someone asked you to compile a list of the nearest stars to the sun based on your own observations, what measurements would you make, and how would you analyze them to detect nearby stars?

2. The sun is sometimes described as an average star. Is that true? What is the average star really like?

Problems

1. If a star has a parallax of 0.050 second of arc, what is its distance in pc? in ly? in AU?

2. If you place a screen of area 1 m² at a distance of 2.8 m from a 100-watt lightbulb, the light flux falling on the screen will be 1 J/sec. To what distance must you move the screen to make the flux striking it equal 0.01 J/sec? (This assumes the lightbulb emits all of its energy as light.)

3. If a star has a prallax of 0.016 seconds of arc and an apparent magnitude of 6, how far away is it, and what is its absolute magnitude?

4. Complete the following table

m	M_v	d(pc)	P (sec of arc)
____	7	10	____
11	____	1000	____
____	−2	____	0.025
4	____	____	0.040

5. The unaided human eye can see stars no fainter than those with an apparent magnitude of 6. If you can see a bright firefly blinking up to 0.5 km away, what is the absolute magnitude of the firefly? (*Hint:* Convert the distance to parsecs and use the formula in By the Numbers 8-2.)

6. If a main-sequence star has a luminosity of 400 $L_\odot$, what is its spectral type? (*Hint:* see Figure 8-7.)

7. If a star is 10 times the radius of the sun and half as hot, what will its luminosity be? (*Hint:* See By the Numbers 8-3.)

8. An 08 V star has an apparent magnitude of +1. Use the method of spectroscopic parallex to find the distance to the star. Why might this distance be inaccurate?

9. Find the luminosity aand spectral type of a 5-$M_\odot$ main-sequence star.

10. In the following table, which star is brightest in apparent magnitude? most luminous in absolute magnitude? largest? least dense? farthest away?

Star	Spectral Type	m
a	G2 V	5
b	B1 V	8
c	G2 lb	10
d	M5 lll	19
e	White dwarf	15

11. If two stars orbit each other with a period of 6 years and a separation of 4 AU, what is their total mass? (*Hint:* See By the Numbers 8-4.)

Critical Inquiries for the Web

1. Hertsprung-Russell diagrams allow astronomers to represent the distribution of stellar properties. Examine H-R diagrams at various Web sites, and determine which types of stars are the most (and least) numerous in our galaxy.

2. What if Algol were oriented such that its stars did not eclipse each other as seen from Earth? Use the Internet to find information about Algol and binary stars in general, and discuss how astronomers could still determine that it is a multipe star system.

 Go to the Brooks/Cole Astronomy Resource Center (www.brookscole.com/astronomy) for critical thinking exercises, articles, and additional readings from InfoTrac College Edition, Brooks/Cole's online student library.

CHAPTER 9

The Formation and Structure of Stars

Jim he allowed [the stars] was made,

but I allowed they happened. Jim

said the moon could'a laid them; well,

that looked kind of reasonable, so I

didn't say nothing against it, because

I've seen a frog lay most as many, so

of course it could be done.

Mark Twain
The Adventures of Huckleberry Finn

Guidepost

We have been building a general understanding of the sky and a facility with the basic tools of astronomy. In this chapter, we begin putting those tools to intensive use. This is the place where we really begin to see how the universe works.

As we study the formation and structure of stars in this chapter, we see how science works. We see how astronomers use the basic properties of light and matter to learn about stars and how they use the basic principles of physics to build theories about the way stars form, age, and die. Even more important, we see how scientists use evidence to test these theories.

Testing theories against evidence is the basic skill required of the scientist, and we will use it over and over in the 11 chapters that follow.

Stars exist because of gravity. They form because gravity makes clouds of gas contract. Once they form, stars spend long, stable lives generating nuclear energy at their centers and balancing their own gravity with the pressure of the hot gas in their interiors. In the end, stars die because they exhaust their fuel supply and can no longer withstand the force of their own gravity.

A star can remain stable only by maintaining great gas pressure in its interior. Gravity tries to make the stellar gas contract, but if the internal temperature is high enough, the pressure of the gas pushes outward just enough to balance gravity. Thus, a star is a battlefield where gas pressure and gravity struggle for dominance. If gas pressure wins, the star must expand, but if gravity wins, the star must contract.

Only by generating tremendous amounts of nuclear energy can a star keep its interior hot enough to maintain the gravity–pressure balance. The sun, for example, generates enough energy in 1 second to power Earth's civilization at its current level for 2 million years.

To understand how a star forms, we must understand how a star generates its energy and how that energy eventually escapes from the star. Thus, we must explore the nuclear reactions that generate energy in the core of a star, and we must explore the star's internal structure, which resists the inward pull of the star's gravity. These explorations will allow us to imagine the slow evolution of stable stars over billions of years.

How can we know what stars are like when we can't see inside them and we don't live long enough to see them evolve? The answer lies in the methods of science. By constructing theories that describe how nature works and then testing those theories against evidence from observations, we can unravel some of nature's greatest secrets. We can even understand the origin and structure of the stars.

9-1 The Birth of Stars

The key to understanding star formation is the correlation between young stars and clouds of gas and dust. Where we find the youngest groups of stars, we also find large clouds of gas illuminated by the hottest

Figure 9-1
Where we find thick clouds of gas and dust, we usually find young stars, a correlation that suggests that stars are born from these clouds. Here, the great dust cloud known as the Horsehead Nebula (0.55 pc from nose to ear) is silhouetted against glowing gas excited by ultraviolet radiation from hot stars imbedded in the dense clouds to the left. *(NOAO and Nigel Sharp)*

and brightest of the new stars (Figure 9-1). This leads us to suspect that stars form from such clouds, just as raindrops condense from the water vapor in a thundercloud. To study the formation of stars, we must examine these cold clouds of gas that float between the stars and understand how these clouds contract, heat up, and become stars.

The Interstellar Medium

Astronomers can describe the gas and dust between the stars, the **interstellar medium,** in some detail. About 75 percent of the mass of the gas is hydrogen, and 25 percent is helium; there are also traces of carbon, nitrogen, oxygen, calcium, sodium, and heavier atoms. Roughly 1 percent of the mass is made up of microscopic dust about the size of the particles in cigarette smoke. The dust seems to be made mostly of carbon and silicates (rocklike minerals) mixed with or coated with frozen water. The average distance between dust grains is about 150 m.

This interstellar material is not uniformly distributed through space; it consists of a complex tangle of cool, dense clouds pushed and twisted by currents of hot, low-density gas. Although the cool clouds contain only 10 to 1000 atoms/cm³ (fewer than any vacuum on Earth), we refer to them as dense clouds in contrast with the hot, low-density gas that fills the spaces between clouds. That thin gas contains only about 0.1 atoms/cm³.

The preceding two paragraphs describe the interstellar medium, but we should not accept these facts blindly. Science is based on evidence, so we should demand to know what observational evidence supports these facts. How do we know there is an interstellar medium, and how do we know its properties?

In some cases, the interstellar medium is easily visible as clouds of gas. Astronomers refer to any cloud of gas and dust as a **nebula** (from the Latin word for "cloud"). Such nebulae (plural) give us clear evidence of an interstellar medium. If a star is hot enough, it can ionize nearby gas, producing an **emission nebula**—a nebula whose spectrum is an emission spectrum. The Balmer lines of hydrogen are strong in the spectra of these nebulae, showing that hydrogen is abundant. In fact, the three visible lines of hydrogen blend together to produce the pink color that is typical of emission nebulae. In contrast, some nebulae look blue, because their spectra are the reflected spectra of nearby stars. This light is scattered by dust in the nebula, and thus the nebulae are called **reflection nebulae.** Because tiny dust grains scatter blue light better than red light, reflection nebulae are typically blue (Figure 9-2). This is clear evidence that the dust grains are tiny—with diameters roughly equal to the wavelength of visible light.

Some nebulae make their presence known by obscuring our view of more distant stars. If a cloud is unusually dense and dusty, it totally obscures our view of the stars beyond, and we see it as a dark cloud (Figure 9-3a). Such clouds are typically not spherical, but torn and twisted like wisps of smoke distorted by a breeze.

If a cloud is less dense, starlight may be able to penetrate it, and we can see stars through the cloud;

a

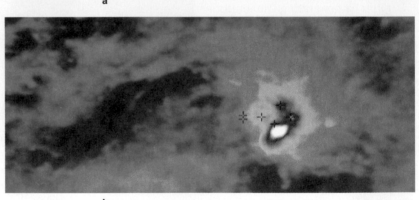

b

Figure 9-2

(a) The Pleiades star cluster, just visible to the naked eye, is surrounded by a faint reflection nebula visible only in long-exposure photographs. The nebula is produced by dust reflecting the starlight. *(Copyright California Institute of Technology and Carnegie Institution of Washington, by permission from Hale Observatories)*
(b) The cloud of gas and dust through which the Pleiades cluster is moving is revealed in this infrared image. The brighter stars of the cluster (marked by crosses) heat the gas and make it expand to produce a low-density cavity (left) tracing the path of the cluster through the cloud. *(Richard E. White; image rendered by Duncan Chesley of American Image, Inc.)*

but the stars look dimmer because the cloud scatters some of the light. The same scattering of blue light by tiny dust grains that makes reflection nebulae appear blue makes stars seen through interstellar dust appear red (as explained in Figure 9-3b). This is called **interstellar reddening,** and it can give us important clues about the dust grains and their sizes.

a

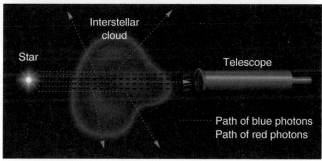

b

Figure 9-3

Some interstellar clouds of gas and dust are so dense we cannot see through them, and they appear as regions nearly free of stars. (a) In the case of Barnard's S Nebula, only a few stars are visible through the dark cloud. *(Courtesy Rudolph E. Schild)* (b) Stars that are visible through interstellar dust look redder than normal because more blue photons, having shorter wavelengths, are scattered by the dust grains. Thus, more red photons reach our telescope, and such stars look redder. Note the reddened stars visible through the dusty nebula in (a).

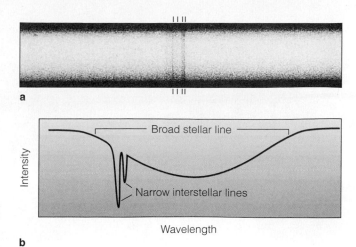

a

b

Figure 9-4

Interstellar absorption lines can be recognized in two ways. (a) The B0 supergiant ε Orionis is much too hot to show spectral lines of once-ionized calcium (CaII), yet this short segment of its spectrum reveals narrow, multiple lines of CaII (tic marks) that must have been produced in the interstellar medium. *(The Observatories of the Carnegie Institution of Washington)* (b) Spectral lines produced in the atmospheres of stars are much broader than the spectral lines produced in the interstellar medium. *(Adapted from a diagram by Binnendijk)* In both (a) and (b) the multiple interstellar lines are produced by separate interstellar clouds with slightly different radial velocities.

The interstellar medium also reveals its presence by forming interstellar absorption lines in the spectra of distant stars (Figure 9-4a). As starlight travels through the interstellar medium, gas atoms of elements such as calcium and sodium absorb photons of certain wavelengths, producing narrow interstellar absorption lines. We can be sure these lines originate in the interstellar medium because they appear in the spectra of O and B stars—stars that are too hot to form calcium and sodium absorption lines in their own atmospheres. Also, the narrowness of the interstellar lines tells us they could not have been formed in the hot atmospheres of the stars. In a hot gas, the atoms move rapidly, and the Doppler shifts of the different atoms smear the spectral line, making it broad. The interstellar medium is very cold, 10 to 50 K, so the atoms are not moving very rapidly at all, and they form extremely sharp spectral lines. In fact, the narrow widths of the spectral lines are clear evidence of the temperature of the interstellar gas. Often multiple interstellar lines appear because the light from the star passed through a number of clouds on its way to Earth (see Figure 9-4b).

Observations at nonvisible wavelengths give us valuable evidence about the interstellar medium. About 1 percent of the interstellar medium is dust, and although it is very cold, the dust can radiate in the far infrared. This would be a negligible amount of radiation except that the interstellar medium contains such vast quantities of dust. Infrared telescopes can map the interstellar medium by its infrared radiation. For example, the Infrared Astronomy Satellite discovered a faint, wispy network of dusty clouds covering the entire sky (Figure 9-5a). This **infrared cirrus** is quite cold, and its wispiness shows that the interstellar medium is in constant motion. In contrast, X-ray observations can detect regions of very hot gas apparently produced by exploding stars (Figure 9-5b), and it appears that the sun lies just inside one of these cavities filled with hot gas. Ultraviolet observations provide information about the distribution of the gas in the interstellar medium, its chemical composition, and its temperature. The interstellar medium is even detectable in the radio part of the spectrum. Radio astronomers have found many different radio signals at the precise frequencies emitted by certain molecules. These are the radio equivalent of emission lines in the spectra of visible light. Thus, the radio observations tell us that some of the atoms in space have linked together to form molecules.

We have no shortage of evidence concerning the interstellar medium. Clearly, space is not empty, and from the correlation between young stars and clouds

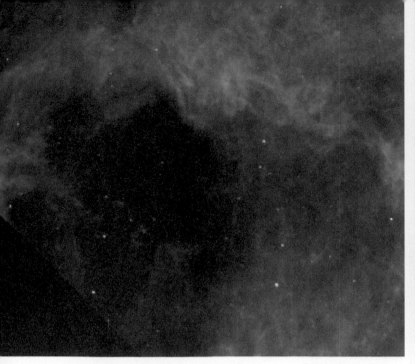

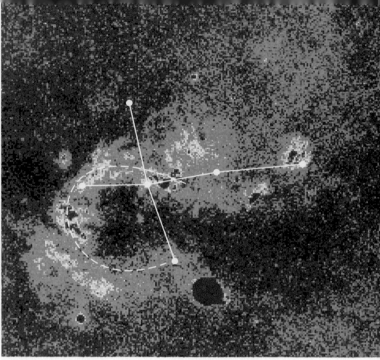

a

b

Figure 9-5
Nonvisible observations of the interstellar medium. (a) The Infrared Astronomy Satellite observed the sky in the far infrared and discovered the glow of the infrared cirrus, a wispy distribution of dust slightly warmed by starlight. At temperatures only slightly above absolute zero, the dust radiates black body radiation in the far infrared. The infrared cirrus is spread all across the sky. *(NASA/IPAC, Courtesy Deborah Levine)* (b) This ROSAT X-ray map of the constellation Cygnus reveals a yellow arch of gas clouds outlining a bubble of high-temperature gas (dashed line) spanning 450 ly. This hot gas has been produced by the explosive deaths of many massive stars. *(Courtesy Steve Snowden and Max-Planck Institute for Extraterrestrial Physics, Germany)*

of gas and dust we can suppose that stars form from these clouds. But how can the icy cold interstellar clouds create hot stars? The answer is gravity.

The Formation of Stars from the Interstellar Medium

To study the formation of stars, we must continue comparing theory with evidence. The theory of gravity predicts that the combined gravitational attraction of the atoms in a cloud of gas will squeeze the cloud, pulling every atom toward the center. Thus, we might expect that every cloud would eventually collapse and become a star; however, the heat in the cloud resists collapse. Heat is a measure of the motion of the atoms or molecules in a material—in a hot gas, the atoms move more rapidly than do those in a cool gas. The interstellar clouds are very cold, but even at a temperature of only 10 K, the average hydrogen atom moves about 0.5 km/sec (1100 mph). This thermal motion would make the cloud drift apart if gravity were too weak to hold it together.

The densest interstellar clouds contain about 1000 atoms/cm³, include from a few hundred thousand to a few million solar masses, and have temperatures as low as 10 K. In such clouds hydrogen can

exist as molecules (H_2) rather than as atoms. Although hydrogen molecules cannot be detected by radio telescopes, the clouds can be mapped by the emission of carbon monoxide molecules (CO) present in small amounts in the gas. Consequently, these clouds are known as **molecular clouds,** and the largest ones are called giant molecular clouds. Stars form in these clouds when the densest parts of the clouds become unstable and contract under the influence of their own gravity.

Most clouds do not appear to be gravitationally unstable, but a stable cloud colliding with a shock wave (the astronomical equivalent of a sonic boom) can be compressed and disrupted into fragments. Theoretical calculations show that some of these fragments can become dense enough to collapse and form stars (Figure 9-6).

Thus, most interstellar clouds may not collapse and form stars until they are triggered by the compression of a shock wave. Happily for this hypothesis, space is filled with shock waves. Supernova explosions (exploding stars described in the next chapter) produce shock waves that compress the interstellar medium, and recent observations show young stars forming at the edges of such shock waves. Another source of shock waves may be the birth of very hot

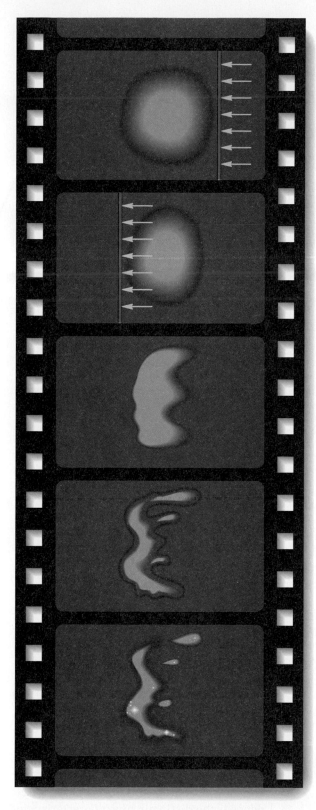

Figure 9-6
A passing shock wave (red line) can compress and fragment an interstellar gas cloud, driving some regions to densities high enough to trigger star formation (bottom frame). The events summarized from first frame to last might span about 6 million years.

stars. If one part of a cloud forms a massive star, the sudden flare of radiation as the star begins emitting radiation could push against other parts of the cloud, compress the gas, and trigger more star formation. Even the collision of two clouds can produce a shock wave and trigger star formation.

Although these are important sources of shock waves, the dominant trigger of star formation in our galaxy may be the spiral pattern itself. In Chapter 1, we noted that our galaxy contains spiral arms, and many astronomers believe the arms are long, curved shock waves traveling through the interstellar medium. If they are, then the collision of an interstellar cloud with a spiral arm could trigger the formation of new stars.

Astronomers have found a number of giant molecular clouds in which stars are forming in a repeating cycle. Both high-mass and low-mass stars form in such a cloud, but when the massive stars form, their intense radiation or eventual supernovae explosions push back the surrounding gas and compress it. This compression in turn can trigger the formation of more stars, some of which will be massive. Thus a few massive stars can drive a continuing cycle of star formation in a giant molecular cloud.

While low-mass stars do form in such clouds along with massive stars, low-mass stars also form in smaller clouds of gas and dust. Because lower-mass stars have lower luminosities and do not develop quickly into supernovae explosions, low-mass stars alone cannot drive a continuing cycle of star formation.

Once part of a cloud is triggered to collapse, gravity draws each atom toward the center. The in-falling atoms pick up speed as they fall until, by the time the gas becomes dense enough for the atoms to collide often, they are traveling very fast. Remember that temperature is a measure of the speed of the atoms in a gas, so as the atoms move faster, the temperature of the gas increases. Thus, the contraction of an interstellar cloud heats the gas by converting gravitational energy into thermal energy.

Such a collapsing cloud of gas does not form a single object; because of instabilities, it fragments, producing an association of 10 to 1000 stars. An **association** is a group of stars that are not gravitationally bound to one another. Thus the association drifts apart within a few million years. The sun probably formed in such a cluster about 5 billion years ago.

The Formation of Protostars

To follow the story of star formation further, we must concentrate on a single fragment of a collapsing cloud as it forms a star. The initial collapse of the material is almost unopposed by any internal pressure. The material falls inward, picking up speed and increasing in density and temperature. This free-fall contraction transforms the cold gas into a warm protostar buried deep in the dusty gas. A **protostar** is an object that will eventually become a star.

When a star or protostar changes, we say it evolves, and we can follow that evolution in the H-R diagram. As a star's temperature and luminosity change, its location in the H-R diagram shifts as well. We can draw an arrow or line called an **evolutionary track** to show how the star moves in the diagram. Of course, the evolutionary motion of the star in the diagram is not related in any way to its motion through space.

Stars begin contracting from the interstellar gas, which is very cold. Thus to trace the contraction of a protostar, we must extend the H-R diagram to the right to include very low temperatures. Figure 9-7, which includes temperatures a low as 32 K, shows the initial collapse of a 1-solar-mass cloud fragment as its temperature rises to more than 1000 K. The exact process is poorly understood because of the complex effect of the dust in the cloud. The dusty cloud also hides the protostar from sight during its contraction. If we could see it, it would be a luminous red object a few thousand times larger than the sun, and we would plot it in the red-giant region of the H-R diagram.

The protostar's original free-fall contraction is slowed by the increasing density and internal pressure, but since the protostar is not yet hot enough to begin nuclear reactions, it must contract further as it radiates energy from its surface. This contraction is much slower than the free-fall contraction.

Throughout its contraction, the protostar converts its gravitational energy into thermal energy. Half of this thermal energy radiates into space, but the remaining half raises the internal temperature. As the internal temperature climbs, the gas becomes ionized, becoming a mixture of positively charged atomic nuclei and free electrons. When the center gets hot enough, nuclear reactions begin generating energy,

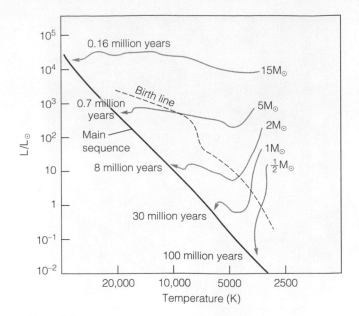

Figure 9-8
The more massive a protostar is, the faster it contracts. A 1-$M_\odot$ star requires 30 million years to reach the main sequence. (Recall that $M_\odot$ means "solar mass.") The dashed line is the birth line, where contracting protostars first become visible as they dissipate their surrounding clouds of gas and dust. Compare with Figure 9-7, which shows the formation of a star of about 1-$M_\odot$ as a dashed line up to the birth line and as a solid line from the birth line to the main sequence. *(Illustration design by author)*

the protostar halts its contraction, and, having blown away its cocoon of gas and dust, it becomes a stable, main-sequence star.

The time a protostar takes to contract from a cool interstellar gas cloud to a main-sequence star depends on its mass. The more massive the star, the stronger its gravity and the faster it contracts (Figure 9-8). The sun took about 30 million years to reach the main sequence, but a 15-solar-mass star can contract in only 160,000 years. Conversely, a star of 0.2 solar mass takes 1 billion years to reach the main sequence.

The story we have told, the story of the formation of a star from the interstellar medium, is based on theory. By understanding what the interstellar medium is like and by understanding how the laws of physics work, astronomers have been able to tell the story of how stars must form. But we can't accept a scientific theory without testing it, and that means we must compare the theory with the evidence. That constant checking of theories against evidence is the distinguishing characteristic of science, so it is time to carefully separate theory from evidence (Window on Science 9-1) and ask how much of this star formation story can be observed.

Figure 9-7
A contracting protostar, hidden deep inside a dust cloud, begins very cool and very faint (arrow). The thermal insulation of the dust makes the exact behavior of the protostar uncertain.

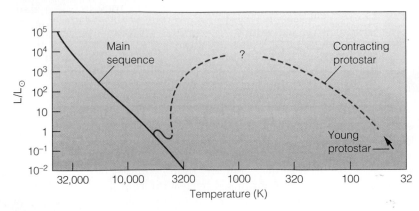

The fundamental work of science is the testing of theories by comparison with facts. As we think about science, we need to clearly distinguish between facts and theories. The facts are the evidence against which we test the theories.

Scientific facts are those observations or experimental results of which we are confident. An astronomer makes observations of stars, a botanist collects samples of related plants, and a chemist performs an experiment to measure the rate of a chemical reaction. A fact could be a precise measurement, such as the mass of a star expressed as a specific number, or a fact could be a simple observation, such as that a certain butterfly no longer visits a certain mountain valley. In each case, the scientist is gathering facts.

A theory, however, is a conjecture as to how nature works. If we are uncertain of the theory, we might call it a hypothesis. In any case, these conjectures are not facts; they are attempts to explain how nature works. In a sense, a theory or a hypothesis is a story that scientists have made up to explain how nature works in some specific case. These stories can be wonderfully detailed and ingenious, but without evidence they are nothing more than hunches.

When one of these stories is tested against the facts and confirmed, we have more confidence that the story is probably right. The more a story is tested successfully, the more confidence we have in it. The facts represent reality, and every theory or hypothesis must be repeatedly tested against reality.

We can't test one theory against another theory. Theories are not evidence; they are conjectures. If we were allowed to test one theory against another theory, we might fall into the trap of circular reasoning. "Elves make the flowers bloom. I know that elves exist because the flowers bloom." Here two theories are used to confirm each other, and it leads us to nonsense. Only facts can be evidence.

Nor can we use a theory to deduce facts. We can use a theory to make predictions that we can test against facts, but the predictions themselves can never be certainties, so they can't be facts. The only way to get facts is to consult nature and make direct measurements or observations.

As we study different problems in astronomy, we must carefully distinguish between facts and theories. Facts are the basic building blocks of all science, and they come only from the careful study of nature. To use the words of Galileo, we must "read the book of nature."

Observations of Star Formation

In astronomy, evidence means observations. Thus, we must ask what observations confirm our theories of star formation. Unfortunately, a protostar is not easy to observe. The protostar stage is less than 0.1 percent of a star's total lifetime, so, although that is a long time in human terms, we cannot expect to find many stars in the protostar stage. Furthermore, protostars form deep inside clouds of dusty gas that absorb any light the protostar might emit. Only when the protostar is hot enough to drive away its enveloping cloud of gas and dust do we see it. The **birth line** in the H-R diagram in Figure 9-8 shows where contracting protostars first become visible, and protostars cross the birth line shortly before they reach the main sequence. Thus the early evolution of a protostar is hidden from our sight.

Although we cannot see protostars at visible wavelengths before they cross the birth line, we can detect them in the infrared. The dust absorbs light from the protostar and grows warm, and warm dust radiates copious infrared. Infrared observations such as those made by the Infrared Astronomy Satellite reveal many bright sources of infrared radiation that are probably protostars buried in dust clouds.

T Tauri stars, named after the first example found, variable star T in Taurus, appear to be protostars just clearing away the surrounding gas and dust.

T Tauri stars vary in brightness, and spectroscopic observations show gas flowing away in strong stellar winds comparable to the sun's solar wind. In some cases, gas appears to be flowing into the star. All T Tauri stars are sources of infrared radiation, which shows that the stars are still accompanied by warm dust. In some cases, evidence suggests the dust and gas surround the star in a rotating disk.

Some very young star clusters such as NGC2264 contain large numbers of T Tauri stars. This cluster formed so recently that the less massive stars, contracting more slowly than the massive stars, have not yet reached the main sequence (Figure 9-9). Scattered throughout the cluster are a number of T Tauri stars. The temperature and luminosity of T Tauri stars place them to the right of the main sequence near the birth line, just where we would expect to find contracting protostars.

T Tauri stars are often accompanied by **Herbig-Haro objects** (named after astronomers George Herbig and Guillermo Haro)—small nebulae that can change their brightness and shape over only a few years. Apparently, Herbig-Haro objects occur where jets of gas flowing away from young stars smash into clouds of interstellar matter. The excited clouds glow, and as the jets of gas fluctuate, the clouds change in brightness and shape. Herbig-Haro objects HH34S and HH34N are clearly related to a young star located

a

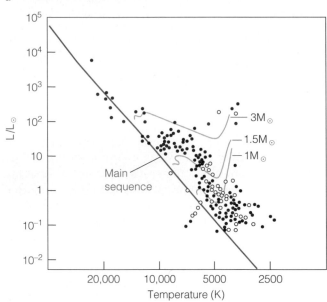

b

Figure 9-9
(a) One of the youngest star clusters known, NGC2264 is only a few million years old. *(Anglo-Australian Telescope Board)* (b) The H-R diagram of the cluster shows that many of its lower-mass stars have not yet reached the main sequence. T Tauri stars (open circles) are common and fall between the birth line and the main sequence (see Figure 9-8). Blue evolutionary tracks are derived from theoretical models of contracting protostars. *(Adapted from data by M. Walker)*

halfway between them (Figure 9-10), and the hottest part of the jet is visible pointing toward HH34S.

When a young star blasts gas away in two jets shooting in opposite directions, astronomers refer to it as a **bipolar flow.** According to recent theoretical work, such beams of gas appear to be caused by the formation of a spinning disk of gas around a forming star. As additional gas falls into the disk, it drags in a

magnetic field from the interstellar medium, and the resulting interaction between the magnetic field and the rotating disk blows beams of gas outward along the axis of rotation (Figure 9-11). We will discover in later chapters that both disks and bipolar flows are common—they occur in dying stars and exploding galaxies as well as in young stars.

We know that many infrared objects, Herbig-Haro objects, and T Tauri stars are young because they are located in regions where hot, blue stars are found. We will see later in this chapter that these bright blue stars have extremely short lives. In fact, they can't live long enough to drift out of the nebulae in which they form, so the presence of these hot, blue stars alerts us that the region is a site of star formation.

Where we find the youngest stars, we often also find objects called **Bok globules** (named after the astronomer Bart Bok)—small clouds only about 1 ly in diameter that contain 10 to 1000 solar masses of gas and dust. Because these clouds produce no light of their own, they are only seen silhouetted against bright nebulae (Figure 9-12). Some astronomers believe the Bok globules are collapsing clouds that will become stars. Infrared observations, including those made by the Infrared Astronomy Satellite, show that some Bok globules do contain infrared sources that appear to be contracting protostars.

Because most of the stages of star formation occur deep inside dense clouds of gas and dust, we can't see what is happening. It would be nice if we could excavate such a cloud the way an archeologist excavates an ancient temple by stripping away the layers of dirt. A Hubble Space Telescope image of part of the Eagle Nebula reveals nature doing exactly that. Hot stars in and near the nebula are evaporating the dust and driving away the gas of the densest parts

Figure 9-10
At the center of this image, a newborn star is emitting a powerful jet toward the left and, presumably, a less powerful jet to the right. Where these jets run into the interstellar medium, they excite the two Herbig-Haro objects HH34S and HH34N. *(Reinhard Mundt, Calar Alto 3.5-m telescope)* (Inset) This Hubble Space Telescope image shows that the jet emitted by the protostar is irregular and turbulent. *(J. Hester, Arizona State University, and NASA)*

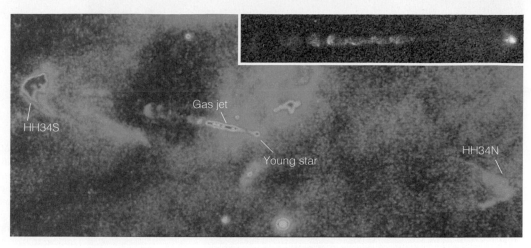

Figure 9-11
(a) Star formation begins in a gas cloud when dense cores begin to form as gravitation draws gas and dust into condensations. (b) The in-falling matter forms a dense, hot, central object destined to become a star at the center of a larger opaque cocoon of gas and dust. Rotation forces some of the matter to fall into a disk rotating around the central body. (c) The central object, now a protostar, and the rotating disk eventually begin to expel gas along the axis of rotation while matter continues to fall into the disk. The bipolar flows, guided and perhaps driven by magnetic fields (red), push out of the surrounding cloud of gas and dust to produce oppositely directed jets.

Figure 9-12
The black objects here are Bok globules—small, dense, dusty clouds that are either contracting to become stars or are the remnants of clouds that have formed stars. These Bok globules are in the nebula IC2944. *(Anglo-Australian Telescope Board)*

Figure 9-13
The Eagle Nebula contains dark clouds of gas and dust that, in this Hubble Space Telescope Image, are revealed as nests of star formation. Hot stars nearby are evaporating the dust and driving the gas away to expose EGGs, evaporating gaseous globules (best illuminated at upper left). Some of these small dense clouds appear to be contracting under their own gravitation to form protostars. At least one seems to contain a newborn star (arrow) exposed by the evaporation of the gaseous globule around it (see Figure 9-6). For a fuller view of the Eagle Nebula, see Figure 6-1. *(Jeff Hester and Paul Scowen, Arizona State University and NASA.)*

of the nebula, revealing small globules of denser gas and dust presumably drawn together by their own gravity. At least some of these seem to be forming stars. In part because these objects were found in the Eagle Nebula, astronomers have enjoyed calling them EGGs — evaporating gaseous globules. Whatever we call them, they are evidently stars exposed in the act of forming (Figure 9-13).

All of these observations confirm our theoretical model of contracting protostars. Although star formation still holds many mysteries, the general process seems clear. In at least some cases, interstellar gas clouds are compressed by passing shock waves, and the clouds' gravity, acting unopposed, draws the matter inward to form protostars.

The planets of our solar system formed with the sun over an interval of no more than a few million years. At some point during the sun's contraction, when it was surrounded by swirling clouds of gas and dust, solid bodies condensed and grew into planets. We will discuss planet building in Chapter 16, but it is important to note here that stars may form planets as they contract.

REVIEW Critical Inquiry

If the interstellar medium is so cold, how can we detect it at infrared wavelengths?

The interstellar medium is certainly very cold. The interstellar clouds have temperatures only 10 to 50 kelvin above absolute zero. Our discussion of black body radiation (Chapter 6) told us that the energy emitted per second by a square meter on the surface of a

black body depends only on the temperature raised to the fourth power (T^4). A dust grain at the temperature of the interstellar medium radiates very little energy per second per square meter, and a dust grain has a tiny surface area. So it seems strange at first that we detect interstellar clouds of gas and dust through the infrared radiation emitted by the dust grains. But recall that the mass of such a cloud might be a million solar masses, and that dust makes up about 1 percent of the cloud, about 10,000 solar masses of dust. An interstellar dust grain has a diameter roughly equal to the wavelength of light, so it has a tiny surface area compared with the sun. But 10,000 solar masses of dust have a surface area roughly 10^{20} times that of the sun. So although a dust grain emits very little infrared energy, the vast number of dust grains in the interstellar medium are very luminous at infrared wavelengths.

The interstellar medium is so cold that we don't really have a word for it. "Cold," "icy," or "frigid" can't express the awful cold of a few degrees above absolute zero. How can such cold material get hot enough to ignite nuclear fusion in the hearts of protostars?

Gravity makes stars form from the interstellar medium, but once a cloud contracts to produce a protostar, something has to stop the contraction and produce a stable star. That stability is reached as the core of the protostar becomes hot enough to ignite thermonuclear fusion. These nuclear reactions are as important to a star's life as gravity is to its formation. Thus, the next step in our study of stellar evolution is the energy source that lies in the hearts of the stars.

9-2 How Stars Generate Nuclear Energy

Stars generate energy through nuclear reactions, which keep the interior of the star hot and the gas pressure high. That pressure balances the force of the star's gravity and prevents further contraction. In a sense, the matter in a star is supported against its own gravity by the nuclear reactions.

The high temperatures inside stars keep the gas entirely ionized. That is, the violent collisions between the atoms strip the electrons away, so the gas is a hot mixture of positively charged nuclei and negatively charged electrons. But how exactly can the nucleus of an atom yield energy? The answer lies in the forces that hold the nuclei together.

Nuclear Binding Energy

Astronomers often use the wrong words to describe energy generation inside stars. Astronomers will say, "The star ignites hydrogen burning." We use the word *ignite* to mean *catch on fire*, and we use *burn* to mean *on fire*. What goes on inside stars isn't really burning in the usual sense.

Stars generate their energy by breaking and reconnecting the bonds between the particles *inside* atomic nuclei. This is quite different from the way we generate energy by burning wood in our fireplaces. The process of burning wood extracts energy by breaking and reconnecting chemical bonds between atoms in the wood. Chemical bonds are formed by the electrons in atoms, and we saw in Chapter 6 that the electrons are bound to the atoms by the electromagnetic force. Thus, chemical energy originates in the electromagnetic force that binds the electrons.

There are only four known forces in nature: the force of gravity (see Chapter 4), the electromagnetic force (see Chapter 6), the **weak force,** and the **strong force**. The weak force comes into play in the radioactive decay of certain kinds of nuclear particles, and the strong force binds together atomic nuclei. Thus nuclear energy originates in the strong force.

Nuclear power plants on Earth generate energy through **nuclear fission** reactions that split uranium nuclei into less massive fragments. A uranium nucleus contains a total of 235 protons and neutrons, and it splits into a range of fragments containing roughly half as many particles. Because the fragments produced are more tightly bound than the uranium nuclei, energy is released during uranium fission.

Stars make energy in **nuclear fusion** reactions that combine light nuclei into heavier nuclei. The most common reaction fuses hydrogen nuclei (single protons) into helium nuclei (two protons and two neu-

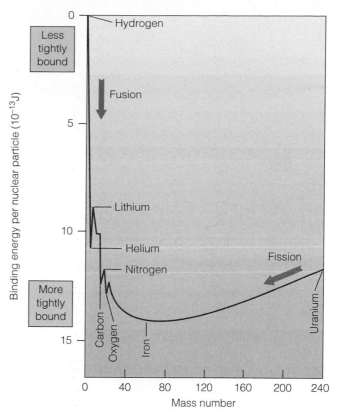

Figure 9-14
The red line in this graph shows the binding energy, the energy that holds an atomic nucleus together, for all the different atoms plotted by their atomic number, the number of protons and neutrons in their nucleus. Both fission and fusion nuclear reactions move downward in the diagram (arrows) toward more tightly bound nuclei. Iron has the most tightly bound nucleus, so no nuclear reactions can begin with iron and release energy.

trons). Because the nuclei produced are more tightly bound than the original nuclei, energy is released. Notice in Figure 9-14 that both fusion and fission reactions move downward in the diagram toward more tightly bound nuclei. Thus, both produce energy by releasing the binding energy of atomic nuclei.

Note from Figure 9-14 that iron has the most tightly bound nucleus. Fusion reactions can release energy only when the nuclei are less massive than iron. This ultimate limitation will be important in the next chapter, when we discuss the deaths of stars.

Hydrogen Fusion

The nuclear reactions that fuse hydrogen in stars join four hydrogen nuclei to make one helium nucleus. Because one helium nucleus has 0.7 percent less mass than four hydrogen nuclei, it seems that some mass vanishes in the process:

$$4 \text{ hydrogen nuclei} = 6.693 \times 10^{-27} \text{ kg}$$
$$\underline{1 \text{ helium nucleus} = 6.645 \times 10^{-27} \text{ kg}}$$
$$\text{difference in mass} = 0.048 \times 10^{-27} \text{ kg}$$

The difference in mass is the mass equivalent to the difference in the binding energy of the nuclei. Recall from Einstein's equation $E = mc^2$ that energy and mass are related. Thus the 0.048×10^{-27} kg is the binding energy released in the fusion reaction:

$$E = mc^2$$
$$= (0.048 \times 10^{-27} \text{ kg})$$
$$(3 \times 10^8 \text{ m/sec})^2$$
$$= 0.43 \times 10^{-11} \text{ J}$$

This is a very small amount of energy, hardly enough to raise a housefly 0.001 inch into the air. Only by concentrating many reactions in a small area can nature produce significant results. A single kilogram (2.2 lb) of hydrogen, for instance, converted entirely to energy, would produce enough power to raise an average-sized mountain 10 km (6 miles) into the air! The sun has a voracious energy appetite and needs 10^{38} reactions per second, transforming 5 million tons of mass into energy every second, just to balance its own gravity. This might sound as if the sun is losing mass at a furious rate, but in its entire 10-billion-year lifetime, the sun will convert less than 0.07 percent of its mass into energy.

These fusion reactions can occur only when the nuclei of two atoms get very close to each other. Since atomic nuclei carry positive charges, they repel each other with an electrostatic force called the Coulomb force (see Chapter 6). Physicists commonly refer to this repulsion between nuclei as the **Coulomb barrier.** To overcome this barrier, atomic nuclei must collide violently. Violent collisions are rare unless the gas is very hot, in which case the nuclei move at high speeds and collide violently. (Remember, an object's temperature is just a measure of the speed with which its particles move.)

Thus nuclear reactions take place only near the centers of stars, where the gas is hot and dense. A high temperature ensures that collisions between nuclei are violent enough to overcome the Coulomb barrier, and a high density ensures that there are enough collisions, and thus enough reactions, to meet the star's energy needs.

We can symbolize this process with a simple nuclear reaction:

$$4\ ^1\text{H} \rightarrow\ ^4\text{He} + \text{energy}$$

In this equation, ^{1}H represents a proton, the nucleus of the hydrogen atom, and ^{4}He represents the nucleus of a helium atom. The superscripts indicate the approximate weight of the nuclei. The actual steps in the process are more complicated than this convenient summary suggests. Instead of waiting for four hydrogen nuclei to collide simultaneously, a highly unlikely event, the process can proceed step-by-step in a chain of reactions that follow either of two pathways—the proton–proton chain or the CNO cycle.

The **proton–proton chain** is a series of three nuclear reactions that builds a helium nucleus by adding to protons one at a time. This process is efficient at temperatures above 10,000,000 K. The sun, for example, manufactures over 90 percent of its energy in this way.

The three reactions in the proton–proton chain are:

$$^1\text{H} + ^1\text{H} \rightarrow\ ^2\text{H} + e^+ + \nu$$
$$^2\text{H} + ^1\text{H} \rightarrow\ ^3\text{He} + \gamma$$
$$^3\text{He} + ^3\text{He} \rightarrow\ ^4\text{He} + ^1\text{H} + ^1\text{H}$$

In the first reaction, two hydrogen nuclei (two protons) combine to form a heavy hydrogen nucleus, emitting a particle called a positron (a positively charged electron, symbolized e^+) and another called a neutrino (ν). In the second reaction, the heavy hydrogen nucleus absorbs another proton and, with the emission of a gamma ray (γ), becomes a lightweight helium nucleus (^{3}He). Finally, two light helium nuclei combine to form a normal helium nucleus and two hydrogen nuclei. Because the last reaction needs two ^{3}He nuclei, the first and second reactions must each occur twice for each ^{4}He produced (Figure 9-15). The net result of this chain reaction is the transformation of four hydrogen nuclei into one helium nucleus plus energy.

The energy appears in the form of gamma rays, positrons, and neutrinos. The gamma rays are photons that are absorbed by the surrounding gas before they can travel more than a few centimeters. This heats the gas. The positrons produced in the first reaction combine with free electrons, and both particles vanish, converting their mass into gamma rays. Thus, the positrons also help keep the center of the star hot. The neutrinos, however, are particles that resemble photons except that they almost never interact with other particles. The average neutrino could pass unhindered through a lead wall 1 ly thick. Thus, the neutrinos do not help heat the gas but race out of the star at the speed of light, carrying away roughly 2 percent of the energy produced.

Like the proton–proton chain, the **CNO cycle** is a series of nuclear reactions that combines four hydrogen nuclei to make one helium nucleus plus energy. The steps in the CNO cycle begin with carbon-12 (^{12}C); pass through stages involving isotopes of carbon, nitrogen (N), and oxygen (O); and finish with the reappearance of a carbon-12 nucleus just like the one that started the process. Thus, we say the carbon is a catalyst; it makes the reactions possible but is not altered in the end (Figure 9-16).

Counting protons in the CNO cycle reactions in Figure 9-16 shows that the net result of the CNO cycle is four hydrogen nuclei combined to make one helium nucleus plus energy.

Because the carbon nucleus has a charge six times that of hydrogen, the Coulomb barrier is high, and much hotter temperatures are necessary to force the pro-

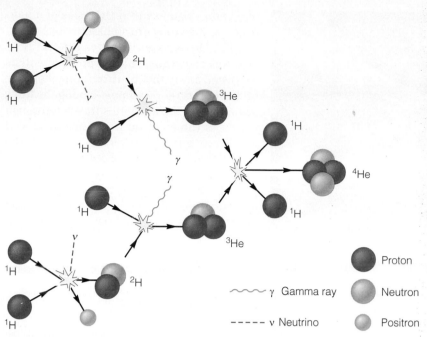

Figure 9-15
The proton–proton chain combines four protons (at left) to produce one helium nucleus (at right). Energy appears as gamma rays and as positrons, which combine with electrons to convert their mass into energy. Neutrinos escape, carrying away about 2 percent of the energy.

〜〜 γ Gamma ray

- - - - - ν Neutrino

● Proton

● Neutron

● Positron

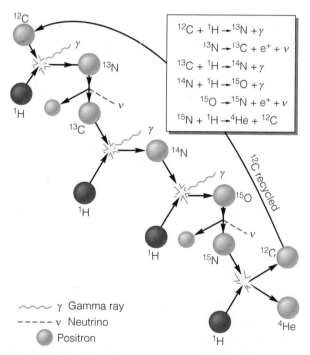

$$^{12}C + ^{1}H \rightarrow ^{13}N + \gamma$$
$$^{13}N \rightarrow ^{13}C + e^{+} + \nu$$
$$^{13}C + ^{1}H \rightarrow ^{14}N + \gamma$$
$$^{14}N + ^{1}H \rightarrow ^{15}O + \gamma$$
$$^{15}O \rightarrow ^{15}N + e^{+} + \nu$$
$$^{15}N + ^{1}H \rightarrow ^{4}He + ^{12}C$$

〜〜 γ Gamma ray
- - - - ν Neutrino
● Positron

Figure 9-16
The CNO cycle uses ^{12}C as a catalyst to combine four hydrogen atoms (^{1}H) to make one helium atom (^{4}He) plus energy. The carbon atom reappears at the end of the process ready to start the cycle over.

ton and carbon nucleus together. Thus the CNO cycle is important only in stars more massive than about 1.1 solar masses. These stars have central temperatures hotter than about 16,000,000 K. Stars less massive than 1.1 solar masses, such as the sun, are not hot enough at their centers to generate much energy on the CNO cycle.

It is time to ask again about evidence. Our discussion of the proton–proton chain and the CNO cycle has been based mostly on theory. The observational facts that show the sun is generating energy by nuclear fusion introduce us to a fascinating problem in modern astronomy.

The Solar Neutrino Problem

The center of a star seems forever hidden from us, but the sun is transparent to **neutrinos** because these subatomic particles almost never interact with normal matter. Nuclear reactions in the sun's core produce floods of neutrinos that rush out of the sun and off into space. If we could detect these neutrinos, we could probe the sun's interior.

Because neutrinos almost never interact with atoms, we never feel the flood of over 10^{12} solar neutrinos that flow through our bodies every second. Even at night, neutrinos from the sun rush through Earth as if it weren't there, up through our beds, through us, and onward into space. Obviously we are lucky to be transparent to neutrinos, but it means that neutrinos are extremely hard to detect. Certain nuclear reactions, however, can be triggered by a neutrino of the right energy; and in the late 1960s, chemist Raymond Davis, Jr., began using such a reaction to detect solar neutrinos.

Davis filled a 100,000-gallon tank with the cleaning fluid perchloroethylene (C_2Cl_4). Theory predicts that about once a day, a solar neutrino will convert a chlorine atom in the tank into radioactive argon, which can be detected later by its radioactive decay. To protect the detector from cosmic rays from space, the tank was buried nearly a mile deep in a South Dakota gold mine (Figure 9-17). Of course, the mile of rock overhead has no effect on the neutrinos.

The result of the Davis experiment startled astronomers. The cleaning fluid detects too few neutrinos—not one neutrino per day as predicted by models of the sun, but about one every three days. The experiment has been refined, tested, and calibrated for three decades; but it has not found the missing neutrinos. In 1988, the Japanese detector, Kamiokande II, buried deep in a lead mine, confirmed the deficiency in solar neutrinos. The detectors count roughly one-third of the expected number of solar neutrinos.

Figure 9-17
The Davis solar neutrino experiment counts neutrinos from the sun using 100,000 gallons of cleaning fluid held in a tank nearly a mile underground. A minuscule fraction of the solar neutrinos that pass through the cleaning fluid convert chlorine atoms into argon atoms that can be counted because they are radioactive. *(Brookhaven National Laboratory)*

The missing neutrinos left astronomers wondering if they misunderstood the nuclear reactions that power the sun. Over the years since the Davis experiment first revealed a deficiency in solar neutrinos, other neutrino detectors have been built, and they confirm the basic observation. We count fewer solar neutrinos than theory predicts.

The solar neutrino problem has been one of the great scientific mysteries of the late twentieth century, and two kinds of solutions were proposed (Window on Science 9-2). Some scientists wondered if there are errors in the models of the sun, while others suggested that we don't yet understand neutrinos well enough to count them with certainty. After decades of research, the second suggestion seems more likely.

Nuclear particle physicists know that neutrinos come in three types, called "flavors." The three flavors of neutrinos, named electron, muon, and tau neutrinos, were all thought to have zero mass, but no one had ever measured the mass of a neutrino because they are so hard to detect.

The Davis experiment can detect only electron neutrinos, so a group of scientists suggested a way that neutrinos might change back and forth—oscillate—among the three flavors while the neutrinos traveled through matter. If the electron neutrinos produced inside the sun could oscillate among the three flavors as they sped out of the sun, we would only detect a third as many as we expected. The other two thirds of the electron neutrinos would have become tau or muon neutrinos, which the Davis experiment could not count.

This neutrino oscillation theory has an interesting consequence. If neutrinos do oscillate, then nuclear particle theory requires that they have some mass. They can't be massless and oscillate.

We could test the theory of neutrino oscillation by trying to measure the mass of neutrinos or by trying to catch neutrinos in the act of oscillating from one flavor to another. Experiments to measure neutrino mass have shown that any mass they have must be very small—roughly 500,000 times less than that of an electron. But some of the experiments have produced tentative evidence that neutrinos do have mass.

To catch neutrinos in the act of oscillating, scientists used the Super Kamiokande neutrino detector, which contains 12.5 million gallons of hyper-pure water and is buried under a mountain in Japan. High-energy particles striking Earth's atmosphere from space should break up atoms and shower Earth with twice as many muon neutrinos as electron neutrinos. From under the mountain, the detector found the predicted numbers of muon and electron neutrinos coming from above, but equal numbers of muon and electron neutrinos coming up through Earth. This suggests that the neutrinos oscillated as they traveled through Earth's dense matter, just as the theory suggests solar neutrinos oscillate as they travel through the dense matter in the sun.

Neutrino oscillation would solve the solar neutrino mystery, but it is also exciting because it requires that neutrinos have mass. Neutrinos are so numerous in the universe that even a very small mass for each one could exert enough gravitational attraction to affect the evolution the universe as a whole. We will discuss this problem in Chapter 15.

Whatever the solution to the solar neutrino problem, we can be sure the sun fuses hydrogen. Some stars, however, fuse heavier elements. To complete our analysis of stellar energy sources, we must consider those forms of nuclear fusion.

Scientists like to claim that every scientific belief is based on evidence, that every theory has been tested, and that the moment a theory fails a test, it is discarded or revised. The truth is much more complicated than that, and the solar neutrino problem is a good illustration. If the detection of solar neutrinos contradicts the theory of stellar structure, why hasn't the theory been abandoned?

While scientists do indeed have tremendous respect for evidence, they also have a faith in theories that have been tested successfully many times. If a theory has been tested and confirmed over and over, they may even begin to call it a natural law, and that means scientists have great faith in its truth. That is, they have confidence that the theory or law is a good description of how nature works.

Nevertheless, it is not unusual for an experiment or an observation to contradict well-established theories. In many cases, the experiments and observations are simple mistakes, or they have not been interpreted correctly. Scientists resist abandoning a well-tested theory even when an observation continues to contradict it. If confidence in the theory is stronger than the evidence, scientists begin testing the evidence. Can it be right? Do we understand it correctly? Of course, if the evidence cannot be impeached and it continues to contradict the theory, scientists must eventually abandon or modify the theory no matter how many times it has previously been tested and confirmed. This ultimate reliance on evidence is the distinguishing characteristic of science.

It is only human nature to hang on to the principles you have come to trust, and that confidence in well-tested scientific principles helps scientists avoid rushing to faulty judgments. For example, claims for perpetual motion machines occasionally crop up in the news, but the world's scientists don't instantly abandon the laws of energy and motion pending an analysis of the latest claim. Of course, if such a claim did prove true, the entire structure of scientific knowledge would come crashing down, but because the known laws of energy and motion have been well tested and no perpetual motion machine has ever been successful, scientists know which way to bet. Like the keel on a ship, confidence in well-tested theories and laws keeps the scientific boat from rocking before every little breeze.

Some forms of faith must be absolute and unshakable—religious faith, for example. But scientific faith may be better described as scientific confidence because it must be open to change. If, ultimately, a single experiment or observation conclusively contradicts our most cherished law of nature, we as scientists must abandon that law and find a new way to understand nature. We can see this scientific confidence at work in many controversies, from the origin of the human race to the meaning of IQ measurements; but one of the best examples is the solar neutrino problem. While we struggle to understand the origin of solar neutrinos, we continue to have confidence that we do indeed understand how stars work.

Heavy-Element Fusion

The hydrogen in a star is the fuel it uses to generate energy. When that fuel is exhausted, some stars can fuse other nuclear fuels. As we learned in the last section, the ignition of these fuels requires high temperatures, because the nuclei have large positive charges. The particles must travel at high velocities to overcome the electrostatic repulsion and force the particles close enough together to react. Helium fusion requires a temperature of at least 100 million K, and carbon fusion requires 600 million K.

We can summarize the helium-fusion process in two steps:

$$^4He + {}^4He \rightarrow {}^8Be + \gamma$$
$$^8Be + {}^4He \rightarrow {}^{12}C + \gamma$$

Because a helium nucleus is called an alpha particle, these reactions are commonly known as the **triple-alpha process.** Helium fusion is complicated by the fact that beryllium-8, produced in the first reaction of the process, is very unstable and may break up into two helium nuclei before it can absorb another helium nucleus. Three helium nuclei can also form carbon directly, but such a triple collision is unlikely.

At temperatures above 600,000,000 K, carbon fuses rapidly in a complex network of reactions illustrated in Figure 9-18, where each arrow represents a different nuclear reaction. The process is complicated because nuclei can react by adding a proton, a neutron, or a helium nucleus or by combining directly with other nuclei. Unstable nuclei can decay by ejecting an electron, a positron, or a helium nucleus or by splitting into fragments. The complexity of this process makes it difficult to determine exactly how much energy will be generated and how many heavy atoms will be produced.

Reactions at still higher temperatures can convert magnesium, aluminum, and silicon into yet heavier atoms. These reactions involving heavy elements will be important in the study of the deaths of massive stars in the next chapter.

The Pressure–Temperature Thermostat

Nuclear reactions in stars manufacture energy and heavy atoms under the supervision of a built-in thermostat that keeps the reactions from erupting out of control. That thermostat is the relation between gas pressure and temperature.

In a star, the nuclear reactions generate just enough energy to balance the inward pull of gravity. If the reactions begin to produce too much energy, the extra energy raises the internal temperature of the star. Because the pressure of the gas depends on its temperature, the pressure also rises. The increased pressure makes the star expand. Expansion of the gas cools the star slightly, slowing the nuclear reactions. Thus the star has a built-in regulator that keeps the nuclear reactions from generating too much energy.

The same thermostat also keeps the reactions from dying down. If the nuclear reactions begin to produce too little energy, the inner temperature decreases, lowering the pressure and allowing gravity to compress the star slightly. As the gas is compressed, it heats up, increasing the nuclear energy generation.

The stability of a star depends on this relation between gas pressure and temperature. If an increase or decrease in temperature produces a corresponding change in pressure, then the thermostat functions correctly and the star is stable. We will see in this chapter how the thermostat accounts for the mass–luminosity relation. In the next chapter, we will see what happens to a star when the thermostat breaks down completely and the nuclear fires rage unregulated.

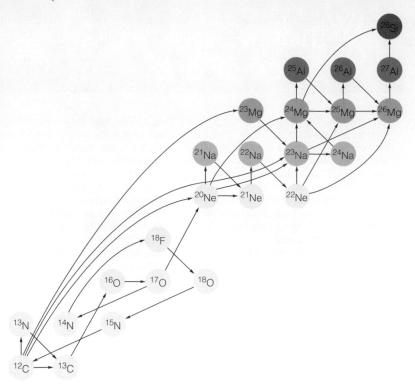

Figure 9-18
Carbon fusion involves many possible reactions that build numerous heavy atoms.

REVIEW Critical Inquiry

Why does nuclear fusion require that the gas be very hot?

Only under certain conditions do the nuclei of atoms fuse together to form a new nucleus. Inside a star, the gas is ionized, which means the electrons have been stripped off the atoms, and the nuclei are bare and have a positive charge. For hydrogen fusion, the nuclei are single protons. These atomic nuclei repel each other because of their positive charges, so they must collide with each other violently to overcome that repulsion and get close enough together to fuse. If the atoms in a gas are moving rapidly, we say it has a high temperature, and so nuclear fusion requires that the gas have a very high temperature. If the gas is cooler than about 10 million K, hydrogen can't fuse because the protons don't collide violently enough to overcome the repulsion of their positive charges.

The ignition of hydrogen fusion requires a minimum temperature. Why does the ignition of more massive nuclei require even higher temperatures?

The law of gravity and the rules of nuclear fusion determine how stars work and allow us to understand what the inside of a star is like. In other words, we can describe the internal structure of a star.

9-3 Stellar Structure

The nuclear fusion at the center of stars heats their interior, creates high gas pressure, and thus balances the inward force of gravity. If there is a single idea in modern astronomy that can be called critical, it is this concept of balance. Stars are simple, elegant, power sources held together by their own gravity and supported by their nuclear fusion.

Having discussed the birth of stars and their generation of energy, we can now consider the structure of a star—that is, the variation in temperature, density, pressure, and so on from the surface of the star to its center. A star's structure depends on how it generates its energy, on four simple laws of structure, and on what it is made of.

It will be easier to think about stellar structure if we imagine that the star is divided into concentric shells like those in an onion. We can then discuss the temperature, density, pressure, and so on in each shell. Of course, these helpful shells do not really exist; stars have no separable layers.

The Laws of Mass and Energy

The first two laws of stellar structure have something in common—they are both conservation laws. They tell us that certain things cannot be created out of nothing or vanish into nothing. Such conservation laws are powerful tools to help us understand how nature works.

The **conservation of mass** law is a basic law of nature, which we can apply to the structure of stars. It says that the total mass of a star must equal the sum of the masses of its shells. This is like saying the weight of a cake must equal the sum of the weight of its layers.

The **conservation of energy** law is another basic law of nature. It says that the amount of energy flowing out of the top of a layer in the star must be equal to the amount of energy coming in at the bottom plus whatever energy is generated within the layer. That means that the energy leaving the surface of the star, its luminosity, must equal the sum of the energies generated in all the layers inside the star. This is like saying that all the new cars driving out of a factory must equal the sum of all the cars made on each of the production lines.

These two laws may seem so familiar and so obvious we hardly need to state them, but they are important clues to the structure of stars. The third law of stellar structure is familiar because we have been using a closely related law in the preceding sections.

Hydrostatic Equilibrium

When we think about a star, it is helpful to think of it as if it were made up of layers. The weight of each layer must be supported by the layer below. The deeper layers must support the weight of all of the layers above. Because the inside of star is made up of gas, the weight pressing down on a layer must be balanced by the gas pressure in the layer. If the pressure is too low, the weight from above will compress the layer, and if the pressure is too high, the layer will expand and lift the layers above.

This balance between weight and pressure is called **hydrostatic equilibrium**. *Hydro* (from the Greek word for water) tells us the material is a fluid, the gases of a star, and *static* tells us the fluid is stable, neither expanding or contracting.

Figure 9-19 shows this hydrostatic balance in the imaginary layers of a star. The weight pressing down on each layer is shown by blue arrows, and they grow larger with increasing depth because the weight grows larger. The pressure in each layer is shown by green arrows, and they must grow larger with increasing depth to support the weight.

The law of hydrostatic equilibrium can tell us something important about the inside of a star. The pressure in a gas depends on the temperature and density of the gas. Near the surface, there is little weight pressing down, so the pressure need not be high. Deeper in the star, the pressure must be higher, and that means that the temperature and density of the gas must also be higher. Hydrostatic equilibrium

Figure 9-19
The law of hydrostatic equilibrium says the pressure in each layer of a stable star balances the weight on that layer. Thus, as the weight increases from the surface of a star to its center, the pressure also increases.

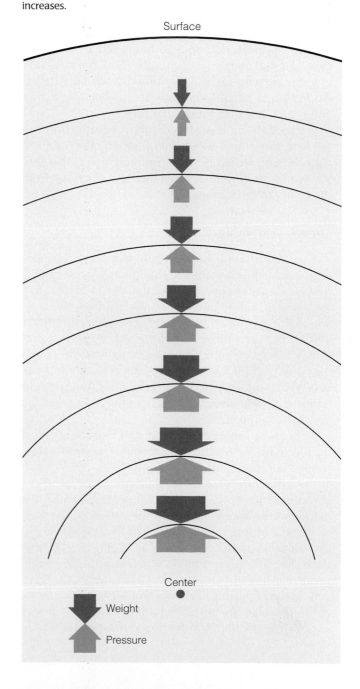

tells us that stars have to be hot inside to maintain the pressure needed to support their own weight. The layers are hot because, as we have seen, energy flows outward from the core of the star where it is made by nuclear fusion.

Now we recognize hydrostatic equilibrium. It is closely related to the pressure–temperature thermostat we have discussed before.

Of course, how hydrostatic equilibrium actually works depends on what an object is made of. Nearly all stars are made of hydrogen and helium gas, with some trace of heavier elements. To understand how a star works, we must describe how that gas responds to changes in temperature and pressure. Hydrostatic equilibrium also applies to planets, including Earth, but Earth is made of rock and metal, so again, understanding how hydrostatic equilibrium supports Earth requires that Earth scientists know how rock and metal respond to changes in temperature and pressure. We will see in later chapters how the structure of a star such as a white dwarf is dramatically different from that of a star because its gas behaves differently from that in most normal stars.

Although the law of hydrostatic equilibrium can tell us some things about the inner structure of stars, we need one more law to completely describe a star. We need a law that describes the flow of energy from the center to the surface.

Energy Transport

The surface of a star radiates light and heat into space and would quickly cool if that energy were not replaced. Because the inside of the star is hotter than the surface, energy must flow outward from the core, where it is generated, to the surface, where it radiates away. This flow of energy through the shells determines their temperature, which, as we saw previously, determines how much weight each shell can balance. To understand the structure of a star, we must understand how energy moves from the center through the shells to the surface.

The law of **energy transport** says that energy must flow from hot regions to cooler regions by conduction, convection, or radiation.

Conduction is the most familiar form of heat flow. If you hold the bowl of a spoon in a candle flame, the handle of the spoon grows warmer. Heat, in the form of motion among the molecules of the spoon, is conducted from molecule to molecule up the handle, until the molecules of metal under your fingers begin to move faster and you sense heat (Figure 9-20). Thus, conduction requires close contact between the molecules. Because the particles (atoms) in most stars are not in close contact, conduction is unimportant. Con-

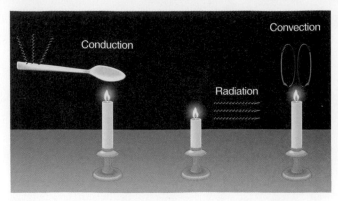

Figure 9-20
The three modes by which energy may be transported from the flame of a candle, as shown here, are the three modes of energy transport within a star.

duction is significant in white dwarfs, which have tremendous internal densities.

The transport of energy by radiation is another familiar experience. Put your hand beside a candle flame, and you can feel the heat. What you actually feel are infrared photons radiated by the flame (Figure 9-20). Because photons are packets of energy, your hand grows warm as it absorbs them.

Radiation is the principal means of energy transport in the sun's interior (Figure 9-21). Photons are absorbed and re-emitted in random directions over and over as they work their way outward. This process breaks the high-energy photons common near the sun's center into large numbers of low-energy photons in the cooler outer layers. Thus, the energy of a single high-energy photon at the center of the sun takes about 1 million years to reach the sun's surface, where it emerges as roughly 1600 photons of visible light.

The flow of energy by radiation depends on how difficult it is for the photons to move through the gas. If the gas is cool and dense, the photons are more likely to be absorbed or scattered, and so the radiation does not get through easily. We would call such a gas opaque. In a hot, thin gas, the photons can get through more easily; such a gas is less opaque. The **opacity** of the gas, its resistance to the flow of radiation, depends strongly on its temperature.

If the opacity is high, radiation cannot flow through the gas easily, and it backs up like water behind a dam. When enough heat builds up, the gas begins to churn as hot gas rises upward and cool gas sinks downward. This heat-driven circulation of a fluid is convection, the third way energy can move in a star. You are familiar with convection; the rising wisp of smoke above a candle flame is carried by convection. Energy is carried upward in these convection currents as rising hot gas (red in Figure 9-20) and also as sinking cool gas (blue in Figure 9-20). In stars, en-

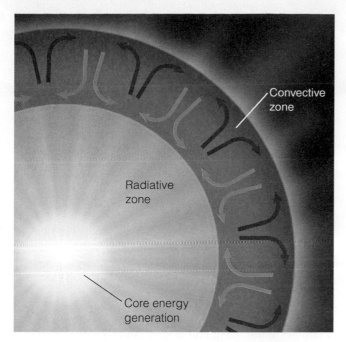

Figure 9-21
A cross section of the sun. Near the center, nuclear fusion reactions generate high temperatures. Energy flows outward through the radiation zone as photons. In the cooler, more opaque outer layers, the energy is carried outward by rising convection currents of hot gas (red) and sinking currents of cooler gas (blue).

Table 9-1 The Four Laws of Stellar Structure

I. Conservation of mass	Total mass equals the sum of of shell masses.
II. Conservation of energies	Total luminosity equals the sum of energy generated in each shell.
III. Hydrostatic equilibrium	The weight on each layer is balanced by the pressure in that layer.
IV. Energy transport	Energy moves from hot to cool by conduction, radiation, or convection.

ergy can be carried upward in convection currents hundreds or thousands of miles in diameter.

Convection is important in stars because it carries energy and because it mixes the gas. Convection currents flowing through the layers of a star tend to homogenize the gas, giving it a uniform composition throughout the convective zone. As you might expect, this mixing affects the fuel supply of the nuclear reactions, just as the stirring of a campfire makes it burn more efficiently.

The four laws of stellar structure are summarized in Table 9-1. These laws, properly understood, can tell us how stars are born, how they live, and how they die.

Stellar Models

The laws of stellar structure, described in general terms in the previous section, can be written as mathematical equations. By solving those equations in a special way, astronomers can build a mathematical model of the inside of a star.

If we wanted to build a model of a star, we would have to divide the star into about 100 concentric shell, and then write down the four equations of stellar structure for each shell. We would then have 400 equations that would have 400 unknowns, namely, the temperature, density, mass, and energy flow in each shell. Solving 400 equations simultaneously is

not easy, and the first such solutions, done by hand before the invention of the electronic computer, took months of work. Now a properly programmed computer can solve the equations in a few seconds and print a table of numbers that represent the conditions in each shell of the star. Such a table is a **stellar model** (Window on Science 9-3).

The table shown in Figure 9-22 is a model of the sun. The bottom line, for radius equal to 0.00, represents the center of the sun, and the top line, for radius equal to 1.00, represents the surface. The other lines tell us the temperature and density in each shell, the mass inside each shell, and the fraction of the sun's luminosity flowing outward through the shell. We can use the table to understand the sun. The bottom line tells us the temperature at the center of the sun is about 15 million Kelvin. At such a high temperature the gas is highly transparent, and energy flows as radiation. Nearer the surface, the temperature is lower, the gas is more opaque, and the energy is carried by convection. This is confirmed by the evidence of convection visible on the sun's surface (Chapter 7).

Stellar models also let us look into a star's past and future. In fact, we can use models as time machines to follow the evolution of stars over billions of years. To look into a star's future, for instance, we use a stellar model to determine how fast the star uses its fuel in each shell. As the fuel is consumed, the chemical composition of the gas changes and the amount of energy generated declines. By calculating the rate of these changes, we can predict what the star will look like at any point in the future.

Although this sounds simple, it is actually a highly challenging problem involving nuclear and atomic physics, thermodynamics, and sophisticated computational methods. Only since the 1950s have electronic computers made the rapid calculation of stellar models possible, and the advance of astronomy since then has been heavily influenced by the use of such models to study the structure and evolution of stars. Our summary of star formation in this chapter

Quantitative Thinking with Mathematical Models

One of the most powerful tools in science is the mathematical model, a group of equations carefully designed to mimic the behavior of the object that scientists want to study. Astronomers build mathematical models of stars using only four equations, but other systems are much more complicated and may require many more equations.

Many sciences use mathematical models. Medical scientists have built mathematical models of the nerves that control the human heart, and physicists have built mathematical models of the inside of an atomic nucleus. Economists have built mathematical models of certain aspects of economic systems, such as the municipal bond market, and Earth scientists have built mathematical models of Earth's atmosphere. In each case, the mathematical model allows the scientists to study something that is difficult to study in reality. The model can reveal regions we cannot observe, speed up a slow process, slow down a fast process, or allow scientists to perform experiments that would be impossible in reality. Astronomers, for example, can change the abundance of different chemical elements in a model star to see how its structure depends on its composition.

Many of these mathematical models require very large and fast computers—in some cases, supercomputers. A modern desktop computer takes only seconds to compute a stellar model; but a model of the motions of the planets of our solar system millions of years into the future would require the largest and fastest computers in the world. Most astronomers are apt computer programmers, and some are experts. A few astronomers have built their own highly specialized computers to compute specific kinds of models. Thus, the mathematical model is one of the most important tools in astronomy.

As is true for any scientific model, a mathematical model is only as reliable as the assumptions that go into its creation. In Window on Science 2-1, we saw that the celestial sphere is an adequate model of the sky for some purposes but breaks down if we extend it too far. So too with mathematical models. We can think of a mathematical model as a numerical expression of one or more theories. While such models can be very helpful, they are always based on theories and assumptions and so must be compared with the real world at every opportunity. As always in science, there is no substitute for a careful comparison with reality.

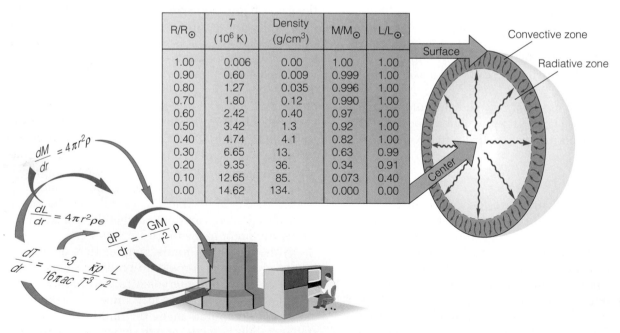

R/R$_\odot$	T (10^6 K)	Density (g/cm^3)	M/M$_\odot$	L/L$_\odot$
1.00	0.006	0.00	1.00	1.00
0.90	0.60	0.009	0.999	1.00
0.80	1.27	0.035	0.996	1.00
0.70	1.80	0.12	0.990	1.00
0.60	2.42	0.40	0.97	1.00
0.50	3.42	1.3	0.92	1.00
0.40	4.74	4.1	0.82	1.00
0.30	6.65	13.	0.63	0.99
0.20	9.35	36.	0.34	0.91
0.10	12.65	85.	0.073	0.40
0.00	14.62	134.	0.000	0.00

$$\frac{dM}{dr} = 4\pi r^2 \rho$$

$$\frac{dL}{dr} = 4\pi r^2 \rho e$$

$$\frac{dP}{dr} = -\frac{GM}{r^2}\rho$$

$$\frac{dT}{dr} = \frac{-3}{16\pi ac}\frac{\bar{\kappa}\rho}{T^3}\frac{L}{r^2}$$

Figure 9-22

A stellar model is a table of numbers that represents conditions inside a star. Such tables can be computed using the four laws of stellar structure, shown here in mathematical form. The table in this figure describes the sun. *(Illustration design by author)*

is based on thousands of stellar models. We will continue to rely on theoretical models as we study main-sequence stars in the next section and the deaths of stars in the next chapter.

R E V I E W Critical Inquiry

What would happen if the sun stopped generating energy?

Stars are supported by the outward flow of energy generated by nuclear fusion in their interiors. That energy keeps each layer of the star just hot enough so that the gas pressure can support the weight of the layers above. Each layer in the star must be in hydrostatic equilibrium; that is, the inward weight is balanced by outward pressure. If the sun stopped making energy in its interior, nothing would happen at first, but over many thousands of years the loss of energy from its surface would reduce the sun's ability to withstand its own gravity, and it would begin to contract. We wouldn't notice much for 100,000 years or so, but eventually, the sun would lose its battle with gravity.

Stars are elegant in their simplicity. Nothing more than a cloud of gas held together by gravity and warmed by nuclear fusion, a star can achieve stability, balancing its weight by generating nuclear energy. But how does the star manage to make exactly the right amount of energy to support its weight?

We have traced the birth of stars from the first instability in the interstellar medium to the final equilibrium of nuclear fusion and gravity. Now we must look carefully at these stable stars and ask two questions. First, what evidence do we have that stars really work this way? And second, how long can these stars remain stable? The answers to these two questions will give us a new insight into the lives of the stars.

9-4 Main-Sequence Stars

When a contracting protostar begins to fuse hydrogen, it stops contracting and becomes stable. These stable stars are located on the main sequence in the H-R diagram. Since 90 percent of all stars, including the sun, are main-sequence stars, it is important for us to understand their structure.

The Mass–Luminosity Relation

Observations of the temperature and luminosity of stars show that main-sequence stars obey a simple rule—the more massive a star is, the more luminous it is. That rule, the mass–luminosity relation discussed in the previous chapter, is the key to understanding the stability of main-sequence stars. In fact, the mass–luminosity relation is predicted by the theory of stellar structure, giving us direct observational confirmation of the theories of stellar structure.

To understand the mass–luminosity relation, we must consider the law of hydrostatic equilibrium, which says that pressure balances weight, and the pressure–temperature thermostat, which regulates energy production. A star that is more massive than the sun has more weight pressing down on its interior, so the interior must have a high pressure to balance that weight. That means the massive star must have its pressure–temperature thermostat turned up to keep the gas in its interior hot and the pressure high. A star less massive than the sun has less weight on its interior, needs less internal pressure, and so has its pressure–temperature thermostat set lower. To sum up, massive stars are more luminous because they must support more weight by making more energy.

This tells us why the main sequence must have a lower end. Masses below about 0.08 solar mass cannot raise their central temperature high enough to begin hydrogen fusion. Called **brown dwarfs**, such objects are only about a dozen times larger than Earth, still warm from contraction, and emit strongly in the infrared. Searches for brown dwarfs have found a few examples. The star GL229, for example, is orbited by an object of about 0.02 $M_\odot$. The low mass of the object and the presence of methane in its spectrum confirm that it cannot be a star. Such objects are difficult to find and study because of their very low temperature, but it seems that nature does indeed make brown dwarfs when a forming star does not have enough mass to begin hydrogen fusion. This observational detection of the lower end of the main sequence further confirms the theories of stellar structure.

Now that we know how stars form, and how they maintain their stability through the mass–luminosity relation, we can predict the evolution of main-sequence stars.

The Life of a Main-Sequence Star

While a star is on the main sequence, it is stable, so we might think its life would be totally uneventful. But a main-sequence star balances its gravity by fusing hydrogen, and as the star gradually uses up its fuel, that balance must change. Thus, even the stable main-sequence stars are changing as they consume their hydrogen fuel.

Recall that hydrogen fusion combines four nuclei into one. Thus, as a main-sequence star fuses its hydrogen, the total number of particles in its interior decreases. Each newly made helium nucleus can exert the same pressure as a hydrogen nucleus, but because

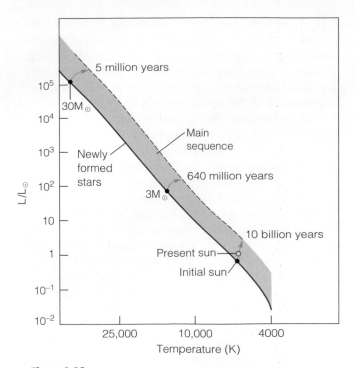

Figure 9-23
The main sequence is not a line but a band (shaded). Newly formed stars begin fusing hydrogen on the lower edge (red). As hydrogen changes their composition, the stars slowly move across the band. Notice that a star of 3 solar masses crosses the band in less than a tenth the time the sun will take, and that a star of 30 solar masses will cross the band in less than a thousandth the time the sun will take.

Table 9-2 Main-Sequence Stars			
Spectral Type	**Mass (Sun = 1)**	**Luminosity (Sun = 1)**	**Years on Main Sequence**
O5	40	405,000	1×10^6
B0	15	13,000	11×10^6
A0	3.5	80	440×10^6
F0	1.7	6.4	3×10^9
G0	1.1	1.4	8×10^9
K0	0.8	0.46	17×10^9
M0	0.5	0.08	56×10^9

the gas has fewer nuclei, its total pressure is less. This unbalances the gravity–pressure stability, and gravity squeezes the core of the star more tightly. As the core contracts, its temperature increases and the nuclear reactions run faster, releasing more energy. This additional energy flowing outward through the envelope forces the outer layers to expand. As the star becomes larger, it becomes more luminous, and eventually the expansion begins to cool the surface.

As a result of these gradual changes in main-sequence stars, the main sequence is not a sharp line across the H-R diagram but rather a band. Stars begin their stable lives fusing hydrogen on the lower edge of this band, but gradual changes in luminosity and surface temperature move the stars upward and slightly to the right as shown in Figure 9-23. By the time they reach the upper edge of the main sequence, they have exhausted nearly all the hydrogen in their centers. Thus we find main-sequence stars scattered throughout the band at various stages of their main-sequence lives.

The sun is a typical main-sequence star, and as it undergoes these gradual changes, Earth will suffer. When the sun began its main-sequence life about 5 billion years ago, it was only about 70 percent as luminous as it is now, and by the time it leaves the main sequence in another 5 billion years, the sun will have twice its present luminosity. This will raise the average tempera-

ture on Earth by at least 19°C (34°F). As this happens over the next few billion years, the polar caps will melt, and Earth's climate will change dramatically. Life on Earth may not survive this change in the sun.

The average star spends 90 percent of its life on the main sequence. This explains why 90 percent of all stars are main-sequence stars—we are most likely to see a star during that long, stable period while it is on the main sequence. To illustrate: Imagine that you photograph, at a single instant, a crowd of 20,000 people. We all sneeze now and then, but the act of sneezing is very short compared to a human lifetime, so we would expect to find very few people in our photograph in the act of sneezing. By contrast, because the main-sequence phase is very long compared to a stellar lifetime, we can expect to see many stars on the main sequence.

The number of years a star spends on the main sequence depends on its mass (Table 9-2). Massive stars consume fuel rapidly and live short lives, but low-mass stars conserve their fuel and shine for billions of years. For example, a 25-solar-mass star will exhaust its hydrogen and die in only about 7 million years. Very-low-mass stars, the red dwarfs, use up their fuel so slowly that they last for 200 to 300 billion years. Because the universe seems to be only 10 to 20 billion years old, red dwarfs must still be in their infancy. (By the Numbers 9-1 explains how we can quickly estimate the life expectancies of stars from their masses.)

Nature makes more low-mass stars than high-mass stars, but this fact is not sufficient to explain the vast numbers of low-mass stars that fill the sky. An additional factor is the stellar lifetime. Because low-mass stars live long lives, there are more of them in the sky than massive stars. Look at Figure 8-22, and notice how much more common the lower-main-sequence stars are than the massive O and B stars. The main-sequence K and M stars are so faint they are difficult to locate, but they are very common. The O and B stars are luminous and easy to locate, but, because of their fleeting lives, there are never more than a few on the main sequence at any one time.

We can estimate the amount of time a star spends on the main sequence—its life expectancy T—by estimating the amount of fuel it has and the rate at which it consumes that fuel:

$$T = \frac{\text{fuel}}{\text{rate of consumption}}$$

The amount of fuel a star has is proportional to its mass M, and the rate at which it uses up its fuel is proportional to its luminosity L. Thus its life expectancy must be proportional to M/L. But we can simplify this equation further because, as we saw in the last chapter, the luminosity of a star depends on its mass raised to the 3.5 power ($L = M^{3.5}$). Thus the life expectancy is:

$$T = \frac{M}{M^{3.5}}$$

or

$$T = \frac{1}{M^{2.5}}$$

If we express the mass in solar masses, the lifetime will be in solar lifetimes.

Example: How long can a 4-solar-mass star live?
Solution:

$$T = \frac{1}{4^{2.5}} = \frac{1}{4 \cdot 4\sqrt{4}}$$

$$= \frac{1}{32} \text{ solar lifetimes}$$

Studies of solar models show that the sun, presently 5 billion years old, can last another 5 billion years. Thus a solar lifetime is approximately 10 billion years, and a 4-solar-mass star will last for about:

$$T = \frac{1}{32} \times (10 \times 10^9 \text{ yr})$$

$$= 310 \times 10^6 \text{ years}$$

R E V I E W   Critical Inquiry

Why are main-sequence O stars so rare compared with stars like the sun (G stars)?

In Section 8-5, we discussed the family of stars and noticed that space contains only about 0.036 main-sequence O stars per million cubic parsecs, but that there are about 2000 main-sequence G stars per million cubic parsecs (see Figure 8-22). Dividing, we discover that main-sequence G stars are over 50,000

times more common than O stars. Our discussion of the lives of main-sequence stars gives us a way to analyze these numbers. O stars live very short lifetimes. Our simple formula predicts a 20-solar-mass O star has a life expectancy of about 6 million years. The sun, however, will live a total of about 10 billion years or about 1600 times longer than an O star. Assuming nature makes equal numbers of different kinds of stars, we would expect to see about 1600 times more main-sequence G stars than O stars. But we see over 50,000 times more, so there must be another factor involved. Nature must be making vastly more lower-mass stars than massive stars. By comparing theory and observation, our analysis has revealed something new about how nature makes stars.

Now we can understand why O stars are so rare. What factors must you take into consideration to explain why main-sequence M stars are even more common than G stars?

Science is a way of understanding nature, and although we use theories, we must always go back to reality and compare our theories with observations. We have constructed an elaborate theory to explain how stars are born and reach stability and checked it a number of times with observations; but there is one fascinating and beautiful nest of newborn stars we should study before we proceed. The Great Nebula in Orion is our final reality check on the formation of stars.

9-5 The Orion Nebula

On a clear winter night, you can see with your naked eye the Great Nebula of Orion as a fuzzy wisp in Orion's sword. With binoculars or a small telescope it is striking, and through a large telescope it is breathtaking. At the center lie four brilliant blue-white stars known as the Trapezium, and surrounding them are the glowing filaments of a nebula more than 8 pc across (Figure 9-24). Like a great thundercloud illuminated from within, the churning currents of gas and dust suggest immense power. A deeper significance lies hidden, figuratively and literally, behind the visible nebula, for radio and infrared astronomers have discovered a vast dark cloud just beyond the visible nebula—a cloud in which stars are now being created.

Evidence of Young Stars

We should not be surprised to find star formation in Orion. Many of the stars in the constellation are hot, blue, upper-main-sequence stars that have short lifetimes. They must have formed recently. The region is

Figure 9-24
The Great Nebula in Orion is a glowing cloud of gas and dust over 8 pc in diameter. The hydrogen glows red because it is ionized by the hottest of the stars of the Trapezium (arrow), a cluster of young, hot stars just visible at the very center of the nebula. Compare with Figure 9-27. *(Daniel Good)*

luminous—nearly 40,000 K and over 300,000 solar luminosities. The other stars are too cool to affect the surrounding gas very much, but the massive star is so hot it radiates large amounts of ultraviolet radiation. These ultraviolet photons have enough energy to ionize hydrogen near the Trapezium (Figure 9-26). Thus, the glowing gas of the nebula tells us that at least one massive star lies inside, and such massive stars are very short-lived. It must be very young.

The Orion Nebula looks impressive, but it is nearly a vacuum, containing a mere 6000 atoms/cm^3. For comparison, the interstellar medium contains about 1 atom/cm^3 and air at sea level about 10^{19} atoms/cm^3. Infrared observations show that the nebula also contains sparsely scattered dust warmed by the stars to about 70 K.

The Molecular Cloud

The significance of the Orion Nebula was established when observations at infrared and radio wavelengths revealed dense clouds of gas and dust just beyond the Great Nebula. At these wavelengths, hot stars and ionized gas are invisible, but cool, dense gas and dust can be detected. In space, ultraviolet photons break up most molecules, but hidden deep in the dark clouds, where ultraviolet photons cannot penetrate, molecules such as carbon monoxide have formed. These molecules emit radio photons of characteristic wavelengths that allow radio astronomers to map these molecular clouds. Dust heated by stars radiates strongly at infrared wavelengths, and thus infrared telescopes can map the location of the dust.

These clouds are significantly different from the Great Nebula. The molecular clouds contain at least 10^6 atoms/cm^3 and large amounts of dust. Because the clouds seem warmest near their centers, astronomers conclude they are contracting.

The Infrared Objects

Infrared observations reveal strong infrared sources hidden within the dusty molecular cloud behind the visible nebula. Invisible at optical wavelengths, these objects are evidently stars in pre-main-sequence stages, wrapped deep in cocoons of dust. Three especially interesting objects lie almost directly behind the visible galaxy (Figure 9-27).

The Becklin-Neugebauer object (BN object), named after its discoverers, was first thought to be a protostar, but infrared and radio observations show that the gas near its center is ionized and flowing out-

also rich in T Tauri stars, which are known pre-main-sequence stars. The Hubble Space Telescope has found jets of gas streaming away from some stars, evidence that the stars are young and surrounded by disks. In fact, other Hubble Space Telescope images reveal that many stars in the region have these disks (Figure 9-25). Furthermore, infrared observations show streamers of gas flowing away from a star that has recently become luminous. All of these observations are evidence of star formation.

In fact, the glowing nebula itself is evidence of recent star formation. The nebula is an ionized hydrogen region excited by young stars, the Trapezium stars, which reached the main sequence no more than a million years ago. The most massive, a star of about 30 solar masses, must consume its fuel at a tremendous rate to support its large mass. Thus, it is hot and

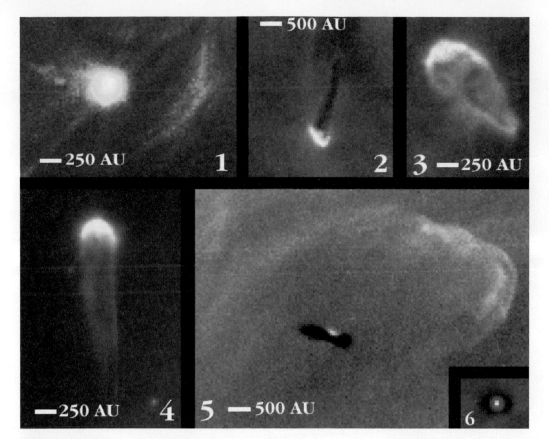

Figure 9-25

Disks of gas and dust around young stars near the hot stars of the Trapezium in the Orion Nebula. The first four images show disks that are being evaporated by the heat from the nearby hot stars with the resulting gas being blown away to produce elongated tails. The last two images show dust disks seen silhouetted against the bright nebula. Bars in images are labeled in astronomical units. The diameter of our solar system taken as the diameter of Pluto's orbit is about 80 AU. *(John Bally, Dave Devine, and Ralph Sutherland and NASA)*

ward. Thus, the object is probably a hot B star that has just reached the main sequence.

The Kleinmann-Low nebula is much cooler than the BN object. The KL nebula is thought to be a contracting cloud of gas in an early stage of star formation.

Near the B-N object and the K-L nebula lies a strong infrared source called IRc2. Radio observations show that the object is surrounded by a thick, expanding disk of gas and is ejecting a powerful bipolar flow. Theory predicts that the object cannot sustain such loss of mass for more than about 1000 years, so it must be a very young protostar. (See Figure 9-25c.)

Telling the Story of Star Formation in Orion

The goal of science is to use theory supported by evidence to describe how nature works. Modern astronomers can now tell the story of the Orion Nebula with some confidence. The history of star formation in Orion is complex, but we can find evidence that the

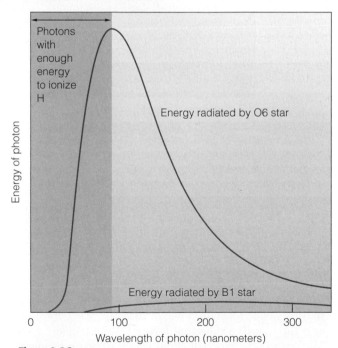

Figure 9-26

Photons with wavelengths shorter than 91.2 nm have enough energy to ionize hydrogen. The O6 star is the only star in the Trapezium hot enough to produce appreciable ionization.

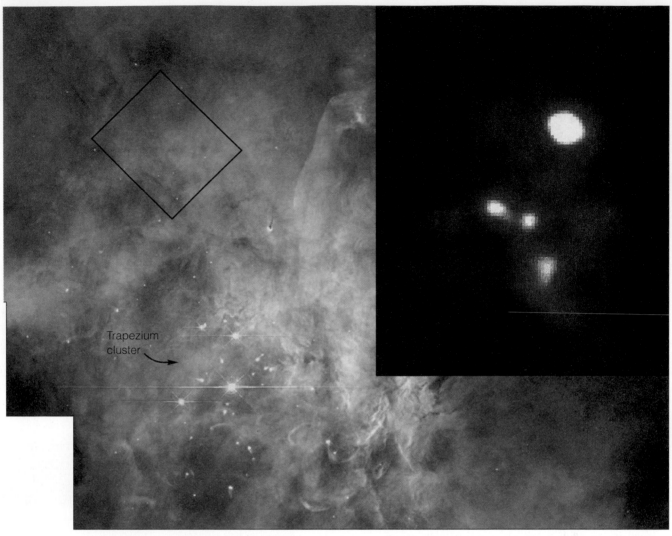

Figure 9-27
The central 2.5 ly of the Great Nebula in Orion shown in this Hubble Space Telescope image contains the four stars of the Trapezium (arrow) and a complex tangle of gas clouds. *(John Bally, Dave Devine, Ralph Sutherland and NASA)* The inset at right shows the boxed area of the nebula at the infrared wavelength of 12.4 microns. Undetectable at visible wavelengths, the infrared sources are evidence of star formation going on in the molecular cloud behind the visible nebula. *(Dan Gezari, Dana Backman, and Mike Werner)*

entire region has been a site of star formation for at least three generations of massive stars. The ages of the stars tell the story. The stars at Orion's west shoulder are about 12 million years old, while the stars of Orion's belt are about 8 million years old. The stars of the Trapezium are no older than 2 million years. Apparently, star formation began near the west shoulder and the massive stars that formed there triggered star formation near the belt. That star formation triggered the formation of the stars we see in the Great Nebula. Like a grass fire, star formation has swept across Orion from northwest to southeast.

We can read the story of the Great Nebula in the Trapezium stars and the molecular cloud that lies behind it. Just a few million years ago, the Trapezium stars were contracting protostars buried within a molecular cloud. As they approached the main sequence,

their temperatures increased and their radiation drove away their cocoons. But not until the 40-solar-mass star turned on was there sufficient ultraviolet radiation to ionize the gas. The ionization transformed the cloud from a cold, dusty cocoon into a hot, transparent bubble of gas expanding away from the Trapezium. We can see evidence of this expansion in the Doppler shifts of spectral lines and in the twisted filaments of gas within the nebula.

Once the hot star ionized the gas, star formation in the Trapezium region stopped. In more distant parts of the cloud, the contraction of the gas into protostars continues. Indeed, the ionized gases of the visible nebula pushing against the remains of the cloud may have triggered the collapse of more protostars. We now see the front of the cloud torn by the expanding ionized gases around the Trapezium cluster, while deeper within the cloud more protostars are forming (Figure 9-28).

In the next few thousand years, the familiar outline of the Great Nebula will change, and a new nebula may form as the protostars in the molecular clouds reach the main sequence and the most massive become hot enough to ionize the surrounding gas. Thus, the Great Nebula in Orion and its parent molecular cloud show how the formation of stars continues even today.

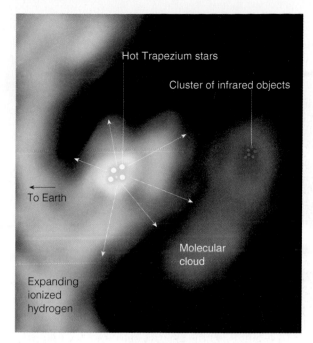

Figure 9-28
This drawing of a side view of the Great Nebula shows the newly formed Trapezium stars heating and driving away the surrounding gas as new stars form in the molecular cloud behind the visible nebula.

REVIEW Critical Inquiry

What did Orion look like to the ancient Egyptians, to the first humans, and to the dinosaurs?

The Egyptian civilization had its beginning only a few thousand years ago, and that is not very long in terms of the history of Orion. The stars we see in the constellation are hot and young, but they are a few million years old, so the Egyptians saw the same constellation we see. (They called it Osiris.) Even the Orion Nebula hasn't changed very much in a few thousand years, and Egyptians may have admired it in the dark skies along the Nile.

Our oldest human ancestors lived about 3 million years ago, and that was about the time that the youngest stars in Orion were forming. Thus, they may have looked up and seen some of the stars we see, but some stars have formed since that time. Also, the Great Nebula is excited by the Trapezium stars, and they are not more than a few million years old, so our early ancestors probably didn't see the Great Nebula.

The dinosaurs saw something quite different. The last of the dinosaurs died about 65 million years ago, long before the birth of the brightest stars in the constellation we see. The dinosaurs, had they had the brains to appreciate the view, might have seen bright stars along the Milky Way, but when we think back millions of years, we must remember that the stars are moving through space and the sun is orbiting the center of our galaxy. The night sky above the dinosaurs contained totally different star patterns.

Of course, a giant molecular cloud can't continue to spawn new stars forever. What processes limit star formation in a molecular cloud and will eventually end star formation in the Orion region?

The ancient Aztecs of central Mexico told the story of how the stars, known as the Four Hundred Southerners, were scattered across the sky when they lost a cosmic battle with their brother, the great war god Huitzilopochtli. Modern astronomy tells a less colorful story of the birth of stars, but the modern story is supported by evidence and leads us to ask further questions. If stars are born, then how do they die? We will begin that story in the next chapter.

Summary

We know there is an interstellar medium because of observational evidence. We can see emission, reflection, and dark nebulae, and we can detect interstellar absorption lines in the spectra of distant stars. The dust in the interstellar medium makes distant stars look fainter and redder than we would expect.

The interstellar medium contains large, dense, cool clouds of gas in which hydrogen exists as molecules rather than as atoms. These molecular clouds are sites of star formation. Such clouds can be triggered into collapse by collision with a shock wave, which compresses and fragments the gas cloud, producing a cluster of protostars.

A contracting protostar is large and cool, and would appear in the red-giant region of the H-R diagram were it not surrounded by a cocoon of dust and gas. The dust in the cocoon absorbs the protostar's light and reradiates it as infrared radiation. Many infrared sources are probably protostars. T Tauri stars may be cocoon stars in later stages as the cocoon is absorbed by the protostar or driven into space. Very young star clusters, like NGC2264, contain large numbers of T Tauri stars. Some protostars have been found emitting bipolar flows of gas as they expel material from their surroundings. Where these flows strike existing clouds of gas, we can see nebulae called Herbig-Haro objects.

As a protostar's center grows hot enough to fuse hydrogen, it settles onto the main sequence to begin its long hydrogen-fusion phase. If it is a low-mass star, it cannot become very hot and must generate energy by the proton–proton chain. But if it is a massive star, it will become hot enough to use the CNO cycle. In either case, the nuclear reactions are regulated by the pressure–temperature thermostat in the star's core.

Almost everything we know about the internal structure of stars comes from mathematical stellar models. The models are based on four simple laws of stellar structure. The first two laws say that mass and energy must be conserved and spread smoothly through the star. The third, hydrostatic equilibrium, says the star must balance the weight of its layers by its internal pressure. The fourth says energy can only flow outward by conduction, convection, or radiation.

The mass–luminosity relation is explained by the requirement that a star support the weight of its layers by its internal pressure. The more massive a star is, the more weight it must support and the higher its internal pressure must be. To keep its pressure high, it must be hot and generate large amounts of energy. Thus the mass of a star determines its luminosity. The massive stars are very luminous and lie along the upper main sequence. The less massive stars are fainter and lie lower on the main sequence.

How long a star can stay on the main sequence depends on its mass. The more massive a star is, the faster it uses up its hydrogen fuel. A 25-solar-mass star will exhaust its hydrogen and die in only about 7 million years, but the sun is expected to last for 10 billion years.

The Great Nebula in Orion is an active region of star formation. The bright stars we see in the center of the nebula formed within the last few million years, and infrared telescopes detect protostars buried inside the molecular cloud that lies behind the visible nebula.

New Terms

interstellar medium	strong force
nebula	nuclear fission
emission nebula	nuclear fusion
reflection nebula	Coulomb barrier
interstellar reddening	proton–proton chain
infrared cirrus	CNO (carbon–nitrogen–oxygen) cycle
molecular cloud	
association	neutrino
protostar	triple-alpha process
evolutionary track	conservation of mass
birth line	conservation of energy
T Tauri star	hydrostatic equilibrium
Herbig-Haro object	energy transport
bipolar flow	opacity
Bok globule	stellar model
weak force	brown dwarf

Review Questions

1. What evidence do we have that the spaces between the stars are not empty?
2. What evidence do we have that the interstellar medium contains both gas and dust?
3. Why would an emission nebula near a hot star look red, but a reflection nebula near its star looks blue?
4. Why do astronomers rely heavily on infrared observations to study star formation?
5. What observational evidence do we have that star formation is a continuous process?
6. How are Herbig-Haro objects related to star formation?
7. How do the proton–proton chain and the CNO cycle resemble each other? How do they differ?
8. Why does the CNO cycle require a higher temperature than the proton–proton chain?
9. What is the solar neutrino problem? How might it be resolved?
10. How does the pressure–temperature thermostat control the nuclear reactions inside stars?
11. Step-by-step, explain how energy flows from the sun's core to Earth.
12. Why is there a mass–luminosity relation?
13. Why is there a lower limit to the mass of a main-sequence star?
14. Why does a star's life expectancy depend on its mass?
15. What evidence do we have that star formation is happening right now in the Orion Nebula?

Discussion Questions

1. When we see distant streetlights through smog, they look dimmer and redder than they do normally. But when we see the same streetlights through fog or falling snow, they look dimmer but not redder. Use your knowledge of the interstellar medium to discuss the relative sizes of the particles in smog, fog, and snow compared to the wavelength of light.
2. If planets form as a natural by-product of star formation, which do you think are more common—stars or planets?

Problems

1. The interstellar medium dims starlight by about 1.9 magnitudes/1000 pc. What fraction of photons survive a trip of 1000 pc? (*Hint:* See By the Numbers 2-1.)

2. A small Bok globule has a diameter of 20 seconds of arc. If the nebula is 1000 pc from Earth, what is the diameter of globule?

3. If a giant molecular cloud has a diameter of 30 pc and drifts relative to neighboring clouds at 20 km/sec, how long will it take to travel its own diameter?

4. If the dust cocoon around a protostar emits radiation most strongly at a wavelength of 30 microns, what is the temperature of the dust? (*Hint:* See By the Numbers 6-1.)

5. The gas in a bipolar flow can travel as fast as 300 km/sec. How long would it take to travel 1 light-year?

6. How much energy is produced when the sun converts 1 kg of mass into energy? (Remember, 1 J/sec = 1 watt.) For how many years would this amount of energy keep a million 100-watt light bulbs burning?

7. How much energy is produced when the sun converts 1 kg of hydrogen into helium? (*Hint:* What fraction of the kilogram of mass vanishes?)

8. Circle all ^{1}H and ^{4}H nuclei in Figures 9-15 and 9-16. Explain how both the proton–proton chain and the CNO cycle can be summarized 4 ^{1}H → ^{4}He + energy.

9. In the model shown in Figure 9-22, how much of the sun's mass is hotter than 12,000,000 K?

10. What is the life expectancy of a 16 $M_\odot$ star?

11. The hottest star in the Orion Nebula has a surface temperature of 30,000 K. At what wavelength does it radiate the most energy? (*Hint:* See By the Numbers 6-1.)

$$\frac{3009999}{30,999} = \frac{300}{3} = 100\ k.$$

Critical Inquiries for the Web

1. If neutrinos are so elusive, how do astronomers detect them? Use an Internet search engine to browse for information on solar neutrino detectors that are currently in operation, under construction, or proposed. Determine similarities and differences between these detectors in terms of method of detection and energy range of detectable neutrinos.

2. Astronomers have been searching for brown dwarfs for years, but few candidates have been identified. Search the Web for information on efforts to locate these substellar objects. Why are they so difficult to detect? List any likely brown dwarfs found so far.

Go to the Brooks/Cole Astronomy Resource Center (www.brookscole.com/astronomy) for critical thinking exercises, articles, and additional readings from InfoTrac College Edition, Brooks/Cole's online student library.

CHAPTER 10

The Deaths of Stars

Natural laws have no pity.

Robert Heinlein
The Notebooks of Lazarus Long

Guidepost

As you read this chapter, take a moment to be astonished and proud that we humans know how stars die. Finding the mass of stars is difficult, and understanding the invisible matter between the stars takes ingenuity; but we human beings have used what we know about stars and what we know about the physics of energy and matter to figure out how stars die.

The death of stars is important to us because our own sun will die, but it is also important because the explosive deaths of massive stars blast atoms heavier than helium into space. The atoms we are made of were made inside stars. If stars didn't die, we would not exist.

In the chapters that follow, we will see that some matter from dying stars becomes trapped in dead ends—neutron stars and black holes—but some escapes into the interstellar medium and is incorporated into new stars and the planets that circle them. Thus, the deaths of stars are part of a cycle of stellar birth and death that includes our sun, our planet, and us.

The stars are dying. Occasionally astronomers see a new star appear in the sky, grow brighter, fade away after a few weeks or a year, and vanish. We will discover that a **nova,** a new star, is in fact a very old dying star, and a **supernova** is a massive star dying in a violent explosion that destroys the star. Modern astronomers find a few novae (plural of nova) each year, but supernovae (plural) are so rare centuries pass between supernovae bright enough to see with the unaided eye.

In the previous chapter, we saw that stars resist their own gravity by generating energy through nuclear fusion. The energy keeps their interiors hot, and the resulting high pressure balances gravity and prevents the star from collapsing. Stars, however, have limited fuel. When they exhaust their fuel, gravity wins, and the stars die.

The mass of a star is critical in determining its fate. Massive stars use up their nuclear fuel at a furious rate and die after only a few million years, while the lowest-mass stars use their fuels sparingly and may be able to live hundreds of billions of years. Massive stars can die in violent explosions that we see as supernovae, but low-mass stars die quiet deaths. To follow the evolution of stars to their graves, we will sort the stars into three categories according to their masses: low-mass red dwarfs, medium-mass sunlike stars, and massive upper-main-sequence stars. These stars lead dramatically different lives and die in different ways (Figure 10-1).

The death of a star leads invariably to one of three final states. Most stars, including stars like the sun, become white dwarfs, stars about the size of Earth with no usable fuels. The most massive stars explode and leave behind either a neutron star or a black hole, exotic objects we will study in detail in the next chapter.

The end states of stellar evolution—white dwarfs, neutron stars, and black holes—represent the final victory of gravity. Understanding stellar evolution and death means we must first understand how the star delays its inevitable collapse. We begin this chapter at the beginning of the end, the evolution of giant stars.

10-1 Giant Stars

A main-sequence star generates its energy by nuclear fusion reactions that combine hydrogen to make helium. The period during which the star fuses hydrogen lasts a long time, and the star remains on the main sequence for 90 percent of its total existence as a star.

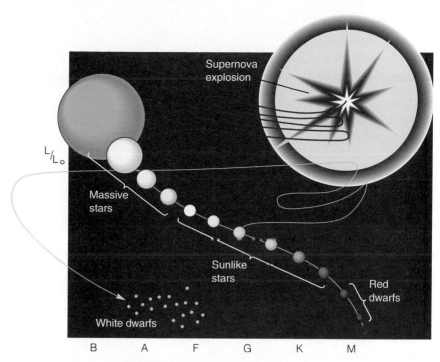

Figure 10-1

Three different modes of stellar death are summarized in this schematic H-R diagram. The lowest-mass stars, the red dwarfs, live long lives and eventually become white dwarfs. Medium-mass stars like the sun become giant stars and then collapse to become white dwarfs. The most massive stars evolve rapidly to become supergiants and explode as supernovae, leaving behind either a neutron star or a black hole.

When the hydrogen is exhausted, however, the star begins to evolve rapidly. It can delay its final collapse by fusing elements heavier than hydrogen and expanding to become a giant star, but that stage is short-lived. It can remain a giant star for only about 10 percent of its total lifetime; then it must die. Thus, the giant-star stage is the first step in the death of a star.

Expansion into a Giant

The nuclear reactions in a main-sequence star's core fuse hydrogen to produce helium. Because the core is cooler than 100,000,000 K, the helium cannot fuse in nuclear reactions, so it accumulates at the star's center like ashes in a fireplace. Initially, this helium ash has little effect on the star, but as hydrogen is exhausted and the stellar core becomes almost pure helium, the star loses the ability to generate nuclear energy. Because it is the energy generated at the center that opposes gravity and supports the star, the core begins to contract as soon as the energy generation starts to die down.

Although the contracting helium core cannot generate nuclear energy, it does grow hotter because it converts gravitational energy into thermal energy (see Chapter 9). The rising temperature heats the unprocessed hydrogen just outside the core, hydrogen that was never before hot enough to fuse. When the temperature of the surrounding hydrogen becomes high enough, hydrogen fusion begins in a spherical layer or shell around the exhausted core of the star. Like a grass fire burning outward from an exhausted campfire, the hydrogen-fusion shell creeps outward, leaving helium ash behind and increasing the mass of the helium core.

The flood of energy produced by the hydrogen fusion shell pushes toward the surface, heating the outer layers of the star, and forcing them to expand dramatically (Figure 10-2). Stars like the sun become giant stars 10 to 100 times the diameter of the sun, and the most massive stars become supergiants some 1000 times larger than the sun. This explains the large diameters and low densities of the giant and supergiant stars. In Chapter 8, we found that giant and supergiant stars were 10 to 1000 times larger in diameter than the sun and from 10 to 10^6 times less dense. Now we understand that these stars were once normal main-sequence stars that expanded to large size and low density when hydrogen shell fusion began.

The expansion of the envelope dramatically changes the star's location in the H-R diagram. As the outer layers of the star expand outward, the surface cools, and the point that represents the star in the H-R diagram moves quickly to the right (in less than a

Figure 10-2
When a star runs out of hydrogen at its center, the core contracts to a small size, becomes very hot, and begins nuclear fusion in a shell (red). The outer layers of the star expand and cool. The red giant star shown here has an average density much lower than the air at Earth's surface. Here $M_\odot$ stands for the mass of the sun and $R_\odot$ stands for the radius of the sun. *(Illustration design by author)*

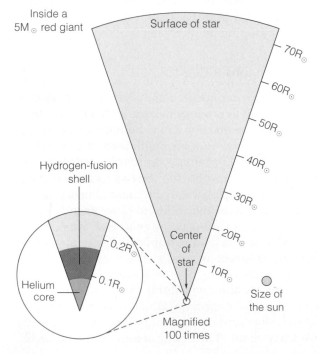

Figure 10-3
The evolution of a massive star moves the point that represents it in the H-R diagram to the right of the main sequence into the region of the supergiants such as Rigel and Betelgeuse. The evolution of a medium-mass star moves its point in the H-R diagram into the region of the giants such as those shown here.

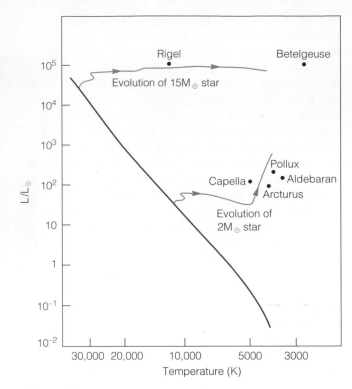

million years for a star of 5 solar masses). A massive star moves to the right across the top of the H-R diagram and becomes a supergiant, while a medium-mass star like the sun becomes a red giant (Figure 10-3). As the radius of a giant star continues to increase, the enlarging surface area makes the star more luminous, moving its point upward in the H-R diagram. Aldebaran, the glowing red eye of Taurus the Bull, is such a red giant, with a diameter 25 times that of the sun but with a surface temperature only half that of the sun.

Degenerate Matter

Although the hydrogen-fusion shell can force the envelope of the star to expand, it cannot stop the contraction of the helium core. Because the core is not hot enough to fuse helium, gravity squeezes it tighter, and it becomes very small. If we represent the helium core of a 5-solar-mass star with a tennis ball, the outer envelope of the star would be about the size of a baseball stadium. Yet the core would contain about 12 percent of the star's mass compressed to very high density. When gas is compressed to such extreme densities, it begins to behave in astonishing ways that can alter the evolution of the star. Thus, to follow the story of stellar evolution, we must consider the behavior of gas at extremely high densities.

Normally, the pressure in a gas depends on its temperature. The hotter the gas is, the faster its particles move and the more pressure it exerts. But the gas inside a star is ionized, so there are two kinds of particles: atomic nuclei and free electrons. If the gas is compressed to very high densities, for example, in the core of a giant star, the difference between these two kinds of particles becomes important.

If the density is very high, the particles of the gas are forced close together, and two laws of quantum mechanics become significant. First, quantum mechanics says that the moving electrons confined in the star's core can have only certain amounts of energy, just as the electron in an atom can occupy only certain energy levels (see Chapter 6). We can think of these permitted energies as the rungs of a ladder. An electron can occupy any rung, but not the spaces between.

The second quantum mechanical law (called the Pauli Exclusion Principle) says that two identical electrons cannot occupy the same energy level. Because electrons spin in one direction or the other, two electrons can occupy an energy level if they spin in opposite directions. That level is then completely filled, and a third electron cannot enter because, whichever way it spins, it will be identical to one or the other of the two electrons already in the level. Thus, no more than two electrons can occupy the same energy level.

A low-density gas has few electrons per cubic centimeter, so there are plenty of energy levels available (Figure 10-4). If a gas becomes very dense, however, nearly all of the lower energy levels may be occupied. In such a gas, a moving electron cannot slow down, because slowing down would decrease its energy, and there are no open energy levels for it to drop down to. It can speed up only if it can absorb enough energy to leap to the top of the energy ladder, where there are empty energy levels.

When a gas is so dense that the electrons are not free to change their energy, astronomers call it **degenerate matter.** Although it is a gas, it has two peculiar properties that can affect the star. First, the degenerate gas resists compression. To compress the gas, we must push against the moving electrons, and changing their motion means changing their energy. That requires tremendous effort, because we must boost them to the top of the energy ladder. Thus, degenerate matter, though still a gas, takes on the consistency of hardened steel.

Second, the pressure of degenerate gas does not depend on temperature but rather on the speed of the electrons, which cannot be changed without tremendous effort. The temperature, however, depends on the motion of all the particles in the gas, both elec-

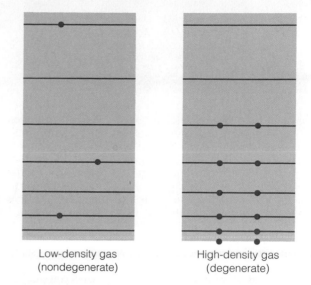

Figure 10-4
Electron energy levels are arranged like rungs on a ladder. In a low-density gas many levels are open, but in a degenerate gas all lower-energy levels are filled.

Low-density gas (nondegenerate)

High-density gas (degenerate)

trons and nuclei. If we add heat to the gas, most of that energy goes to speed up the motions of the nuclei, and only a few electrons can absorb enough energy to reach the empty energy levels at the top of the energy ladder. Thus changing the temperature of the gas has almost no effect on the pressure.

These two properties of degenerate matter become important when stars end their main-sequence lives and approach their final collapse (Window on Science 10-1). Eventually, many stars collapse into white dwarfs, and we will discover that these tiny stars are made of degenerate matter. But long before that, the cores of many giant stars become so dense that they are degenerate, a situation that can produce a cosmic bomb.

Helium Fusion

Hydrogen fusion in main-sequence stars leaves behind helium ash, which cannot begin fusing. Helium nuclei have a positive charge twice that of a proton, and so to overcome the repulsion between nuclei, they must collide at a velocity higher than that at which hydrogen fuses. The temperature for hydrogen fusion is too cool to fuse helium. As the star becomes a giant star fusing hydrogen in a shell, the inner core of helium contracts and grows hotter. It may even become degenerate, but when it finally reaches a temperature of 100,000,000 K, it begins to fuse helium nuclei to make carbon (see Chapter 9).

How a star begins helium fusion depends on its mass. Stars more massive than about 3 solar masses contract rapidly, their helium-rich cores heat up, and helium fusion begins gradually. But less massive stars evolve more slowly, and their cores contract so much

One of the most interesting lessons of science is that the behavior of very small things often determines the structure and behavior of very large things. In degenerate matter, the quantum-mechanical behavior of electrons helps determine the evolution of giant stars. Such links between the very small and the very large are common in astronomy, and we will see in later chapters how the nature of certain subatomic particles may determine the fate of the entire universe.

This is a second reason why astronomers study atoms. Recall that the first reason is that atoms interact with light and astronomers must understand that interaction in order to analyze the light. The second reason is the link between the very small and the very large. It would be impossible to understand the evolution of the largest stars if we failed to understand how degenerate electrons behave. Similarly, astronomers studying galaxies must understand how microscopic dust in space reddens starlight, and astronomers studying planets must know how atoms form crystals.

As you study any science, look for the way very small things affect very large things. Sociologists and psychologists know that the mass behavior of very large groups of people can depend on the behavior of a few key individuals. Biologists study the visible consequences of atomic bonds in molecules we call genes, and meteorologists know that tiny changes in temperature in one part of the world can affect worldwide climate weeks or months later.

Nature is the sum of many tiny parts, and science is the study of nature. Thus, one way to do science is to search out the tiny causes that determine the way nature behaves on the grandest scales.

---

the gas becomes degenerate. On Earth, a teaspoon of the gas would weigh as much as an automobile. In this degenerate matter, the pressure does not depend on temperature, and that means the pressure–temperature thermostat does not regulate energy production. When the temperature becomes hot enough, helium fusion begins to make energy, the temperature rises, but pressure does not increase because the gas is degenerate. The higher temperature increases the helium fusion even further, and the result is a runaway explosion called the **helium flash** in which, for a few minutes, the core of the star can generate more energy per second than does an entire galaxy.

Although the helium flash is sudden and powerful, it does not destroy the star. In fact, if you were observing a giant star as it experienced the helium flash, you would probably see no outward evidence of the eruption. The helium core is quite small (Figure 10-2), and all of the energy of the explosion is absorbed by the distended envelope. In addition, the helium flash is a very short-lived event in the life of a star. In a matter of minutes to hours, the core of the star becomes so hot it is no longer degenerate, the pressure–temperature thermostat brings the helium fusion under control, and the star proceeds to fuse helium steadily in its core (Figure 10-5).

The sun will experience helium fusion, but stars less massive than about 0.4 solar mass never get hot enough to ignite helium. Stars more massive than 3 solar masses ignite helium before their contracting cores become degenerate.

Whether a star experiences a helium flash or not, the ignition of helium in the core changes the structure of the star. The star now makes energy in its helium-fusion core and in its hydrogen-fusion shell. The energy flowing outward from the core can halt the contraction of the core, and the distended envelope of the star contracts and grows hotter. Thus, the point that represents the star in the H-R diagram moves downward and to the left toward the hot side of the H-R diagram.

Helium fusion produces carbon and oxygen atoms in the core, atoms that require much higher temperatures to fuse. Thus, as the helium fuel is used up, carbon and oxygen atoms accumulate in an inert

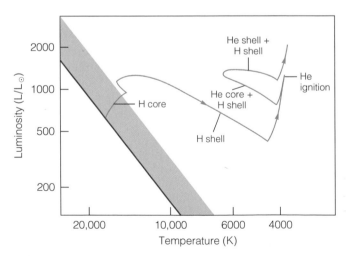

Figure 10-5

An expanded region of the H-R diagram shows the evolution of a star through helium fusion. After exhausting hydrogen in its core, the star ignites a hydrogen-fusion shell and leaves the main sequence as it expands to become a giant star. It eventually ignites helium in its core (in a helium flash for some stars) and then settles down to fuse helium in its core and hydrogen in a shell. Later it will fuse helium in a shell inside the hydrogen-fusion shell (top).

core. Once again, the core contracts and heats up, and a helium-fusion shell ignites below the hydrogen-fusion shell. The star now makes energy in two fusion shells, it quickly expands, and its surface cools once again. The point that represents the star in the H-R diagram moves back to the right, completing a loop (Figure 10-5).

The evolution of a star after helium exhaustion is uncertain, but the general course is clear. The inert carbon–oxygen core contracts and becomes hotter. Stars more massive than about 3 solar masses can reach temperatures of 600,000,000 K and ignite carbon fusion. Subsequent contraction may be able to fuse oxygen and silicon as well as other heavy elements.

Eventually, the star must exhaust all of its fuels and die in a final collapse. Exactly how the star collapses depends on which fuels it has consumed and on the strength of its gravity. Both of these factors depend on the star's mass, so we must group the stars into low-mass stars and high-mass stars.

Now we can understand why giant stars are so rare (see Figure 8-22). A star spends about 90 percent of its lifetime on the main sequence and only 10 percent as a giant star. During any particular human lifetime, only a fraction of the visible stars will be passing through the giant stage.

REVIEW Critical Inquiry

How are giant stars different from main-sequence stars?

We have seen that main-sequence stars begin the death process by expanding and cooling to become giant or supergiant stars, and in our discussion we discovered that these bloated stars are fundamentally different from main-sequence stars. Stars like the sun fuse hydrogen in their cores, and the energy flows outward to support the weight of the star. The pressure–temperature thermostat regulates the rate of energy production of a stable main-sequence star. But a giant star has used up the hydrogen in its core, and it can no longer maintain that main-sequence stability. Hydrogen fusion continues in a shell, but the core of helium ash contracts while the envelope of the star expands and cools. Later, helium fusion will begin in the core of the star and will produce an inert carbon–oxygen core and a helium-fusion shell below the hydrogen-fusion shell. Thus, the internal structure of a giant star is dramatically different from the simple structure of a main-sequence star. One of the pitfalls of critical analysis is to assume that similar consequences must arise from similar causes; in this case, main-sequence stars and giants look very similar, but they are quite different inside.

Consider a 5-solar-mass star and the sun. They look quite similar, and both will become giant stars, yet their internal evolution includes an important difference, the helium flash. What makes the difference?

We have now explored the structure of giant stars, so we are ready to group the stars by mass and tell the story of each kind of star. We begin with the low- and medium-mass stars.

10-2 The Deaths of Low- and Medium-Mass Stars

Because contracting stars heat up by converting gravitational energy into thermal energy, low-mass stars cannot get very hot. This limits the fuels they can ignite. In Chapter 9, we saw that protostars less massive than 0.08 solar mass cannot get hot enough to ignite hydrogen. Consequently, this section will concentrate on stars more massive than 0.08 solar mass but no more than a few times the mass of the sun.

Structural differences divide the low-mass stars into two subgroups — very-low-mass red dwarfs and medium-mass stars such as the sun. The critical difference between the two groups is the extent of interior convection. If the star is convective, fuel is constantly mixed, and the resulting evolution is drastically altered.

Red Dwarfs

Stars less massive than about 0.4 solar mass — the red dwarfs — have two advantages over more massive stars. First, they have very small masses, and thus they have very little weight to support. Their pressure–temperature thermostats are set low, and they consume their hydrogen fuel very slowly. Our discussion of the life expectancies of stars in Chapter 9 predicted that the red dwarfs could live very long lives.

The red dwarfs have a second advantage because they are totally convective. That is, they are stirred by circulating currents of hot gas rising from the interior and cool gas sinking inward. This means the stars are mixed like a pot of soup that is constantly stirred as it cooks. Hydrogen is consumed and helium accumulates uniformly throughout the star, and that means the star is not limited to the fuel in its core. It can use all of its hydrogen to prolong its life on the main sequence.

Because a red dwarf is mixed by convection, it cannot develop an inert helium core surrounded by unprocessed hydrogen. Then it can never ignite a hydrogen shell and thus cannot become a giant star. Rather, nuclear fusion converts hydrogen into helium, which cannot fuse because the star cannot get hot enough. Thus, what we know about stellar evolution tells us that these red dwarfs should use up nearly all of their hydrogen and live very long lives on the lower main sequence. They could survive for a hundred billion years—longer than the present age of the universe. When they finally run out of hydrogen, they should gradually contract, grow hotter, and become white dwarfs.

This discussion is fully consistent with what we know about the evolution of stars, but it leaves out one critical issue—mass loss. The sun slowly loses mass as the solar wind carries gas away into space, and some stars have stronger stellar winds. A study from the early 1990s showed that red dwarfs should have strong stellar winds carrying mass away so rapidly that a typical red dwarf might evaporate completely in just a few billion years. Thus mass loss can dramatically affect the evolution of stars.

Medium-Mass Stars

Stars with masses between roughly 4 solar masses and 0.4 solar mass,* including the sun, evolve in the same way. They can ignite hydrogen and helium and be-

*This mass limit is uncertain, as are many of the masses quoted here. The evolution of stars is highly complex, and such parameters are difficult to specify.

come giants, but they cannot get hot enough to ignite carbon, the next fuel after helium. When they reach that impasse, they collapse and become white dwarfs. There are two keys to the evolution of these sunlike stars: the lack of complete mixing and mass loss.

The interiors of medium-mass stars are not completely mixed (Figure 10-6). Stars with a mass of 1.1 solar masses or less have no convection near their centers, so they are not mixed at all. Stars with a mass greater than 1.1 solar masses have small zones of convection at their centers, but this mixes no more than about 12 percent of the star's mass. Thus medium-mass stars, whether they have convective cores or not, are not mixed, and the helium ash accumulates in an inert helium core surrounded by unprocessed hydrogen. When this core contracts, the unprocessed hydrogen ignites in a shell and swells the star into a giant.

In the giant stage, the core of the star contracts and the envelope expands. The star fuses helium in its core and then in a shell surrounding a core of carbon and oxygen. This core contracts and grows hotter; but, because the star has too low a mass, the core cannot get hot enough to ignite carbon fusion. Thus the carbon–oxygen core is a dead end for these medium-mass stars.

All of this discussion is based on theoretical models of stars and our general understanding of how stars evolve. Does it really happen? We need observational evidence to confirm our theoretical discussion, and the

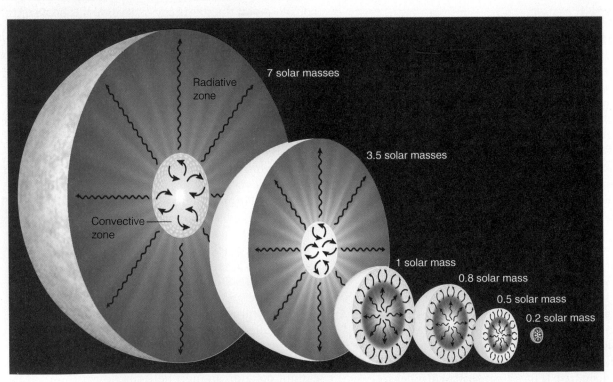

Figure 10-6
Inside stars. The more massive stars have small convective interiors and radiative envelopes. Stars like the sun have radiative interiors and convective envelopes. The lowest-mass stars are convective throughout. The "cores" of the stars where nuclear fusion occurs (not shown) are smaller than the interiors.
(Illustration design by author)

gas that is lost from these giant stars gives us visible evidence that sunlike stars do indeed die in this way.

Planetary Nebulae

When a medium-mass star like the sun becomes a distended giant, it can expel its outer atmosphere to form one of the most interesting objects in astronomy, a **planetary nebula,** so called because it looks like the greenish-blue disks of planets such as Uranus and Neptune. In fact, a planetary nebula has nothing to do with a planet. It is composed of ionized gases expelled by a dying star.

Basic observations tell us a great deal about planetary nebulae. In photographs, they look like roughly spherical shells of gas centered on hot stars (Figure 10-7). From estimates of their distances and their angular diameters, we can find their typical radii—from 0.2 ly to 3 ly. The spectra of the nebulae contain emission lines, which tells us the gas is low-density, excited gas. Doppler shifts show that the nebulae are expanding away from their central stars at 10 to 20 km/sec. If we divide the radius of a nebula by its velocity of expansion, we get its age, and planetary nebulae tend to be 1000 to 10,000 years old. Evidently, older nebulae become mixed into the interstellar medium and disappear.

Some planetary nebulae appear to be nearly spherical, and that helps us understand how they are produced. Just as the sun expels its solar wind, a giant star expels a strong stellar wind, which astronomers can detect as blue shifted lines in the star's spectrum. During the final centuries of its life, a giant star expels even more of its surface layers, but this gas is not easily visible in photographs because it is not ionized. Soon, however, it expels the last of its surface gases and exposes deeper, hotter layers. The intense radiation drives the gas near the star violently outward to overtake and sweep up the gas that was lost earlier, much as a snowplow sweeps up snow. The short-wavelength radiation from the hot surface ionizes the expanding shell of gas, and, as if someone flipped a switch, it lights up like a great neon sign, a planetary nebula.

The Hubble Space Telescope has revealed that many planetary nebulae are not spherical (Figure

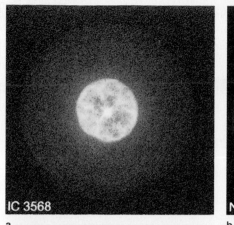

IC 3568

a

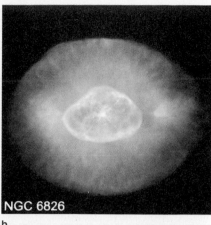

NGC 6826

b

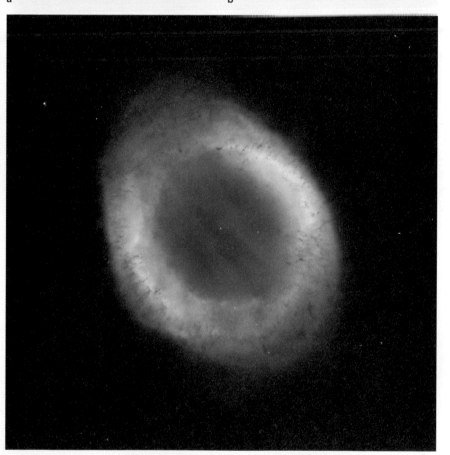

c

Figure 10-7
Planetary nebulae are shells of gas expelled from dying stars. (a) IC 3568 is highly symmetric and appears to be a spherical shell. (b) The red regions of NGC 6826 are produced by some asymmetry in the ejected gas. (c) About a light-year in diameter, the Ring Nebula is bright enough to be visible in small telescopes. Although it looks round, it is believed to be a cylinder of gas that happens to be pointed nearly at Earth. *(a and b, H. Bond, B. Ballick, and NASA; c, Hubble Heritage Team, AURA/STScI/NASA)*

10-8). A number of things could cause asymmetry in a planetary nebula. Some of the gas expelled from the star before it exposes its hot inner layers can form a disk around its equator. Later, when the star

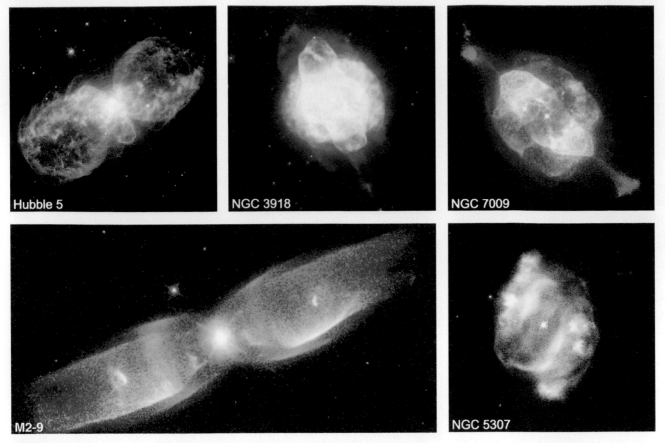

Figure 10-8
When the outflowing gas creating a planetary nebula is focused in opposite directions it can produce the complex structures seen here. *(H. Bond, B. Ballick, and NASA)*

expels high-speed gas to form the planetary nebula, the disk can focus the expanding gases into two opposite directed flows. Also, a second star orbiting the aging giant star or even planets can have a similar effect. Some of the planetary nebulae are very complex, which suggests there are undiscovered processes at work.

Planetary nebulae don't survive long, but we see roughly 1500 of them, and that must mean they are a common event. Most medium-mass stars end their lives by producing planetary nebulae. The sun will die in this way within about 5 billion years. Stars a few times more massive than the sun can also produce planetary nebulae if they have strong stellar winds and expel much of their mass before they end their lives.

Once an aging giant star blows its surface into space, the remaining hot interior collapses into a small, intensely hot star. It consists of the carbon–oxygen core surrounded by the helium- and hydrogen-fusion shells and topped by a shallow atmosphere of hydrogen. Such a central star in a planetary nebula is very hot at first—in some cases, over 100,000 K—and we find it plotted far to the upper left in the H-R diagram (Figure 10-9). As the nucleus cools, the point that represents it

moves toward the lower right in the diagram and eventually reaches the region of the white dwarfs. Thus it seems that white dwarfs are the remains of sunlike stars that ejected their surfaces to produce planetary nebulae.

This suggests a second way we can search for evidence of the deaths of medium mass stars: We can study white dwarfs.

White Dwarfs

In Chapter 8, we surveyed the stars and discovered that white dwarfs are the second most common kind of star (Figure 8-22). Only red dwarfs are more abundant. Now we can recognize these white dwarfs as the remains of medium-mass stars that fused hydrogen and helium, failed to ignite carbon, drove away their outer layers to form planetary nebulae, and collapsed and cooled to form white dwarfs. The billions of white dwarfs in our galaxy must be the remains of medium-mass stars.

The first white dwarf discovered was the faint companion to Sirius. In that visual binary system, the bright star is Sirius A. The white dwarf, Sirius B, is 10,000 times fainter than Sirius A. The orbital mo-

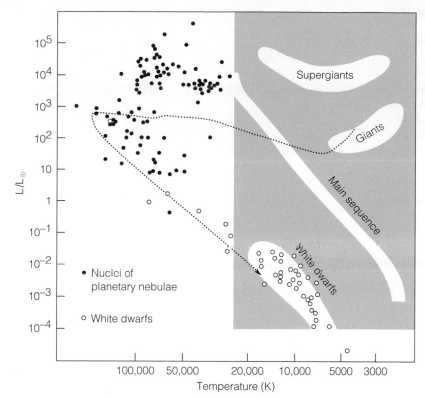

Figure 10-9
Our customary H-R diagram (shaded) must be expanded to show the evolution of a star after it ejects a planetary nebula. Stars at the centers of planetary nebulae (filled circles) are hotter and more luminous than most white dwarfs (open circles). The dotted line shows the evolution of a 0.8 M$_\odot$ star as it collapses toward the white-dwarf region. *(Adapted from diagrams by C. R. O'Dell and S. C. Vila)*

tions of the stars (shown in Figure 8-12) tell us that the white dwarf's mass is about 1 solar mass, and its blue-white color tells us that its surface is hot, about 25,000 K. Although it is very hot, it has a very low luminosity, so it must have a small surface area (see By the Numbers 8-3)—in fact, its diameter is about 92 percent of Earth's. Dividing its mass by its volume reveals that it is very dense—over 3×10^6 g/cm³. On Earth, a teaspoonful of Sirius B material would weigh more than 15 tons. Thus, basic observations and simple physics lead us to the conclusion that white dwarfs are astonishingly dense.

A normal star is supported by energy flowing outward from its core, but a white dwarf has no internal energy source. It has exhausted its hydrogen and helium and cannot get hot enough to ignite carbon. Thus, there is no energy flow to balance gravity. The star contracts until it becomes degenerate. A white dwarf is supported not by energy flowing outward but by the refusal of its electrons to pack themselves into a smaller volume.

The interior of a white dwarf is mostly carbon and oxygen ions immersed in a whirling storm of degenerate electrons. Theory predicts that as the star cools these ions will lock together to form a crystal lattice,

so there may be some truth in thinking of aging white dwarfs as great crystals of carbon and oxygen. Near the surface, where the pressure is lower, a layer of ionized gases makes up a hot atmosphere, which, because of the strong gravity, is pulled down into a shallow layer. If Earth's atmosphere were equally shallow, people on the top floors of skyscrapers would have to wear space suits.

Clearly, a white dwarf is not a normal star. It generates no nuclear energy, is almost totally degenerate, and, except for a thin layer at its surface, contains no gas. Instead of calling a white dwarf a "star," we can call it a "compact object." In the next chapter, we will discuss two other compact objects—neutron stars and black holes.

A white dwarf's future is bleak. As it radiates energy into space, its temperature gradually falls, but it cannot shrink any smaller because its degenerate electrons cannot get closer together. This degenerate matter is a very good thermal conductor, so heat flows to the surface and escapes into space, and the white dwarf gets fainter and cooler, moving downward and to the right in the H-R diagram. Because the white dwarf contains a tremendous amount of heat, it needs billions of years to radiate that heat through its small surface area. Eventually, such objects may become cold and dark, so called **black dwarfs.** Our galaxy is not old enough to contain black dwarfs. The coolest white dwarfs in our galaxy are about the temperature of the sun.

Perhaps the most interesting thing we have learned about white dwarfs has come from mathematical models. The equations predict that if we added mass to a white dwarf, its radius would *shrink* because added mass would increase its gravity and squeeze it tighter. If we added enough to raise its total mass to about 1.4 solar masses, its radius would shrink to zero (Figure 10-10). This is called the **Chandrasekhar limit** after Subrahmanyan Chandrasekhar, the astronomer who discovered it. It seems to imply that a star more massive than 1.4 solar masses could not become a white dwarf unless it shed mass in some way.

Stars do lose mass. Young stars have strong stellar winds, and aging giants also lose mass (Figure 10-11). This suggests that stars more massive than the Chandrasekhar limit can eventually die as white dwarfs if they reduce their mass. Theoretical models show that an 8-solar-mass star should be able to reduce its mass to 1.4 solar masses before it collapses, and slightly more massive stars may also be able to get under the limit. Thus, a wide range of medium-mass stars eventually die as white dwarfs.

How will the sun die?

The sun's death process will begin in a few billion years, when it exhausts the hydrogen in its core. It will then expand to become a giant star fusing helium. Although the surface of the sun will become cooler, it will expand until it has such a large surface that it will be about 100 times more luminous than it is now. (Incidentally, this will evaporate Earth's atmosphere and oceans.) As a giant star, the sun will have a strong solar wind carrying gas into space and pushing back the interstellar medium. Eventually, the loss of mass will expose deeper layers in the sun. These extremely hot layers will ionize the gas around the sun and propel it outward to scoop up previously expelled gases and produce a planetary nebula. Soon the last remains of the sun will collapse inward and slowly cool to form a small, dense white dwarf. The surface will be very hot but so small that the sun will be about 100 times fainter than it is now (and any cinder that remains of Earth will fall into a deadly deep freeze).

This story of stellar evolution tells us that our sun will someday create a planetary nebula and a white dwarf. At that point, what will have become of all the hydrogen the sun once contained in its core?

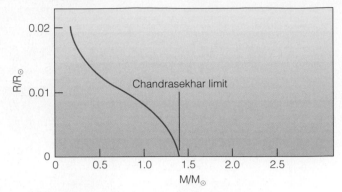

Figure 10-10
The more massive a white dwarf is, the smaller its radius. Stars more massive than the Chandrasekhar limit of 1.4 $M_\odot$ cannot be white dwarfs.

Medium-mass stars die by producing beautiful planetary nebulae and tiny white dwarfs, but we have discovered a limit—1.4 solar masses. What happens to massive stars that can't shed mass fast enough to die this way? We will address that question in the next section.

Figure 10-11
Mass loss from stars takes many forms. (a) This far-infrared map of the Orion constellation shows clouds of gas and dust heated and pushed away by hot young stars. The Orion Nebula is at lower right, and the large ring at upper right is expanding away from λ Orionis, an O8 giant. *(NASA/IPAC, Courtesy Deborah Levine)* (b) The Cat's Eye planetary nebula is formed by an inner and outer cavity inflated by a strong stellar wind carrying mass away from the central star. *(J. P. Harrington and K. J. Borkowski, University of Maryland and NASA)*

a

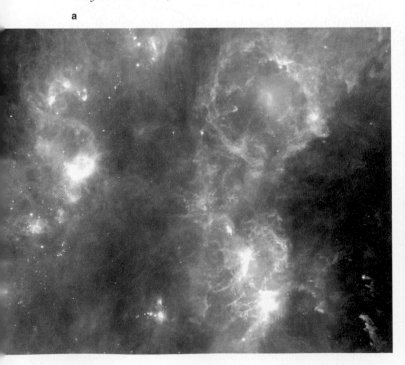

b

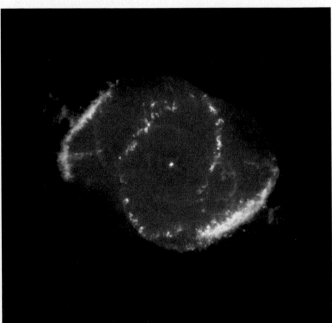

10-3 The Deaths of Massive Stars

We have seen that low- and medium-mass stars die relatively quietly as they exhaust their hydrogen and helium and then eject their surface layers to form planetary nebulae. In contrast, massive stars live spectacular lives (Figure 10-12a) and destroy themselves in violent explosions.

Nuclear Fusion in Massive Stars

Stars on the upper main sequence have too much mass to die as white dwarfs, but their evolution begins much like that of their lower-mass cousins. They consume the hydrogen in their cores and ignite hydrogen shells; as a result, they expand and become giants or, for the most massive stars, supergiants. Their cores contract and fuse helium first in the core and then in a shell, producing a carbon–oxygen core.

Unlike medium-mass stars, the massive stars are able to ignite carbon fusion at a temperature of about 1 billion Kelvin. For some stars, the core becomes degenerate before the carbon ignites, and thus the carbon ignites explosively in a **carbon detonation,** an event similar to but more violent than the helium flash. Carbon fusion produces more oxygen and neon, but as soon as the carbon is exhausted in the core, the core contracts and carbon ignites in a shell. This pattern of core ignition and shell ignition continues with fuel after fuel, and the star develops a layered structure (Figure 10-12b), with a hydrogen-fusion shell above a helium-fusion shell above a carbon-fusion shell above. . . . After carbon fuses, oxygen, neon, and magnesium fuse to make silicon and sulfur, and then the silicon fuses to make iron.

The fusion of these nuclear fuels goes faster and faster as the massive star evolves rapidly. Recall that massive stars must consume their fuels rapidly to support their great weight, but other factors also cause the heavier fuels like carbon, oxygen, and silicon to fuse at increasing speed. For one thing, the amount of energy released per fusion reaction decreases as the mass of the fusing atom increases. To support its weight, a star must fuse oxygen much faster than it fused hydrogen. Also, there are fewer atoms in the core of the star by

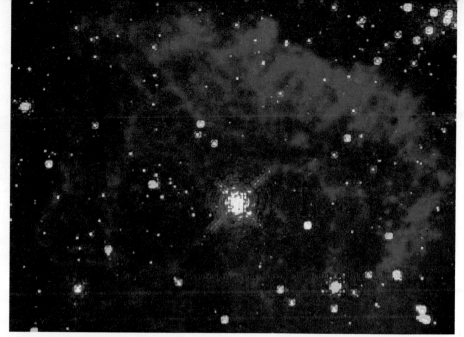

a

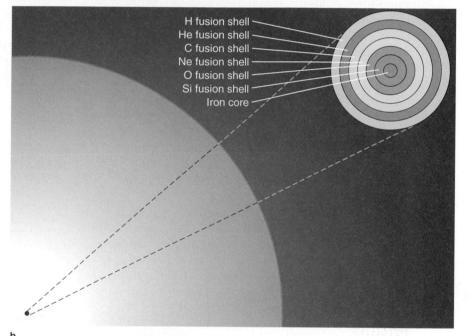

b

Figure 10-12
(a) The most massive star known is about 10 million times more luminous than the sun and contains about 100 solar masses. Only a few million years old, it has already expelled about half its mass to form the nebula shown in this infrared image. *(Don F. Figer, UCLA, and NASA)* (b) This aging massive star contains an Earth-size core (magnified 100,000 times in this figure) composed of concentric layers of gases undergoing nuclear fusion. The iron at the center is unable to produce energy.

the time heavy atoms begin to fuse. Four hydrogens made a helium atom, and three heliums made a carbon, so there are 12 times fewer atoms of carbon available for fusion than there were hydrogen. This means the heavy elements are used up and fusion goes very quickly in massive stars (Table 10-1). Hydrogen fusion

Table 10-1 Heavy-Element Fusion in a 25 M$_\odot$ Star

Fuel	Time	Percentage of Lifetime
H	7,000,000 years	93.3
He	500,000 years	6.7
C	600 years	0.008
O	0.5 years	0.000007
Si	1 day	0.00000004

can last 7 million years in a 25-solar-mass star, but that same star will fuse its oxygen in 6 months and its silicon in a day.

Supernova Explosions

Theoretical models of evolving stars combined with nuclear physics allow astronomers to describe what happens inside a massive star when the last nuclear fuels are exhausted. It begins with iron atoms and ends in cosmic violence.

Silicon fusion produces iron, the most tightly bound of all atomic nuclei (see Figure 9-14). Nuclear fusion is able to release energy by combining less tightly bound nuclei into a more tightly bound nucleus, but iron is the limit. Once the gas in the core of the star has been converted to iron, there are no nuclear reactions that can combine iron atoms and release energy. Thus, the iron core is a dead end in the evolution of a massive star.

As a star develops an iron core, energy production begins to decline, and the core contracts. For atoms less massive than iron, such contraction heats the gas and ignites new fusion fuels, but nuclear reactions involving iron tend to remove energy from the core in two ways. First, the iron atoms begin capturing electrons and breaking into smaller nuclei. The degenerate electrons helped support the core, so the loss of the electrons allows the core to contract even faster. Second, temperatures are so high that the average photon is a high-energy gamma ray, and these gamma rays are absorbed by atomic nuclei, which break into smaller fragments. The removal of the gamma rays further cools the core and allows it to contract more. With the loss of the electrons and the gamma rays, the core of the star collapses inward in less than a tenth of a second.

This collapse happens so rapidly that our most powerful computers are unable to predict the details. Thus models of supernova explosions contain many approximations. Nevertheless, the models predict that the collapsing core of the star must quickly become a neutron star or a black hole, the subjects of the next chapter, while the envelope of the star is blasted apart in a supernova explosion.

To understand how the inward collapse of the core can produce an outward explosion, we can think about a traffic jam. The collapse of the innermost part of the degenerate core allows the rest of the core to fall inward, and this creates a tremendous traffic jam as all of the nuclei fall toward the center. It is as if every car owner in Indiana suddenly tried to drive into downtown Indianapolis. There would be a traffic jam not only downtown but also in the suburbs; and as more cars arrived, the traffic jam would spread outward. Similarly, as the inner core falls inward, a shock wave (a "traffic jam") develops and begins to move outward.

The shock wave moves outward through the star aided by two additional sources of energy. First, when the iron atoms in the core are disrupted, they produce a flood of neutrinos. In fact, for a short time the core produces more energy per second than all of the stars in all of the visible galaxies in the universe; and 99 percent of that energy is in the form of neutrinos. This flood of neutrinos rushes out of the core and helps heat the gas and accelerate the outward-bound shock wave. Second, the tremendous heat in the core of the star triggers giant convection currents, and intensely hot gas rushes outward from the interior (Figure 10-13). Again, this convection carries energy out into the envelope and helps drive the shock wave outward.

Figure 10-13
The inside of an exploding star is shown in this mathematical model of a supernova explosion. The image here represents a region 150 km on a side near the center of the star (lower left). Black arrows indicate the motion of the gas. Turbulent material (blue) has bounced off the core and is rushing outward to meet the rest of the star, which is falling inward (red at upper right). The innermost part of the star is forming a neutron star (red at lower left). To show the entire star at this scale, this figure would have to be 140 km in diameter. (*Courtesy Adam Burrows, John Hayes, and Bruce Fryxell*)

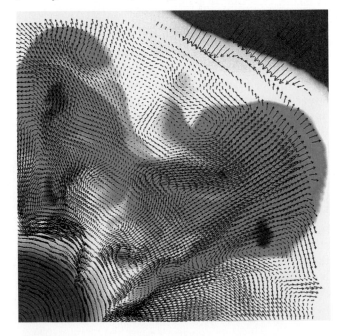

Within a few hours, the shock wave bursts through the surface of the star and blasts it apart.

The supernova we see from Earth is the brightening of the star as its distended envelope is blasted outward by the shock wave. As months pass, the cloud of gas expands, thins, and fades, but the rate at which some supernovae fade tells astronomers that the gas is enriched with short-lived radioactive nuclei such as nickel-56. The gradual decay of these nuclei can keep the gas hot and prevent the supernova from fading rapidly. The presence of these radioactive elements tells us that the density and temperature inside the exploding star were so high that nuclear reactions took place and built large numbers of heavy atoms.

Astronomers working with the largest and fastest computers are using modern theory to try to understand supernova explosions. The companion to theory is observation, so we should ask what observational evidence supports this story of the deaths of massive stars.

Observations of Supernovae

In A.D. 1054, Chinese astronomers saw a "guest star" appear in the constellation we know as Taurus the Bull. The star quickly became so bright it was visible in the daytime, and then, after a month, it slowly faded, taking almost 2 years to vanish from sight. When modern astronomers turned their telescopes to the location of the "guest star," they found a cloud of gas about 1.35 pc in radius expanding at 1400 km/sec. Projecting the expansion back in time, they concluded the expansion must have begun about 900 years ago, just when the "guest star" made its visit. Thus, the nebula, now called the Crab Nebula because of its shape (Figure 10-14a), marks the site of the A.D. 1054 supernova.

Supernovae are rare. Only a few have been seen with the naked eye in recorded history. Arab astronomers saw one in A.D. 1006, and the Chinese saw one in A.D. 1054. European astronomers observed two—one in A.D. 1572 (Tycho's supernova) and one in A.D. 1604 (Kepler's supernova). In addition, the guest stars of A.D. 185, 386, 393, and 1181 may have been supernovae.

In the centuries following the invention of the astronomical telescope in 1609, no supernova was bright enough to be visible to the naked eye. Supernovae are sometimes seen in distant galaxies, but they are faint and hard to study. Then, in the early morning hours of February 24, 1987, astronomers around the world were startled by the discovery of a naked-eye supernova still growing brighter in the southern sky (Figure 10-14b). The supernova, known officially as SN1987A, is only 53,000 pc away in the Large Magellanic Cloud, a small satellite galaxy to our own Milky Way galaxy. This first naked-eye supernova in 383

a

b

Figure 10-14
(a) The Crab Nebula is the remains of the supernova of A.D. 1054. At visual wavelengths, the glow of the nebula is woven through a shell of expanding filaments. *(Copyright California Institute of Technology and Carnegie Institution of Washington)* (b) Supernova 1987A is the bright image at the right. This supernova, which is in the Large Magellanic Cloud, was the first supernova visible to the naked eye since 1604. The Tarantula Nebula is at upper left. *(NOAO)*

years has given astronomers a ringside seat for the most spectacular event in stellar evolution.

One obseravtion of SN1987A is critical in that it confirms the theory of core collapse. At 2:35 A.M. EST on February 23, 1987, nearly 4 hours before the supernova was first seen, a blast of neutrinos swept through Earth, including perhaps 20 trillion that passed harmlessly through your body. Instruments buried in a salt mine under Lake Erie and in a lead mine in Japan, though designed for another purpose, recorded 19 neutrinos in less than 15 seconds. Neutrinos are so difficult to detect that the 19 neutrinos actually detected mean that some 10^{17} neutrinos must have passed through the detectors in those 15 seconds. Furthermore, the neutrinos were arriving from the direction of the supernova. Thus, astronomers conclude that the burst of neutrinos was released when the iron core collapsed, and the supernova was seen hours later when the shock wave blasted its surface into space.

Most supernovae are seen in distant galaxies (Figure 10-15), and careful observations show that there are fundamentally two kinds. Type I supernovae are more luminous, reaching a maximum luminosity about 4 billion times that of the sun. They decline rapidly at first and then more slowly. Type II supernovae become only about 0.6 billion times more luminous than the sun and decline in a more irregular way. Spectra of Type II supernovae show hydrogen lines, but spectra of Type I do not. These observational clues make sense when we consider an alternative way a star can collapse.

Type II supernovae are apparently produced by the collapse of the iron core in a massive star.

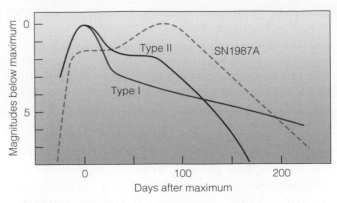

Figure 10-16
Type I supernovae decline rapidly at first and then more slowly, but Type II supernovae pause for about 100 days before beginning a steep decline. Supernova 1987A was odd in that it did not rise directly to maximum brightness. These light curves have been adjusted to the same maximum brightness. Generally, Type II supernovae are about 2 magnitudes fainter than Type I.

SN1987A was a Type II supernova, although its light curve is not typical (Figure 10-16). It was produced by the explosion of a hot, blue supergiant rather than the usual cool, red supergiant. Evidently, the star was a red supergiant a few thousand years ago but had contracted and heated up slightly, becoming smaller, hotter, and bluer before it exploded. Scientists believe that most Type II supernovae are caused by the collapse of red supergiants. Spectra of Type II supernovae contain hydrogen lines because the envelopes of giant and supergiant stars are rich with unfused hydrogen.

Type I supernovae are a related phenomenon in that they too are caused by the collapse of a star, but in this case the star is a white dwarf. A typical white dwarf has a mass of 0.6 solar mass and is supported by the degenerate matter in its interior. A white dwarf in a binary star system can gain mass from its companion star (a process we will discuss in detail later in this chapter). It is also possible for a pair of white dwarfs orbiting each other to merge. In either case, a white dwarf could find itself with more mass than the Chandrasekhar limit—more mass than the degenerate matter can support—and then it must collapse. The sudden compression and heating of the degenerate matter triggers violent nuclear fusion, leading to a supernova explosion that destroys the star. Spectra of Type I supernovae lack hydrogen lines—evidently because white dwarfs contain very little hydrogen.

Although the supernova explosion fades to obscurity in a year or two, an expanding shell of gas marks the site of the explosion. The gas, originally expelled at 10,000 to 20,000 km/sec, may carry away a fifth of the mass of the star. The collision of that expanding gas with the surrounding interstellar medium can sweep up even more gas and excite it to produce a **supernova remnant,** the nebulous remains of a supernova explosion (Figure 10-17a).

Figure 10-15
A supernova (arrow) is visible in this Hubble Space Telescope photograph of the Whirlpool Galaxy. (Dark marks across the image are caused by missing data.) *(Robert Kirshner, Harvard and NASA)*

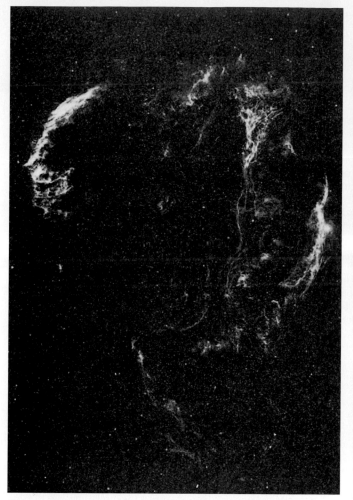

a

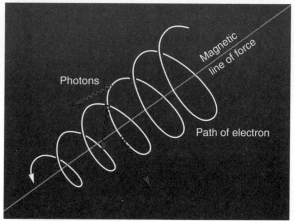

b

c

Figure 10-17

(a) The wispy sphere of the supernova remnant called the Cygnus Loop fills this photograph. It was formed when a massive star exploded about 20,000 years ago. *(Palomar Obs/Caltech)* (b) This image of the 300-year-old remnant Cas A was made with a radio telescope. *(NRAO)* (c) Some of the energy emitted by supernova remnants is synchrotron radiation, produced when an electron spirals around a magnetic field and radiates its energy away as photons. Usually synchrotron radiation is detected at radio wavelengths, but the Crab Nebula also produces synchrotron radiation at visible wavelengths.

Supernova remnants look quite delicate and do not survive very long—a few tens of thousands of years—before they gradually mix with the interstellar medium and vanish. The Crab Nebula is a young remnant, only about 900 years old and about 2.7 pc in diameter. The Cygnus Loop (see Figure 10-17a) is an estimated 20,000 years old, with a diameter of 40 pc. Some supernova remnants are visible only at radio and X-ray wavelengths (Figure 10-17b and Window on Science 10-2). They have become too tenuous to emit detectable light, but the collision of the expanding gas shell with the interstellar medium can generate radio and X-ray radiation. We saw in Chapter 9 that the compression of the interstellar medium by supernova remnants can also trigger star formation.

The radio emission coming from the Crab Nebula is **synchrotron radiation** (Figure 10-17c). This form of radiation is produced by rapidly moving electrons spiraling through magnetic fields; and in the case of the Crab Nebula, the electrons are so energetic that they also emit visible light. This leaves us with a puzzle. The Crab Nebula is 900 years old, so the electrons should have radiated away their energy long ago. The Crab Nebula must contain a powerful energy source to maintain the synchrotron radiation. We will search for that energy source in the next chapter.

Astronomers use computers to add false color to images and reveal things our eyes cannot see. Radio, infrared, ultraviolet, and X-ray radiation are not visible to our eyes, so data collected at these wavelengths can be displayed in images that are given colors visible to our eyes. The choice of colors is usually arbitrary, but the variation in color across an image does have meaning. A radio astronomer might choose shades of red through blue to display a radio map of a supernova remnant, and although the choice of colors was arbitrary, we would be able to identify the regions of strongest radio emission as the reddest regions on the map.

Astronomers also use false color to enhance visible-wavelength photographs. By converting the image to digital form and assigning colors to different levels of intensity, the astronomer can bring out subtle details that might not otherwise be visible. This is a common technique in other fields as well. Physicians often apply false color to medical X rays and CAT scans. Biologists use false color to analyze microscope photographs, and geologists use false color to study images of the earth recorded by earth satellites.

Even a simple photograph of an astronomical object such as the Crab Nebula is a kind of false-color image. The nebula is too faint to see with the human eye, and even if we flew in a spaceship to within a short distance of the nebula, it would still be too faint to see without special equipment. We could not see its shape or its colors. The photograph we see is a time exposure made using special filters, and thus it reveals a part of nature that is invisible to our eyes. Even a black-and-white photograph of a nebula shows us an aspect of the universe we can never see with our unaided eyes. In that sense, all astronomical photographs are false-color images. Their virtue is not that they are colorful, but that they are meaningful.

Evolution and Star Clusters

Before we rush ahead with our story of stellar evolution, we should pause to test our theories against reality. We can now arrange the stars by mass and describe the different stages in their lives (Figure 10-18), but do stars really evolve this way? We live such short lives compared with the stars that it is hard to test our theories. Nevertheless, all of science is based on evidence, so we should demand evidence to support these theories. Star clusters give us dramatic confirmation that stars do indeed evolve as our theories predict.

The stars in a star cluster all form at about the same time, but they include a range of masses. Like entrants in a foot race, the stars quickly sort themselves out. The most massive stars evolve and die quickly, while lower-mass stars evolve more slowly. When we look at the H-R diagram of a star cluster, we can see stellar evolution made visible.

Suppose we follow the evolution of a star cluster by making H-R diagrams like frames in a film (Figure 10-19). Our first frame shows the cluster only 10^6 years after it began forming, and already the most massive stars have reached the main sequence, con-

Figure 10-18
How a star evolves depends on its mass. The lowest-mass stars become white dwarfs, while stars like the sun become giants, produce planetary nebulae, and then collapse into white dwarfs. Mass loss can move stars to the right in this diagram so a star of 6 or more solar masses may be able to shed enough mass to die as a planetary nebula and a white dwarf. A more massive supergiant may be able to eject enough mass to leave behind a neutron star instead of a black hole. *(Illustration design by author)*

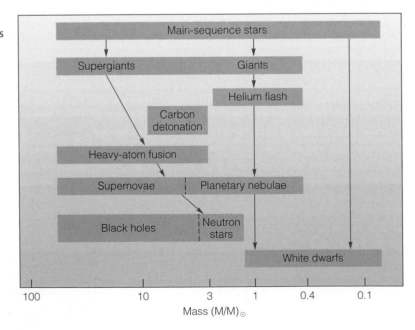

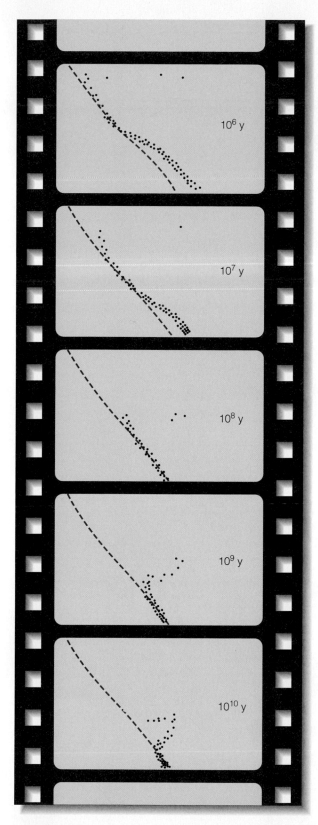

10^6 y

10^7 y

10^8 y

10^9 y

10^{10} y

Figure 10-19
A series of theoretical H-R diagrams, like frames in a film, illustrates the evolution of a cluster of stars. Recall that the most massive stars are the most luminous and are plotted at the top of the H-R diagram.

sumed the hydrogen in their cores, and moved off to become supergiants. The medium- to low-mass stars have not yet reached the main sequence.

Because evolution is such a slow process, we cannot make the time step between frames equal, or we would fill over 1000 pages with nearly identical diagrams. Instead, we increase the time step by a factor of 10 with each frame. Thus, the second frame shows the cluster after 10^7 years and the third after 10^8 years.

By the third frame, all massive stars have died, and stars slightly more massive than the sun are beginning to leave the main sequence. Notice that the lowest-mass stars have finally begun to fuse hydrogen. Only after 10^{10} years will the sun begin to swell into a giant.

These five frames were made from theoretical models of stellar evolution, but they compare very well with H-R diagrams of real star clusters. That is, our theory compares well with reality. For example, NGC2264 (Figure 10-20a) is only a few million years old and still has many of its lower-mass stars contracting toward the main sequence. The Pleiades (Figure 10-20b), a cluster visible to the naked eye in Taurus, is older than NGC2264, dating back about 10^8 years. Compare these younger clusters with M67 (Figure 10-20c), a faint cluster of stars about 4×10^9 years old whose more massive stars have died.

We can estimate the age of a star cluster by noting the point at which its stars turn off the main sequence and move toward the red-giant region. The masses of the stars at this **turnoff point** tell us the age of the cluster because those stars are on the verge of exhausting their hydrogen-fusion cores. Thus the time the stars at the turnoff point have spent on the main sequence equals the age of the cluster (see By the Numbers 9-1).

Star clusters confirm by their H-R diagrams that astronomers do understand how stars are born, evolve, and die. The theory of stellar evolution is one of the great accomplishments of 20th-century astronomy. It gives us insights into how nature converts the cold, dark interstellar medium into heat and light.

REVIEW Critical Inquiry

What causes a Type II supernova explosion?

A Type II supernova occurs when a massive star reaches the end of its usable fuel and develops an iron core. The iron is the final ash produced by nuclear fusion, and it cannot fuse to produce energy because iron is the most stable element. When energy generation begins to fall, the star contracts; but because iron can't ignite, there is no new energy source to stop the contraction. In a fraction of a second, the core of the star falls inward and a shock wave moves outward. Aided by a flood of neutrinos and sudden

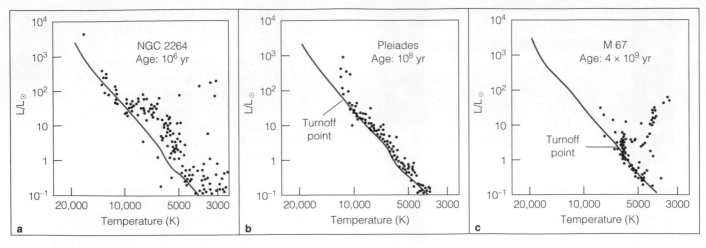

Figure 10-20

H-R diagrams of real star clusters compare well with H-R diagrams computed from theoretical models as shown in Figure 10-19.

convection, the shock wave blasts the star apart, and we see it brighten as its surface gases expand into space.

Type II supernova explosions are easy to recognize because their spectra contain hydrogen lines. Use what you know about Type I supernova explosions to explain why their spectra do not contain visible hydrogen lines.

Supernovae mark the death of massive stars, and white dwarfs mark the death of less massive stars. But so far we have discussed only single stars. At least half of all stars are binary. How do stars that are members of binary systems die? That is the subject of the next section.

10-4 The Evolution of Binary Systems

Stars in binary systems can evolve independently of each other if they orbit at a large distance from each other. In this situation, one of the stars can swell into a giant and collapse without disturbing its companion. But some binary stars are as close to each other as 0.1 AU, and when one of those stars begins to swell into a giant, its companion can suffer in peculiar ways.

These interacting binary stars are interesting in their own right. The stars share a complicated history and can evolve to experience strange and violent phenomena. But such systems are also important because they can help us understand the ultimate fate of stars and certain observed phenomena, such as the temporary appearance of new stars in the sky. In the next chapter, we will use interacting binary stars as tools to help us find black holes.

Mass Transfer

Binary stars can sometimes interact by transferring mass from one star to the other. Of course, the gravitational field of each star holds its mass together, but the gravitational fields of the two stars, combined with the rotation of the binary system, define a teardrop-shaped volume around each star called the **Roche lobe** (Figure 10-21). Matter inside a star's Roche lobe is gravitationally bound to the star. The size of the Roche lobes depends on the mass of the

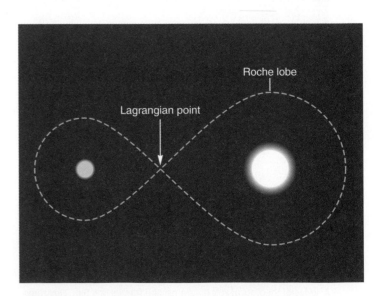

Figure 10-21

The Roche lobes around a pair of binary stars represent the two volumes of space that the stars can control with their gravity. The Roche lobes meet at the Lagrangian point. If matter can leave a star and reach the Lagrangian point, it can overflow into the other Roche lobe and eventually fall into the other star.

stars and on the distance between the stars. If the stars are far apart, the lobes are very large, and the stars easily control their own mass.

Like the lobes of a giant dumbbell, the Roche lobes in a binary star meet at the **Lagrangian point** somewhere between the stars. The Lagrangian point connects the two Roche lobes, and, if matter can leave a star and reach the Lagrangian point, it can flow into the other star. Thus, the Lagrangian point is the connection through which the stars can transfer matter.

In general, there are only two ways matter can escape from a star and reach the Lagrangian point. First, if a star has a strong stellar wind, some of the gas blowing away from the star can pass through the Lagrangian point and be captured by the other star. Second, if an evolving star expands so far that it fills its Roche lobe, which can occur if the stars are close together and the lobes are small, then matter can overflow through the Lagrangian point onto the other star. Mass transfer driven by a stellar wind tends to be slow, but mass can be transferred rapidly by an expanding star.

Evolution with Mass Transfer

Mass transfer between stars can affect evolution of the stars in surprising ways. In fact, this explains a problem that puzzled astronomers for many years.

In some binary systems, the less massive star has become a giant, while the more massive star is still on the main sequence. If higher-mass stars evolve faster than lower-mass stars, how do the lower-mass stars in such binaries manage to leave the main sequence first? This is called the Algol paradox, after the binary system Algol.

Mass transfer explains how this could happen. Imagine a binary system that contains a 5-solar-mass star and a 1-solar-mass companion. The two stars formed at the same time, so we expect the higher-mass star to evolve faster and leave the main sequence first. When it expands into a giant, however, it can fill its Roche lobe and transfer matter to the low-mass companion. Thus the higher-mass star could evolve into a lower-mass star, and the companion could gain mass and become a high-mass star still on the main sequence. Thus, we might find a system such as Algol containing a 5-solar-mass main-sequence star and a 1-solar-mass giant.

The first four frames of Figure 10-22 show mass transfer producing a system like Algol. The last frame shows an additional stage in which the giant star collapses to form a white dwarf, and the massive companion evolves to become a giant transferring matter back onto the white dwarf. Such systems can become the site of tremendous explosions. To see how this can happen, we must consider how mass falls into a star.

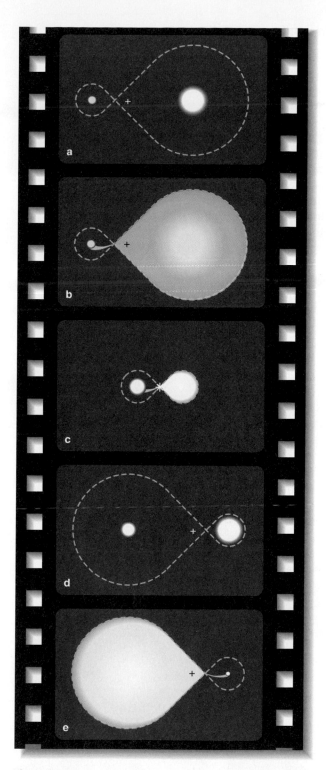

Figure 10-22
The evolution of a close binary system. As the more massive star (at the right) evolves (a), it fills its Roche lobe (b) and begins to transfer mass through the Lagrangian point to its companion. The companion grows more massive (c), and the star losing mass shrinks (d). Eventually the star on the right in this illustration can collapse into a white dwarf, while the star on the left evolves into a giant and begins transferring mass back to the white dwarf (e).

Accretion Disks

Matter flowing from one star to another cannot fall directly into the star. Rather, because of conservation of angular momentum, it must flow into a whirling disk around the star.

Angular momentum refers to the tendency of a rotating object to continue rotating. All rotating objects possess some angular momentum, and in the absence of external forces, an object maintains (conserves) its total angular momentum. An ice skater takes advantage of conservation of angular momentum by starting a spin slowly with arms extended and then drawing them in. As her mass becomes concentrated closer to her axis of rotation, she spins faster (Figure 10-23). The same effect causes the slowly circulating water in a bathtub to spin in a whirlpool as it approaches the drain.

Mass transferred through the Lagrangian point in a binary system toward a white dwarf must conserve its angular momentum. Thus it must flow into a rapidly rotating whirlpool called an **accretion disk** around the white dwarf (Figure 10-24).

Two important things happen in an accretion disk. First, the gas in the disk grows very hot due to friction and tidal forces, the same kind of gravitational forces that make Earth's oceans ebb and flow (Chapter 3). The disk also acts as a brake, ridding the gas of its angular momentum and allowing it to fall into the white dwarf. The temperature of the gas in the inner parts of the accretion disk can exceed a million Kelvin, causing the gas to emit intense X rays. In addition, the matter falling inward from the accretion disk can cause a violent explosion if it accumulates on a white dwarf.

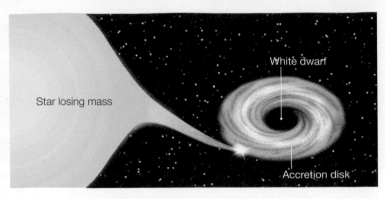

Figure 10-24
Matter falling into a compact object forms a whirling accretion disk.

Novae

At the beginning of this chapter we saw that the word **nova** refers to a new star that appears in the sky for a while and then fades away. The evidence tells us something different. The star is not new; it is old. After a nova fades, astronomers can photograph the spectrum of the remaining star. Invariably, they find a closely spaced binary star containing a normal star and a white dwarf. A nova is evidently an explosion involving a white dwarf.

Observational evidence can tell us how nova explosions occur. As the explosion begins, spectra show blue-shifted absorption lines, which tells us the gas is dense and coming toward us at a few thousand kilometers per second. After a few days, the spectral lines change to emission lines, which tells us the gas has thinned, but the blue shifts remain, showing that a cloud of debris has been ejected into space.

We can understand nova explosions as the result of mass transfer from a normal star through the Lagrangian point into an accretion disk around the white dwarf. As the matter loses its angular momentum in the accretion disk, it settles inward onto the surface of the white dwarf and forms a layer of unused nuclear fuel—mostly hydrogen. As the layer deepens, it becomes denser and hotter until the hydrogen fuses in a sudden explosion that blows the surface off of the white dwarf. Although the expanding cloud of debris contains less than 0.0001 solar mass, it is hot and its expanding surface area makes it very luminous. Nova explosions can become 100,000 times more luminous than the sun. As the debris cloud expands, cools, and thins over a period of months and years, the nova fades from view.

The explosion of its surface hardly disturbs the white dwarf and its companion star. Mass transfer quickly resumes, and a new layer of fuel begins to accumulate. How fast the fuel builds up depends on the rate of mass transfer. Accordingly, we can expect novae to repeat each time an explosive layer accumu-

Figure 10-23
A skater demonstrates conservation of angular momentum when she spins faster by drawing her arms and legs closer to her axis of rotation.

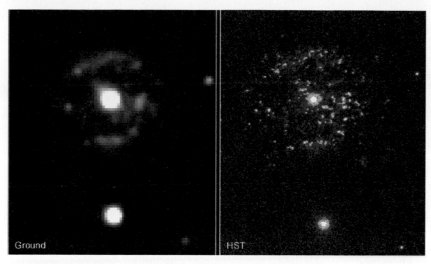

Figure 10-25
Nova T Pyxidis erupts about every two decades, expelling shells of gas into space. The shells of gas are visible from ground-based telescopes, but Hubble Space Telescope reveals much more detail. The shell consists of knots of excited gas that presumably form when a new shell collides with a previous shell. *(M. Shara and R. Williams, STScI; R. Gilmozzi, ESO; and NASA)*

lates. Many novae take thousands of years to build an explosive layer, but some take only decades (Figure 10-25).

R E V I E W Critical Inquiry

How does modern astronomy explain the Algol paradox?

When we encounter a paradox in nature, it is usually a warning that we don't understand things as well as we thought we did. The Algol paradox is a good example. The binary star Algol contains a lower-mass giant star and a more massive main-sequence star. Because the two stars must have formed together, they are the same age; and the more massive star should have evolved first and left the main sequence. But in binary systems such as Algol, the lower-mass star has left the main sequence, or so it seems.

We can understand this paradox if we add mass transfer to our theory of stellar evolution. The first star to leave the main sequence must be the more massive star; but if the stars are close together, the star that is initially more massive can fill its Roche lobe and transfer mass back to its companion. The companion can grow more massive and the giant can grow less massive. Thus, the lower-mass giant star we see in the system may have originally been more massive and may have evolved away from the main sequence, leaving its companion behind to increase in mass as it decreased.

Mass transfer can explain the Algol paradox, and it can also explain the violent explosions called novae. How does mass transfer explain why novae can explode over and over in the same binary system?

The study of the deaths of stars has led us to discover astonishing objects of unbelievable density, temperature, and violence—all consequences of the victory of gravity over matter. But we have not considered the strangest circumstance of all. What happens when degenerate matter can't support the weight of a dying star—that is, when the mass of the compact object exceeds the Chandrasekhar limit? The answer is in the title of the next chapter.

Summary

When a star's central hydrogen-fusion reactions cease, its core contracts and heats up, and hydrogen fusion begins in a spherical layer around the core—a hydrogen-fusion shell. Energy from this shell swells the star into a cool giant. The contraction of the star's core ignites helium first in the core and later in a shell. If the star is massive enough, it can eventually fuse carbon and other elements.

If a star's mass lies between about 0.4 and 3 solar masses, its helium core becomes degenerate before the helium ignites. In degenerate gas, pressure does not depend on temperature, so there is no pressure–temperature thermostat to control the reactions. As a result, the core explodes in a helium flash. All of the energy produced is absorbed by the star. A similar thing happens when carbon ignites in more massive stars. This carbon detonation is much more powerful than the helium flash.

How a star evolves depends on its mass. Stars less massive than about 0.4 solar mass are completely mixed and will have very little hydrogen left when they die. They cannot ignite a hydrogen-fusion shell or a helium-fusion core, so they will eventually become white dwarfs. Medium-mass stars between about 0.4 and 3 solar masses become giants and fuse helium but cannot fuse carbon. They produce planetary nebulae and become white dwarfs.

Stars as massive as 8 solar masses may lose enough mass to eject planetary nebulae and die as white dwarfs, but more massive stars suffer a different fate. The most massive stars fuse nuclear fuels up to iron but cannot generate further nuclear energy because iron is the most tightly bound of all atomic nuclei. When an iron core forms in a massive star, the core collapses and triggers a supernova explosion that expels the outer layers of the star to form an expanding supernova remnant. The first supernova visible to the naked eye since 1604 was seen in February 1987.

We can see evidence of stellar evolution in the H-R diagrams of clusters of stars. Beginning their evolution at about the same time, the stars evolve in different ways, depending on their masses. The most massive leave the main sequence first and are followed later by progressively less massive stars. Thus, we can estimate the age of a star cluster from the turnoff point in its H-R diagram.

Close binary stars evolve in complex ways because they can transfer mass from one star to the other. This explains why some binary systems contain a main-sequence star more massive than its giant companion—the Algol paradox. Also, mass transfer into an accretion disk around a white dwarf can produce X rays from the hot disk and can trigger nova explosions.

New Terms

nova	supernova remnant
supernova	synchrotron radiation
degenerate matter	turnoff point
helium flash	Roche lobe
planetary nebula	Lagrangian point
black dwarf	angular momentum
Chandrasekhar limit	accretion disk
carbon detonation	

Review Questions

1. Why does helium fusion require a higher temperature than hydrogen fusion?

2. How can the contraction of an inert helium core trigger the ignition of a hydrogen-fusion shell?

3. Why does the expansion of a star's envelope make it cooler and more luminous?

4. Why is degenerate matter so difficult to compress?

5. How does the presence of degenerate matter in a star trigger the helium flash?

6. Why don't red dwarfs become giant stars?

7. What causes an aging giant star to produce a planetary nebula?

8. Why can't a white dwarf contract as it cools? What is its fate?

9. Why can't a white dwarf have a mass greater than 1.4 solar masses?

10. How can a star of as much as 8 solar masses form a white dwarf when it dies?

11. How can the inward collapse of the core of a massive star produce an outward explosion?

12. What is the difference between Type I and a Type II supernovae?

13. How can star clusters confirm our theories of stellar evolution?

14. How can we understand the Algol paradox?

15. What is the difference between a supernova explosion and a nova explosion?

Discussion Questions

1. How do we know the helium flash occurs if it cannot be observed? Can we accept something as real if we can never observe it?

2. False-color radio images and time-exposure photographs of astronomical images show us aspects of nature we can never see with our unaided eyes. Can you think of common images in newspapers or on television that reveal phenomena we cannot see?

Problems

1. About how long would a 0.4 $M_\odot$ star spend on the main sequence? (*Hint:* See By the Numbers 9-1.)

2. The Ring Nebula in Lyrae is a planetary nebula with an angular diameter of 72 seconds of arc and a distance of 5000 ly. What is its linear diameter? (*Hint:* See By the Numbers 3-1.)

3. If the Ring Nebula is expanding at a velocity of 15 km/sec, typical of planetary nebulae, how old is it?

4. Suppose a planetary nebula is 1 pc in radius. If the Doppler shifts in its spectrum show it is expanding at 30 km/sec, how old is it? (*Hints:* 1 pc equals 3×10^{13} km, and 1 year equals 3.15×10^7 seconds.)

5. If a star the size of the sun expands to form a giant 20 times larger in radius, by what factor will its average density decrease? (*Hint:* The volume of a sphere is $(\frac{4}{3})\pi r^3$.)

6. If a star the size of the sun collapses to form a white dwarf the size of Earth, by what factor will its density increase? (*Hints:* The volume of a sphere is $(\frac{4}{3})\pi r^3$. See Appendix A for radii of sun and Earth.)

7. The Crab Nebula is now 1.35 pc in radius and is expanding at 1400 km/sec. About when did the supernova occur? (*Hint:* 1 pc equals 3×10^{13} km.

8. If the Cygnus Loop is 40 pc in diameter and is 40,000 years old, with what average velocity has it been expanding? (*Hints:* 1 pc equals 3×10^{13} km, and 1 year equals 3.15×10^7 seconds.)

9. Observations show that the gas ejected from SN1987A is moving at about 10,000 km/sec. How long will it take to travel one astronomical unit? one parsec? (*Hints:* 1 AU equals 1.5×10^8 km, and 1 pc equals 3×10^{13} km.)

10. If the stars at the turnoff point in a star cluster have masses of about $4M_\odot$, how old is the cluster?

Critical Inquiries for the Web

1. As seen in Figure 10-7, there is an incredible diversity of appearance for planetary nebulae. Browse the Web for images and information on these dying stars, and discuss why there is a range of shapes of planetary nebulae that we see.

2. Naked-eye supernovae in our galaxy are are, but astronomers have noted supernovae in other galaxies for years. Look for summaries of observations of recent supernovae. Are similar numbers of Type I and Type II being seen? Compare the number of supernovae seen during the last few years with that of two decades ago. Why are we finding so many more supernovae in recent years than in the past?

Go to the Brooks/Cole Astronomy Resource Center (www.brookscole.com/astronomy) for critical thinking exercises, articles, and additional readings from InfoTrac College Edition, Brooks/Cole's online student library.

CHAPTER 11

Almost anything is easier to get

into than out of.

Agnes Allen

Neutron Stars and Black Holes

Guidepost

The preceding two chapters have traced the story of stars from their birth as clouds of gas in the interstellar medium to their final collapse. This chapter finishes that story by discussing the two most extreme of the three end states of stellar evolution: neutron stars and black holes.

How strange and wonderful that we humans can talk about places in the universe where gravity is so strong it bends space, slows time, and curves light back on itself! To carry on these discussions, astronomers have learned to use the language of relativity. Throughout this chapter, remember that only the most sophisticated application of general relativity allows us to understand these strange bodies.

This chapter ends the story of individual stars, but it does not end the story of stars. The next chapter extends that story to the giant communities that stars live in—the galaxies.

Gravity always wins. However a star lives, it must eventually die in one of three final states—white dwarfs, neutron stars, or black holes. These objects, often called compact objects, are small monuments to the power of gravity. Almost all of the energy available has been squeezed out of compact objects, and we find them in their final, high-density states.

In this chapter, we must compare evidence and hypothesis with great care. Theory predicts the existence of these objects; but by their nature, they are difficult to detect. Is theory right? Do these objects really exist? To confirm the theories, astronomers have searched for real objects that could be identified as having the properties predicted by theory. That is, they have tried to find real neutron stars and real black holes. No medieval knight ever rode off on a more difficult quest, but modern astronomers have found objects that do seem to be neutron stars and black holes. We can judge their success by critically analyzing the evidence to see if it does confirm the theory.

11-1 Neutron Stars

A **neutron star** is a star of a few solar masses compressed to a radius of about 10 km. Its density is so high that the matter is stable only as a fluid of neutrons. Theory predicts that such an object would spin a number of times a second, be nearly as hot at its surface as the inside of the sun, and have a magnetic field a billion times stronger than Earth's. Two questions should occur to us immediately. First, how could any theory predict such a wondrously unbelievable star? And second, do such neutron stars really exist?

Theoretical Prediction of Neutron Stars

The neutron was discovered in the laboratory in 1932, and its properties suggested something fantastic. Neutrons spin in much the way that electrons do, which means that neutrons must obey the Pauli exclusion principle. In that case, if neutrons are packed together tightly enough, they can become degenerate just as electrons do. White dwarfs are supported by degenerate electrons, and, in 1932, the Russian physicist Lev Landau predicted that neutron stars might exist supported by degenerate neutrons. Of course, the inside of a

neutron star would have to be much denser than the inside of a white dwarf.

Only two years later, in 1934, Walter Baade and Fritz Zwicky provided a pedigree for neutron stars. They suggested that some of the most luminous novae in the historical record were not true novae but were caused by the collapse of a massive star in an explosion they called a supernova. What was left of the core of the star, they proposed, was a small, high-density neutron star.

Atomic physics gives us an explanation of how the collapsing core of a massive star could form a neutron star. As the core of the star begins to collapse, the density quickly reaches that of a white dwarf, but the weight is too great to be supported by degenerate electrons. The collapse of the core continues, and the atomic nuclei are broken apart by gamma rays. Almost instantly, the increasing density forces the freed protons to absorb electrons and become neutrons. In a fraction of a second, the collapsing core becomes a contracting ball of neutrons. As we saw in the previous chapter, the envelope of the star is blasted away in a supernova explosion. The remaining core of the star is left behind as a neutron star.

Theoretical calculations predict that a neutron star will be only 10 or so kilometers in radius (Figure 11-1) and will have a mass of about 10^{14} g/cm^3. On Earth, a sugar-cube-sized lump of this material would weigh 100 million tons. This is roughly the density of the atomic nucleus, and we can think of a neutron star as matter with all of the empty space squeezed out of it.

Figure 11-1
A tennis ball and a road map illustrate the relative size of a neutron star. Such an object, containing slightly more than the mass of the sun, would fit with room to spare inside the beltway around Washington, D.C. *(Photo by author)*

How massive can a neutron star be? That is a critical question, and a difficult one to answer because we don't know the strength of pure neutron material. Scientists can't make such matter in the laboratory, so its properties must be predicted theoretically. The most widely accepted calculations suggest that a neutron star cannot be more massive than 2 to 3 solar masses. If a neutron star were more massive than that, the degenerate neutrons would not be able to support the weight, and the object would collapse (presumably into a black hole).

Simple physics, the physics we have used in previous chapters to discuss normal stars, predicts that neutron stars should be hot, spin rapidly, and have strong magnetic fields. We have seen that contraction heats the gas in a star. As the gas particles fall inward, they pick up speed, and when they collide, their high speeds become heat. The sudden collapse of the core of a massive star to a radius of 10 km should heat it to millions of degrees. Furthermore, neutron stars should cool slowly because the heat can escape only from the surface, and neutron stars are so small they have little surface from which to radiate. Thus, basic theory predicts that neutron stars should be very hot.

The conservation of angular momentum predicts that neutron stars should spin rapidly. All stars rotate to some extent because they form from swirling clouds of interstellar matter. As such a star collapses, it must rotate faster because it conserves angular momentum. Recall that we see this happen when ice skaters spin slowly with their arms extended and then speed up as they pull their arms closer to their bodies (see Figure 10-23). In the same way, a collapsing star must spin faster as it pulls its matter closer to its axis of rotation. If the sun collapsed to a radius of 10 km, its period of rotation would decrease from 25 days to about 0.001 second. We might expect the collapsed core of a massive star to rotate 10 or 100 times a second.

Basic theory also predicts that a neutron star should have a powerful magnetic field. Whatever magnetic field a star has is frozen into the star. The gas of the star is ionized, and that means the magnetic field cannot move easily through the gas. When the star collapses, the magnetic field is carried along and squeezed into a smaller area, which could make the field a billion times stronger. Because some stars have magnetic fields over 1000 times stronger than the sun's, we might expect a neutron star to have a magnetic field as much as a trillion times stronger than the sun's. For comparison, that is about 10 million times stronger than any magnetic field ever produced in the laboratory.

Basic physics predicted the properties of neutron stars, but it also predicted that such objects should be difficult to observe. Neutron stars are very hot, so from our understanding of black-body radiation we can predict they will radiate most of their energy in the X-ray part of the spectrum, radiation that could not be observed in the 1940s and 1950s because astronomers could not put their telescopes above Earth's atmosphere. Also, the small surface areas of neutron stars mean that they will be faint objects. Thus astronomers of the mid-20th century were not surprised that none of the newly predicted neutron stars were found. Neutron stars were, at that point, entirely theoretical objects. Progress came not from theory but from observation.

The Discovery of Pulsars

In November 1967, Jocelyn Bell, a graduate student at Cambridge University in England found a peculiar pattern on the paper chart from a radio telescope. Unlike other radio signals from celestial bodies, this was a series of regular pulses (Figure 11-2). At first she and the leader of the project, Anthony Hewish, thought the signal was interference, but they found it day after day in the same place in the sky. Clearly, it was celestial in origin.

Another possibility, that it came from a distant civilization, led them to consider naming it LGM for Little Green Men. But within a few weeks, the team found three more objects in other parts of the sky pulsing with different periods. The objects were clearly natural, and the team dropped the name LGM in favor of **pulsar**—a contraction of *pulsing star*. The pulsing radio source Bell had observed with her radio telescope was the first known pulsar (Figure 11-2).

As more pulsars were found, astronomers argued over their nature. Periods ranged from 0.033 to 3.75 seconds and were nearly as exact as an atomic clock. Months of observation showed that many of the periods were slowly growing longer by a few billionths of a second per day. Thus, whatever produced the regular pulses had to be highly precise, nearly as exact as an atomic clock, but it had to gradually slow down.

It was easy to eliminate possibilities. Pulsars could not be stars. A normal star, even a small white dwarf, is too big to pulse that fast. Nor could a star

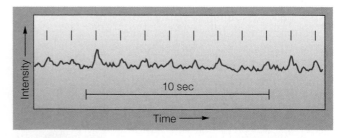

Figure 11-2
The 1967 detection of regularly spaced pulses in the output of a radio telescope led to the discovery of pulsars. This record of the radio signal from the first pulsar, CP1919, contains regularly spaced pulses (marked by ticks). The period is 1.33730119 seconds.

with a hot spot on its surface spin fast enough to produce the pulses. Even a small white dwarf would fly apart if it spun 30 times a second.

The pulses themselves gave the astronomers a clue. The pulses lasted only about 0.001 second. This places an upper limit on the size of the object producing the pulse. If a white dwarf blinked on and then off in that interval, we would not see a 0.001-second pulse. The near side of the white dwarf would be about 6000 km closer to us, and light from the near side would arrive 0.022 seconds before the light from the bulk of the white dwarf. Thus its short blink would be smeared out into a longer pulse. This is an important principle in astronomy—an object cannot change its brightness appreciably in an interval shorter than the time light takes to cross its diameter. If pulses from pulsars are no longer than 0.001 second, then the objects cannot be larger than 300 km (190 miles) in diameter.

Only a neutron star is small enough to be a pulsar. In fact, a neutron star is so small, it can't vibrate slowly enough, but it can spin as fast as 1000 times a second without flying apart. The missing link between pulsars and neutron stars was found in late 1968, when astronomers discovered a pulsar at the heart of the Crab Nebula. The Crab Nebula is a supernova remnant, and theory predicts that some supernovae leave behind a neutron star (Figure 11-3). Since 1968, over 1000 pulsars have been found, of which only a few are located inside supernova remnants.

The short pulses and the discovery of the pulsar in the Crab Nebula are strong hints that pulsars are neutron stars. If we combine theory and observation, we can devise a model of a pulsar.

A Model Pulsar

Pulsar is a misnomer. The periodic flashing of pulsars is linked to rotation, not pulsation. The spinning neutron star emits beams of radiation that sweep around the sky. When one of these beams sweeps over us, we detect a pulse, just as sailors see a pulse of light when the beam from a lighthouse sweeps over their ship. In fact, this model is called the **lighthouse theory.**

The lighthouse theory is generally accepted, and astronomers are becoming more confident of the mechanism that produces the beams. The theory suggests that the neutron star spins so fast and its magnetic field is so strong that it acts as a generator and creates an electric field around itself. This field is so intense that it rips charged particles, mostly electrons, out of the surface near the magnetic poles and accelerates them to high velocity. These accelerated elec-

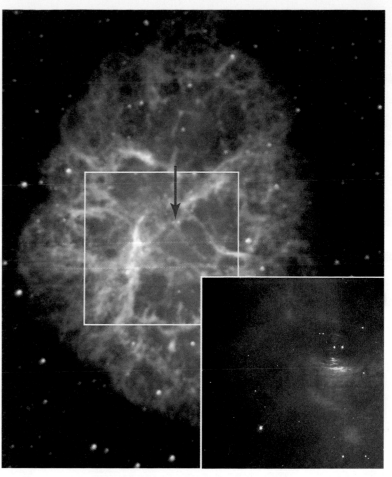

Figure 11-3
The Crab Nebula (Figure 10-14) is a hollow shell of filaments enclosing a hazy nebula. A pulsar blinking at the center (arrow) produces energy to power the synchrotron radiation from the nebula. Jets from the pulsar (inset) excite gas near the pulsar, producing changing wisps and waves in the nebula. *(Palomar Observatory/Caltech; inset, Jeff Hester and Paul Scowen, Arizona State University, NASA)*

trons emit photons traveling in the same direction as the electrons. Thus, the photons leave the neutron star in narrow beams emanating from the magnetic poles. If the magnetic axis is inclined with respect to the axis of rotation, as is the case with Earth and most of the planets in the solar system that have magnetic fields, then the neutron star will sweep the beams around the sky (Figure 11-4).

Our model describes the general properties of a pulsar, but a detailed analysis of the Crab Nebula warns us that spinning neutron stars are much more complex than our model. In Chapter 10, we noted that the blue glow of the Crab Nebula is produced by synchrotron radiation, which is produced by high-speed electrons spiraling through a magnetic field. In the 900 years since the Crab supernova, such high-speed electrons should have radiated away their energy and slowed, but the Crab Nebula continues to glow. Astronomers have recognized that the Crab pulsar is the source of the energy that keeps the nebula glowing.

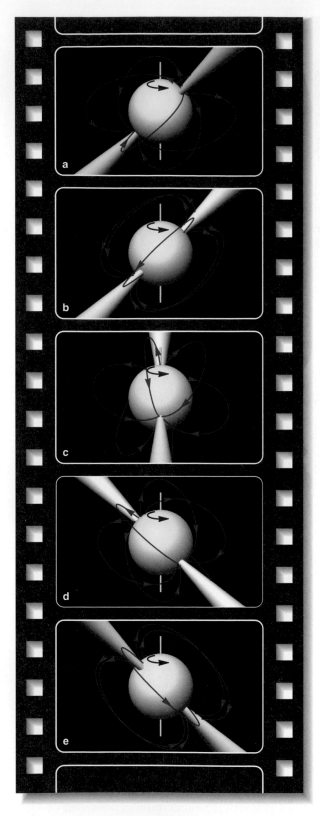

Figure 11-4
Schematic diagram of a neutron star (yellow green) with its powerful magnetic field (red). Beams of electromagnetic radiation blast out of the magnetic poles, and the rotation of the neutron star sweeps them around the sky like beams of light from a lighthouse.

Images recorded by the Hubble Space Telescope reveal the complexity of the phenomena associated with the pulsar (Figure 11-3 inset). Circular wisps of excited gas move away from the pulsar in its equatorial plane at about half the speed of light and jets of high-energy particles blast away from its poles. Both of these phenomena appear to be produced as the spinning neutron star accelerates subatomic particles such as electrons and positrons to nearly the speed of light and flings them away into the nebula. Where these streams of particles encounter the gas of the inner supernova remnant, they excite the gas, and we see glowing regions that shift position in times as short as days. Clearly, the 900-year-old neutron star in the Crab Nebula is still an active source of energy for the nebula.

Since the discovery of the first pulsar, astronomers have found over a thousand, and there are probably many more. Only when a pulsar's beams sweep over Earth do we detect its presence. In most cases, the beams of radio energy never point to Earth, and the pulsar remains almost undetectable (Figure 11-5).

The Evolution of Pulsars

Astronomers are beginning to understand how a pulsar ages. When it first forms, it is spinning very fast, perhaps nearly 100 times a second, and it contains a

Figure 11-5
A solitary neutron star (arrow) glows faintly in this Hubble Space Telescope image. With a temperature of nearly 700,000 K, the neutron star cannot be larger than about 28 km in diameter. It is not known to pulse, perhaps because its beams do not sweep over Earth. *(Fred Walter, State University of New York at Stony Brook, NASA)*

strong magnetic field. As it converts its energy of rotation into radiation, it gradually slows, and its magnetic field grows weaker. The average pulsar is apparently only a few million years old, and the oldest is about 10 million years old. By the time a pulsar gets older than that, it is rotating too slowly to generate detectable radio beams.

If a pulsar is young, we would expect it to spin fast, have a strong magnetic field, and consequently emit very strong beams of radiation. The Crab Nebula gives us an example of such a system. Only about 900 years old, the Crab pulsar is so powerful it emits photons of radio, infrared, visible, X-ray, and gamma-ray wavelengths (Figure 11-6). Careful measurements of its brightness with high-speed instruments show that it blinks twice for every rotation. One beam sweeps almost directly over us and we detect a strong pulse. Half a rotation later, the edge of the other beam sweeps past us, and we detect a weaker pulse.

We would expect only the most energetic pulsars to produce short-wavelength photons and thus pulse at visible wavelengths. The Crab Nebula pulsar is young and powerful, and it produces visible pulses, and so does a pulsar called the Vela pulsar (located in the Southern Hemisphere constellation Vela). The Vela pulsar is fast, pulsing about 11 times a second

and, like the Crab Nebula pulsar, is located inside a supernova remnant. Its age is estimated at about 11,000 years, young in terms of the average pulsar. Thus we suspect that pulsars are capable of producing optical pulses only when they are young.

We might expect to find the pulsars inside supernova remnants, but the statistics must be examined with care. Not every supernova remnant contains a pulsar, and not every pulsar is located inside a supernova remnant. Many supernova remnants probably contain pulsars whose beams never sweep over Earth, and it will be difficult to detect such pulsars. Also, some pulsars move through space at high velocity, which suggests that supernova explosions can occur asymmetrically, perhaps because of the violent convection in the exploding core. Such a violent, off-center explosion could give the neutron star a high velocity through space. Also, some supernovae probably occur in binary systems and fling the two stars apart at high velocity. In any case, pulsars are known to have such high velocities that many probably escape the disk of our galaxy. We should not be surprised that many neutron stars quickly leave their supernova remnant behind. And finally, we must remember that a pulsar can remain detectable for 10 million years or so, but a supernova remnant cannot survive more than about

Figure 11-6
High-speed images of the Crab Nebula pulsar show it pulsing at visual wavelengths and at X-ray wavelengths *(© AURA, Inc., NOAO, KPNO)* The period of pulsation is 33 milliseconds, and each cycle includes two pulses as its two beams of unequal intensity sweep over Earth. *(Courtesy F. R. Harnden, Jr., from* The Astrophysical Journal, *published by the University of Chicago Press; © 1984 The American Astronomical Society)*

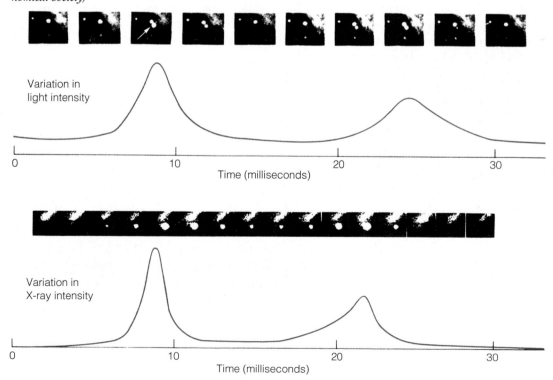

50,000 years before it is mixed into the interstellar medium. Thus, we should not be surprised that most pulsars are not in supernova remnants, and most supernova remnants do not contain pulsars.

The explosion of Supernova 1987A in February 1987 apparently formed a neutron star. We can draw this conclusion because a burst of neutrinos was detected passing through Earth, and theory predicts that the collapse of a massive star's core into a neutron star would produce such a burst of neutrinos. At first the neutron star would be hidden at the center of the expanding shells of gas ejected into space; but as the gas expands and thins we might be able to see it. Or, if its beams don't sweep over Earth, we might be able to detect it from its X-ray and gamma-ray emission. Although, years after the explosion, no neutron star has been detected, astronomers continue to watch the site, hoping to see the youngest pulsar known.

Our theory predicts that newborn pulsars should blink rapidly and old pulsars should blink slowly, but the handful that blink the fastest may be quite old. One of the fastest known pulsars is catalogued as PSR 1937 + 21 in the constellation Vulpecula. It pulses 642 times a second and is slowing down only slightly. The energy stored in the rotation of a neutron star at this rate is equal to the total energy of a supernova explosion, so it seemed difficult at first to explain this pulsar. It now appears that PSR 1937 + 21 is an old neutron star that has gained mass and rotational energy from a companion in a binary system. Like water hitting a mill wheel, the matter falling on the neutron star has spun it up to 642 rotations per second. With its old, weak magnetic field, it slows down very slowly and will continue to spin for a very long time.

A number of other very fast pulsars have been found and they are known generally as **millisecond pulsars** because their pulse periods are almost as short as a millisecond (0.001 sec). This produces some fascinating physics because the pulse period of a pulsar equals the rotation period of the neutron star. If a neutron star 10 km in radius spins 642 times a second, as does PSR 1937 + 21, then the period is 0.0016 seconds, and the equator of the neutron star must be traveling about 40,000 km/sec. That is fast enough to flatten the neutron star into an ellipsoidal shape and is nearly fast enough to break it up.

One reason pulsars are so fascinating is the extreme conditions we find in spinning neutron stars. To see natural processes of even greater violence, we have only to look at pulsars in binary systems.

Binary Pulsars

Of the hundreds of pulsars now known, a few are located in binary systems. These pulsars are of special interest because we can learn more about the neutron star by studying the orbital motions of the binary.

Also, in some cases, mass can flow from the companion star onto the neutron star, and that produces high temperatures and X rays.

The first binary pulsar was discovered in 1974 when astronomers noticed that the pulse period of the pulsar PSR 1913 + 16 was changing. The period first grew longer and then grew shorter in a cycle that took 7.75 hours. Thinking of the Doppler shifts seen in spectroscopic binaries, the radio astronomers realized that the pulsar had to be in a binary system with a period of 7.75 hours. When the orbital motion of the pulsar carries it away from Earth, we see the pulse period slightly lengthened, just as the wavelength of light emitted by a receding source is lengthened. That is, we see a red shift. Then, when the pulsar rounds its orbit and approaches Earth, we see the pulse period slightly shortened — a blue shift. From these changing Doppler shifts, the astronomers could calculate the radial velocity of the pulsar around its orbit just as if it were a spectroscopic binary star (Chapter 8). The resulting graph of radial velocity versus time could be analyzed to find the shape of the pulsar's orbit (Figure 11-7).

A few dozen binary pulsars have been found; and by analyzing the Doppler shifts in their pulse periods and making reasonable assumptions about the mass of the companion stars, astronomers can estimate the mass of the neutron stars. Such masses lie between 1.5 and 2 solar masses, in good agreement with our theory of neutron stars.

Binary pulsars are relatively quiet systems now, but they may once have been violently active. The typical companion star to a binary pulsar was probably once a giant star losing mass to the neutron star, and that matter falling into the neutron star would have liberated tremendous energy. A single marshmallow dropped onto the surface of a neutron star from a distance of 1 AU would hit with the impact of a 3-megaton nuclear bomb. Even a small amount of matter flowing into a neutron star can generate high temperatures and release X rays and gamma rays.

As an example of such an active system, we can examine Hercules X-1. It emits pulses of X rays with a period of about 1.2 seconds, but every 1.7 days the pulses vanish for a few hours (Figure 11-8a). Astronomers can understand this system by comparing it to an eclipsing binary star. Hercules X-1 seems to contain a 2-solar-mass star and a neutron star that orbit each other with a period of 1.7 days. Matter flowing from the normal star into the neutron star becomes trapped in the neutron star's magnetic field and generates X rays that the spinning neutron star beams into space (Figure 11-8b). We receive a pulse of X rays every time a beam points our way. The X rays shut off completely every 1.7 days when the neutron star is eclipsed behind the normal star. Hercules X-1 is a complex system with many different

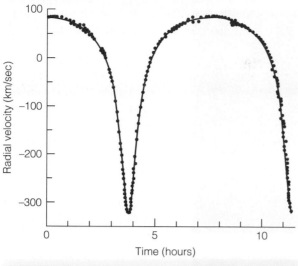

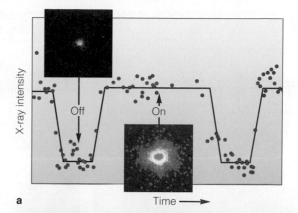

a

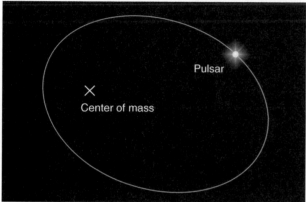

Figure 11-7
The radial velocity of a pulsar orbiting another star can be calculated from the Doppler shifts in its rate of pulsation. Here, the radial velocity of pulsar PSR 1913 + 16 varies over an interval of 7.75 hours. Analyzing this changing radial velocity allows astronomers to determine the size and shape of the pulsar's orbit as it revolves around the center of mass of the binary system.

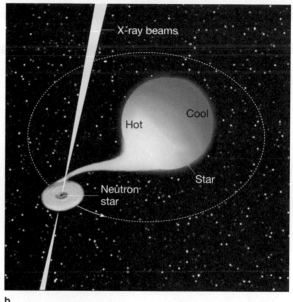

b

Figure 11-8
(a) X-ray observations of Hercules X-1 show that the X-ray pulses disappear periodically and then reappear (insets), and a graph of X-ray intensity versus time looks like the light curve of an eclipsing binary. *(Insets: J. Trümper, Max-Planck Institute)* (b) In the Hercules X-1 system, matter flows from a stellar companion into an accretion disk around a neutron star producing X rays, which heat the near side of the star to 20,000 K compared to only 7000 K on the far side.

high-energy processes going on simultaneously, and our quick analysis only serves to illustrate how complex and powerful such binary systems are during mass transfer.

The X-ray source 4U1820-30 illustrates another way neutron stars can interact with normal stars. In this system, a neutron star orbits a white dwarf with a period of only 11 minutes (Figure 11-9a and b). The separation between the two objects is only about a third the distance between Earth and the moon. To explain how such a very close pairing could originate, theorists suggest that a neutron star collided with a giant star and went into an orbit *inside* the star. (Recall the low density of the outer envelope of giant stars.) The neutron star would have gradually eaten away the giant star's envelope from the inside, leaving the white dwarf behind. Matter still flows from the white dwarf into an accretion disk and then down to the surface of the neutron star (Figure 11-9c), where it accumulates until it ignites to produce periodic

bursts of X rays. Notice the similarity between this mechanism and that responsible for novae.

In the previous section, we concluded that millisecond pulsars are old neutron stars that have been spun up to high-speed rotation by matter flowing from a companion star. A few millisecond pulsars that are not members of binary systems seem to challenge this hypothesis, but a millisecond pulsar called the Black Widow may help explain this seeming contradiction. The Black Widow pulsar is a fast pulsar (622 pulses per second) with a low-mass companion. Spectroscopic observations reveal that the blast of radiation and high-energy particles from the neutron star is boiling away the surface of the companion. The Black Widow pulsar is destroying its companion and will eventually be left as a solitary millisecond pulsar.

Figure 11-9
(a) At visible wavelengths, the center of star cluster NGC6624 is crowded with stars. *(Ivan King and NASA/ESA)* (b) In the ultraviolet, one object stands out, an X-ray source consisting of a neutron star orbiting a white dwarf. *(Ivan King and NASA/ESA)* (c) An artist's conception shows matter flowing from the white dwarf into an accretion disk around the neutron star. *(Dana Berry/STScI)*

Neutron stars were first predicted by theory, and, although the evidence can never prove absolutely that neutron stars exist (Window on Science 11-1), modern astronomers have found many examples of objects for which no other explanation seems adequate. Thus, we can have great confidence that neutron stars are real. Other theories that describe how they emit beams of radiation and how they form and evolve are less certain, but continuing observations at many wavelengths are expanding our understanding of these last embers of massive stars. In fact, observations of one pulsar have turned up objects no one predicted.

Pulsar Planets

Finding planets orbiting stars other than the sun is very difficult, and only a few are known. Oddly, the first such planets were found orbiting a neutron star.

Because a pulsar's period is so precise, astronomers can detect tiny variations by comparison with atomic clocks. When astronomers checked pulsar PSR 1257 + 12, they found variations in the period of pulsation much like those caused by the orbital motion of the binary pulsar (Figure 11-10a). However, in the case of PSR 1257 + 12, the variations were much smaller, and when they were interpreted as Doppler shifts, it became evident that the pulsar was being orbited by at least two objects with planet-

No scientific theory or hypothesis can be proved correct. We can test a theory over and over by performing experiments or making observations, but we can never prove that the theory is absolutely true. It is always possible that we have misunderstood the theory or the evidence, and the next observation we make might disprove the theory. In that sense, we never learn anything new until we disprove a theory or hypothesis. Only at that point do we discover something we did not know before.

For example, we might propose the theory that the sun is mostly iron. We might test the theory by looking at the iron lines in the solar spectrum, and the strength of the lines would suggest our theory is right. Although our observation has confirmed our theory, we know nothing more than we did before. We might continue to confirm the theory over and over, but we still know nothing new and there is always the danger that we have misunderstood the evidence. Finally, we might study the formation of atomic spectra and discover that iron atoms are very good absorbers of photons, so the strong iron lines in the solar spectrum do not mean that the sun is made mostly of iron. In fact, the hydrogen lines mean that the sun is mostly hydrogen atoms. Now that we have disproved our hypothesis, we have learned something new.

The nature of scientific hypothesis testing can lead to two kinds of mistakes. Sometimes nonscientists will say, "You scientists just want to tear everything down —you don't believe anything." It is the nature of science to test every hypothesis, not because the scientist wants to tear things down, but because the scientist wants to know what is the most dependable description of nature. Others will say, "You scientists are never sure of anything." Again, the scientist knows that no theory can ever be proved correct. That the sun will rise tomorrow is very likely, but in the end it is still a theory.

People will say of an idea they dislike, "That is only a theory," as if a theory were simply a random guess. In fact, a theory can be a well-tested truth in which all scientists have great confidence. Yet we can never prove that any theory is absolutely true.

---

like masses of 3.4 and 2.8 Earth masses. The gravitational tugs of the planets make the pulsar wobble about the center of mass of the system by no more than 800 km, and that produces the tiny changes in period (Figure 11-10b).

Astronomers greeted this discovery with both enthusiasm and skepticism. As usual, they looked for ways to test the hypothesis. Simple gravitational theory predicts that the planets should interact and slightly modify each other's orbit. When the data were analyzed, that interaction was found, further confirming the hypothesis that the variations in the period of the pulsar are caused by planets. In fact, further data suggest the presence of two more planets orbiting the pulsar.

Astronomers wonder how a neutron star can have planets. The innermost planet orbiting PSR 1257 + 12 is only 0.36 AU from the pulsar. Any planets that orbit a star would be lost or vaporized when the star exploded. Furthermore, a star about to explode as a supernova would be a large giant or a supergiant, and planets only a few AU distant would be inside such a large star and could not survive. It seems more likely that these planets are the remains of a stellar companion that was devoured by the neutron star. In fact, the pulsar is very fast (162 pulses per second), suggesting that it was spun up in a binary system.

We can imagine what these worlds might be like. Formed from the remains of dying stars, they might have chemical compositions richer in heavy elements than Earth. We can imagine visiting these worlds, landing on their surfaces, and hiking across

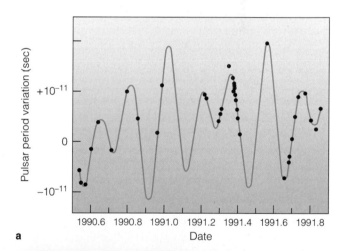

a

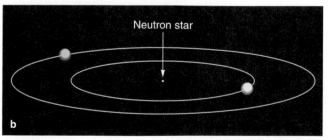

b

Figure 11-10
(a) The dots in this graph are observations showing that the period of pulsar PSR 1257 + 12 varies from its average value by a fraction of a billionth of a second. The blue line shows the variation that would be produced by two planets orbiting the pulsar. (b) As the planets orbit the pulsar, they cause it to wobble by less than 800 km, a distance that is invisibly small in this diagram.

their valleys and mountains. Above us, the neutron star would glitter in the sky, a tiny point of light.

Gamma Ray Bursts from Neutron Stars

In 1968, satellites that could detect gamma rays were put in orbit around Earth, and they immediately began picking up bursts of gamma rays coming from space. Typically one burst a day was seen lasting for seconds or minutes. These gamma-ray bursts have been a mystery for decades, but new observations suggest at least some of the bursts may be coming from neutron stars.

A 1992 theory proposes that some young neutron stars can have magnetic fields 100 times stronger than the average neutron star. Dubbed **magnetars,** these objects would be tremendously hot because the magnetic field would heat the surface of the neutron star to 10 million Kelvin. Shifts in the magnetic field would break the rigid crust of the neutron star in a violent starquake—equivalent to an earthquake on Earth. Such a starquake would release a great burst of high-energy photons.

Observations of two repeating gamma-ray bursters confirm that they are very hot and have powerful magnetic fields. This has been taken as confirmation that they are magnetars. One of the objects was the source of a burst of gamma rays that reached Earth on August 27, 1998, and temporarily ionized Earth's upper atmosphere, disrupting radio communication worldwide.

Astronomers estimate the distance to these two magnetars as a few times 10,000 ly. They are within our Milky Way galaxy. Other gamma-ray bursters are more distant, but they too may be related to neutron stars.

Because a gamma-ray burst is over so quickly and most do not appear to repeat, it has been hard to trace them to their source, but coordination between orbiting space telescopes and ground-based observatories has revealed that at least some gamma-ray bursters are located in very distant galaxies. To be detectable at such great distances, they must produce tremendous amounts of energy. One theory proposes that the gamma-ray bursts occur when two neutron stars orbiting each other in a binary system spiral together and collide. Such an event would release about the right amount of energy to be observed from Earth as a gamma-ray burst.

Gamma ray bursts occur daily, but exotic magnetars and colliding neutron stars seem unusual. In fact, neutron stars are very common objects in the universe, and neutron stars in binaries must occur in every galaxy. These peculiar neutron stars only seem unusual because they are difficult to observe.

The study of gamma-ray bursts lies at the very frontier of modern astronomy. The solution to the mystery seems likely to involve neutron stars.

REVIEW Critical Inquiry

How can a neutron star be found at X-ray wavelengths?

First, we should remember that a neutron star is very hot because of the heat released when it contracts to a radius of 10 km. It could easily have a surface temperature of 1,000,000 K, and Wien's law (By the Numbers 6-1) tells us that such an object will radiate most intensely at a very short wavelength typical of X rays. However, we know that the total luminosity of a star depends on its surface temperature and its surface area, and a neutron star is so small it can't radiate much energy. It would be hard to find even with an X-ray telescope.

There is, however, a second way a neutron star can radiate X rays. If a normal star in a binary system loses mass to a neutron star companion, the inflowing matter will hit with so much energy that it will be heated to very high temperatures. It may form a very hot accretion disk that can radiate intense X rays easily detectable by X-ray telescopes orbiting above Earth's atmosphere.

If you discovered a pulsar, what observations would you make to determine whether it was young or old, single or a member of a binary system, alone or accompanied by planets?

Perhaps the strangest planets in the universe are those orbiting pulsars. But however strange a pulsar planet may be, we can imagine going to one. Our next topic, in contrast, seems beyond the reach of even our imaginations.

11-2 Black Holes

We have now studied white dwarfs and neutron stars, two of the three end states of dying stars. Now we turn to the third—black holes.

Although the physics of black holes is difficult to discuss without sophisticated mathematics, simple logic is sufficient to predict that they should exist. Our problem is to consider their predicted properties and try to confirm that they exist. What objects observed in the heavens could be real black holes? More difficult than the search for neutron stars, the quest for black holes is nevertheless meeting with success.

To begin our discussion of black holes, we must consider a simple question. How fast must an object travel to escape from the surface of a celestial body? The answer will lead us to black holes.

Escape Velocity

Suppose we threw a baseball straight up. How fast must we throw it if it is not to come down? Of course, gravity would pull back on the ball, slowing it, but if the ball were traveling fast enough to start with, it would never come to a stop and fall back. Such a ball would escape from Earth. The **escape velocity** is the initial velocity an object needs to escape from a celestial body (Figure 11-11).

Whether we are discussing a baseball leaving Earth or a photon leaving a collapsing star, the escape velocity depends on two things: the mass of the celestial body and the distance from the center of mass to the escaping object. If the celestial body has a large mass, its gravity is strong and we need a high velocity to escape; but if we begin our journey farther from the center of mass, the velocity needed is less. For example, to escape from Earth, a spaceship would have to leave Earth's surface at 11 km/sec (25,000 mph), but if we could launch spaceships from the top of a tower 1000 miles high, the escape velocity would be only 8.8 km/sec (20,000 mph). If we could make an object massive enough or small enough, its escape velocity could be greater than the speed of light. Such an object could never be seen because light could never leave it.

Rev. John Mitchell, a British amateur astronomer, was the first person to realize that Newton's laws of gravity and motion implied the existence of black holes. In 1783, he pointed out that an object 500 times the radius of the sun but of the same density would have an escape velocity greater than the speed of light. Then, "all light emitted from such a body would be made to return towards it." Mitchell had discovered the black hole.

Schwarzschild Black Holes

If the core of a star collapses and contains more than about 3 solar masses, no known force can stop it. The object cannot stop collapsing when it reaches the size of a white dwarf because degenerate electrons cannot support the weight. It cannot stop when it reaches the size of a neutron star because degenerate neutrons cannot support the weight. No known force remains to stop the object from collapsing to zero radius.

As an object collapses, its density and the strength of its surface gravity increase; and if an object collapses to zero radius, its density and gravity become infinite. Mathematicians call such a point a **singularity;** but in physical terms it is difficult to imagine an object of zero radius. Some theorists believe that a singularity is impossible and that when we better understand the laws of physics, we will discover that the collapse halts before the radius zero. Astronomically, it seems to make little difference.

If the contracting core of a star becomes small enough, the escape velocity in the region around it is so

Figure 11-11
Escape velocity, the velocity needed to escape from a celestial body, depends on mass. The escape velocity at the surface of a very small body would be so low we could jump into space. Earth's escape velocity is much larger, about 11 km/sec (25,000 mph).

large that no light can escape. We can receive no information about the object or about the region of space near it, and we refer to this region as a **black hole** (Figure 11-12). The boundary of the black hole is called the **event horizon,** because any event that takes place inside the event horizon is invisible to an outside observer (Figure 11-13). To see how a black hole can exist, we must consider general relativity.

In 1916, Albert Einstein published a mathematical theory of space and time that became known as the general theory of relativity. Einstein treated space and time as a single entity—space–time. His equations showed that gravity could be described as a curvature of space–time, and almost immediately the astronomer Karl Schwarzschild found a way to solve the equations to describe the gravitational field around a single, nonrotating, electrically neutral lump of matter. That solution contained the first general relativistic description of a black hole; nonrotating, electrically neutral black holes are now known as Schwarzschild black

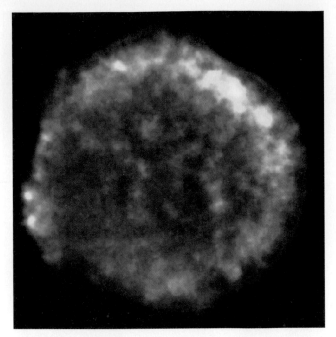

Figure 11-12
The collapse of a massive star causes a supernova explosion that ejects the outer layers of the star in an expanding shell while the core of the star may become so small it forms a black hole. Tycho's super-nova occurred in 1572, and the expanding shell is shown in this X-ray image. No pulsar has been found, so the collapsing star may have formed a black hole. *(John P. Hughes, Rutgers Univ.)*

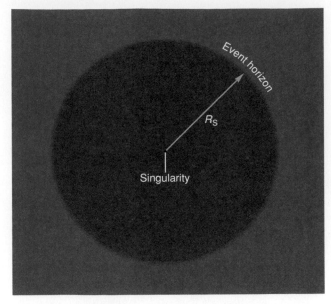

Figure 11-13
A black hole forms when an object collapses to a small size (perhaps to a singularity) and the escape velocity becomes so great light cannot escape. The boundary of the black hole is called the event horizon because any event that occurs inside is invisible to outside observers. The radius of the black hole R_s is the Schwarzschild radius.

holes. In recent decades, theorists such as Stephen W. Hawking have found a way to apply the mathematical equations of the general theory of relativity to charged, rotating black holes. For our discussion the differences are minor, and we may proceed as if all black holes were Schwarzschild black holes.

Schwarzschild's solution shows that if matter is packed into a small enough volume, then space–time curves back on itself. Objects can still follow paths that lead into the black hole, but no path leads out, so nothing can escape, not even light. Thus, the inside of the black hole is totally beyond the view of an outside observer. The event horizon is the boundary between the isolated volume of space–time and the rest of the universe, and the radius of the event horizon is called the **Schwarzschild radius, R_s**—the radius within which an object must shrink to become a black hole (Figure 11-13).

Although Schwarzschild's work was highly mathematical, his conclusion is quite simple. The Schwarzschild radius (in meters) depends only on the mass of the object (in kilograms):

$$R_s = \frac{2GM}{c^2}$$

In this simple formula, G is the gravitational constant, M is the mass, and c is the speed of light. A bit of arithmetic shows that a 1-solar-mass black hole will have a Schwarzschild radius of 3 km, a 10-solar-mass black hole will have a Schwarzschild radius of 30 km, and so on (Table 11-1). Even a very massive black hole would not be very large.

Every object with mass has a Schwarzschild radius, but not every object is a black hole. For example, Earth has a Schwarzschild radius of about 1 cm, but it could become a black hole only if we squeezed it inside that radius. Fortunately, Earth will not collapse spontaneously into a black hole because its mass is less than the critical mass of about 3 solar masses. Only exhausted stellar cores more massive than this can form black holes under the sole influence of their own gravity. In this chapter, we are interested in black holes that might originate from the deaths of massive stars. These would have masses larger than 3 solar masses. In later chapters, we will

Table 11-1 The Schwarzschild Radius

	Mass ($M_\odot$)	R_s
Star	10	30 km
Star	3	9 km
Star	2	6 km
Sun	1	3 km
Earth	0.000003	0.9 cm

encounter black holes whose masses might exceed a million solar masses.

Do not think of black holes as giant vacuum cleaners that will pull in everything in the universe. A black hole is just a gravitational field, and at a reasonably large distance its force is quite small. If the sun were replaced by a 1-solar-mass black hole, the orbits of the planets would not change at all. The gravity of a black hole becomes extreme only when we approach close to it. Earth's gravitational field is not dangerous because we can never come closer to its center than 6400 km (4000 miles). But were Earth compressed into a black hole only a centimeter in radius, we could come much closer to the center of the gravitational field, and then we would feel extreme gravitation. We can imagine that the universe contains numerous black holes. So long as we and other objects stay a safe distance from the black holes, they will have no catastrophic effects.

A Leap into a Black Hole

Before we can search for real black holes, we must understand what theory predicts about the appearance of a black hole. To explore that idea, we can imagine that we leap, feet first, into a Schwarzschild black hole.

If we were to leap into a black hole of a few solar masses from a distance of an astronomical unit, the gravitational pull would not be very large, and we would fall slowly at first. Of course, the longer we fell and the closer we came to the center, the faster we would travel. Our wristwatches would tell us that we fell for about 65 days before we reached the event horizon.

Our friends who stayed behind would see something different. They would see us falling more slowly as we came closer to the event horizon because, as explained by general relativity, clocks slow down in curved space–time. This is known as **time dilation.** In fact, our friends would never actually see us cross the event horizon. To them we would fall more and more slowly until we seemed hardly to move. Generations later, our descendants could focus their telescopes on us and see us still inching closer to the event horizon. We, however, would have sensed no slowdown and would conclude that we had crossed the event horizon after only about 65 days.

Another relativistic effect would make it difficult to see us with normal telescopes. As light travels out of a gravitational field, it loses energy, and its wavelength grows longer. This is known as the **gravitational red shift.** Although we would notice no effect as we fell toward the black hole, our friends would need to observe at longer and longer wavelengths in order to detect us.

While these relativistic effects seem merely peculiar, other effects would be quite unpleasant. Imagine again that we are falling feet first toward the event horizon of a black hole. We would feel our feet, which would be closer to the black hole, being pulled in more strongly than our heads. This is a tidal force, and at first it would be minor. But as we fell closer, the tidal force would become very large. Another tidal force would compress us as our left side and our right side both fell toward the center of the black hole. For any black hole with a mass like that of a star, the tidal forces would crush us laterally and stretch us longitudinally long before we reached the event horizon. The friction from such severe distortions of our bodies would heat us to millions of degrees, and we would emit X rays and gamma rays (Figure 11-14).

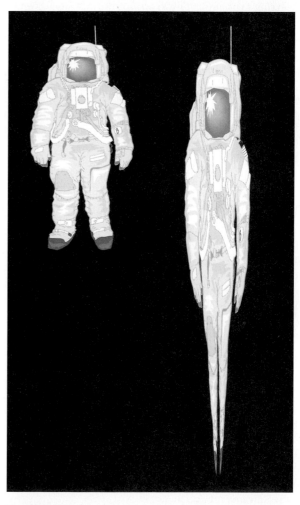

Figure 11-14
Leaping feet first into a black hole. A person of normal proportions (left) would be distorted by tidal forces (right) long before reaching the event horizon around a typical black hole of stellar mass. Tidal forces would stretch the body lengthwise while compressing it laterally. Friction from this distortion would heat the body to high temperatures.

Fraud is actually quite rare in science. The nature of science makes fraud difficult, and the way scientists publish their research makes it almost impossible. In fact, we can think of science as a set of unwritten rules of behavior that have evolved to prevent scientists from lying to one another or to themselves, even by accident.

Suppose for a moment that we wanted to commit scientific fraud. We would have to invent data supposedly obtained from experiment or observation. We might invent X-ray data supposedly obtained by observing an X-ray binary star. Or if we were interested in theory, we might invent a fraudulent mathematical calculation of the physics going on in an X-ray binary. We might get away with it for a short time, but one of the most important rules in science is that good results must be reproducible. Other people must be able to repeat our observations, experiments, and calculations. In fact, most scientists routinely repeat other scientists' work as a way of getting started on a research topic. As soon as someone tries to repeat our fraudulent research, we will be caught. The more important a scientific result is, the sooner other scientists will repeat it, so we don't have much of a chance of getting away with scientific fraud. In this way, science is self-correcting.

Even if we could invent some convincing scientific research, we would probably have difficulty publishing it. When a scientist submits an article to a scientific journal, it is subject to peer review. That is, the editor of the journal sends the article to one or two other experts in the field for comment and suggestions. These reviewers often make helpful suggestions, but they may also point out errors that have to be fixed before the journal can publish the article. In some cases, an article may be so flawed the editor will refuse to publish it at all. If we submitted our fraudulent research on X-ray binaries, the reviewers would almost certainly notice things wrong with it, and it would never get into print.

Scientists know the rules, and they use them. If someone makes a big discovery and is interviewed by the press, scientists will begin asking, "Has this work been published in a peer-reviewed journal yet?" That is, they want to know if other experts have checked the work. Until research is published, it isn't really official, and most scientists would treat the results with care.

Fraud isn't impossible in science. Some cases have even been in the national news. Big grant money is a terrible temptation. But in science, "the truth will out." Because of the way scientists reproduce research and because of the way research is published, scientific fraud is quite rare.

Some years ago a popular book suggested that we could travel through the universe by jumping into a black hole in one place and popping out of another somewhere far across space. That might make for good science fiction, but tidal forces would make it an unpopular form of transportation even if it worked.

Our imaginary leap into a black hole is not entirely frivolous. We now know how to find a black hole: Look for a strong source of X rays. It may be a black hole into which matter is falling.

The Search for Black Holes

Do black holes really exist? Beginning in the 1970s, astronomers searched for observational proof that their theories were correct. They tried to find one or more objects that were obviously black holes. That very difficult search is a good illustration of how the unwritten rules of science help us understand nature (Window on Science 11-2).

A black hole alone is totally invisible because nothing can escape from the event horizon. But a black hole into which matter is flowing would be a source of X rays. Of course, X rays can't escape from inside the event horizon, but X rays emitted by the heated matter flowing into the black hole could escape. An isolated black hole in space will not have much matter flowing into it, but a black hole in a binary system might receive a steady flow of matter transferred from the companion star. Thus, we can search for black holes by searching among X-ray binaries.

Some X-ray binaries such as Hercules X-1 contain a neutron star, and they will emit X rays much as would a binary containing a black hole. We can tell the difference, however, if we can find the mass of the compact object. If the mass is greater than 3 solar masses, the object can't be a neutron star, and we can conclude it must be a black hole.

The first X-ray binary suspected of harboring a black hole was Cygnus X-1, the first X-ray object discovered in Cygnus. It contains a supergiant O star and a compact object orbiting each other with a period of 5.6 days. Matter flows from the O star as a strong stellar wind, and some of that matter enters a hot accretion disk around the compact object (Figure 11-15). The compact object is invisible, but Doppler shifts in the spectrum reveal the motion of the O star around the center of mass of the binary. By making assumptions about the mass of the O star and the geometry of the orbit, astronomers can calculate the mass of the compact object—between 10 and 15 solar masses, well above the maximum for a neutron star.

To confirm that black holes exist, we need to find a conclusive example, an object that can't be anything else. Cygnus X-1 fails. Astronomers were quick to point out that the O star might not be a normal star

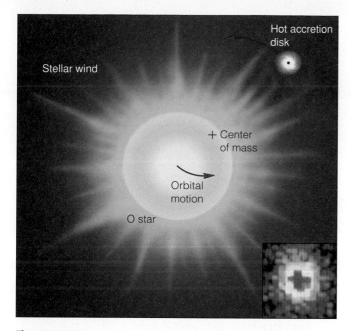

Figure 11-15

The X-ray source Cygnus X-1 is a supergiant O star and a compact object orbiting each other. Gas from the O star's stellar wind flows into the hot accretion disk, and the X rays we detect come from the disk. An X-ray image is shown in the inset. The masses can be estimated from the Doppler shifts produced in the spectrum of the O star as it circles the center of mass. *(J. Trümper, Max-Planck Institute)*

and assumptions about its mass might be far off. Also, there may be a third star in the system, and that would confuse the analysis. The compact object in Cygnus X-1 cannot be conclusively shown to be more massive than 3 solar masses, and thus Cygnus X-1 is not a conclusive example of a black hole.

As X-ray telescopes have found more X-ray objects, the list of black hole candidates has grown to a few dozen. A few of these objects, such as the first two in Table 11-2, contain massive stellar companions, either giants, supergiants, or massive main-sequence stars. This makes such systems difficult to analyze because such massive companions dominate the system. A good example of this sort is LMC X-3 (LMC refers to the Large Magellanic Cloud, a small galaxy near our own). The compact object in LMC X-3 has a mass be-

tween 4 and 11 solar masses. One reason astronomers think the compact object is so massive is that it distorts the shape of the B main-sequence star into an egg shape; and as the system rotates, the light from the B star varies because we see the side of the egg and then the end. By analyzing the light change, astronomers can determine the shape of the B star and from that find the mass of the compact object.

Many of the black hole candidates are binary systems in which the normal star is a lower-mass main-sequence star. Such systems do not remain X-ray sources continuously, but suffer X-ray nova outbursts as matter flows rapidly into the accretion disk. A year or so after an outburst, the flow has stopped, the accretion disk has dimmed, and astronomers can detect the spectrum of the main-sequence star. The spectral type and Doppler motions of the main-sequence companion reveal the mass of the compact object with only limited uncertainty. The lower-mass stellar companion makes these X-ray nova systems easier to analyze than are systems with massive companions.

Four examples of these X-ray nova systems are shown in Table 11-2. V616 Mon is an old nova that erupted again in 1975. It contains an ordinary main-sequence K star and a compact object that orbit each other with a period of 7.75 hours. From the orbital motion and the distortion of the K star, astronomers conclude the compact object must have a mass between 3.3 and 4.2 solar masses. Three of the best understood examples are V404 Cygni; J1655-40, also known as Nova Scorpii 1994; and QZ Vul. The compact objects in these systems seem much too massive to be anything but black holes.

The growing list of X-ray binaries with compact objects exceeding 3 solar masses has convinced nearly all astronomers that black holes really do exist. The problem now is to understand how these objects interact with matter flowing into them through accretion disks to produce X rays, gamma rays, and jets of matter.

Jets of Energy from Compact Objects

Our impression of a black hole might suggest that it is impossible to get any energy out of such an object. In later chapters, we will meet galaxies that extract vast amounts of energy from massive black holes, so we should pause here to see how a compact object can produce energy.

Whether a compact object is a black hole or a neutron star, it has a strong gravitational field. Any matter flowing into that field is accelerated inward, and because it must conserve angular momentum, it flows into an accretion

Table 11-2 Six Black Hole Candidates

Object	Location	Companion Star	Orbital Period	Mass of Compact Object
Cygnus X-1	Cygnus	O supergiant	5.6 days	10–15 $M_\odot$
LMC X-3	Dorado	B3 main-sequence	1.7 days	4–11 $M_\odot$
V616 Mon	Monocerotis	K main-sequence	7.75 hours	3.3–4.2 $M_\odot$
V404 Cygni	Cygnus	K main-sequence	6.47 days	8–15 $M_\odot$
J1655-40	Scorpius	F-G main-sequence	2.61 days	4–5.2 $M_\odot$
QZ Vul	Vulpecula	K main-sequence	8 hours	5–14 $M_\odot$

disk made so hot by friction that the inner regions can emit X rays and gamma rays. Thermal and magnetic processes can cause the inner parts of the accretion disk to eject high-speed jets of matter in opposite directions along the axis of rotation. This process recalls the bipolar flows ejected by protostars, but it is much more powerful. As examples, the Vela pulsar is known to have a jet visible at X-ray wavelengths, and the black hole candidate J1655-40 is observed at radio wavelengths to be ejecting oppositely directed jets at 92 percent the speed of light.

One of the most powerful examples of this process is an X-ray binary called SS433. Its optical spectrum shows sets of spectral lines that are Doppler-shifted by about one-fourth the speed of light, with one set shifted to the red and one set shifted to the blue. The object is both receding and approaching at a fantastic speed. Furthermore, the two sets of lines shift back and forth across each other with a period of 164 days.

Apparently, SS433 is a binary system in which a compact object (a neutron star) pulls matter from its companion star and forms an extremely hot accretion disk. Jets of high-temperature gas blast away in beams aimed in opposite directions. As the disk precesses, it sweeps these beams around the sky once every 164 days, and we see light from gas trapped in both beams. One beam produces a red shift, and the other produces a blue shift (Figure 11-16). SS433 is a prototype that illustrates how the gravitational field around a compact object can produce powerful beams of radiation and matter. We will meet this phenomenon again when we study peculiar galaxies.

Figure 11-16
The generally accepted model of SS433 includes a compact object with a very hot accretion disk producing beams of radiation with embedded blobs of gas traveling at 25 percent the velocity of light. The precession of the disk swings the beams around a conical path every 164 days.

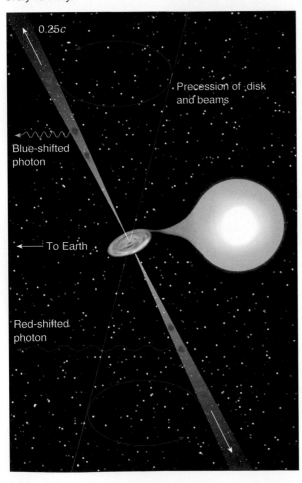

0.25c

Precession of disk and beams

Blue-shifted photon

To Earth

Red-shifted photon

Compact objects emitting X rays and producing precessing jets of radiation and gas may not be as unusual as they seem. Many stars collapse to form black holes or neutron stars in binary systems, but these are objects of only a few solar masses. We will see similar phenomena many times more powerful when we explore the galaxies in Chapters 12, 13, and 14.

Summary

If the remains of a star collapse with a mass greater than the Chandrasekhar limit of 1.4 solar masses, then the object cannot reach stability as a white dwarf. It must collapse to the neutron-star stage, with a radius of about 10 km and a density equal to that of an atomic nucleus. Such a neutron star can be supported by the pressure of its degenerate neutrons. But if the mass is greater than 2 to 3 solar masses, then the degenerate neutrons cannot stop the collapse, and the object must become a black hole.

Theory predicts that a neutron star should rotate very fast, be very hot, and have a strong magnetic field. Such objects have been identified as pulsars, sources of pulsed radio energy. Pulsars are evidently spinning neutron stars that emit beams of radiation from their magnetic poles. As they spin, they sweep the beams around the sky; and if the beams sweep over Earth, we detect pulses. This is known as the lighthouse theory. The spinning neutron star slows as it radiates energy into space.

A few pulsars have been found in binary systems, which allows astronomers to estimate the masses of the pulsars. Such masses are consistent with the predicted masses of neutron stars. In some binary systems, such as Hercules X-1, mass flows into a hot accretion disk around the neutron star and causes the emission of X rays.

If a collapsing star has a mass greater than 2 to 3 solar masses, then it must contract to a very small size—perhaps to a singularity, an object of zero radius. Near such an object, gravity is so strong that not even light can escape, and we term the region a black hole. The surface of this region, called the event horizon, is the boundary of the black hole. The Schwarzschild radius is the radius of this event horizon, amounting to only a few kilometers for black holes of stellar mass.

If we were to leap into a black hole, our friends who stayed behind would see two relativistic effects. They would see our clock slow relative to their own clock because of time dilation. Also, they would see our light red-shifted to longer wavelengths. We would not notice these effects, but we would feel powerful tidal forces that would deform and heat our mass until we grew hot enough to emit X rays. Any X rays we emitted before crossing the event horizon could escape.

To search for black holes, we must look for binary star systems in which mass flows into a compact object and emits X rays. If the mass of the compact object is greater than about 3 solar masses, then the object is presumably a black hole. A number of such objects have been located.

SS433 is an X-ray binary that illustrates how rapid mass transfer can generate tremendous energy. Apparently, mass flows rapidly into an accretion disk around a compact object, and the hot accretion disk ejects beams of radiation and gas in opposite directions at about 25 percent the speed of light.

New Terms

neutron star

pulsar

lighthouse theory

millisecond pulsar

magnetar

escape velocity

singularity

black hole

event horizon

Schwarzschild radius (R_s)

time dilation

gravitational red shift

Review Questions

1. How are neutron stars and white dwarfs similar? How do they differ?

2. Why is there an upper limit to the mass of neutron stars?

3. Why do we expect neutron stars to spin rapidly?

4. If neutron stars are hot, why aren't they very luminous?

5. Why do we expect neutron stars to have a powerful magnetic field?

6. Why did astronomers conclude that pulsars could not be pulsating stars?

7. What does the short length of pulsar pulses tell us?

8. How does the lighthouse theory explain pulsars?

9. What evidence do we have that pulsars are neutron stars?

10. Why would astronomers at first assume that the first millisecond pulsar was young?

11. How can a neutron star in a binary system generate X rays?

12. If the sun has a Schwarzschild radius, why isn't it a black hole?

13. How can a black hole emit X rays?

14. What evidence do we have that black holes really exist?

15. How can mass transfer into a compact object produce jets of high-speed gas?

Discussion Questions

1. In your opinion, has the existence of neutron stars been sufficiently tested to be called a theory, or should it be called a hypothesis? What about the existence of black holes?

2. Why wouldn't an accretion disk orbiting a giant star get as hot as an accretion disk orbiting a compact object?

Problems

1. If a neutron star has a radius of 10 km and rotates 642 times a second, what is the speed of the surface at the neutron star's equator in terms of the speed of light? (*Hint:* The circumference of a circle is $2\pi R$.)

2. A neutron star and a white dwarf have been found orbiting each other with a period of 11 minutes. If their masses are typical, what is the average distance between them? (*Hint:* See By the Numbers 8-4.)

3. If Earth's moon were replaced by a typical neutron star, what would the angular diameter of the neutron star be, as seen from Earth? (*Hint:* See By the Numbers 3-1.)

4. What is the Schwarzschild radius of Jupiter (mass = 2×10^{27} kg)? of a human adult (mass = 75 kg)? (*Hint:* See Appendix C for the values of G and c.)

5. If the inner accretion disk around a black hole has a temperature of 10^6 K, at what wavelength will it radiate the most energy? (*Hint:* See By the Numbers 6-1.)

6. What is the orbital period of a bit of matter in an accretion disk 2×10^5 km from a 10 $M_\odot$ black hole? (*Hint:* See By the Numbers 8-4.)

7. If SS433 consists of a 20 $M_\odot$ star and a neutron star orbiting each other every 13.1 days, then what is the average distance between them? (*Hint:* See By the Numbers 8-4.)

8. What is the orbital velocity at a distance of 7400 meters from the center of a 5-solar-mass black hole? What kind of particles could orbit at this distance? (*Hint:* See By the Numbers 4-1.)

9. Compare the orbit in Problem 8 with an orbit having the same velocity around a 2-solar-mass neutron star. Why is this orbit impossible? (*Hint:* See By the Numbers 4-1.)

Critical Inquiries for the Web

1. Imagine that you are on a mission to explore one of the pulsar planets noted in the chapter. What would you find there? Look for information about pulsars and the known pulsar planets on the Web and describe what you might encounter on such a mission.

2. What would you experience if you were to pilot a spacecraft near a black hole? Visit black-hole-related Internet sites to determine what the gravitational effects and general environment would be. Also use the Internet to find the limits of human tolerance to strong gravitational forces. (*Hint:* Look for information about astronaut training and find out how many Gs a human can withstand.) Use these sources to give a brief account about what your voyage would be like.

Go to the Brooks/Cole Astronomy Resource Center (www.brookscole.com/astronomy) for critical thinking exercises, articles, and additional readings from InfoTrac College Edition, Brooks/Cole's online student library.

CHAPTER 12

The Milky Way Galaxy

A hypothesis or theory is clear, deci-

sive, and positive, but it is believed by

no one but the man who created it.

Experimental findings, on the other

hand, are messy, inexact things,

which are believed by everyone except

the man who did that work.

Harlow Shapley
Through Rugged Ways to the Stars

Guidepost

This chapter plays three parts in our cosmic drama. First, it expands our understanding of our place in nature by introducing us to our home, the Milky Way Galaxy. Second, by helping us analyze our own galaxy, it gives us a preview of galaxies in general, the subject of the next chapter. Third, this chapter elaborates our study of stars by showing us how stars are born, evolve, and die in these vast communities we call galaxies.

This chapter contains many examples of scientists using evidence and theory. If in some cases the theories seem incomplete and the evidence contradictory, we should not be disappointed. Rather, we must conclude that the adventure of discovery is not yet over.

We struggle to understand our own galaxy as an example. We will extend our study in the next two chapters, and then we will apply our understanding to the study of the universe as a whole.

223

We are all fantastically wealthy. We live inside one of the largest star systems in the universe. Our Milky Way Galaxy is over 75,000 ly in diameter and contains over 100 billion stars. If Earth is the only inhabited planet in the galaxy, then the galaxy belongs to us; and, sharing equally, each person on Earth owns 50 stars plus assorted planets, moons, comets, and so on. Even if we must share with a few other inhabited planets, we are all rich beyond any dream. Of course, we lack transportation to visit our dominions, but we can still admire our galaxy. It is, after all, ours.

Almost every celestial object visible to our naked eyes is part of our Milky Way Galaxy. The only exception visible from Earth's northern hemisphere is the Andromeda Galaxy, just visible to our unaided eyes as a faint patch of light in the constellation Andromeda.* Our galaxy probably looks much like the Andromeda Galaxy (Figure 12-1a).

*Consult the star charts at the end of this book to locate the Milky Way and the Andromeda Galaxy in your night sky.

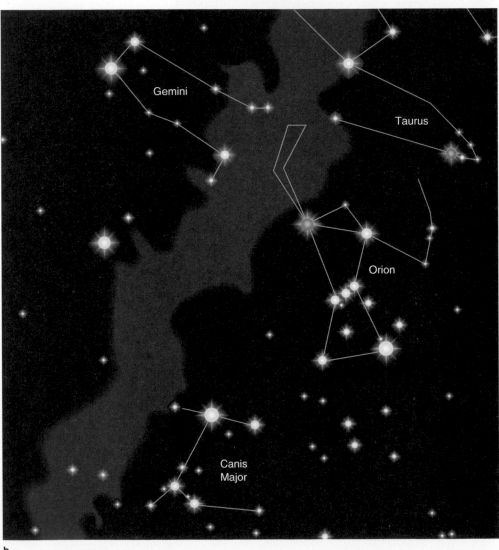

Figure 12-1
(a) This photograph of the Great Galaxy in Andromeda, a spiral galaxy about 2.3 million ly from us, shows us what our Milky Way Galaxy would look like if we could view it from such a great distance. *(Bill Schoening and Vanessa Harvey, NSF REU, NOAO, © AURA, Inc.)* (b) From inside, we see our galaxy as scattered nearby stars and distant clouds of stars in a faintly luminous path that circles the sky. Invisible from the brightly lit skies of cities, the band of the Milky Way is easily visible from a dark site. This artwork shows the location of a portion of the Milky Way near a few bright winter constellations. (See Figure 12-4 and the star charts at the end of this book to further locate the Milky Way in the sky.)

Gemini

Taurus

Orion

Canis Major

a

b

It isn't obvious that we live in a galaxy. We are inside, and we see nearby stars scattered all over the sky, while the more distant clouds of stars in our galaxy make a faint band of light circling the sky (Figure 12-1b). The ancient Greeks named that band *galaxies kuklos,* the "milky circle." The Romans changed the name to *via lactea,* "milky road" or "milky way." It was not until early in this century that astronomers understood that we live inside a great wheel of stars and that the universe is filled with other such star systems. Drawing on the Greek word for milk, we call them galaxies.

Our goal is to understand our galaxy—how it formed and evolved and why it takes the form it does. As always, we begin by gathering evidence for comparison with theories. In this case, our evidence includes the size, mass, and shape of our galaxy and the distribution of stars within it.

12-1 The Discovery of the Galaxy

It seems odd to say astronomers discovered something that is all around us; but until the last century, although everyone could see the faint band of the Milky Way, no one knew what it was. Two hundred years ago, astronomers generally assumed that stars were uniformly scattered through space. One hundred years ago, astronomers understood that the stars we see lie in a great wheel-shaped star system, but they thought the sun was near the center and that the star system was not very large. Then, about one lifetime ago, a woman studying variable stars and a man studying star clusters unlocked one of nature's biggest secrets—that we live in a galaxy.

We follow the story of their discovery for two reasons. First, it is an important historical moment in human history—as dramatic as the Copernican realization that we live on a planet. Second, the story of the discovery of our galaxy illustrates how scientists build on the work of their predecessors and step by step refine their ideas about the natural world we live in.

Variable Stars

Although we think of stars as unchanging, astronomers have known for centuries that certain stars change their brightness, growing brighter, fainter, and brighter with periods that range from less than a day to hundreds of days (Figure 12-2a). These **variable stars** were not understood until the 20th century.

In Chapter 9 we saw that the layers of a star are supported against gravity by the flow of energy out of their interiors. Most stars reach a stable balance between gravity and energy, but in a certain region of the H-R diagram called the **instability strip,** the temperature and size of the star are just right to make it unstable (Figure 12-2b). A layer just below the surface of the star absorbs and releases energy like a spring, and the star pulsates like a beating heart—expanding and contracting and thus changing its brightness. Modern astronomers understand that the evolution of some giant stars carries them through the instability strip, and thus they become variable stars while they cross the region of instability.

There are many kinds of variable stars, but two kinds are important to our story of the discovery of the Milky Way. The **RR Lyrae variable stars,** named after variable star RR in the constellation Lyrae, have periods of one-half to one day and are a bit less than 100 times as luminous as the sun. A related kind of variable star is the **Cepheid,** named after δ Cephei, the first star of this type to be found. Cepheids have periods ranging from 1 to 60 days, but they do not all have the same luminosity. More massive stars cross the instability strip higher in the diagram and are thus more luminous. Because they are larger in radius, they pulsate more slowly than do the lower-mass stars. Thus, there is a relation between the period and luminosity of Cepheid variable stars: the longer the period, the greater the luminosity.

In the early part of the 20th century, astronomers knew none of the stellar evolution we have summarized in the preceding paragraphs. Nevertheless, in 1912 the Harvard astronomer Henrietta Leavitt (1868–1921) was photographing the sky and discovering large numbers of Cepheid and RR Lyrae variables when she discovered that the longer-period Cepheid variables were brighter than the short-period Cepheids (Figure 12-2c). This **period–luminosity relation** has become a powerful tool for finding distance. It was the key to the Milky Way Galaxy.

The Size of the Galaxy

By the beginning of the 20th century, astronomers studying the distribution of the stars in the sky concluded that the faint band of the Milky Way was evidence that we lived near the center of a disk-shaped cloud of stars that was not very large. Many astronomers believed that the star system, as it was called, was isolated in an otherwise empty universe.

A young astronomer named Harlow Shapley (1885–1972) began the discovery of the true nature of the Milky Way when he noticed that different kinds of star clusters have different distributions in the sky. **Open star clusters** (Figure 12-3a) contain 100 to 1000 stars in a region 10 to 60 ly in diameter and are concentrated along the band of the Milky Way. The Pleiades (See Figure 9-2a) is a well-known open cluster. **Globular star clusters** (Figure 12-3b) contain

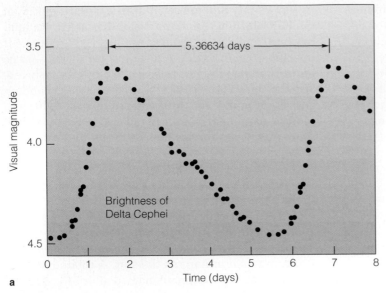

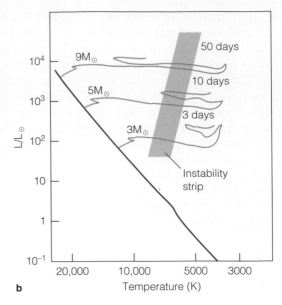

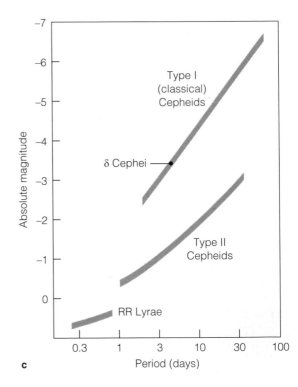

Figure 12-2

Variable stars: (a) A graph of the brightness of the star Delta Cephei versus time shows that it is varying in brightness with a period slightly more than 5 days. (b) The instability strip is a region of the H-R diagram in which stars become unstable and pulsate as variable stars. More massive stars evolve across the instability strip higher in the diagram, are larger, and have longer periods of pulsation. (c) A modern diagram shows the period–luminosity relation for RR Lyrae stars and two kinds of Cepheid variable stars.

100,000 to 1,000,000 stars crowded into a region about 75 ly in diameter. Shapley noticed that globular clusters are highly concentrated in one part of the sky. Almost half of all known globular clusters are located in or near the constellation Sagittarius (Figure 12-4).

Shapley assumed that this great cloud of globular clusters was controlled by the combined gravitational field of the entire star system, in which case he could study the size and extent of the star system by studying the globular clusters. To do that, he needed to measure the distances to as many globular clusters as possible.

Globular clusters are much too far away to have measurable parallaxes, but they do contain variable stars. Shapley knew of Henrietta Leavitt's work on these stars, but she knew only their apparent magnitudes. Cepheids and RR Lyrae stars are relatively rare, and there are none near enough to have measurable parallaxes. Consequently, Leavitt could not find the absolute magnitude of any of the variable stars she discovered, but only their apparent magnitudes. Shapley knew that he could find the distance to the globular clusters if he could find the absolute magnitude of the variable stars in the clusters, so he turned his attention to variable stars.

All stars are moving through space, and over periods of a few years these motions, known as **proper motions,** can be detected as small shifts in the positions of the stars in the sky. The more distant stars

have undetectably small proper motions, but the nearer stars have larger proper motions. Clearly, proper motions contain clues to distance. Although no Cepheids are close enough to have measurable parallaxes, Shapley found 11 with measured proper motions. Through a statistical process, he was able to find their average distance and thus their average absolute magnitude. That meant he could replace Leavitt's apparent magnitudes with absolute magnitudes on the period–luminosity diagram (see Figure 12-2c). That is, he knew how bright the variable stars really were.

Finally, he could find the distance to the globular clusters. He could identify the variable stars he

b

Figure 12-3

(a) The "Jewel Box" is an open star cluster containing only a few hundred stars in a region about 8 pc in diameter. *(NOAO)* (b) The object known as 47 Tucanae is a large globular cluster containing over 1 million stars in a region about 50 pc in diameter—slightly larger than the typical globular cluster. *(NOAO)*

a

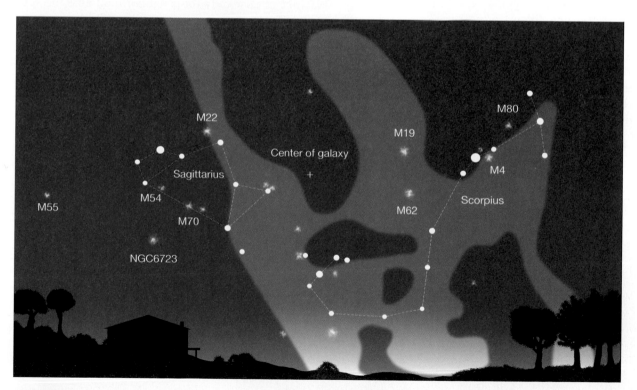

Figure 12-4

The globular clusters form a great swarm concentrated in the sky toward the constellation Sagittarius. A few of the brighter globular clusters are shown here as fuzzy spots. The brightest are marked with their catalog designations and can be seen with binoculars on a dark night. Constellations are shown as they appear above the southern horizon on a summer night as seen from latitude 40° N, typical for most of the United States.

Astronomers often say that Shapley "calibrated" the Cepheids for the determination of distance, meaning that he did all the detailed background work so that the Cepheids could be used to find distances. Other astronomers could use the Cepheids without repeating the detailed calibration.

Calibration is actually very common in science. Chemists, for instance, have carefully calibrated the colors of certain compounds against acidity. They can quickly measure the acidity of a solution by dipping into it a slip of paper containing the indicator compound and looking at the color. They don't have to repeat the careful calibration every time they measure acidity.

Engineers in steel mills have calibrated the color of molten steel against its temperature. They can use a handheld device to measure the color of a ladle of molten steel and then look up the temperature in a table. They don't have to repeat the calibration every time. Astronomers have made the same kind of color–temperature calibration for stars.

As you read about any science, notice how calibrations are used to simplify common measurements. But notice, too, how important it is to get the calibration right. An error in calibration can throw off every measurement made with that calibration. Some of the biggest errors in science have been errors of calibration.

---

found in the clusters (RR Lyrae stars), and he could find their apparent magnitude from his photographs. The comparison of apparent and absolute magnitude gave him the distance to the star cluster using the process we identified in Chapter 8 as spectroscopic parallax. Notice how Shapley proceeded step by step in his research; astronomers say he **calibrated** the variable stars for distance determination (Window on Science 12-1).

Shapley later wrote that it was late at night when he plotted the direction and distance to the globular clusters and found that, just as he had supposed, they formed a great swarm. But it was not centered on the sun. The center of the swarm of clusters lay many thousands of light-years in the direction of Sagittarius. Evidently the center of the star system was in Sagittarius (Figure 12-5). He called the only other person in the building, a cleaning lady, and the two stood looking at his graph as he explained that they were the only two people on Earth who understood that humanity lives, not at the center of a small star system, but in the suburbs of a vast wheel of stars, a galaxy.

Why did astronomers before Shapley think we lived near the center of a small star system? Space is filled with gas and dust that dims our view of distant stars. When we look toward the band of the Milky Way, we can see only the neighborhood near the sun. Most of the star system is invisible and, like travelers in a fog, we seem to be at the center of a small region. Shapley was able to see the globular clusters at greater distances because they lie outside the plane of the star system and are not dimmed very much by the gas and dust.

Building on Shapley's work, other astronomers began to suspect that some of the faint patches of light visible through telescopes were other galaxies like our

Figure 12-5
A photograph of the Milky Way near Sagittarius shows no evidence that this is the center of the galaxy. Gas and dust block our view, and we cannot see all the way to the center (boxed area). *(Daniel Good)*

own. Shapley and astronomer Heber Curtis met in 1920 in a famous debate known as the **Shapley–Curtis debate.** Curtis claimed the faint objects were other galaxies, and Shapley argued they were not other galaxies but were part of our star system. It isn't clear who won the debate; but, in 1923, Edwin Hubble photographed individual stars in the Andromeda Galaxy, and in 1924 he identified Cepheids there. Clearly the faint patches were other star systems like our own.

How big is our galaxy? Modern studies tell us the disk is at least 75,000 ly in diameter, but there is more to our galaxy than just the disk. To get a true impression, we must conduct a careful analysis.

An Analysis of the Galaxy

Our galaxy, like many others, contains two primary components—a disk and a sphere. Figure 12-6 shows these components and other features we will discuss in this section as we analyze the structure of our galaxy.

75,000 ly

Sun   Nuclear bulge   Disk

Halo

Globular cluster

Figure 12-6
An artist's conception of our Milky Way Galaxy seen face-on and edge-on. Note the position of the sun and the distribution of globular clusters.

The **disk component** consists of all matter confined to the plane of rotation—that is, everything in the disk itself. This includes stars, open star clusters, and nearly all of the galaxy's gas and dust. Because the disk contains lots of gas and dust, it is the site of most of the star formation in the galaxy, and so the disk is illuminated by the brilliant, blue, massive stars. Consequently, the disk of the galaxy tends to be blue.

The dimensions of the disk are uncertain for a number of reasons. The thickness is uncertain because the disk does not have sharp boundaries. Stars become less crowded as we move away from the plane. Also, the thickness depends on what kind of object we study—O stars lie within a narrow disk only about 300 ly thick, but sunlike stars are more widely spread. The diameter of the disk and the position of the sun are also difficult to determine. Gas and dust block our view in the plane of the galaxy so we cannot see to the center or to the edge, and the outer edge of the disk is not well defined. The best studies suggest the sun is about 8.5 kpc from the center, where 1 kpc is a **kiloparsec,** or 1000 pc. We seem to be about two-thirds of the way from the center to the edge, so the diameter of our galaxy appears to be about 25 kpc. This is often approximated as 75,000 ly, which is a bit smaller than 25 kpc, but this illustrates the uncertainty in this number. We don't know it to better than 10 percent. This is the diameter of the luminous part of our galaxy, the part we would see from a distance. We will learn later that strong evidence suggests that our galaxy is much larger than this but that the outer parts are not luminous.

One way to explore our galaxy is to use radio telescopes to map the distribution of un-ionized hydrogen. Fortunately for astronomers, neutral hydrogen in space is capable of radiating photons with a wavelength of 21 cm. This happens because the spinning proton and electron act like tiny magnets as shown in Figure 12-7. When the spinning electron in the hydrogen atom reverses its spin, the magnetic fields reverse, and the atom is left with a tiny amount of excess energy, which it radiates as a photon with a wavelength of 21 cm. Radio astronomers can map our entire galaxy at a wavelength of 21 cm because these long-wavelength photons are unaffected by the microscopic dust grains that scatter photons of light. Thus, radio telescopes can "see" our entire galaxy, while optical telescopes cannot.

Observations made at other wavelengths can also let us see through the dust clouds. Infrared photons have wavelengths long enough to be unaffected by the dust. Thus, a map of the sky at long infrared wavelengths reveals the disk of our galaxy (Figure 12-8).

The most striking features of the disk component are the **spiral arms**—long curves of bright stars, star clusters, gas, and dust. Such spiral arms are easily

Figure 12-7
Both the proton and electron in a neutral hydrogen atom spin and consequently have small magnetic fields. When they spin in the same direction, their magnetic fields are reversed, and when they spin in opposite directions, their magnetic fields are aligned. As explained in the text, this allows cold, neutral hydrogen in space to emit radio photons with a wavelength of 21 cm.

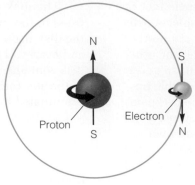
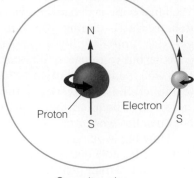

Same spins
Magnetic fields reversed

Opposite spins
Magnetic fields the same

visible in other galaxies, and we will see later that our own galaxy has a spiral pattern.

The second component of our galaxy is the **spherical component,** which includes all matter in our galaxy scattered in a roughly spherical distribution around the center. This includes a large halo and the nuclear bulge.

The **halo** is a spherical cloud of thinly scattered stars and globular clusters. It contains only about 2 percent as many stars as the disk of the galaxy and contains very little gas and dust. Thus, no new stars are forming in the halo. In fact, the vast majority of the halo stars are old, cool, lower-main-sequence stars

and giants much like the stars in globular clusters. Although we can map the halo of our galaxy by studying these stars, the halos of other galaxies are too faint to detect.

The **nuclear bulge** is the dense cloud of stars that surrounds the center of our galaxy. It has a radius of about 2 kpc and is slightly flattened. We cannot observe it easily because thick dust in the disk scatters radiation of visible wavelengths, but observations at longer wavelengths can penetrate the dust. The bulge seems to contain little gas and dust, and there is thus little star formation. Most of the stars are old, cool stars like those in the halo.

Figure 12-8
Just as the inset map projects the entire surface of Earth onto a flat oval, this infrared image from the IRAS satellite projects the entire sky onto a flat map. The plane of our galaxy runs horizontally from left to right with the center of the galaxy at the center of the oval. The red glow in this false-color image is produced by dust in the disk of the galaxy warmed by starlight. The thickness of the distribution of dust is evident in this image. (Warmer dust in our solar system is located along the ecliptic and produces the faint blue s-shaped curve.) *(NASA)*

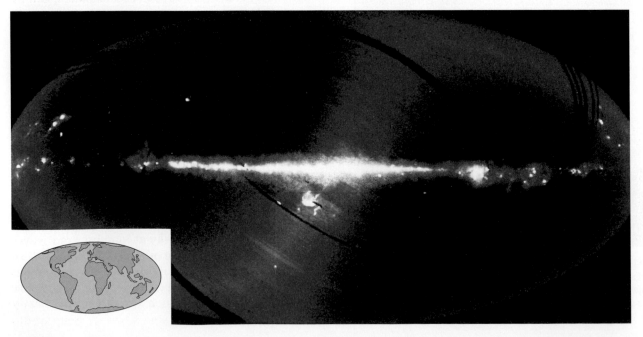

The vast numbers of stars in the disk, halo, and nuclear bulge lead us to raise a basic question. How massive is the galaxy?

The Mass of the Galaxy

When we needed to find the masses of stars, we studied the orbital motions of the pairs of stars in binary systems. To find the mass of the galaxy, we must look at the orbital motions of the stars within the galaxy. Every star in our galaxy follows an orbit around the center of mass of the galaxy, and thus we say the galaxy rotates. That rotation can reveal the total mass of the galaxy. Consequently, our discussion of the mass of the galaxy is also a discussion of the rotation of the galaxy and the orbits of the stars within the galaxy.

We can find the orbits of stars by finding how they move. Of course, we could use the Doppler effect to find a star's radial velocity (see By the Numbers 6-2). In addition, if we could measure the distance to and proper motion of a star, we could find the velocity of the star perpendicular to the radial direction. Combining all of this information, we could then find the shape of the star's orbit.

The orbital motions of the stars in the halo are strikingly different from those in the disk (Figure 12-9). In the halo, each star and globular cluster follows its own randomly tipped elliptical orbit. These orbits carry the stars and clusters far out into the spherical halo, where they move slowly, but when they fall

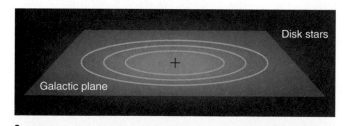

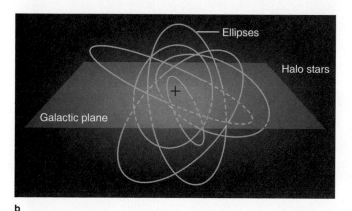

Figure 12-9
(a) Stars in the galactic disk have nearly circular orbits that lie in the plane of the galaxy. (b) Stars in the halo have randomly oriented, elliptical orbits.

back into the inner part of the galaxy, their velocities increase. Thus, motions in the halo do not resemble a general rotation but are more like the random motions of a swarm of bees. In contrast, the stars in the disk of the galaxy move in the same direction in nearly circular orbits that lie in the plane of the galaxy. The sun is a disk star and follows a nearly circular orbit around the galaxy that never carries it out of the disk.

We can use the orbital motion of the sun to find the mass of the galaxy. By observing the radial velocity of other galaxies in various directions around the sky, astronomers can tell that the sun moves about 220 km/sec in the direction of Cygnus, carrying Earth and the other planets of our solar system along with it. Because its orbit is a circle with a radius of 8.5 kpc, we can divide the circumference of the orbit by the velocity and find that the sun completes a single orbit in about 240 million years. Obviously, we notice no change in a time as short as a human lifetime.

If we think of the sun and the center of mass of our galaxy as two objects orbiting each other, we can find the mass of the galaxy (see By the Numbers 8-4). We need to divide the separation in AU cubed by the period in years squared to find the mass in solar masses. The radius of the sun's orbit is about 8.5 kpc, and the orbital period is 240 million years. Converting kiloparsecs to AU and doing the arithmetic tells us the galaxy must have a mass of about 100 billion solar masses.

This estimate is uncertain for a number of reasons. First, we don't know the radius of the sun's orbit with great certainty. We estimate the radius as 8.5 kpc, but we could be wrong by 10 percent or more, and this radius gets cubed in the calculation, so it makes a big difference. Second, this estimate of the mass includes only the mass inside the sun's orbit. Mass spread uniformly outside the sun's orbit will not affect its orbital motion. Thus 100 billion solar masses is a lower limit for the mass of the galaxy, but we don't know how much to increase our estimate to include the rest of the galaxy.

This estimate is also uncertain because it depends on our knowledge of the rotation of the galaxy, and that rotation is very complex. The motion of the stars near the sun shows that the disk does not rotate as a solid body. Each star follows its own orbit, and stars in some regions have shorter or longer orbital periods than the sun. This is called **differential rotation** (Figure 12-10) and is different from the rotation of a solid body—a record on a turntable, for instance. Whereas three spots lined up on a record would stay together as the record turned, three stars lined up in the galaxy will draw apart because their orbital periods differ.

A graph of the orbital velocity of stars at various orbital radii in the galaxy is called a **rotation curve** (Figure 12-11). If all of the mass in the galaxy were concentrated at its center, then orbital velocity would

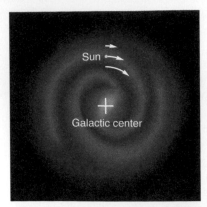

Figure 12-10

The differential rotation of the galaxy means that stars at different distances from the center have different orbital periods. In this example, the star just inside the sun's orbit has a shorter period and gains on the sun, while the star outside falls behind.

be high near the center and would decline as we move outward. This kind of motion has been called *Keplerian motion* because it follows Kepler's third law, as in the case of our solar system, where nearly all of the mass is in the sun. Of course, the galaxy's mass is not all concentrated at its center, but if most of the mass is inside the orbit of the sun, then we would expect to see orbital velocities decline at greater distances. Many observations confirm, however, that velocities actually increase at greater distance; and this fact tells us that these larger orbits are enclosing more mass. In other words, significant amounts of matter in our galaxy lie in its outer parts. Although it is difficult to determine a precise edge to the visible galaxy, it seems

clear that large amounts of matter are located beyond the traditional edge of the galaxy—the edge we would see if we journeyed into space and looked back at our galaxy. Thus, this matter must not be producing light, and astronomers refer to it as **dark matter.**

Some of this dark matter is evidently located in the disk of our galaxy, but the shape of the rotation curve and theoretical models of the rotation of the disk suggest that much of the dark matter is located in an extended **dark halo,** also called the **galactic corona.** Estimates suggest this dark halo extends as far as seven times the visible radius of our galaxy and contains 10 to 100 times the mass of the visible part of the Milky Way Galaxy.

Astronomers debate the nature of this dark matter. It cannot be normal stars, because they would produce light and we would see them. We can also eliminate neutron stars and black holes because we would be able to detect the X rays such objects emit. Remember the amount of dark matter is 10 to 100 times the amount of visible matter, so such a large mass of normal matter should be detectable. Many astronomers suspect that the dark matter is not normal matter, but is composed of some exotic atomic particle whose properties are not understood.

We will return to the mysterious dark matter in the following chapters when we discuss other galaxies and the universe as a whole. For now it is sufficient to consider the implications of this dark matter for understanding our own Milky Way Galaxy. Simple observations of the motion of stars in our galaxy and a simple theory of orbital motion lead us to the conclusion that the visible part of our galaxy may

Figure 12-11

The rotation curve of our galaxy is plotted here as orbital velocity versus radius. Data points show measurements made by radio telescopes with error bars showing the uncertainty in the measurements. The fitted rotation curve (solid line) does not decline outside the sun's orbit as we would expect for Keplerian motion (dashed line). Rather, the curve is approximately flat at great distances, suggesting that the galaxy contains significant mass outside the orbit of the sun. *(Adapted from a diagram by Françoise Combes)*

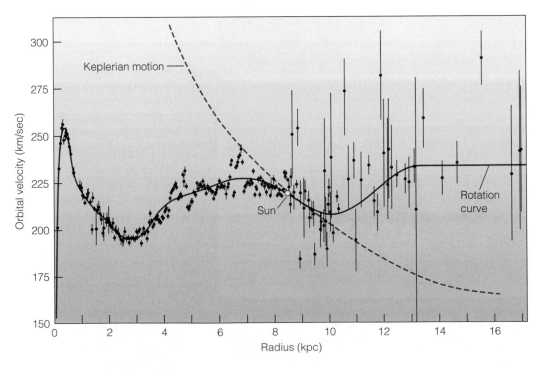

make up only a small percentage of a much larger object. This is disturbing; it is as if we looked at a ham sandwich and saw only the mayonnaise. Furthermore, we are led to suspect that there are forms of matter unknown to us at present. Our simple quest to find the mass of our galaxy has opened a door to an exciting puzzle—the dark matter.

REVIEW Critical Inquiry

If clouds of gas and dust block our view, how do we know how big our galaxy is?

The gas and dust in our galaxy block our view only in the plane of the galaxy. When we look away from the plane of the galaxy, we look out of the gas and dust and can see to great distances. The globular clusters are scattered through the halo with a strong concentration in the direction of Sagittarius. When we look above or below the plane of the galaxy, we can see those clusters, and careful observations reveal variable stars in the clusters. By using Shapley's calibration of the period–luminosity diagram, we can find the distance to those clusters. If we assume that the distribution of the clusters is controlled by the gravitation of the galaxy as a whole, then we can find the distance to the center of the galaxy by finding the center of the distribution of globular clusters.

In the critical analysis of ideas, it is important to ask, "How do we know?" and that is especially true in science. For example, how do we know the mass of our galaxy?

The disk component and spherical component are quite different from each other. In the next section, we will discover that they also differ in chemical composition; and that clue, combined with our knowledge of stellar evolution, suggests a model for the formation of the galaxy.

12-2 The Origin of the Milky Way

Just as paleontologists reconstruct the history of life on Earth from the fossil record, astronomers try to reconstruct our galaxy's past from the fossil it left behind as it formed and evolved. That fossil is the spherical component of the galaxy. The stars we see in the halo formed when the galaxy was young. The chemical composition and the distribution of these stars can give us clues to how our galaxy formed.

The Age of the Milky Way

To begin, we should ask ourselves how old our galaxy is. That question is easy to answer because we already know how to find the age of star clusters. But there are uncertainties that make our easy answer hard to interpret.

The oldest open clusters have ages of about 7 billion years. These ages are determined by analyzing the turnoff point in the cluster H-R diagram (see Chapter 10), but three things make these ages uncertain. First, finding the age of a star cluster becomes more difficult for older clusters because they change more slowly. Also, the location of their turnoff points and giant regions depends on their chemical composition, which can differ from cluster to cluster. Finally, open clusters are not strongly bound by their gravity, so there may have been older open clusters that dissipated as their stars wandered away. In any case, from open clusters, we can get a rough age for the disk of our galaxy of at least 7 billion years.

The H-R diagrams of globular clusters tell us their age, and conventional data suggest that they are from 13 to 17 billion years old. Data from the Hipparcos satellite, however, provides more accurate distances and luminosities, suggesting that globular clusters may be as young as 11 billion years. Part of the uncertainty reflects the difficulty in measuring the ages of old star clusters, but part of the variation in age seems real. Some globular clusters are older than others. From this information, it seems that the halo is at least 11 billion years old.

Our study of star clusters reveals an important fact. The two components of our galaxy have different ages. The spherical component seems to have formed first and the disk later. To get further clues, we have to look at the subtle differences in chemical composition between these two components.

Stellar Populations

In the 1940s, astronomers realized that there were two types of stars in the galaxy. The type they were accustomed to studying was that located in the disk, such as the stars near the sun. These they called **population I** stars. The second type, called **population II** stars, are usually found in the halo, in globular clusters, or in the central bulge. In other words, the two stellar populations are associated with the two components of the galaxy.

The stars of the two populations are very similar. They burn nuclear fuels and evolve in nearly identical ways. They differ only in their abundance of atoms heavier than helium, atoms that astronomers refer to collectively as **metals.** (Note that this is not the way the word *metal* is commonly used by nonastronomers.)

Population I stars are metal rich, containing 2 to 3 percent metals, while population II stars are metal poor, containing only about 0.1 percent metals or less. The metal content of the star defines its population.

Population I stars belong to the disk component of the galaxy and are sometimes called *disk population stars*. They have circular orbits in the plane of the galaxy and are relatively young stars that formed within the last few billion years. The sun is a population I star.

Population II stars belong to the spherical component of the galaxy and are sometimes called the *halo population stars*. These stars have randomly tipped, elliptical orbits and are old stars. The metal-poor globular clusters are part of the halo population.

To see how important this difference is, refer back to Figure 12-2c. We can now understand the two kinds of Cepheid variables shown in the period–luminosity diagram. Type I Cepheids are population I stars with metal abundance similar to that of the sun. The Type II Cepheids are population II stars. The difference in metal abundance affects the ease with which radiation flows through the gas, and that changes the balance between gravity and energy flowing outward through the star. A few percent difference in metal abundance may seem like a small detail, but it makes a big difference to a star.

Since the discovery of stellar populations, astronomers have realized that there is a gradation between populations (Table 12-1). The most metal-rich stars, such as the stars in Orion, are found only in the spiral arms and are called *extreme population I stars*. Slightly less metal-rich population I stars, called *intermediate population I stars*, are located throughout the disk. The sun is such a star. Stars even less metal rich, such as stars in the nuclear bulge, are intermediate population II stars. The most metal-poor stars are those in the halo and in globular clusters. These are extreme population II stars.

Why do the disk and halo stars have different metal abundances? The two types of star must have formed at different stages in the life of the galaxy, at stages when the chemical composition of the galaxy differed. This is a clue to the history of our galaxy, but to use the clue, we must discuss the cycle of element building.

The Element-Building Cycle

The atoms of which we are made were created in a process that spanned a number of generations of stars. Natural processes are all around us, and we must learn to recognize and understand them if we are to understand nature (Window on Science 12-2). The process that built the chemical elements may be one of the most important processes in the history of our galaxy.

We saw in Chapter 10 how elements heavier than helium but lighter than iron are built up by nuclear reactions inside evolving stars. Atoms more massive than iron are made by short-lived nuclear reactions that occur during a supernova explosion. This explains why lower-mass atoms like carbon, nitrogen, and oxygen are so common and why atoms more massive than iron—such as gold, silver, platinum, and uranium—are so rare. Figure 12-12a shows the abundance of the chemical elements, but notice that the graph has an exponential scale. To get a feeling for the true abundance of the elements, we should draw this graph using a linear scale as in Figure 12-12b. Then we see how rare the elements heavier than helium are.

Most of the matter in stars is hydrogen and helium, and the metals (including carbon, nitrogen, oxygen, and so on) were cooked up inside stars. When the galaxy first formed, there should have been no metals because stars had not yet manufactured any. The gas from which the galaxy condensed must have been almost pure hydrogen (80 percent) and helium (20 percent). (Where the hydrogen and helium came from is a mystery we will save until Chapter 15.)

The first stars to form from this gas were metal poor, and now, 15 billion years later, their spectra still show few metal lines. Of course, they may have manufactured some atoms heavier than helium, but because the stars' interiors are not mixed, those heavy atoms stay trapped at the centers of the stars where they were produced and do not affect the spectra (Figure 12-13). The population II stars that we see today

Table 12-1 Stellar Populations

	Population I		Population II	
	Extreme	Intermediate	Intermediate	Extreme
Location	Spiral arms	Disk	Nuclear bulge	Halo
Metals (%)	3	1.6	0.8	Less than 0.8
Shape of orbit	Circular	Slightly elliptical	Moderately elliptical	Highly elliptical
Average age (yr)	100 million and younger	0.2–10 billion	2–10 billion	10–14 billion

One of the most common organizing themes in science is the concept of a process—a sequence of events that leads to some result or condition. Scientists try to discover the natural processes that shape the universe. Telling the stories of these processes explains why the universe is the way it is. At the same time, recognizing these processes and learning to tell their stories makes science easier to understand.

For example, astronomers try to assemble the sequence of events that led to the formation of the chemical elements. We know that the first stars formed from nearly pure hydrogen and helium and that they made certain heavier elements in their cores. Supernova explosions made other elements and mixed the elements back into the interstellar medium. New stars formed from this gas, and generation after generation of stars added to the heavier elements that we are made of. Telling that story explains much of the science of the stars.

As you study any science, be alert for processes as organizing themes. Science, at first glance, seems to be nothing but facts, but one of the ways to organize the facts is to fit them into the story of a process. If you study biology and learn the process by which plant cells know whether to become roots, stems, or leaves, you will have organized a great deal of factual information into a story that not only is easy to remember but also makes sense.

When you see a process in science, ask yourself a few basic questions. What conditions prevailed at the beginning of the process? What sequence of steps occurred? Can some steps occur simultaneously, or must one step occur before another can occur? What is the final state that this process produces? Most important, what evidence do we have that nature really works this way?

Not everything in science is a process, but identifying a process will help you remember and understand a great many details and principles.

Figure 12-12

The abundance of the elements in the universe. (a) When the elements are plotted on an exponential scale, we see that elements heavier than iron are about a million times less common than iron and that all elements heavier than helium (the metals) are quite rare. (b) The same data plotted on a linear scale provide a more realistic impression of how rare the metals are. Carbon, nitrogen, and oxygen make small peaks near atomic mass 15, and iron is just visible in the graph.

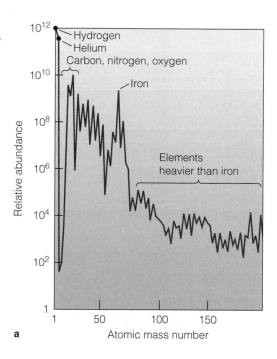
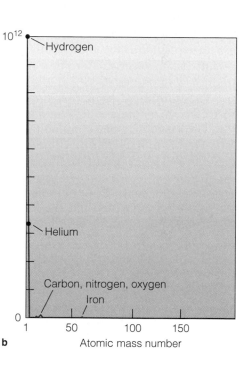

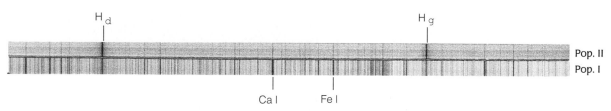

Figure 12-13

Spectra of population II and population I stars of similar spectral type show hydrogen lines of equal strength, but the lines of heavier atoms are conspicuously weak in the population II star's spectrum. Labeled lines are calcium (Ca) and iron (Fe). *(Lick Observatory photograph)*

in the halo are the survivors of an earlier generation of stars to form in the galaxy.

Most of the first stars evolved and died, and the most massive became supernovae and enriched the interstellar gas with metals. Succeeding generations of stars formed from gas clouds that were more enriched, and each generation added to the enrichment through the supernovae of massive stars. By the time the sun formed, 5 billion years ago, the element-building process had added about 1.6 percent metals. Since then, the metal abundance has increased further, and stars forming now, in the Orion Nebula for example, incorporate 2 to 3 percent metals and become extreme population I stars. Thus metal abundance varies between populations because of the production of heavy atoms in successive generations of stars.

The lack of metals in the spherical component tells us it is very old, a fossil left behind by the galaxy when it was young and drastically different from its present disk shape. Thus, the study of element building and stellar populations leads us to the fundamental question, "How did our galaxy form?"

The History of the Milky Way Galaxy

In the 1950s, astronomers began to develop a hypothesis to explain the formation of our galaxy. Recent observations, however, are forcing a reevaluation of that traditional hypothesis.

The traditional hypothesis says that the galaxy formed from a single large cloud of gas 10 to 15 billion years ago. As gravity pulled the gas together, the cloud began to fragment into smaller clouds, and because the gas was turbulent, the smaller clouds had random velocities. Stars and star clusters that formed from these fragments went into randomly shaped and randomly tipped orbits. Of course, these first stars were metal poor because no stars had existed earlier to enrich the gas with metals. Thus, the contraction of the large gas cloud produced the spherical component of the galaxy.

The second stage of this hypothesis accounts for the disk component. The contracting gas cloud was roughly spherical at first, but the turbulent motions canceled out, as do eddies in recently stirred coffee, leaving the cloud with uniform rotation. A rotating, low-density cloud of gas cannot remain spherical. A star is spherical because its high internal pressure balances its gravity, but in a low-density cloud, the pressure cannot support the weight. Like a blob of pizza dough spun in the air, the cloud must flatten into a disk (Figure 12-14).

This contraction into a disk took billions of years, with the metal abundance gradually increasing as generations of stars were born from the gradually flattening gas cloud. The stars and globular clusters of

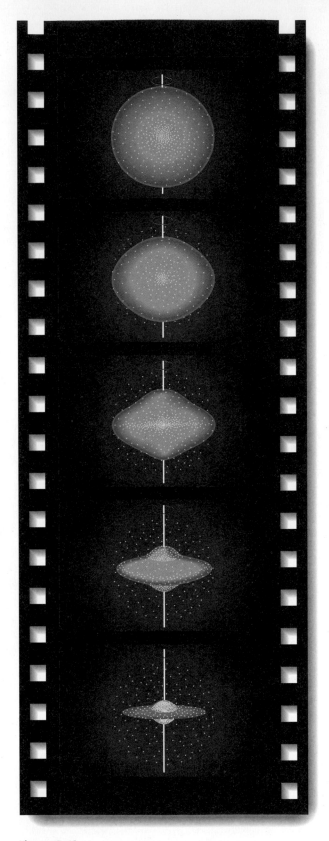

Figure 12-14
According to the traditional theory, the galaxy began as a spherical cloud of gas (shaded) in which stars and star clusters (dots) formed. As the rotating gas cloud collapsed into a disk, the halo stars were left behind as a fossil of the early galaxy.

the halo were left behind by the cloud when it was spherical, and subsequent generations of stars formed in flatter distributions. The gas distribution in the galaxy now is so flat that the youngest stars are confined to a disk only about 100 parsecs thick. These stars are metal rich and have nearly circular orbits.

This traditional hypothesis accounts for many of the Milky Way's properties. Advances in technology, however, have improved astronomical observation, and beginning in the 1980s, contradictions arose. For example, many halo stars have metal abundances similar to those of globular clusters and may have escaped from such clusters, but some stars are even more metal poor. One star contains 30,000 times less metals than the sun. These stars may be older than the globular clusters. Also, some of the oldest stars in our galaxy are in the central bulge, not in the halo. Furthermore, not all globular clusters have the same age, and the younger clusters seem to be in the outer halo. The traditional hypoethesis says that the halo formed first and that the clusters within it should have a uniform age.

Ages are a key problem for the traditional hypothesis. The youngest globular clusters are about 11 billion years old, but the disk seems much younger. The oldest open star clusters in the disk are only about 7 billion years old. Also, white dwarfs tend to accumulate over the eons, and there aren't as many white dwarfs in the disk as there should be if the disk were 11 billion years old. A successful hypothesis should account for these ages.

Another problem is that the oldest stars are metal poor but not metal free. There must have been at least a few massive stars to create these metals before the formation of the oldest stars we see in the halo.

Can we modify the traditional hypothesis to explain these observations? Perhaps the galaxy began with the contraction of a gas cloud to form the central bulge and the later accumulation of the halo from gas clouds that had been slightly enriched in metals by an early generation of massive stars. This would explain the age of the central bulge and the metals in the oldest stars.

The disk could have formed later as the gas that was already in the galaxy flattened and as more gas fell into the galaxy and settled into the disk. Perhaps entire galaxies were captured by the growing Milky Way galaxy. (We will see in the next chapter that such mergers do occur.) If the galaxy absorbed a few small but partially evolved galaxies, then some of the globular clusters we see in the halo may be hitchhikers. This would explain the range of globular cluster ages.

The problem of the formation of our galaxy is frustrating because the theories are incomplete. We prefer certainty, but astronomers are still gathering observations and testing hypotheses. We see an older theory that has proven to be inadequate to explain the observations, and we see astronomers attempting to refine the observations and devise new theories. The metal abundances and ages of the stars in our galaxy seem to be important clues, but metal abundance and age do not tell the whole story. Astronomer Bernard Pagel was thinking of this when he said, "Cats and dogs may have the same age and metallicity, but they are still cats and dogs."

REVIEW Critical Inquiry

Why do metal-poor stars have the most elliptical orbits?

Of course, the metal abundance of a star cannot affect its orbit, so our analysis must not confuse cause and effect with the relationship between these two factors. Both chemical composition and orbital shape depend on a third factor—age. The oldest stars are metal poor because they formed before there had been many supernova explosions to create and scatter metals into the interstellar medium. Those stars formed long ago when the galaxy was young and motions were not organized into a disk. In the top-down hypothesis, the first stars formed before the collapsing galaxy could form a disk; in the bottom-up hypothesis, the oldest stars date from the assembly of the halo from smaller clouds of gas and stars. In either case, those first stars tended to take up randomly shaped orbits, and many of those orbits are quite elliptical. Thus, today we see the most metal-poor stars following the most elliptical orbits.

Nevertheless, even the oldest stars we can find in our galaxy contain some metals. They are metal poor, not metal free. How does that affect our theories?

The origin of our galaxy is an astronomical work in progress, but that is not the only mystery hidden in the Milky Way. The next two sections discuss two special problems that astronomers are trying to understand—the nature of spiral arms and the secret of the galactic nucleus.

12-3 Spiral Arms and Star Formation

The most striking feature of galaxies like the Milky Way is the system of spiral arms that wind outward through the disk. These arms contain swarms of hot, blue stars, clouds of dust and gas, and young star clusters. These young objects hint that the spiral arms involve star formation. As we try to understand the spiral arms, we face two problems. First, how can we be sure our galaxy has spiral arms when our view is obscured by dense clouds of dust? Second, why doesn't the differential rotation of the galaxy destroy the arms? The solution to both problems involves star formation.

Tracing the Spiral Arms

Studies of other galaxies show us that spiral arms contain hot, blue stars (Figure 12-15a). Thus, one way to study the spiral arms of our own galaxy is to locate these stars. Fortunately, this is not difficult, since O and B stars are often found in associations and, being very bright, are easy to detect across great distances. Unfortunately, at these great distances their parallax is too small to measure, so their distances must be found by other means, usually by spectroscopic parallax (see Chapter 8).

O and B associations in the sky are not located randomly but reveal parts of three spiral arms near the sun, which have been named for the prominent constellations through which they pass (Figure 12-15b). If we could penetrate the dust clouds, we could locate other O and B associations and trace the spiral arms farther, but like a traveler in a fog, we see only the region near us.

Objects used to map spiral arms are called **spiral tracers.** O and B associations are good spiral tracers because they are bright and easy to see at great distances. Other tracers include young open clusters, clouds of hydrogen ionized by hot stars (emission nebulae), and certain higher-mass variable stars.

Notice that all spiral tracers are young objects. O stars, for example, live for only a few million years. If their orbital velocity is about 250 km/sec, they cannot have moved more than about 500 pc since they formed. This is less than the width of a spiral arm. Because they don't live long enough to move away from the spiral arms, they must have formed there.

The youth of spiral tracers gives us an important clue about spiral arms. Somehow they are associated with star formation. Before we can follow this clue, however, we must extend our map of spiral arms to show the entire galaxy.

a

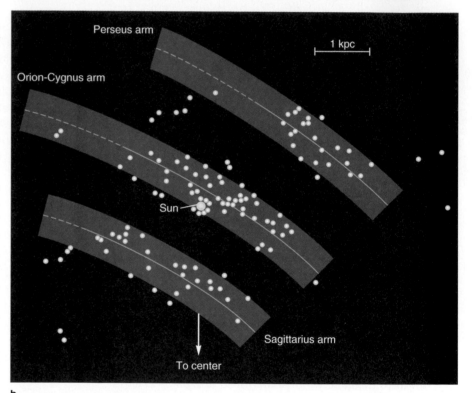

b

Figure 12-15

(a) Images of other galaxies, such as NGC1232 shown here, show the spiral arms illuminated by clouds of young stars. The hot O and B stars in these clouds give the arms a blue tint. *(European Southern Observatory)* (b) Segments of spiral arms near the sun appear when we plot the location of O and B associations and the ionized nebulae that often surround such associations. These groups of massive stars have very short lives, so we find them only in regions of recent star formation. Dust clouds block our view, and we can see only a few kiloparsecs in the plane of the Milky Way, but in the region we can see near the sun, O and B associations trace out segments of three spiral arms as diagrammed here.

Radio Maps of Spiral Arms

The dust clouds that block our view at visual wavelengths are transparent at radio wavelengths, because radio waves are much longer than the diameter of the dust particles. When we point a radio telescope at a section of the Milky Way, we receive 21-cm radio signals coming from cool hydrogen in spiral arms at various distances across the galaxy. Fortunately, the signals can be unscrambled by measuring the Doppler shifts of the 21-cm radiation, except in the direction toward the nucleus. There, the orbital motions of gas clouds are perpendicular to the line of sight, and all of the radial velocities are zero. Thus, the radio map shown in Figure 12-16 reveals spiral arms throughout the disk of the galaxy, except in the wedge-shaped region toward the center.

Radio astronomers can also use the strong radio emission from carbon monoxide (CO) to map the location of giant molecular clouds in the plane of the galaxy. Recall from Chapter 9 that these clouds are sites of active star formation. Maps constructed from such observations reveal that the giant molecular clouds are located in large structures that resemble segments of spiral arms (Figure 12-17).

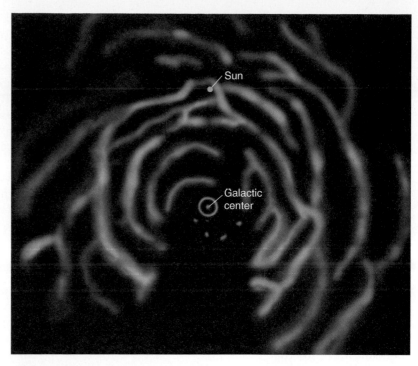

Figure 12-16
This map of our galaxy was made using 21-cm radio observations, and it confirms that our galaxy has spiral arms. Branches and spurs in the spiral arms are visible. The conical region around and beyond the center of the galaxy is the region where 21-cm radiation from gas clouds at different distances cannot be unscrambled. *(Adapted from a radio map by Gart Westerhout)*

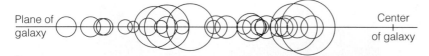

a

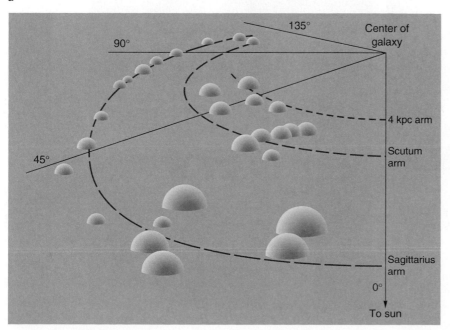

b

Figure 12-17
Mapping spiral arms using giant molecular clouds. (a) At the wavelength emitted by the CO molecule we find many overlapping giant molecular clouds along the Milky Way. By finding the distance to each cloud from its radial velocity and the rotation of the galaxy, radio astronomers can use the clouds to map spiral arms. (b) This diagram shows the location of giant molecular clouds in the plane of our galaxy as hemispheres seen from a location 2 kpc directly above the sun. Now we can see that the molecular clouds are located along spiral arms. Angles in this diagram are galactic longitudes measured clockwise from the sun. *(Adapted from a diagram by T. M. Dame, B. G. Elmegreen, R. S. Cohen, and P. Thaddeus)*

Radio maps combined with optical data reveal the spiral pattern of our galaxy. The segments we see near the sun are part of a spiral pattern that continues throughout the disk. But the spiral arms are rather irregular and are interrupted by branches, spurs, and gaps. The stars we see in Orion, for example, appear to be a detached segment of a spiral arm, a spur. There are significant sources of error in the radio mapping method, but many of the irregularities along the arms seem real, and photographs of nearby spiral galaxies show similar features. One study compared all of the available data on our galaxy's spiral pattern with the patterns seen in other galaxies and constructed a model of what our galaxy probably looks like from a distance (Figure 12-18).

The most important feature in the radio maps is easy to overlook—spiral arms are regions of higher gas density. Spiral tracers told us that the arms contain young objects, and we suspected active star formation. Now radio maps confirm our suspicion by telling us that the material needed to make stars is abundant in spiral arms.

The Density Wave Theory

Having mapped the spiral pattern, we can ask, "What are spiral arms?" We can be sure that they are not physically connected structures such as bands of a magnetic field holding the gas in place—a theory that was considered about three decades ago. Like a kite string caught on a spinning hubcap, such arms would be wound up and pulled apart by differential rotation within a few tens of millions of years. Yet spiral arms are common in disk-shaped galaxies and must be reasonably permanent features.

The most popular theory since the 1950s is called the **density wave theory.** It proposes that spiral arms are waves of compression rather like sound waves, which move around the galaxy, triggering star formation. Because these waves move slowly, orbiting gas clouds overtake the spiral arms from behind and create a moving traffic jam within the arms.

We can imagine the density wave as a traffic jam behind a truck moving slowly along a highway. Seen from an airplane overhead, the jam seems a permanent, though slow-moving, feature. But individual cars move up through the jam behind the truck, await their opportunity, and pass the truck. So too do clouds of gas overtake the spiral density wave, become compressed in the "traffic jam," and eventually move out in front of the arm, leaving the slower-moving density wave behind.

Mathematical models of this process have been very successful at generating spiral patterns that look like our own and other spiral galaxies. In each case, the density wave takes on a regular two-armed spiral pattern winding outward from the nuclear bulge to the edge of the disk. In addition, the theory predicts that the spiral arms will be stable over a period of a billion years. The differential rotation does not wind them up because they are not physically connected structures.

a

Figure 12-18
(a) An observer a few million light-years away might have this view of our galaxy. Note that the nucleus is elongated and that the sun, marked by a cross, is located on the inside edge of a separate branch of a spiral arm. *(Painting by M. A. Seeds based on a study by G. De Vaucouleurs and W. D. Pence)*
(b) Spurs and branches in spiral arms are common. The galaxy M83, shown here, appears to have a strong two-armed spiral pattern but also has spurs and branches. *(Anglo-Australian Telescope Board)*

b

Of course, star formation will occur where the gas clouds are compressed. Stars pass through the spiral arms unaffected, like bullets passing through a wisp of fog, but large clouds of gas slam into the spiral density wave from behind and are suddenly compressed (Figure 12-19). We saw in Chapter 9 that sudden compression could trigger the formation of stars in a gas cloud. Thus, new star clusters should form along the spiral arms.

The brightest stars, the O and B stars, live such short lives that they never travel far from their birthplace and are found only along the arms. Their presence is what makes the spiral arms glow so brightly. Lower-mass stars, like the sun, live longer and have time to move out of the arms and continue their journey around the galaxy. The sun may have formed in a star cluster about 5 billion years ago when a gas cloud smashed into a spiral arm. Since that time, the sun has escaped from its cluster and made about 20 trips around the galaxy, passing through spiral arms many times.

The density wave theory is very successful in explaining the properties of spiral galaxies, but it has two problems. First, what stimulates the formation of the spiral pattern? Of course, some process must generate the spiral arms in the first place, but the pattern must be restimulated or it would die away over a period of a billion years or so. One possibility is that the galaxy is naturally unstable to certain disturbances, just as a guitar string is unstable to certain vibrations. Any sudden disturbance—the rumble of a passing truck, for example—can set the string vibrating at its natural frequencies. Similarly, minor fluctuations in the galaxy's disk or gravitational interactions with passing galaxies might generate a density wave.

The second problem with the density wave theory is that it does not account for the branches and spurs in the spiral arms of our own and other galaxies. Computer models of density waves produce regular, two-armed spiral patterns. Some galaxies, called grand-design galaxies, do indeed have symmetric two-armed patterns, but others do not. Other galaxies have a great many short spiral segments, giving them a fluffy appearance. These galaxies have been termed **flocculent,** meaning "woolly." Our galaxy is probably an intermediate galaxy. How can we explain these variations if the density wave theory always produces two-armed, grand-design, spiral arms? Perhaps the solution lies in a process that sustains star formation once it begins.

Star Formation in Spiral Arms

Star formation is a critical process in the creation of the spiral patterns that we see. It makes spiral arms visible and may shape the spiral pattern itself.

Figure 12-19
According to the density wave theory, gas clouds overtake the spiral arm from behind and smash into the density wave. The compression triggers the formation of stars. The massive stars are so short-lived that they die before they can leave the spiral arm. The less massive stars emerge from the front of the arm with the remains of the gas cloud.

The spiral density wave creates spiral arms by the gravitational attraction of the stars and gas flowing through the arms. Even if there were no star formation at all, rotating disk galaxies could form spiral arms. But without star formation to make young, hot, luminous stars, the spiral arms would be difficult to see. It is the star formation that lights up the spiral arms and makes them so prominent. Thus, star formation helps determine what we see when we look at a spiral galaxy.

But star formation can also control the shape of spiral patterns if the birth of stars in a cloud of gas can renew itself and continue making more new stars. Consider a newly formed star cluster with a single massive star. The intense radiation from that hot star can compress nearby parts of the gas cloud and trigger further star formation (Figure 12-20). Also, massive stars evolve so quickly that their lifetimes are only an instant in the history of a galaxy. When they explode as supernovae, the expanding gasses can compress neighboring clouds of gas and trigger more star formation. Examples of such a **self-sustaining star formation** have been found. The Orion complex consisting of the Great Nebula in Orion and the star formation buried deep in the dark interstellar clouds behind the nebula is such a region.

Self-sustaining star formation can produce growing clumps of new stars, and the differential rotation of the galaxy can drag the inner edge ahead and let the outer edge lag behind to produce a cloud of star formation shaped like a segment of a spiral arm, a spur. Mathematical models of galaxies filled with such segments of spiral arms do have a spiral appearance, but they lack the bold, two-armed spiral that astronomers refer to as the grand-design spiral pattern (Figure 12-21). Rather, astronomers suspect that self-sustaining star formation can produce the branches and spurs so prominent in flocculent galaxies, but only the spiral density wave can generate the beautiful two-armed spiral patterns.

While the spiral density wave theory has been very successful in explaining spiral arms in galaxies, astronomers still do not understand all of the details. For example, Figure 12-22 shows the spiral pattern in a galaxy extending inward very close to the center of the galaxy. Such examples warn us that we do not yet understand all there is to know about spiral arms.

Our discussion of star formation in spiral structure illustrates the importance of natural processes. The spiral density wave creates graceful arms, but it is the star formation in the arms that makes them stand out so prominently. Self-sustaining star formation can act in some galaxies to modify the spiral arms and produce branches and spurs. In some galaxies, it can make the spiral pattern flocculent. By searching out and understanding the details of such natural processes, we can begin to understand the overall structure and evolution of the universe around us.

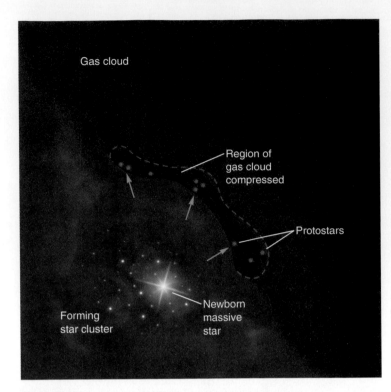

Figure 12-20
The energetic birth of a massive star compesses nearby gas clouds, triggering the gravitational contraction of the gas to form new stars.

REVIEW Critical Inquiry

Why can't we use solar-type stars as spiral tracers?

Sometimes the timing of events is the critical factor in an analysis. In this case, we must think about the evolution of stars and their orbital periods around the galaxy. Stars like the sun live about 10 billion years, but the sun's orbital period around the galaxy is 240 million years. The sun almost certainly formed when a gas cloud passed through a spiral arm, but since then the sun has circled the galaxy many times and has passed through spiral arms often. Thus, the sun's present location has nothing to do with the location of any spiral arms. An O star, however, lives only a few million years. It is born in a spiral arm and lives out its entire lifetime before it can leave the spiral arm. We find short-lived stars such as O stars only in spiral arms, but G stars are found all around the galaxy.

The spiral arms of our galaxy must make it beautiful in photographs taken from a distance, but we are trapped inside it. How do we know that the spiral arms we can trace out near us actually extend across the disk of our galaxy?

Spiral arms are a puzzle of great beauty, but the center of our galaxy hides a puzzle of great power.

a

b

Figure 12-21

(a) Some galaxies are dominated by two spiral arms, but even in these galaxies, minor spurs and branches are common. The spiral density wave can generate the two-armed, grand-design pattern, but self-sustained star formation may be responsible for the irregularities. (b) Many spiral galaxies do not appear to have two dominant spiral arms. Spurs and branches suggest that star formation is proceeding rapidly in such galaxies. *(Anglo-Australian Telescope Board)*

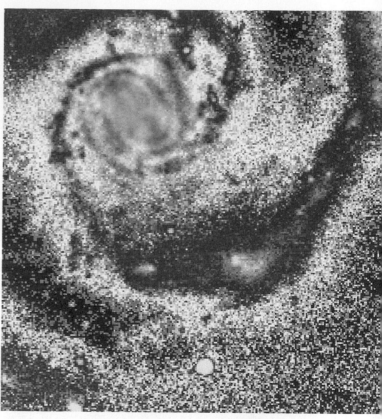

Figure 12-22

This near-infrared image of the center of the Whirlpool Galaxy has been computer enhanced and given false colors to reveal the spiral pattern. Hidden by dust at visible wavelengths, the innermost spiral arms are visible at longer wavelengths. The image shows the spiral pattern wrapped much closer to the center than current theories predict they should be. (See Figure 13-16.) *(Courtesy Dennis F. Zaritsky and Hans-Walter Rix)*

12-4 The Nucleus

The most mysterious region of our galaxy is its very center, the nucleus. At visual wavelengths, this region is totally hidden from us by gas and dust that dim the light by 30 magnitudes (Figure 12-23). If a trillion (10^{12}) photons of light left the center of the galaxy on a journey to Earth, only one would make it through the gas and dust. Thus, visual-wavelength photos tell us nothing about the nucleus. Observations at radio and infrared wavelengths paint a picture of tremendously crowded stars orbiting the center at high velocity. To understand what is happening at the center of our galaxy, we must carefully compare observations and theories.

a

tronomers turned their telescopes toward Sagittarius, they found a collection of radio sources with one, **Sagittarius A*** (abbreviated Sgr A* and usually pronounced "sadge A-star"), lying at the expected location of the galactic core (Figure 12-24). Radio astronomers can resolve (detect the diameter of) this central object, and it seems to be a bit less than 4 AU in diameter. Yet it is emitting tremendous amounts of energy. Astronomers suspect that Sgr A* is the geometrical center of our galaxy.

Infrared observations show that the stars are very close together in the central region of the galaxy. Infrared radiation at wavelengths shorter than 2 microns (2000 nm) comes almost entirely from cool stars. Observations at these wavelengths show that the stars are only about 1000 AU apart. Near the sun, stars are about 330,000 AU (1.6 pc) apart. Infrared photons with wavelengths longer than 4 microns (4000 nm) come almost entirely from interstellar dust warmed by stars, and radiation at these wavelengths from the central region is intense.

Radio observations show that the region around Sgr A* is disturbed. A ring of gas clouds dense enough to contain molecules surrounds the center, extending from a distance of about 25 pc inward to about 6 pc. The inner edge of this ring contains gas ionized by bright ultraviolet radiation from the stars crowded in the central area. Inside this ring of gas lies a cavity in which the gas is 10 to 100 times less dense. Because differences in gas density should smooth out within about 100,000 years, astronomers suspect that a central eruption drove most of the gas outward, forming the cavity within the last 100,000 years.

During 1998, astronomers announced two infrared studies of the motions of stars near the center of the Milky Way Galaxy. One group of astronomers was able to measure the motions of stars only 860 AU from the center and found them moving so fast they would orbit with a period a bit less than 16 years. A simple calculation of the mass using a^3 divided by P^2 (see By the Numbers 8-4) tells us the central mass must be about 2.6 million solar masses. Another team of astronomers detected stars only 275 AU from the center moving so fast they would orbit in only 2.8 years. This implies a central mass of 2.7 million solar masses.

b

Figure 12-23
(a) In this visual-wavelength photograph of the Milky Way from Cassiopeia at left to Sagittarius at right, gas and dust hide the center of the galaxy at extreme right (arrow). *(Steward Observatory)* (b) An infrared photo penetrates the dust clouds and reveals the crowded stars near the galactic center. This image was made by combining images made at the wavelengths 1350 nm, 1680 nm, and 2260 nm and coloring them blue, green, and red. Only at the longest of the three wavelengths can we see the star clouds near the center, and thus they look red in this composite images. *(Courtesy Bradford W. Greeley, Rhodes College)*

Observations

If we observe the Milky Way on a dark night, we might notice a slight thickening in the direction of the constellation Sagittarius, but nothing specifically identifies this as the direction of the heart of the galaxy. Even Shapley's study of globular clusters identified the center only approximately. When radio as-

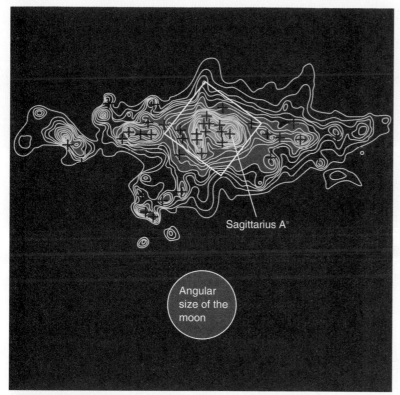

Figure 12-24
A radio map of the Sagittarius region of the Milky Way (see box in Figure 12-5) reveals an intense radio source, Sagittarius A*, at the expected location of the center. Crosses mark far-infrared sources associated with star formation. Box marks boundaries of Figure 12-25. Angular size of the moon shown for comparison. *(Adapted from observations by W. L. Altenhoff, D. Downes, T. Pauls, and J. Schrami, and by S. F. Odenwalk and G. G. Fazio)*

ter it could draw inward. Also, other energy sources are limited. A supernova, for example, converts very little of its mass into energy before blowing itself apart, and there is a limit to the number of luminous, massive stars capable of producing supernovae that can occupy a small region. For steady energy production from a very small region, a massive black hole works very well.

Such a massive black hole would not be the remains of a single collapsed star but would have developed gradually as gas clouds and stars lost orbital velocity and drifted toward the center of the galaxy. We will see in later chapters that other galaxies may contain massive black holes.

Certain characteristics of Sgr A* are compatible with its being a massive black hole within an accretion disk. Studies of Sgr A* show that it does not seem to be moving at all. While other stars in the central regions rush about their orbits, Sgr A* remains fixed, suggesting that it must have a mass of at least a few hundred solar masses. It can't be a typical compact object of a few solar masses. Radio astronomers using the highest resolution find that Sgr A* is less than a few astronomical units in diameter and elongated, as we might expect of a massive black hole surrounded by an inclined accretion disk.

Consider how small Sgr A* is in comparison with the entire galaxy. If the Milky Way galaxy were shrunk to the size of Texas, Sgr A* would be only 0.02 inches in diameter—about the size of a mosquito's eye. Yet Sgr A* seems to be a powerful energy source. For example, X-ray observations show clouds of highly ionized gas extending 100 pc on either side of the center, and this suggests that the core erupted with the power of several thousand supernovae only a few centuries ago. The combination of small size and great power suggests accretion into a massive black hole.

Some observations suggest the existence of high-energy jets coming from the core. Radio maps reveal a jet extending about 30 pc toward the galaxy's south pole. Such jets are characteristic of accretion disks around black holes, as in the case of SS433, and the strong magnetic field found near the core (Figure 12-25a) suggests that the center of the galaxy contains a rapidly spinning object. Also, the orbiting Gamma Ray Observatory has found gamma rays coming from a region extending several thousand light years from the central region out into the halo. The gamma rays have the energy produced when an electron collides with its antiparticle, the positron. It isn't known what is making these positrons, but they may be coming from high-energy processes in the core, such as a hot accretion disk.

What could be so massive and so small? Since the mid-1970s, astronomers have been asking themselves that question. The most interesting possibility is a gigantic black hole.

The Massive Black Hole Hypothesis

To understand the center of our galaxy, astronomers must explain how a large mass can occupy a small space, and a massive black hole fills the bill quite well. A black hole of 3 million solar masses, for example, would be only about 13 times larger than the sun—quite small in terms of the galaxy. Also, a massive black hole could produce energy if matter dribbled into it through an accretion disk, and it could produce occasional eruptions if an entire star fell in. Before we can decide, we must examine the hypothesis more carefully, compare it with observations, and consider alternatives.

A massive black hole could supply almost unlimited energy. There is no known limit to the mass of a black hole, and the more massive it is, the more mat-

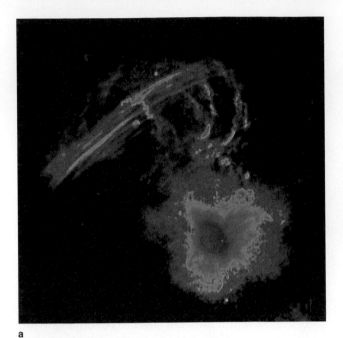

a

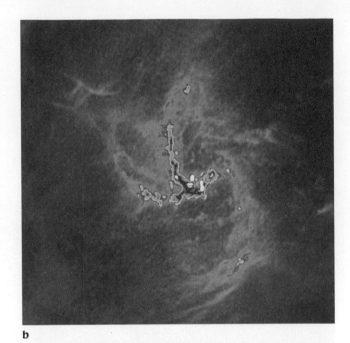

b

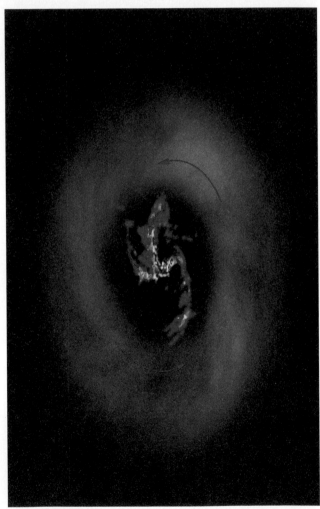

c

Figure 12-25
Sagittarius A* (a) This VLA map shows filaments of excited gas 50 pc long arching out of the center of our galaxy as if constrained by a magnetic field. Compare with boxed region in Figure 12-24. *(NRAO)* (b) A detailed map of the center, shown roughly by the box in (a), reveals the excited inner edge of a disk of gas and dust with gas flowing into a central object. *(N. Killeen and Kwok-Yung Lo)* (c) The central spiral appears to lie in the open center of the larger disk of neutral gas. It is believed to show matter flowing into a central object.

At the highest resolution, the VLA radio telescope can detect a spiral swirl of matter in the central 3 pc of the galaxy, the hole in the disk of gas and dust that whirls around the center. This spiral is not related to the spiral arms but appears to be matter spiraling inward toward the central object (Figure 12-25b, 12-25c). This inflow appears to add a few tenths of a solar mass per year to the central mass, and accretion at that rate onto a massive black hole could easily provide the observed energy.

A supermassive black hole is an exciting idea, but we must notice the difference between adequacy and necessity. A supermassive black hole is adequate to explain the observations, but it is not necessary. A few astronomers have thought of other explanations. For example, gas flowing into the core could form a compact cluster of very massive, highly energetic stars. Thus, the observational evidence we have does not require a black hole at the center of our galaxy. However, astronomers are finding more and more evidence that other galaxies contain supermassive black holes in their nuclei (Chapter 18), so most astronomers take the black hole hypothesis very seriously.

REVIEW Critical Inquiry

Why do we think the center of our galaxy contains a large mass?

The best way to measure the mass of an astronomical object is to watch something orbit around it. We can then use Kepler's third law to find the mass inside the orbit. In the case of the nucleus of our galaxy, we can't see individual stars orbiting the center, but we can detect the effects of the circulation of matter there. Ionized neon radiates an infrared emission line at the specific wavelength of 12.8 microns (12,800 nm), so infrared astronomers can measure the observed wavelength of this emission and use the Doppler shift to calculate the velocity with which gas is moving near the center of the galaxy. That velocity is very high at small distances from the center, and Kepler's third law tells us that the mass inside these small orbits must be millions of solar masses.

In addition to these orbital motions, the large amount of infrared radiation at wavelengths longer than 4 microns (4000 nm) also implies that vast numbers of stars are crowded into the center. Why?

Despite all the observations and theories, we cannot yet be sure what lies at the center of the Milky Way Galaxy. One way to extend our research is to compare our galaxy with others. We begin that strategy in the next chapter.

Summary

The Milky Way Galaxy is a star system roughly 75,000 ly in diameter, containing about 200 billion stars. Its true dimensions were not known until early in this century when Harlow Shapley began studying globular clusters. In the clusters, he found a type of variable star that had been studied some years before by Henrietta Leavitt. She had found that the period of variation of these Cepheid variables was related to their luminosity. Shapley used these variable stars to find the distance to the globular clusters and thus discover that the Milky Way Galaxy was much larger than anyone had previously supposed.

The Milky Way Galaxy contains two components, the disk component and the spherical component. Disk stars are metal-rich population I stars moving in circular orbits that lie in the plane of the disk. Because stars that are farther from the center move more slowly in their orbits, the disk rotates differentially.

The spherical component consists of a nuclear bulge at the center and a halo of thinly scattered stars and globular clusters that completely envelops the disk. Halo stars are metal-poor population II stars moving in random, elliptical orbits.

Studies of the motions of stars and star clusters in the outer parts of the galaxy show that it is rotating more rapidly than was expected. This seems to mean that the galaxy is surrounded by a galactic corona extending roughly seven times farther than the traditional edge of the galaxy. Because this mass is nonluminous, astronomers have dubbed it dark matter.

The distribution of populations through the galaxy suggests a way the galaxy could have formed from a spherical cloud of gas that gradually flattened into a disk. The younger the stars, the more metal rich they are, and the more circular and flat their orbits are.

The very youngest objects lie along spiral arms within the disk. They live such short lives they don't have time to move from their place of birth in the spiral arms. Maps of these spiral tracers and cool hydrogen clouds reveal the spiral pattern of our galaxy.

The density wave theory suggests that the spiral arms are regions of compression that move through the disk. When an orbiting gas cloud smashes into the compression wave, the gas cloud forms stars. Another process, self-sustaining star formation, may act to modify the arms as the birth of massive stars triggers the formation of more stars by compressing neighboring clouds.

The nucleus of the galaxy is invisible at visual wavelengths, but radio, infrared, X-ray, and gamma-ray radiation can penetrate the dust clouds. These wavelengths reveal crowded central stars and heated clouds of dust. Some of the central features are expanding outward.

The very center of the Milky Way is marked by a radio source, Sagittarius A*, that is also a source of infrared radiation, X rays, and gamma rays. The core must be less than 20 AU in diameter and must contain a few million solar masses. Many astronomers believe that this central object is a black hole.

New Terms

variable star	nuclear bulge
instability strip	differential rotation
RR Lyrae variable star	rotation curve
Cepheid	dark matter
period–luminosity relation	dark halo
open star cluster	galactic corona
globular star cluster	population I
proper motion	population II
calibration	metals
Shapley–Curtis debate	spiral tracer
disk component	density wave theory
kiloparsec (kpc)	flocculent galaxy
spiral arm	self-sustaining star formation
spherical component	Sagittarius A*
halo	

Review Questions

1. Why isn't it possible to tell from the appearance of the Milky Way that the center of our galaxy is in Sagittarius?

2. Why is there a period–luminosity relation?

3. How can astronomers use variable stars to find distance?

4. Why is it difficult to specify the thickness or diameter of the disk of our galaxy?

5. Why didn't astronomers before Shapley realize how large the galaxy is?

6. How do we know how old our galaxy is?

7. Why do we conclude that metal-poor stars are older than metal-rich stars?

8. How can astronomers find the mass of the galaxy?

9. What evidence do we have that our galaxy has an extended corona of dark matter?

10. How do the orbits of stars around the Milky Way Galaxy help us understand its origin?

11. What evidence contradicts the traditional theory for the origin of our galaxy?

12. Why do spiral tracers have to be short-lived?

13. What evidence do we have that the density wave theory is not fully adequate to explain spiral arms in our galaxy?

14. What evidence do we have that the center of our galaxy is a powerful source of energy?

15. Why is the lack of motion of Sgr A* important evidence in our study of the center of our galaxy?

Discussion Questions

1. How would this chapter be different if interstellar matter didn't absorb starlight?

2. Are there any observations you could make with the Hubble Space Telescope that would allow you to better understand the nature of Sgr A*?

Problems

1. Make a scale sketch of our galaxy in cross section. Include the disk, sun, nucleus, halo, and some globular clusters. Try to draw the globular clusters to scale size.

2. Because of dust clouds, we can see only about 5 kpc into the disk of the galaxy. What percentage of the galactic disk can we see? (*Hint:* Consider the area of the entire disk and the area we can see.)

3. If the fastest passenger aircraft can fly 1600 km/hr (1000 mph), how long would it take to reach the sun? the galactic center? (*Hint:* 1 pc = 3×10^{13} km.)

4. If the RR Lyrae stars in a globular cluster have apparent magnitudes of 14, how far away is the cluster? (*Hint:* See By the Numbers 8-2.)

5. If interstellar dust makes an RR Lyrae star look 1 magnitude fainter than it should, by how much will we overestimate its distance? (*Hint:* See By the Number 8-2.)

6. If a globular cluster is 10 minutes of arc in diameter and 8.5 kpc away, what is its diameter? (*Hint:* Use the small-angle formula from By the Numbers 3-1.)

7. If we assume that a globular cluster 4 minutes of arc in diameter is actually 25 pc in diameter, how far away is it? (*Hint:* Use the small-angle formula from By the Numbers 3-1.)

8. If the sun is 5 billion years old, how many times has it orbited the galaxy?

9. If the true distance to the center of our galaxy is found to be 7 kpc and the orbital velocity of the sun is 220 km/sec, what is the minimum mass of the galaxy? (*Hints:* Find the orbital period of the sun, and then see By the Numbers 8-4.)

10. Infrared radiation from the center of our galaxy with a wavelength of about 2×10^{-6} m (2000 nm) comes mainly from cool stars. Use this wavelength as λ_{max} and find the temperature of the stars. (*Hint:* See By the Numbers 6-1.)

Critical Inquiries for the Web

1. How is our view of other galaxies affected by our position in the Milky Way? Search the Internet or other sources for sky maps that trace the path of the Milky Way through the constellations. Which constellations does the Milky Way run through? Choose one of these constellations and another constellation that lies far from the band of the Milky Way. Consult the Web for lists of galaxies visible in these two constellations. In which one have more galaxies been found? Why is this so?

2. Massive stars like Eta Carinae can play a significant role in the formation of stars along a spiral arm. Look for information on the Web related to Eta Carinae, which is on of the most carefully studied stars in the heavens. Briefly summarize the observed behavior of this star, and discuss why it is a good candidate for contributing to self-sustaining star formation in our galaxy.

Go to the Brooks/Cole Astronomy Resource Center (www.brookscole.com/astronomy) for critical thinking exercises, articles, and additional readings from InfoTrac College Edition, Brooks/Cole's online student library.

Galaxies

Guidepost

We can no more understand galaxies by understanding a single example, the Milky Way, than we could understand humanity by understanding a single person. Whereas the last chapter examined a single galaxy, this chapter generalizes to examine the entire range of objects called galaxies.

We take two lessons from this chapter for use in future chapters. First, galaxies interact with each other in complicated ways, and that will become critical in the next chapter when we study peculiar galaxies. The second lesson is that 90 to 99 percent of the matter in the universe is invisible. That astonishing conclusion comes from the study of galaxies, but we will understand its impact better when we get to Chapter 19, where we will see that it determines the structure and fate of the entire universe.

Science fiction heroes flit effortlessly between the stars, but almost none voyage between the galaxies. As we leave our home galaxy, the Milky Way, behind, we voyage out into the vast depths of the universe, out among the galaxies, space so deep it is unexplored even in fiction.

Less than a century ago, astronomers did not understand that there were galaxies. Nineteenth-century telescopes revealed faint nebulae scattered among the stars, and some were spiral. Astronomers argued about the nature of these nebulae, but it was not until the 1920s that astronomers understood that some were other galaxies much like our own, and it was not until recent decades that astronomical telescopes could reveal the tremendous beauty and intricacy of the galaxies (Figure 13-1).

In this chapter, we will try to understand how galaxies form and evolve, and we will discover that the amount of gas and dust in a galaxy is a critical clue. We will also discover that interactions between galaxies can influence their structure and evolution.

Before we can build theories, however, we must gather some basic data concerning galaxies. We must classify the different kinds of galaxies and discover their basic properties—diameter, luminosity, and mass. Once we know what galaxies are like, once we can characterize them by describing their typical properties, we will be ready to theorize about their origin and evolution.

13-1 Three Types of Galaxies

Often a new branch of science begins with a simple classification system. The shapes of animals can tell a biologist how a species evolved, and the shapes of galaxies can help modern astronomers understand

Figure 13-1
Modern photographs of spiral galaxies, such as this image of M100, show bright stars and star clusters clearly. Until early in the 20th century, astronomers could see only faint nebulosity and vague spiral patterns. Not until the 1920s were telescopes large enough and photographic plates sensitive enough to reveal that many of the faint nebulae in the sky are other galaxies. *(Anglo-Australian Telescope Board)*

how galaxies evolve. Our discussion of galaxy shapes begins in the 1920s with Mount Wilson astronomer Edwin Hubble (namesake of the Hubble Space Telescope), who obtained some of the first good photographs of other galaxies. He classified the galaxies according to their shapes (Window on Science 13-1) and summarized the classification in a diagram known today as the **tuning fork diagram** (Figure 13-2). We will consider each main shape category in detail.

As we study galaxies, recall that star formation occurs in clouds of gas and dust. The presence or absence of gas and dust in different kinds of galaxies will help us understand their shapes and histories.

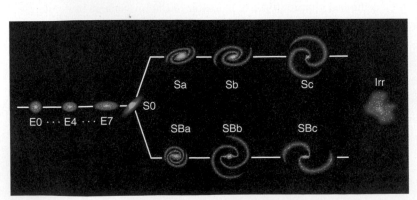

Figure 13-2
Modern astronomers use the tuning fork diagram as an organizing framework for galaxy classification. Elliptical (E), spiral lacking dust and gas (S0), normal spiral (S), barred spiral (SB), and irregular (Irr) galaxies are placed from left to right in approximate order of increasing gas, dust, and star formation.

Classification is one of the most basic and most powerful of scientific tools. It is often a way to begin studying a topic, and it can lead to dramatic insights. Charles Darwin, for example, sailed around the world with a scientific expedition aboard the ship HMS *Beagle.* Everywhere he went, he studied the living things he saw and tried to classify them. For example, he studied the tortoises, mockingbirds, and finches he saw on the Galápagos Islands. His classifications of these and other animals eventually led him to think about evolution and natural selection. Scientists in many fields depend on carefully designed systems of classification.

Classification is an everyday mode of thought. We all use classifications when we order lunch, buy shoes, catch a bus, and so on. In science, classifications reveal the relationships between different kinds of objects and at the same time save the scientist valuable time by bringing order out of seeming disorder. An economist can classify different kinds of businesses and does not have to analyze each of the millions of businesses as if it were unique. Classifications of minerals, plants, psychological learning styles, modes of transportation, or sandwiches bring order to the world and help us deal with information.

Astronomers use classifications of galaxies, stars, moons, telescopes, and more. Whenever you encounter a scientific discussion, look for the classifications on which it is based. Classifications are the orderly framework on which much of science is built.

Elliptical Galaxies

Galaxies that appear to be round or elliptical in shape; have almost no visible gas or dust; lack hot, bright stars; and have no spiral pattern are called **elliptical galaxies** (Figure 13-3). The stars are most crowded toward the center. The largest, the giant ellipticals, are a number of times larger than our galaxy, and the smallest, the dwarf ellipticals, are hardly more luminous than a globular cluster.

Elliptical galaxies are classified according to shape. If we measure on a photograph the largest diameter through the center of an elliptical galaxy and call this distance a and also measure the smallest diameter b, then the equation

$$\frac{10(a - b)}{a}$$

gives an index of its shape. This index, rounded to the nearest integer, ranges from 0, for circular outlines, to

Figure 13-3

Elliptical galaxies contain little gas and dust and few bright stars. Thus, they are nearly featureless clouds of stars. (a) M87 is a giant elliptical galaxy surrounded by a swarm of over 500 globular clusters. Objects to the lower right are smaller galaxies. *(NOAO)* (b) The Leo I dwarf elliptical is a small, nearby galaxy and is resolved into individual stars. Note the lack of dust, gas, and hot, bright stars. *(Anglo-Australian Telescope Board)*

a

b

7, for the most elliptical outlines observed. Thus, elliptical galaxies are classified into subgroups ranging from E0 to E7.

The true shape of an elliptical galaxy is not evident from its outline. An elliptical galaxy with a true shape like a rounded football might appear elliptical if we saw it from the side or round if we saw it from one end. Statistical studies of different types of elliptical galaxies reveal a range of true shapes. Some are truly spherical, but others are more ellipsoidal. None are known to be more elliptical than E7.

The spectra of elliptical galaxies reveal that they are composed mainly of population II stars. In fact, the best images of giant ellipticals reveal that they are surrounded by a swarm of globular clusters. A few studies have detected small dust and gas clouds where a few new stars are forming, but for the most part, these galaxies contain a single, old generation of stars. Lack of gas and dust means there is little star formation, so elliptical galaxies cannot make new stars.

Spiral Galaxies

Galaxies that contain an obvious disk component are called **spiral galaxies.** To varying degrees, these galaxies contain gas, dust, and hot, bright, population I stars. As in our own galaxy, the gas, dust, and youngest stars, which include the bright O and B stars, are confined to a thin disk (Figure 13-4). The dust clouds in the disks of spiral galaxies are clearly visible when they are silhouetted against bright clouds of stars or a more distant galaxy (Figure 13-4b). We also see in these galaxies nuclear bulges and, in the nearest spiral galaxies, halos filled with population II stars and globular clusters. Thus, spiral galaxies contain both old and young stars.

Among spiral galaxies we identify three distinct types: S0 galaxies, normal spirals, and barred spirals. Unlike our galaxy, the S0 galaxies show no obvious spiral arms, have very little gas and dust, and contain very few hot, bright stars (Figure 13-5). However, they have an obvious disk component with a large nuclear bulge at the center. They appear to be intermediate between elliptical and spiral galaxies.

Normal spiral galaxies can be further subclassified into three groups according to the size of their nuclear bulge and the degree to which their arms are wound up. Spirals that have little gas and dust, larger nuclear bulges, and tightly wound arms are classified Sa (Figure 13-6a). Sc galaxies (Figure 13-6c) have large clouds of gas and dust, small nuclear bulges, and very loosely wound arms. The Sb galaxies (Figure 13-6b) are intermediate between Sa and Sc. Because there is more gas and dust in the Sb and Sc galaxies than in S0 and Sa galaxies, we find more gas and dust in the Sb and Sc galaxies, and more young, hot, bright stars along their arms. The Andromeda Galaxy (see Figure 12-1a) and our own Milky Way Galaxy are Sb galaxies.

A minority of spirals have an elongated nuclear bulge with spiral arms springing from the ends of the

Figure 13-4

(a) The disk of the spiral galaxy NGC253 is filled with gas, dust, and clouds of bright young stars that are forming from the gas and dust. The nucleus of the galaxy is at upper left. *(Hubble Heritage Team, AURA/STScI/NASA)* (b) The dust in this spiral galaxy is dramatically revealed where it is silhouetted against a slightly more distant galaxy. *(W. Keel and R. White, NASA)*

a

b

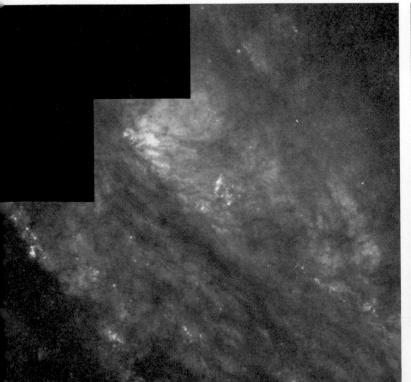

Figure 13-5
S0 galaxies have a disk but have no bright O and B stars to illuminate spiral arms. Even when we see an S0 galaxy edge-on, as in this image, no dust is visible. Compare the lack of dust in this galaxy with the dust visible in the spiral galaxy shown in Figure 13-4b. *(Rudolph E. Schild)*

bar. These **barred spiral galaxies** are classified SBa, SBb, or SBc, according to the same criteria listed for normal spirals (Figure 13-7). Astronomers working with computer models have succeeded in imitating the rotating bar structure. It appears to occur when an instability develops in the stellar distribution within the rotating galaxy, altering the orbits of the inner stars and generating a stable, elongated nuclear bulge. Thus, except for the peculiar rotation in the nuclear bulge, barred spiral galaxies are quite similar to normal spirals. In fact, some studies suggest that most spiral galaxies, including our own Milky Way Galaxy, are barred.

Irregular Galaxies

Certain galaxies that have a chaotic appearance, with large clouds of gas and dust mixed with both young and old stars, are known as **irregular galaxies.** They have no obvious nuclear bulge or spiral arms. The prototypical irregular galaxies are the **Large** and the **Small Magellanic Clouds,** which were first noted by Magellan's men as they sailed through the southern seas in the early 1500s. For observers in the Southern Hemisphere, these nearby galaxies are visible to the naked eye as hazy patches in the sky (Figure 13-8a).

It is difficult to decide how common irregular galaxies are because they are faint (Figure 13-8b). In catalogues of galaxies, about 70 percent are spiral, but that is the result of a selection effect (Window on Science 13-2). Spiral galaxies contain hot, bright stars and are thus very luminous and easy to see. Many ellipticals are faint and harder to notice. Small galaxies, such

a

b

c

Figure 13-6
(a) The galaxy NGC3623 is an Sa galaxy with a large nuclear bulge and tightly wound spiral arms. (b) NGC3627 is an Sb galaxy with a smaller nuclear bulge and more open arms. (c) NGC2997 is an Sc galaxy with a small nuclear bulge and loosely wound spiral arms. *(Anglo-Australian Telescope Board)*

a

b

Figure 13-7

Barred spiral galaxies. (a) The SBb galaxy NGC4639 has a bar-shaped nuclear bulge with spiral arms springing from the ends of the bar (see inset). Notice the bright blue stars and dust lanes in the spiral arms. Notice also the fainter galaxies visible, including a distant barred spiral near the bottom edge of the frame. *(Sandage, Saha, Tammann, Labhardt, Macchetto, Panagia, and NASA)* (b) NGC1365 is an SBc galaxy with a pronounced bar and very open spiral arms (see inset). *(Anglo-Australian Telescope Board)*

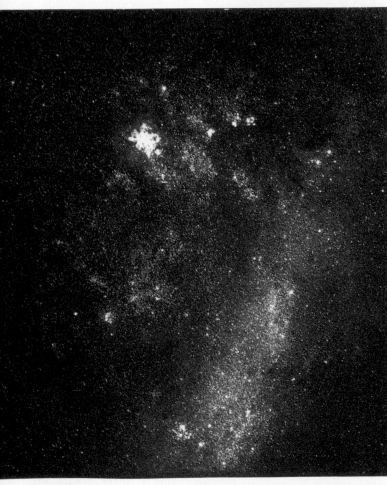

a

b

Figure 13-8

(a) The Large Magellanic Cloud is an irregular galaxy that is only about 50 kpc away. Bright pink regions are ionized gas. The bright nebula at upper left is the Tarantula Nebula, which is also shown in Figure 10-14b. *(Copyright R. J. Dufour, Rice University)* (b) Galaxy IC4182 is typical of the irregular galaxies in that it is small and faint and contains both red and blue stars but no obvious spiral arms. The two bright stars are foreground stars in our own galaxy. *(George Jacoby and Mike Pierce, NOAO, ©AURA, Inc.)*

Many different kinds of science depend on selecting objects to study. Scientists studying insects in the rain forest, for example, must choose which ones to catch. They can't catch every insect they see, so they might decide to catch and study any insect that is red. If they are not careful, a selection effect could bias their data and lead them to incorrect conclusions without their ever knowing it. The reason selection effects are dangerous in science is that they can have powerful influences without being evident. Only by carefully designing a research project can scientists avoid selection effects.

For example, suppose we decide to measure the speed of cars on a highway. There are too many cars to measure every one, so we reduce the workload and measure only the speed of red cars. It is quite possible that this selection criterion will mislead us because people who buy red cars may be more likely to be younger and drive faster. Should we measure only brown cars? No, because we might suspect that only older, more sedate people would buy a brown car. Should we measure the speed of any car in front of a truck? Perhaps we should pick any car following a truck? Again, we may be selecting cars that are traveling a bit faster or a bit slower than normal. Only by very carefully designing our experiment can we be certain that the cars we measure are traveling at representative speeds.

Astronomers face the danger of selection effects quite often. Iron meteorites are easier to find than stone meteorites. Very luminous stars are easier to see at great distances than faint stars. Spiral galaxies are brighter, bluer, and more noticeable than elliptical galaxies. What we see through a telescope depends on what we notice, and that is powerfully influenced by selection effects.

Scientists engaged in observation must spend considerable time designing their experiments. They must be careful to observe an unbiased sample if they expect to make logical deductions from their results. The scientists in the rain forest, for example, should not catch and study only the red insects. Often, the most brightly colored insects are poisonous. If they are good scientists, they will plan their work with great care and avoid any possible selection effects.

---

as irregular galaxies and dwarf ellipticals, are very hard to spot if they are not nearby, so compiling a census of galaxies is difficult. It seems that ellipticals are more common than spirals and that irregulars make up only about 25 percent of all galaxies.

REVIEW Critical Inquiry

What color are galaxies?

What we know about stars and populations allows us to analyze this question. Different kinds of galaxies have different colors, depending mostly on how much gas and dust they contain. If a galaxy contains large amounts of gas and dust, it probably contains lots of young stars, and a few of those young stars will be massive, hot, luminous O and B stars. They will give the galaxy a distinct blue tint. In contrast, a galaxy that contains little gas and dust will probably contain few hot stars. Its most luminous stars will be red giants, and they will give the galaxy a red tint. Because the light from a galaxy is a blend of the light from billions of stars, the colors are only tints. Nevertheless, the most luminous stars in a galaxy determine its color. From this we can conclude that elliptical galaxies tend to be red and the disks of spiral galaxies tend to be blue.

Of course, the halo of a spiral galaxy would be both dimmer and redder than the disk. Why?

We end our discussion of the classification of galaxies with the irregulars, and that is convenient because it leads us to two important questions. First, how do we know the properties of galaxies? In the next section, we will try to find the basic properties of galaxies. Second, we must wonder why the irregular galaxies look as if they have been stirred with a big spoon. In the last section of this chapter, we will try to answer that question and see what is happening to the unfortunate Magellanic clouds.

13-2 Measuring the Properties of Galaxies

Looking beyond the edge of our Milky Way galaxy, astronomers find many billions of galaxies. Great clusters of galaxies, some containing thousands of galaxies, fill space as far as we can see. We can recognize the different types of galaxies, but what are the properties of these star systems? What are the diameters, luminosities, and masses of galaxies? Just as in our study of stellar characteristics (Chapter 8), the first step in our study of galaxies is to find out how far away they are. Once we know a galaxy's distance from us, its size and luminosity are relatively easy to find. Later in this section, we will see that finding mass is more difficult.

Figure 13-9
The vast majority of spiral galaxies are too distant for Earth-based telescopes to detect Cepheid variable stars. The Hubble Space Telescope, however, can locate Cepheids in some of these galaxies, as it has in the bright spiral galaxy M100. From a series of images taken on different dates, astronomers can locate Cepheids (inset), determine the period of pulsation, and measure the average apparent brightness. They can then deduce the distance to the galaxy—51 million ly for M100. *(J. Trauger, JPL; Wendy Freedman, Observatories of the Carnegie Institution of Washington; and NASA)*

Distance

The distances to galaxies are so large that it is not convenient to measure them in light-years, parsecs, or even kiloparsecs. Instead, we will use the unit **megaparsec (Mpc),** or 1 million pc. One Mpc equals 3.26 million ly, or approximately 2×10^{19} miles.

To find the distance to a galaxy, we must search among its stars for a familiar object whose luminosity or diameter we know. Such objects are called **distance indicators.**

Because their period is related to their luminosity (see Figure 12-2c), Cepheid variable stars are reliable distance indicators. If we know the period of the star's variation, we can use the period–luminosity diagram to learn its absolute magnitude. By comparing its absolute and apparent magnitudes, we can find its distance (see By the Numbers 8-2). Figure 13-9 shows a galaxy in which the Hubble Space Telescope detected Cepheids.

Even with the Hubble Space Telescope, Cepheids are not visible in galaxies beyond about 50 million ly (15 Mpc), so astronomers must search for less common but brighter distance indicators and calibrate them using nearby galaxies containing visible Cepheids. For example, by studying nearby galaxies with distances known from Cepheids, astronomers have found that the brightest globular clusters have absolute magnitudes of about −10. If we find globular clusters in a more distant galaxy, we can assume that the brightest have similar absolute magnitudes and thus calculate the distance.

Another distance indicator is the cloud of ionized hydrogen called an **H II region** that forms around a very hot star. Astronomers have attempted to calibrate these, but they have not been as reliable as Cepheids because some are larger and brighter than others. Also, the size of H II regions seems to depend on the kind of galaxy they occupy, and so astronomers have not been able to calibrate them as accurately as other distance indicators.

Planetary nebulae have proven to be very useful distance indicators, which is surprising considering how faint their central star is. However, the central stars of planetary nebulae seem faint because they are very hot and radiate most of their energy in the ultraviolet. The planetary nebula absorbs this ultraviolet radiation and reradiates it as visible emission lines, and astronomers have been able to calibrate the brightest planetary nebulae by studying nearby galaxies, such as the Andromeda Galaxy. Like brightly colored paper lanterns illuminated from within, planetary nebulae in other galaxies can be identified by their emission at the wavelength of 500.7 nm, the wavelength of an emission line of oxygen; and that identification allows the calculation of the distance to the galaxy.

When a supernova explodes in a distant galaxy, astronomers rush to observe it. Studies show that Type Ia supernovae, those caused by the collapse of a white dwarf, all reach about the same absolute magnitude at maximum. Because these supernovae are much brighter than Cepheids, they can be seen at distances as great as 200 Mpc, and thus astronomers can calculate the distances to these galaxies. The drawback to this distance indicator is that supernovae are rare, and none may occur during our lifetime in a galaxy we might be studying.

At the greatest distances, astronomers must calibrate the total luminosity of the galaxies themselves. For example, studies of nearby galaxies show that an average galaxy like our Milky Way Galaxy has a luminosity about 16 billion times the sun's. If we see a similar galaxy far away, we can measure its apparent magnitude and calculate its distance. Of course, it is important to recognize the different types of galaxies, and that is difficult to do at great distances (Figure 13-10). Averaging the distances to the brightest galaxies in a cluster can reduce the uncertainty in this method.

Notice how astronomers use calibration (Window on Science 12-1) to build a distance scale reaching

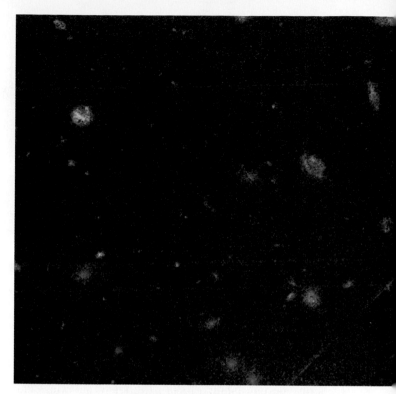

Figure 13-10
Only the general characteristics of the galaxies are recognizable in this cluster of galaxies 4 to 6 billion light-years away. The bright blue galaxy at upper left is probably a foreground galaxy much closer than the cluster. The larger galaxies in this cluster are about the size of our Milky Way Galaxy. The diagonal line at lower right was produced by an artificial Earth satellite. *(K. Ratnatunga, R. Griffiths, Carnegie Mellon University, and NASA)*

from the nearest galaxies to the most distant visible galaxies. Often astronomers refer to this as the distance pyramid or the distance ladder because each step depends on the steps below it. Of course, the foundation of the distance scale rests on the Cepheid variable stars and our understanding of the luminosities of the stars in the H-R diagram, which ultimately rest on measurements of the parallax of stars within a few hundred light-years of Earth.

The most distant visible galaxies are roughly 10 billion ly (3000 Mpc) away, and at such distances we see an effect akin to time travel. When we look at a galaxy millions of light-years away, we do not see it as it is now, but as it was millions of years ago when its light began the journey toward Earth. Thus, when we look at a distant galaxy, we look back into the past by an amount called the **look-back time,** a time in years equal to the distance to the galaxy in light-years.

The look-back time to nearby objects is usually not significant. The look-back time to the moon is 1.3 seconds, to the sun only 8 minutes, and to the nearest star about 4 years. The Andromeda Galaxy has a look-back time of about 2 million years, a mere eye blink in the lifetime of a galaxy. But when we look at more distant galaxies, the look-back time becomes an appreciable part of the age of the universe. We will see evidence in Chapter 15 that the universe began no more than 20 billion years ago. Thus, when we look at the most distant visible galaxies, we are looking back 10 billion years to a time when the universe may have been significantly different. This effect will be important in our discussions in this and the next chapter.

The Hubble Law

Although astronomers find it difficult to measure the distance to a galaxy, they often estimate such distances using a simple relationship.

Early in this century, astronomers noticed that the spectral lines in galaxy spectra were shifted slightly toward longer wavelengths—what astronomers call *red shifts*. Interpreted as a consequence of the Doppler shift, these red shifts implied that the galaxies had large radial velocities and were receding from us. (Review the Doppler effect in Chapter 6.)

In 1929, the American astronomer Edwin Hubble published a graph that plotted the velocity of recession versus distance for a number of galaxies. The points in the graph fell along a straight line (Figure 13-11). The straight line relation from Hubble's diagram can be written as the simple equation

$$V_r = Hd$$

That is, the velocity of recession V_r equals the distance d in millions of parsecs times the constant H. This relation between red shift and distance is known as the **Hubble law,** and the constant H is known as the **Hubble constant.**

Modern attempts to measure H have been difficult because of the uncertainty in the distances to galaxies. For the past few decades, astronomers have narrowed the range of possible values down to somewhere between 50 and 100 km/sec/Mpc,[*] and the

newest results suggest a value of 70 to 80. In Chapter 15 we will see that the value of H has important consequences for our understanding of the history of the universe, but here we should recognize that the Hubble constant is not precisely known.

The Hubble law is important because it is commonly interpreted to show that the universe is expanding. In Chapter 15, we will discuss this expansion, but here we will use the Hubble law as a practical way to estimate the distance to a galaxy. Simply stated, a galaxy's distance equals its velocity of recession divided by H. For example, if a galaxy has a radial velocity of 700 km/sec, and H is 70 km/sec/Mpc, then the distance to the galaxy is 700 divided by 70, or about 10 Mpc. This makes it relatively easy to estimate the distances to galaxies because a large telescope can photograph the spectrum of a galaxy and determine its velocity of recession even when it is too distant to have visible distance indicators. We will refer to such distances as estimates because H is not known accurately.

Diameter and Luminosity

The distance to a galaxy is the key to finding its diameter and its luminosity. We can easily photograph a galaxy and measure its angular diameter in seconds of arc. If we know the distance, we can use the small-angle formula (see By the Numbers 3-1) to find its linear diameter. Also, if we measure the apparent magnitude of a galaxy, we can use the distance to find its absolute magnitude and thus its luminosity (see By the Numbers 8-2).

The results of such observations show that galaxies differ dramatically in size, mass, and luminosity. Irregular galaxies tend to be small, 1 to 25 percent the size of our galaxy, and of low luminosity. Although they are common, they are easy to overlook. Our Milky Way Galaxy is large and luminous compared with most spiral galaxies, though we know of a few spiral galaxies that are even larger and more luminous—at least 50 percent larger in diameter and about 10 times more luminous. Elliptical galaxies cover a wide range of diameters and luminosities. The largest, called giant ellipticals, are five times the size of our Milky Way. But many elliptical galaxies are very small, dwarf ellipticals, only 1 percent the diameter of our galaxy.

To put galaxies in perspective, we can use an analogy. If our galaxy were an 18-wheeler, the smallest dwarf galaxies would be the size of pocket-size toy cars, and the largest giant ellipticals would be the size of jumbo jets. Even among spiral galaxies that looked similar, the sizes could range from that of a go-cart to that of a small yacht.

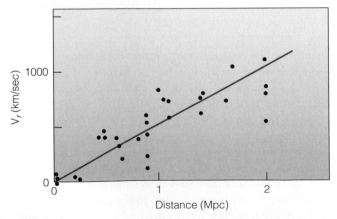

Figure 13-11
Edwin Hubble's first diagram of the velocities and distances of galaxies did not probe very deeply into space. It did show, however, that the galaxies are receding.

[*]H has the units of a velocity divided by a distance. These are usually written as km/sec/Mpc, meaning km/sec per Mpc.

Clearly, the diameter and luminosity of a galaxy do not determine its type. Some small galaxies are irregular, and some are elliptical. Some large galaxies are spiral, and some are elliptical. We need more data before we can build a theory for the origin and evolution of galaxies.

Of the three basic parameters that describe a galaxy, we have found two—diameter and luminosity. The third, as was the case for stars, is more difficult to discover.

Mass

Although the mass of a galaxy is difficult to determine, it is an important quantity. It tells us how much matter the galaxy contains, which gives us clues to the galaxy's origin and evolution. In this section, we will examine two fundamental ways to find the masses of galaxies.

One way to find the mass of a galaxy is to watch it rotate. We know the stars in the outer parts of the galaxy are in orbit, so we can use Newton's laws to find the mass. All we need to know is the size of the galaxy in meters and the velocity with which its outer stars orbit, and then we can use the equation for circular velocity (By the Numbers 4-1).

It is easy to find the size of the galaxy if we know the distance. We just measure its angular size and use the small angle formula (By the Numbers 3-1).

We can find the orbital velocity of the stars from the Doppler effect. If we focused the image of the galaxy on the slit of a spectrograph, we would see a bright spectrum formed by the bright nucleus of the galaxy, but we would also see fainter emission lines produced by ionized gas in the disk of the galaxy. Because the galaxy rotates, one side moves away from us and one side moves toward us, so the emission lines would be red shifted on one side of the galaxy and blue shifted on the other side. We could measure those changes in wavelength, use the Doppler formula to find the velocities, and plot a diagram showing the velocity of rotation at different distances from the center of the galaxy—a diagram called a **rotation curve.** The artwork in Figure 13-12a shows the process of creating a rotation curve, and Figure 13-12b shows a real galaxy, its spectrum, and its rotation curve.

The rotation curve is the best method for finding the masses of galaxies, but it works only for the nearer galaxies, whose rotation curves can be observed. More distant galaxies appear so small we cannot measure the radial velocity at different points along the galaxy. Also, recent studies of our own galaxy and others show that the outer parts of the rotation curve do not decline to lower velocities (Figure 13-12b). This shows that the outermost visible parts of some galaxies do not travel more slowly and tells us that the galaxies contain large amounts of mass outside this radius, perhaps in extended galactic coronas like the one that seems to surround our own galaxy (see Chapter 12). Because the rotation curve method can be applied only to nearby galaxies and because it cannot determine the masses of galactic coronas, we must look at the second way to find the masses of galaxies.

The **cluster method** of finding galactic mass depends on the motions of galaxies within a cluster. If we measure the radial velocities of many galaxies in a cluster, we find that some velocities are larger than others because of the orbital motions of the individual galaxies in the cluster. Given the range of velocities and the size of the cluster, we can ask how massive a cluster of this size must be to hold itself together with this range of velocities. Dividing the total mass of the cluster by the number of galaxies in the cluster yields the average mass of the galaxies. This method contains the built-in assumption that the cluster is not flying apart. If it is, our result is too large. Since it seems likely that most clusters are held together by their own gravity, the method is probably valid.

A related way of measuring a galaxy's mass is called the **velocity dispersion method.** It is really a version of the cluster method. Instead of observing the motions of galaxies in a cluster, we observe the motions of matter within a galaxy. In the spectra of some galaxies, broad lines indicate that stars and gas are moving at high velocities. If we assume the galaxy is bound by its own gravity, we can ask how massive it must be to hold this moving matter within the galaxy. This method, like the one before, assumes that the system is not coming apart.

The masses of galaxies cover a wide range. The smallest contain about 10^{-6} as much mass as the Milky Way, and the largest contain as much as 50 times more than the Milky Way. The structure and evolution of a star is determined by its mass, but this is clearly not so for galaxies. We must search further for an explanation of the different types of galaxies we see.

Dark Matter in Galaxies

Given the size and luminosity of a galaxy, we can make a rough guess as to the amount of matter it should contain. We know how much light stars produce, and we know about how much matter there is between the stars, so it is quite possible to estimate very roughly the mass of a galaxy from its luminosity. But when astronomers measure the masses of galaxies, they often find that the measured masses are much too large. We discovered this effect in Chapter 12 when we studied the rotation curve of our own galaxy and concluded that it must contain large amounts of dark matter. This seems to be true of most galaxies. Measured masses of galaxies amount to 10 to 100 times more mass than we can see.

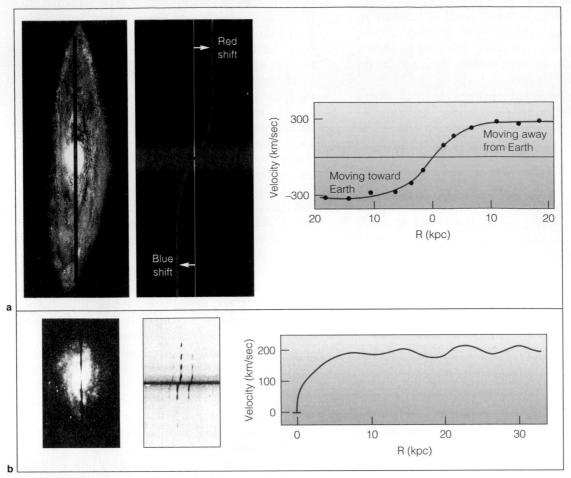

Figure 13-12
(a) In this diagram, the astronomer has placed the galaxy over the slit of the spectrograph (left). A short segment of the resulting spectrum includes an emission line red shifted on one side of the galaxy and blue shifted on the other. Converting these shifts to radial velocities, the astronomer can plot the galaxy's rotation curve (right). *(Lick Observatory Photo)* (b) A short segment of the spectrum of galaxy NGC2998 includes three emission lines. The rotation curve is plotted for only half of the galaxy. *(Courtesy Vera Rubin)*

The rotation curves of galaxies are strong evidence for dark matter, but astronomers have found even more evidence that most of the mass in the universe is invisible. For example, in many clusters of galaxies, the velocities of the galaxies are so high the clusters would fly apart if they contained only the mass we see. These clusters must contain large amounts of dark matter.

X-ray observations reveal more evidence of dark matter. X-ray images of galaxy clusters show that many of them are filled with very hot, low-density gas that would easily leak away if it were not held by a strong gravitational field. To provide that gravity, the cluster must contain much more matter than we can see. The galaxies near the center of the Coma cluster, for instance, amount to only 11 to 35 percent of the total mass (Figure 13-13).

Dark matter is not an insignificant issue. Observations of galaxies and clusters of galaxies tell us that 90 to 99 percent of the matter in the universe is dark matter. The universe we see—the luminous matter—has been compared to the foam on an invisible ocean. Dark matter affects the formation and evolution of galaxies and can affect the evolution of the universe as a whole (Chapter 15).

Dark matter is difficult to detect, and it is even harder to explain. Some astronomers have suggested that dark matter consists of low-luminosity stars and planets scattered through an extended halo around each galaxy. Searches for such objects have not been successful. For example, if the dark matter in our galaxy is hidden in cool, low-mass stars in the halo, then careful searches using the Hubble Space Telescope should reveal many of these stars. Actual photographs looking out into the halo reveal almost none of these low-mass objects, so it seems unlikely that cool, low-mass stars can explain the dark matter in galaxies.

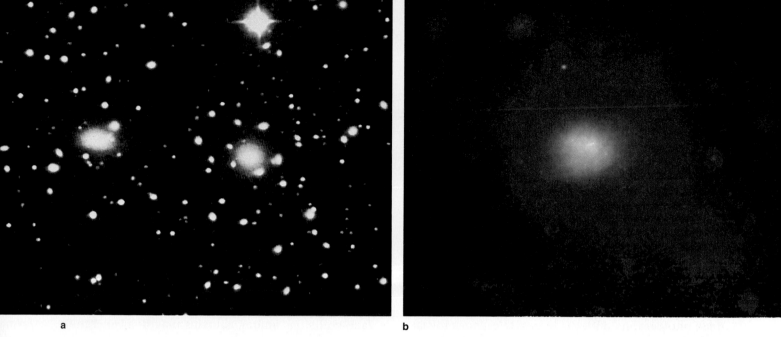

a   b

Figure 13-13
(a) The Coma cluster of galaxies contains at least 1000 galaxies and is especially rich in E and S0 galaxies. Two giant galaxies lie near its center. If the cluster were visible, it would span 8 times the diameter of the full moon in the sky. *(L. Thompson © NOAO)* (b) In false colors, this X-ray image of the Coma cluster shows it filled with hot gas. The clumpy distribution suggests that the cluster formed by the merger of smaller clusters, and the temperature of the gas reveals the presence of up to 90 percent dark matter in the cluster. *(Simon D. M. White, Ulrich G. Briel, and J. Patrick Henry)*

The dark matter can't be hidden in vast numbers of black holes and neutron stars, because we don't see the X rays these objects would emit. Remember, we need 10 to 100 times more dark matter than visible matter, and that many black holes should produce X rays that are easy to detect.

Because observations seem to tell us the dark matter can't be hidden in normal objects, some theorists have suggested that the dark matter is made up of unexpected forms of matter. Until recently neutrinos were thought to be massless, but studies now suggest they have a very small mass. Thus they may represent part of the dark matter.

Dark matter remains one of the fundamental unresolved problems of modern astronomy. We will return to this problem in Chapter 15 when we try to understand how dark matter affects the nature of the universe, its past, and its future.

REVIEW Critical Inquiry

Why does the measured mass of a galaxy depend on its distance?

Often the critical analysis of a problem is complicated by interrelationships between factors, but a bit of care is usually enough to reveal these connections. For example, to find the mass of a galaxy, we compare the motions of stars in the outer parts of a galaxy with velocities predicted by Kepler's third law. Measuring the velocity of stars in a galaxy is simply a matter of observing the Doppler shift in the spectrum, but we also need to know the size of a star's orbit around the galaxy. We can observe the radius of the orbit, the distance from the stars to the center of the galaxy we are observing, in seconds of arc, but Kepler's third law requires we know that distance in some linear unit of distance such as AU. That is where the distance enters. If we know the distance to the galaxy, we can use the small-angle formula to convert the orbital radius from seconds of arc to AU. Then we can use Kepler's third law to find the mass.

Many different measurements in astronomy depend on the calibration of the distance scale. For example, what would happen to our estimate of the diameter and luminosity of galaxies if astronomers discovered that the Cepheid variable stars were slightly more luminous than had been believed?

We have gathered the basic data to help us characterize galaxies. That is, we can describe their general characteristics. Now we turn to the more difficult and more interesting question. How did galaxies get to be the way they are?

13-3 The Evolution of Galaxies

Our goal in this chapter has been to build a theory to explain the evolution of galaxies. In Chapter 12, we developed a theory that described the origin of our own Milky Way Galaxy; and, presumably, other galaxies formed similarly from the contraction of large clouds of gas. But why did some galaxies become spiral, some elliptical, and some irregular? Clues to that mystery lie in the clustering of galaxies.

Clusters of Galaxies

The distribution of galaxies is not entirely random. Galaxies tend to occur in clusters ranging from a few to thousands. Furthermore, modern astronomers have catalogued over 2700 clusters within 4 billion ly, and deep photographs with the largest telescopes reveal even more clusters of galaxies at the limit of our vision. Not only are these vast clusters fascinating, but they can also help us understand the lives of the galaxies.

For our discussion, we can sort clusters into two groups: rich and poor. **Rich galaxy clusters** contain over a thousand galaxies, mostly elliptical, scattered through a spherical volume about 3 Mpc (10^7 ly) in diameter. Such a cluster is very crowded, with the galaxies more concentrated toward the center. Rich clusters often contain one or more giant elliptical galaxies at their centers.

The Coma cluster (located in the constellation Coma Berenices) is an example of a rich cluster (Figure 13-13). It lies over 100 Mpc from us and contains at least 1000 galaxies, mostly E and S0 galaxies. Its galaxies are highly crowded around a central giant elliptical and a large S0.

One of the nearest clusters to us, the Virgo cluster, contains over 2500 galaxies and is, by our definition, a rich cluster. It does contain a giant elliptical galaxy, M87, near its center, but it is not very crowded and contains mostly spiral galaxies.

Poor galaxy clusters contain fewer than 1000 galaxies and are irregularly shaped. Instead of being crowded toward the center, poor clusters tend to have subcondensations, small groupings within the clusters.

Our own **Local Group,** which contains the Milky Way, is a good example of a poor cluster (Figure 13-14). It contains a few dozen members scattered irregularly through a volume slightly over 1 Mpc in diameter. Of the brighter galaxies, 14 are elliptical, 3 are spiral, and 4 are irregular.

The total number of galaxies in the Local Group is uncertain because some lie in the plane of the Milky Way Galaxy and are difficult to detect. For example, the two Maffei galaxies, a giant elliptical and a spiral, are detectable in the infrared but totally hidden at visible wavelengths. A small dwarf galaxy, known as the Sagittarius Dwarf, has been found on the far side of our own galaxy where it is almost totally hidden behind the star clouds of Sagittarius (Figure 13-14b).

We have classified galaxy clusters into rich and poor clusters to reveal a fascinating and suggestive clue to the evolution of galaxies. In general, rich clusters tend to have a larger proportion of E and S0 galaxies than do poor clusters. This suggests that a galaxy's environment is important in determining its structure and has led astronomers to suspect that the secrets to galaxy evolution lie in collisions between galaxies.

Colliding Galaxies

Astronomers are finding more and more evidence to show that galaxies collide, interact, and merge. In fact, collisions among galaxies may dominate their evolution.

We should not be surprised that galaxies collide with each other. The average separation between galaxies is only about 20 times their diameter. Like two elephants blundering about under a circus tent, galaxies should bump into each other once in a while. Stars, on the other hand, almost never collide. In the region of our galaxy near the sun, the average separation between stars is about 10^7 times their diameter. Thus a collision between two stars is about as likely as a collision between two gnats flitting about in a baseball stadium.

Even if two galaxies pass through each other, the stars never collide. The interstellar gas and dust in the galaxies do interact as gas clouds collide, and the sudden compression can trigger rapid star formation. It is also possible for one galaxy to strip gas and dust away from the other galaxy. Thus, bursts of star formation and galaxies stripped of gas and dust are clear consequences of galaxy collisions.

Because collisions between galaxies can last hundreds of millions of years, we cannot see the galaxies change, but computer models of collisions can speed up the process and show how colliding galaxies are twisted and deformed by their gravitational fields to produce peculiar shapes, tails, and bridges. It is not even necessary for the galaxies to penetrate each other —a close encounter is enough to allow tidal forces to distort the galaxies. One of the most famous pairs of colliding galaxies is called the Mice because of the tail-like deformities (Figure 13-15). These tails are flung out as the galaxies whip around their common center of mass.

Both observation and theory tell us that collisions between galaxies can trigger rapid changes in a galaxy. Colliding galaxies are often driven to make stars rapidly, but some galaxies contain plenty of gas and

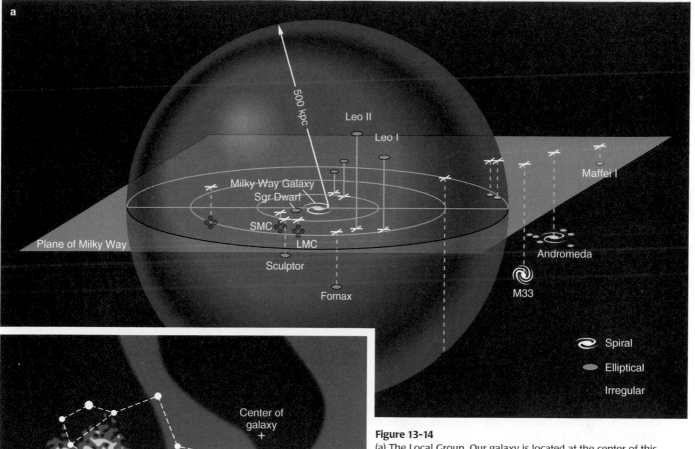

a

Leo II

Leo I

500 kpc

Milky Way Galaxy
Sgr Dwarf

SMC

LMC

Plane of Milky Way

Sculptor

Fornax

Maffei I

Andromeda

M33

🌀 Spiral

⬭ Elliptical

Irregular

Figure 13-14

(a) The Local Group. Our galaxy is located at the center of this diagram. The vertical lines giving distances from the plane of the Milky Way are solid above the plane and dashed below. (b) The Sagittarius Dwarf galaxy (Sgr Dwarf) lies on the other side of our galaxy. If we could see it in the sky, it would be 17 times larger than the full moon.

Center of
galaxy
+

Sagittarius

Scorpius

b

dust but are not making new stars very rapidly. Such galaxies tend to be isolated, with no nearby galaxies that might collide with them and trigger star formation. Theoretical models confirm that collisions can trigger spiral density waves (Figure 13-16), and observations tell us that flocculent galaxies (Chapter 12), which lack strong two-armed spiral patterns, tend to be isolated from neighboring galaxies.

As two galaxies approach each other, they each carry large amounts of orbital momentum. Computer simulations show that colliding galaxies experience

tidal forces that convert some of their orbital motion into random motion among their stars and gas clouds and thus rob them of orbital momentum. It is possible for galaxies to collide, interact, and then go their separate ways, but it is also possible for colliding galaxies to lose enough momentum that they fall together and merge.

A small galaxy merging with a larger galaxy will be gradually pulled apart, and its stars will spread through the larger galaxy in a process called **galactic cannibalism.** Computer simulations of such mergers

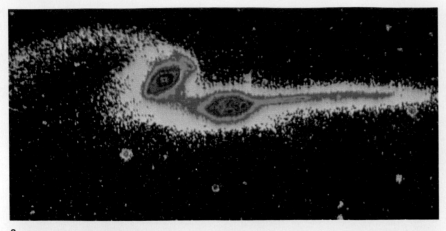

a

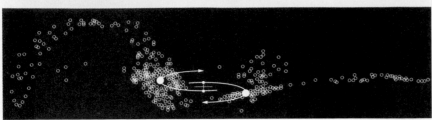

b

Figure 13-15

(a) The Mice are a pair of colliding galaxies with peculiar tails (shown here in a false-color image). *(NOAO)* (b) A computer simulation of a close encounter between normal galaxies produces similar tails. *(Illustration by Allen Beechel)*

show that the galaxies throw off shells of stars as the nuclei of the two galaxies spiral into each other, and such shells of stars have been observed around real galaxies (Figure 13-17).

Evidence for galaxy mergers is all around us. The Large Magellanic Cloud (Figure 13-8a) and the Small Magellanic Cloud are orbiting our galaxy and passed through the outer disk of our galaxy about 200 million years ago. They are being pulled apart and will eventually be merged with our galaxy. The Sagittarius Dwarf (Figure 13-14b) will meet a similar fate. The Hubble Space Telescope reveals many galaxies that are in the process of merging with all of the symptoms that collisions produce (Figure 13-18).

Mergers long past leave traces in the centers of galaxies. For example, the peculiar galaxy NGC7252 is highly distorted and rapidly making new stars (Figure 13-19a). When the Hubble Space Telescope looked into the center of the galaxy, it found a small spiral of gas and young stars rotating in the direction opposite to the rest of the galaxy. Evidently, the distorted galaxy was created about a billion years ago by the merger of two galaxies that were rotating in opposite directions. Another backward galaxy is M64, a bright galaxy visible through small telescopes. High-resolution VLA radio maps reveal that the central disk of the galaxy is

rotating backward (Figure 13-19b), and this suggests a past merger.

In contrast to mergers, high-speed head-on crashes between galaxies can produce dramatic results. **Ring galaxies** consist of a bright nucleus surrounded by a ring (Figure 13-20). Computer models show that they could be produced by a galaxy passing roughly perpendicularly through the disk of a larger galaxy. Observations show that many ring galaxies have nearby companions.

We have seen that collisions and mergers between galaxies are common and can produce dramatic changes in the structure of the galaxies. We can now propose a hypothesis to help us understand how galaxies evolve.

The Origin and Evolution of Galaxies

The test of any scientific understanding is whether or not we can put all the evidence and theory together to tell the history of the objects we study. Can we describe the origin and evolution of the galaxies? Just a few decades ago, it would have been impossible, but new evidence is helping astronomers understand the history of the galaxies.

We can eliminate a few older ideas immediately. The tuning fork diagram tempts us to think that galaxies evolve from left to right, from elliptical to spiral to irregular. The evidence, however, shows that elliptical galaxies contain no gas and dust to make new stars, and they contain lots of old stars. Ellipticals can't be young. Furthermore, galaxies can't evolve from right to left in the tuning fork diagram, from irregular to spiral to elliptical; irregular galaxies contain both young and old stars and thus can't be young. The tuning fork diagram is not an evolutionary diagram. It tells us that the galaxies have had different histories of star formation, but it does not tell us how galaxies evolve.

Another old idea held that galaxies that form from rapidly rotating clouds of gas would have lots of angular momentum and would contract slowly to form disk-shaped spiral galaxies. Clouds of gas that rotated less rapidly would contract more rapidly, form stars quickly, use up all the gas and dust, and become elliptical galaxies. The evidence clearly shows that galaxies didn't form and evolve in isolation. Collisions and mergers dominate the history of the galaxies.

The ellipticals appear to be the product of galaxy mergers, which triggered star formation that used up the gas and dust. In fact, we see many **starburst galaxies** that are very luminous in the infrared because a

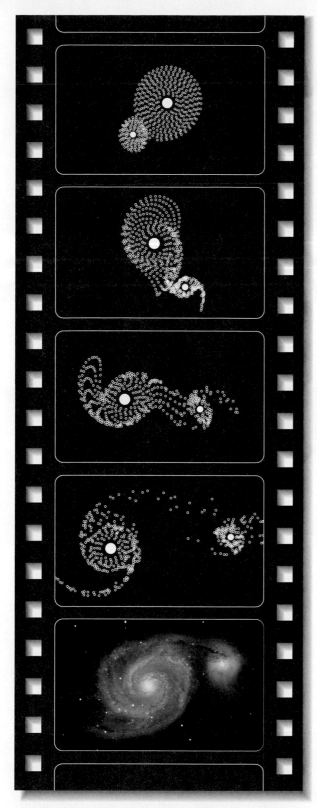

Figure 13-16
A computer model of two colliding disk galaxies produces a strong two-armed spiral pattern that resembles the well-known Whirlpool Galaxy (bottom). *(Illustration by Allen Beechel; photo NOAO)*

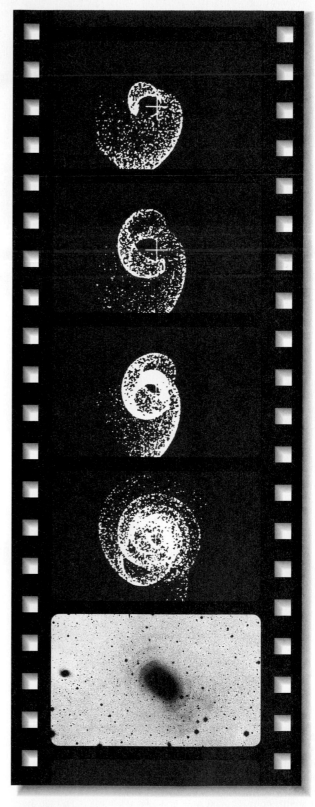

Figure 13-17
Computer models of merging galaxies show that mergers can produce shells of stars as the two galaxies whirl around their common center of mass. To illustrate the production of shells, these diagrams show only one of two galaxies circling their common center of mass (cross). Such shells are seen around galaxies such as NGC3923 (bottom), shown here as a negative image enhanced to reveal low-contrast features. *(Model courtesy François Schweizer and Alan Toomre; photo courtesy David Malin, Anglo-Australian Telescope Board)*

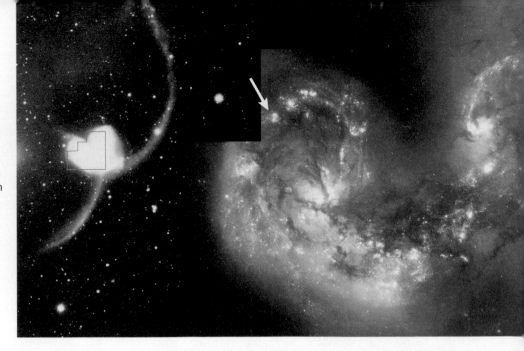

Figure 13-18
The colliding galaxies NGC4038 and NGC4039 are known as The Antennae because their long curving tails resemble the antennae of an insect. Earth-based photos (left) show little detail, but a Hubble Space Telescope image (right) reveals the collision of two galaxies producing thick clouds of dust and raging star formation, creating roughly a thousand massive star clusters such as the one at the top (arrow). Such collisions between galaxies are common. *(Brad Whitmore, STScI, and NASA)*

a
Ground view

HST view

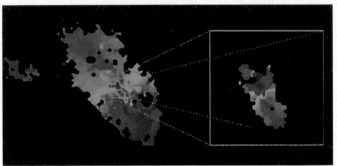

b

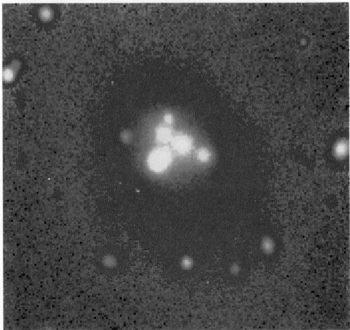

c

Figure 13-19
Evidence of mergers. (a) The twisted shape of NGC7252 suggests a collision, and a Hubble Space Telescope image reveals a small spiral of young stars spinning backward in the heart of the larger galaxy. This is thought to be the remains of two oppositely rotating galaxies that merged about a billion years ago. *(François Schweizer, Carnegie Institution of Washington, and Brad Whitmore, STScI)* (b) The apparently normal galaxy M64 is shown at left in a radio map that shows radial velocity as color. Red regions are receding from us and blue regions approaching. At the center of the galaxy, the rotation is reversed, evidently the consequence of a merger between counter-rotating galaxies. *(Robert Braun, NRAO)* (c) Giant elliptical galaxies in rich clusters sometimes have multiple nuclei, thought to be the densest parts of smaller galaxies that have been absorbed and partly digested. *(Michael J. West)*

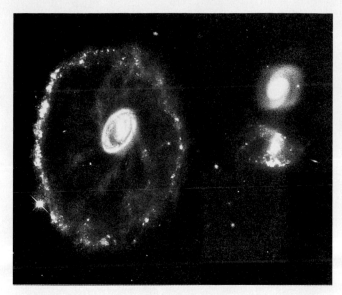

Figure 13-20

The Cartwheel Galaxy was once a normal spiral galaxy, but a few hundred thousand years ago one or the other of the two smaller galaxies at the right passed through the center of the larger galaxy. Tidal forces generated an expanding wave of star formation and a central core of newborn stars heavily reddened by dust. Such ring galaxies appear to be the fleeting consequences of bull's-eye collisions. Soon after the collision, the violent star formation exhausts itself, and normal spiral arms may reform. This is suggested in the Cartwheel Galaxy by the faint spiral structure inside the outer ring. *(Kirk Borne, STScI, and NASA)*

collision has triggered a burst of star formation that is heating the dust. The warm dust reradiates the energy in the infrared. The Antennae (Figure 13-18) contain over 15 billion solar masses of hydrogen gas and will become a starburst galaxy as the merger triggers rapid star formation. Supernovae in a starburst galaxy may eventually blow away any remaining gas and dust that doesn't get used up making stars. A few collisions and mergers could leave a galaxy with no gas and dust from which to make new stars. Astronomers now suspect that most ellipticals are formed by the merger of at least two or three galaxies.

In contrast, spirals seem never to have suffered major collisions. Their thin disks are delicate and would be destroyed by tidal forces in a collision with a massive galaxy. Also, they retain plenty of gas and dust and continue making stars.

Other processes can alter galaxies. The S0 galaxies may have lost much of their gas and dust in a burst of star formation, but they still managed to remain disk shaped. Also, galaxies moving through the gas trapped in dense clusters of galaxies may have their own gas and dust blown away. For example, X-ray observations show that the Virgo cluster contains thin, hot gas between the galaxies. A galaxy orbiting throught that gas would encounter a tremendous wind blowing its gas and dust away. This could explain the dwarf ellipticals, which are too small to be made of merged spirals. In contrast, the irregular galaxies may be small fragments of galaxies ripped apart by collisions.

The evidence for galaxy evolution by merger is quite strong. Observations with the largest telescopes take us to great distances and great look-back times. We see the galaxies as they were long ago, and we discover that there were spirals then and fewer ellipticals. Also, the ellipticals were smaller than they are now. We can even see that galaxies were closer together long ago;

about a third of all distant galaxies are in close pairs, but only 7 percent of nearby galaxies are in pairs. The observational evidence clearly supports the hypothesis that galaxies have evolved by merger.

The Hubble Space Telescope has recorded very long exposure images called *deep fields* at two seemingly empty spots in the northern and southern sky. Each of these frames shows over a thousand galaxies at such large look-back times that we see them as they were when the universe was very young (Figure 13-21). The entire sky is filled with many billions of faint galaxies.

Among the smallest objects in such deep photos are small, blue clouds of stars and gas containing about 10 times the mass of large globular clusters (Figure 13-21c). These distorted clouds may be the objects that first fell together to form galaxies when the universe was young. More-developed galaxies in the Deep Field can now be studied, such as the one shown in Figure 13-22, which shows at visual wavelengths characteristics typical of a disk-shaped galaxy and shows at infrared wavelengths the presence of cool stars and warm dust thought to be typical of a young galaxy.

REVIEW Critical Inquiry

How did elliptical galaxies get that way?

A growing body of evidence suggests that elliptical galaxies have been subject to collisions in their past and that spiral galaxies have not. During collisions, a galaxy can be driven to use up its gas and dust in a burst of star formation may be stripped of gas and dust by an encounter. Thus, elliptical galaxies now contain little star-making material. The beautiful disk typical of spiral galaxies is very orderly, with all the stars following

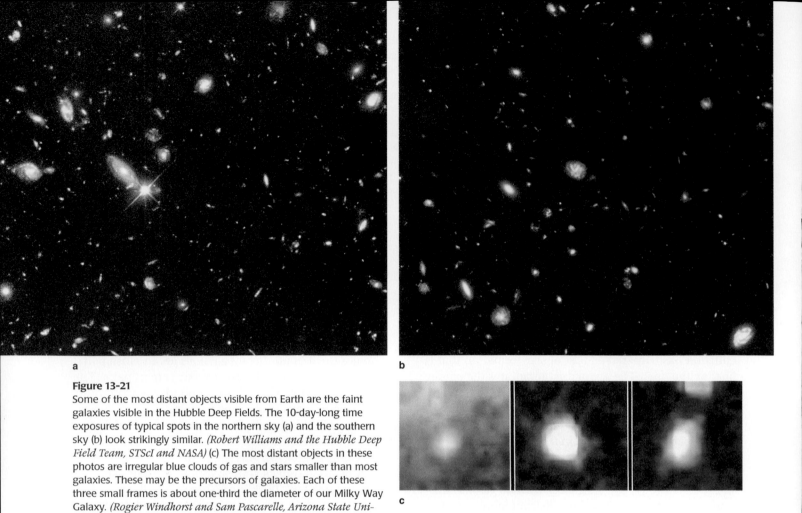

a

b

c

Figure 13-21
Some of the most distant objects visible from Earth are the faint galaxies visible in the Hubble Deep Fields. The 10-day-long time exposures of typical spots in the northern sky (a) and the southern sky (b) look strikingly similar. *(Robert Williams and the Hubble Deep Field Team, STScI and NASA)* (c) The most distant objects in these photos are irregular blue clouds of gas and stars smaller than most galaxies. These may be the precursors of galaxies. Each of these three small frames is about one-third the diameter of our Milky Way Galaxy. *(Rogier Windhorst and Sam Pascarelle, Arizona State University, and NASA)*

Figure 13-22
A very distant galaxy selected from the Hubble Deep Field (Figure 13-21a) is shown here at visual wavelengths (left) and at infrared wavelengths (right). Presumably we see the galaxy at an early age. Bright clumps in the galaxy in the visual images are believed to be regions of active star formation. The infrared image shows that the disk of the galaxy is filled with cool stars and warm dust. Other objects in the field of view are probably other distant galaxies. *(Rodger I. Thompson, University of Arizona, and NASA.)*

a

b

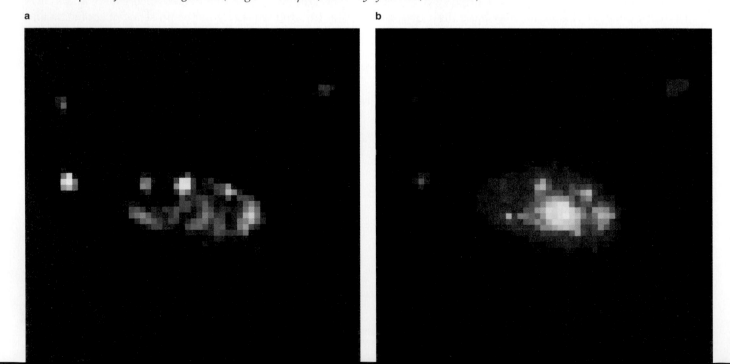

similar orbits. When galaxies collide, the stellar orbits get scrambled, and an orderly disk galaxy could be converted into a chaotic swarm of stars typical of elliptical galaxies. Thus, it seems likely that elliptical galaxies have had much more complex histories than spiral galaxies have had.

What evidence do we have to support the story in the preceding paragraph?

Before we can consider the universe as a whole, we must examine the galaxies from a different perspective. Some galaxies are suffering tremendous eruptions in their centers. In the next chapter, we will discover that these peculiar galaxies are closely related to collisions and mergers between galaxies.

Summary

We can divide galaxies into three classes—elliptical, spiral, and irregular—with subclasses specifying the galaxy's shape. The elliptical galaxies contain little gas and dust and few bright, young stars. Spiral and irregular galaxies have large amounts of gas and dust and are actively making new stars.

To measure the properties of galaxies, we must first find out how far away they are. For the nearer galaxies, we can judge distances using distance indicators, objects whose luminosity or diameter is known. The most accurate distance indicators are Cepheid variable stars. Other distance indicators are bright giants and supergiants, globular clusters, planetary nebulae, and novae. In addition, we can estimate the distance to the farthest galaxy clusters using the average luminosity of the brightest galaxies.

The Hubble law shows that the radial velocity of a galaxy is proportional to its distance. Thus, we can use the Hubble law to estimate distances. The galaxy's radial velocity divided by the Hubble constant equals its distance in megaparsecs.

The masses of galaxies can be measured in two basic ways—the rotation curve method and the velocity dispersion method. The rotation curve method is more accurate but can be applied only to nearby galaxies. Both methods suggest that galaxies contain 10 to 100 times more dark matter than visible matter.

Galaxies occur in clusters. Our own galaxy is a member of the Local Group, a small cluster. A galaxy in a rich cluster may collide with other galaxies more often than a galaxy in a poor cluster, and such collisions can force a galaxy to form new stars and use up its gas and dust. Collisions can also strip gas out of a galaxy. This may explain why elliptical and S0 galaxies are more common in rich clusters than

in poor clusters. Spiral galaxies may be star systems that have not experienced many collisions.

New Terms

tuning fork diagram	Hubble law
elliptical galaxy	Hubble constant (H)
spiral galaxy	rotation curve
barred spiral galaxy	cluster method
irregular galaxy	velocity dispersion method
Large Magellanic Cloud	rich galaxy cluster
Small Magellanic Cloud	poor galaxy cluster
megaparsec (Mpc)	Local Group
distance indicator	galactic cannibalism
H II region	ring galaxy
look-back time	starburst galaxy

Review Questions

1. Why didn't astronomers at the beginning of the 20th century recognize galaxies for what they are?

2. How can a classification system aid a scientist?

3. Why are the true shapes of elliptical galaxies uncertain?

4. What is the difference between an SBa and an SBb galaxy? between an SBb and an Sb?

5. Why can't galaxies evolve from E to S to Irr? Why can't they evolve from Irr to S to E?

6. How do selection effects make it difficult to decide how common E, S, and Irr galaxies are?

7. Why are Cepheid variable stars good distance indicators? What about planetary nebulae?

8. Why is it difficult to measure the Hubble constant?

9. How is the rotation curve method related to binary stars and Kepler's third law?

10. What evidence do we have that galaxies contain dark matter?

11. What evidence do we have that galaxies collide and merge?

12. Why are the shells visible around some elliptical galaxies significant?

13. Ring galaxies often have nearby companions. What does that suggest?

14. Propose an explanation for the lack of gas, dust, and young stars in elliptical galaxies.

15. How do deep images by the Hubble Space Telescope confirm our hypothesis about galaxy evolution?

Discussion Questions

1. Why do we believe that galaxy collisions are likely, but star collisions are not?

2. Should an orbiting infrared telescope find irregular galaxies bright or faint in the far infrared? Why? What about elliptical galaxies?

Problems

1. If a galaxy contains a Type I (classical) Cepheid with a period of 30 days and an apparent magnitude of 20, what is the distance to the galaxy?

2. If you find a galaxy that contains globular clusters that are 2 seconds of arc in diameter, how far away is the galaxy? (*Hints:* Assume that a globular cluster is 25 pc in diameter, and see By the Number 3-1.)

3. If a galaxy contains a supernova that at its brightest has an apparent magnitude of 17, how far away is the galaxy? (*Hints:* Assume that the absolute magnitude of the supernova is −19, and see By the Numbers 8-2.)

4. If we find a galaxy that is the same size and mass as our Milky Way Galaxy, what orbital velocity would a small satellite galaxy have if it orbited 50 kpc from the center of the larger galaxy? (*Hint:* See By the Numbers 4-1.)

5. Find the orbital period of the satellite galaxy described in Problem 4. (*Hint:* See By the Numbers 8-4.)

6. If a galaxy has a radial velocity of 2000 km/sec and the Hubble constant is 70 km/sec/Mpc, how far away is the galaxy? (*Hint:* Use the Hubble law.)

7. If you find a galaxy that is 20 minutes of arc in diameter, and you measure its distance to be 1 Mpc, what is its diameter? (*Hint:* See By the Numbers 3-1.)

8. We have found a galaxy in which the outer stars have orbital velocities of 150 km/sec. If the radius of the galaxy is 4 kpc, what is the orbital period of the outer stars? (*Hints:* 1 pc = 3.08×10^{13} km, and 1 yr = 3.15×10^7 sec.)

9. A galaxy has been found that is 5 kpc in radius and whose outer stars orbit the center with a period of 200 million years. What is the mass of the galaxy? On what assumptions does this result depend? (*Hint:* See By the Numbers 8-4.)

Critical Inquiries for the Web

1. How far out into the universe can we see Cepheid variables? Research sources on the Internet to find other galaxies whose distances have been found through observation of Cepheids. List the galaxies in which Cepheids have been identified and the distances determined from these data.

2. How does the Milky Way stack up against the other galaxies in the Local Group? Look for information on the other galaxies in our cluster, and rank the top six members in order of total mass.

3. Locate a Web page dedicated to the Messier Catalogue —a list of galaxies, clusters, and nebulae that is often used as a list of targets for small telescopes. Be sure that your destination includes images of the objects. For each of the galaxies in the Messier list, determine its Hubble classifications (see Figure 13-2). You may be given this information at the site, but examine the images to see if the features of these galaxies conform to a particular Hubble type.

Go to the Brooks/Cole Astronomy Resource Center (www.brookscole.com/astronomy) for critical thinking exercises, articles, and additional readings from InfoTrac College Edition, Brooks/Cole's online student library.

CHAPTER 14

Galaxies with Active Nuclei

Virtually all scientists have the bad habit of displaying feats of virtuosity in problems in which they can make some progress and leave until the end the really difficult central problems.

M. S. Longhair
High Energy Astrophysics

Guidepost

This chapter is important for two reasons. First, it draws together ideas from many previous chapters to show how nature uses the same basic rules on widely different scales. Matter flowing into a protostar, into a white dwarf, into a neutron star, or into the heart of a galaxy must obey the same laws of physics, so we see the same geometry and the same phenomena. The only difference is the level of violence.

Second, this chapter is important because the most distant objects we can see in the universe are the most luminous galaxies, and many of those are erupting in outbursts and are thus peculiar. By studying these galaxies, our attention is drawn out in space to the edges of the visible universe and back in time to the earliest stages of galaxy formation. In other words, we are led to think of the origin and evolution of the universe, the subject of the next chapter.

Explosion? Eruption? Detonation? Colossal? Titanic? Words fail us when we try to describe the violence that astronomers find at the center of some galaxies. If we say a massive star explodes as a supernova, what word shall we use for the eruption of an entire galaxy? Astonishing it may be, but violence in the centers of galaxies is easily detected across billions of light-years, and our double goal in this chapter is to explain the origin of these eruptions and relate them to the formation and evolution of galaxies.

The adventure of studying active galaxies began when radio astronomers first noticed such objects in the 1950s. In the decades that followed, astronomers used all parts of the electromagnetic spectrum to study these galaxies and piece together the puzzle. Today, astronomers generally agree on how a galaxy can develop an active core. Thus, our study of active galaxies not only reveals how galaxies can erupt but also illustrates how scientists can use evidence, theory, and their own ingenuity to learn about the natural world.

A number of astronomers object to the quotation that opens this chapter, perhaps because it seems to belittle scientists. But it is a realistic description of how scientists attack a problem. Big problems are made of little problems, and scientists begin by solving the little problems, learning from those solutions, and moving ever closer to the center of the big problem. Perhaps no better example can be found than modern astronomy's attack on "the really difficult central problem" in the nuclei of active galaxies.

14-1 Active Galactic Nuclei

In the 1950s, radio astronomers began to notice that some galaxies were powerful sources of radio energy. Of course, all galaxies, including our own Milky Way, emit some radio energy, but some emit over 10 million times more radio energy than does a normal galaxy. Such objects were first known as **radio galaxies.** But newer observations reveal that many of these galaxies also radiate at infrared, ultraviolet, and X-ray wavelengths, so these objects have become known as **active galaxies.** In many cases, the source of the activity lies in the nucleus of the galaxy, and such objects are known as **active galactic nuclei (AGN).** In some cases, however, the radio energy is detected coming from two areas (lobes) on opposite sides of the visible galaxy. We will eventually develop a model that relates these seemingly different kinds of objects.

Double-Lobed Radio Galaxies

Radio astronomers have found many examples of radio sources in which the radio energy comes from two regions, called lobes, located near each other in the sky. Photographs of the region between these lobes reveal a galaxy, usually a giant elliptical galaxy. Many of these **double-lobed radio galaxies** are peculiar in some way, with distorted shapes, close companion galaxies, dust clouds, or jets; but some of these galaxies are not themselves unusually bright at radio wavelengths.

Radio lobes are generally much larger than the galaxy they accompany (Figure 14-1). One of the largest-known radio galaxies is 3C236 (the 236th object in the *Third Cambridge Catalogue of Radio Sources*). In fact, it is one of the largest objects in the universe, with radio lobes that reach across 5.8 Mpc $(19 \times 10^6$ ly).

Radio lobes have two properties that hint at their origin: They radiate synchrotron radiation, and they are often most intense on the side away from the central galaxy. Recall that synchrotron radiation is produced when high-speed electrons spiral through a magnetic field (see Chapter 10). Though the field is at least 1000 times weaker than Earth's, it fills a tremendous volume and represents vast stored energy. In addition, the high-speed electrons also contain energy. The total energy in a radio lobe equals about 10^{53} J, approximately the energy equivalent of converting 10^6 solar masses directly into energy. Clearly, the process that creates radio lobes is powerful.

The second hint to the nature of the radio lobes is that many radio lobes emit more intensely from the side away from the galaxy (Figure 14-2a). Not all radio lobes have these **hot spots,** but many do.

Synchrotron radiation and hot spots suggest an explanation for double-lobed radio galaxies now known as the **double-exhaust model.** According to this model, if an eruption in the core of the galaxy drove jets of high-energy gas outward in opposite directions, they could blow up a pair of radio lobes like bubbles in the thin gas of the intergalactic medium. The ejected gas would be ionized and would drag along part of the galaxy's magnetic field, and the high-speed electrons in the gas would produce synchrotron radiation. The beams could produce hot spots where they pushed against the far side of each radio lobe (Figure 14-2b).

The radio map of Cygnus A shown in Figure 14-1 contains a striking detail: A long narrow jet of high-velocity gas leads from the center of the galaxy out into the radio lobe at the right. Another jet, slightly harder to see in this map, leads into the left radio lobe. Notice also that the radio lobes contain hot spots at their outer edges. These jets give us direct evidence that the vast structure of the radio lobes is produced by high-energy jets springing from the nucleus of the galaxy.

Another fascinating galaxy is NGC5128. It looks like a giant elliptical galaxy, but it is surrounded by a

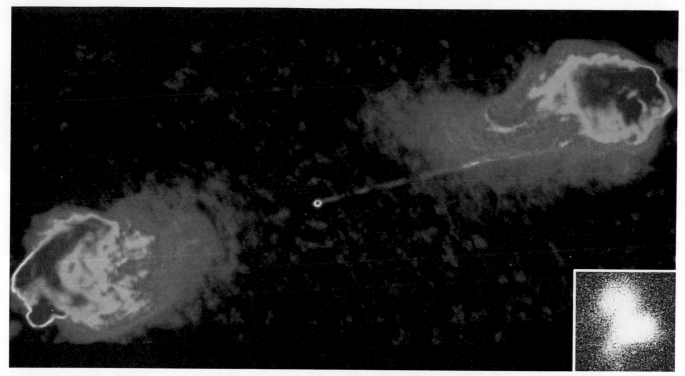

Figure 14-1
Cygnus A is a large, powerful double-lobed radio source. This radio map shows regions of strongest radio signal as red. From tip to tip, this pair of radio lobes spans over 100 kpc, about 6 times the diameter of the disk of our Milky Way Galaxy. *(NRAO)* Inset: In a visible light photograph, the deformed central galaxy is roughly the size of the region between the radio lobes. *(Palomar Obs/Caltech)*

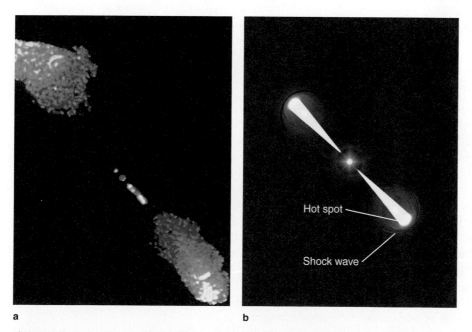

a   b

Hot spot

Shock wave

Figure 14-2
(a) The radio map of galaxy 3C219 displays regions of strongest radio signal as red and yellow. The yellow areas on the radio lobes mark hot spots. *(NRAO; Observers A. H. Bridle and R. A. Perley)* (b) The double-exhaust model proposes that jets of high-speed particles from an active galaxy nucleus inflate radio lobes and produce hot spots where they run into intergalactic gas. Notice the jet in part (a) extending from the nucleus toward the lower right radio lobe.

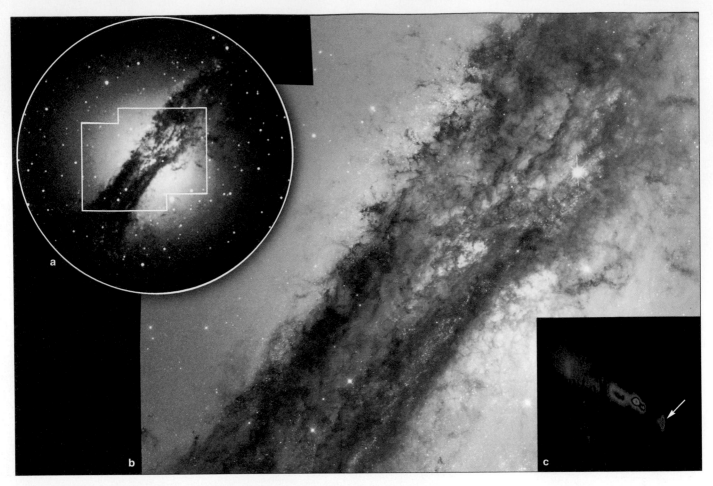

Figure 14-3
(a) The peculiar galaxy NGC5128 appears to be an elliptical galaxy colliding with a spiral. *(NOAO)* (b) A photo by the Hubble Space Telescope reveals young, bright, blue stars, evidence of star formation triggered by a collision, scattered through the dust disk. *(E. Schreier, STScI, and NASA)* (c) High-resolution radio maps of the center of the galaxy reveal an energetic central object (arrow) and a jet pointing toward one of the radio lobes that lie on either side of the galaxy. *(NRAO)*

belt of dust, and astronomers believe we are seeing a collision between a giant elliptical and a small spiral (Figure 14-3a). The evidence for this interpretation is strong. Doppler shifts show that the bright, spherical part of the galaxy is rotating about an axis that lies in the plane of the dust ring, but that the dust ring itself is rotating about a different axis that is perpendicular to the plane of the ring. Also, the best photos of this galaxy reveal bright blue stars in the dust ring, stars that must have formed recently (see Figure 14-3b). Collisions are known to trigger star formation, so the presence of young stars supports the collision hypothesis.

Radio astronomers know this object by the name Centaurus A, two giant radio lobes that span 10° in the sky, bigger than the bowl of the Big Dipper. Radio inteferometers reveal a small jet leading from the center of the galaxy toward the northern lobe (Figure 14-3c). In fact, at the highest resolution, radio astronomers see a jet only 1 pc long pointing toward the northern lobe. Clearly the center of this galaxy is

somehow ejecting jets of high-velocity gas to inflate the radio lobes. Because this is a collision between two galaxies, we must wonder if the collision could have triggered the eruption in the core.

We can't see the core of NGC5128 at visible wavelengths because of dust, but an infrared photograph made by the Hubble Space Telescope penetrates the dust. The photo, shown in Figure 14-4, reveals a small, brilliant object at the very center of the galaxy surrounded by a disk of hot gas 130 ly in diameter. Evidently, this disk and the brilliant object at its center are creating the jets and lobes that radio astronomers detect.

The radio lobes themselves are very tenuous. They do not contain sufficient gravity or magnetic fields strong enough to hold themselves together. Apparently, the lobes are held together by their collision with the intergalactic gas.

What other evidence can we find to show that our double-exhaust model is a valid way to think about

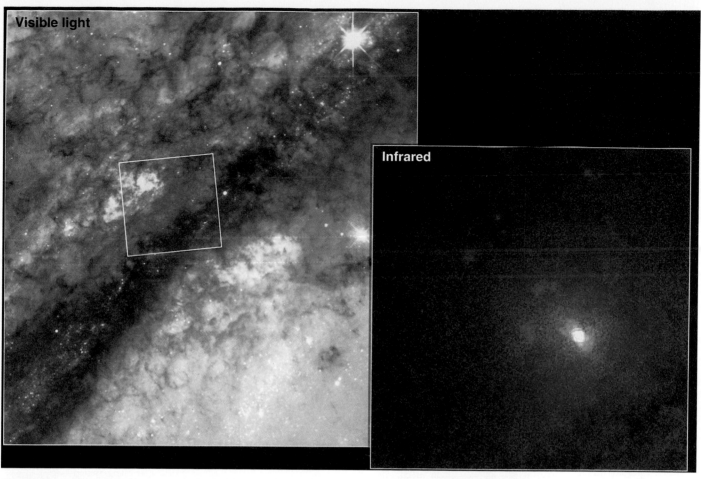

Figure 14-4

An infrared photograph of the center of peculiar galaxy NGC5128 (position shown on the visible light photo at left) reveals a small bright disk of hot gas (white) surrounding the central object. Red objects in the photo are clouds of gas ionized by the radiation from the center. *(E. Schreier, STScI, and NASA)*

these galaxies? Radio astronomers have found examples of double-lobed radio galaxies in which the jets of gas flowing out of the galaxy are distorted in some way. For example, the jets flowing out of the galaxy 3C449 (Figure 14-5a) are twisted, perhaps because the source of the jets is orbiting around the center of the galaxy. Some of these galaxies have complex structure in their jets, such as separate blobs or spirals along the two streams (Figure 14-5b). This suggests that the jets may not remain "on" all the time, and the spirals suggest that the axis along which the beams shine may be precessing and spewing relativistic particles (subatomic particles such as electrons and protons traveling near the speed of light) out the galaxy in swirling streams, like someone shaking a garden hose. A subclass of these galaxies is called **head–tail radio galaxies** because they are moving relative to the intergalactic gas, and the hot gas spewing outward in their jets is swept away like smoke from smokestacks to form tails. 3C129 in Figure 14-5b is an example of a head–tail radio galaxy. Radio astronomers have confidence in this in-

terpretation of the radio images because these head–tail galaxies are most common in clusters of galaxies, where the intergalactic gas is relatively dense.

All of the evidence suggests that radio lobes are produced by two oppositely directed jets of high-energy gas flowing out of the central galaxy. The double-exhaust model helps us understand the overall geometry, but it does not help us understand the central problem: How does the center of a galaxy produce such eruptions? To pursue that part of the problem, we must consider active galaxies that lack radio lobes.

Galaxies with Erupting Nuclei

Some galaxies have active nuclei but lack radio lobes. **Seyfert galaxies,** for example, are spiral galaxies with unusually bright, tiny cores that fluctuate into brightness (Figure 14-6). Most Seyfert galaxies are powerful sources of infrared radiation, and some are radio sources. Of the roughly 120 known, 25 emit X rays and one is known to emit gamma rays.

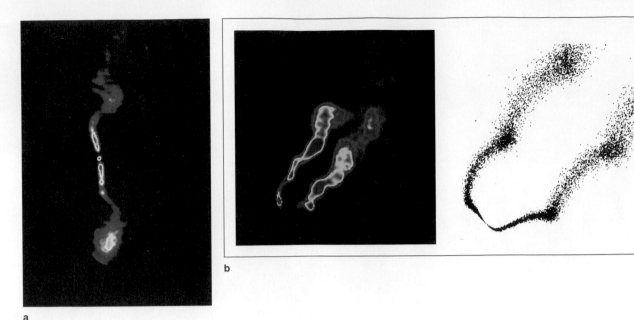

b

Figure 14-5

(a) Active galaxy 3C449 has spiral patterns in its radio lobes, thought to be produced by the orbital motion of the source of the jets. At visual wavelengths, the galaxy is revealed to be a giant elliptical about 10 times the size of the central spot in this false-color image *(Richard A. Perley)* (b) Head–tail radio galaxy 3C129 has its jets swept back by the motion of the galaxy through the intergalactic medium. Twists in the jets, shown in the mathematical model at right, seem to be produced by the precession of the jets. *(Laurence Rudnick, University of Minnesota, and Jack Burns, University of New Mexico; model courtesy Vincent Icke, University of Minnesota and* The Astrophysical Journal, *published by the University of Chicago Press, © 1981 American Astronomical Society)*

Figure 14-6

The spiral galaxy NGC7742 has a small, brilliant central core identifying it as a Seyfert galaxy. The thick lumpy ring surrounding the yellow glow of the core is a region of active star formation about 3000 ly from the center of the galaxy. *(Hubble Heritage Team, AURA, STScI, and NASA)*

Detailed studies reveal two different kinds of Seyfert galaxies. Both kinds have emission lines in their spectra, but Type 1 Seyferts have broad spectral lines typical of gas swirling around at velocities over 1000 km/sec. These Type 1 Seyferts are also very bright at ultraviolet and X-ray wavelengths. Type 2 Seyferts have narrow emission lines in their spectra, suggesting the gas emitting the light is not moving so violently, and these galaxies are bright at infrared wavelengths. The differences among Seyfert galaxies will become critical evidence when we try to devise a model to explain what is happening in the hearts of these active galaxies.

Statistics can give us further insights (Window on Science 14-1). Since 2 percent of all spiral galaxies are Seyferts, we can conclude either of two things: Either active cores occur in 2 percent of all spiral galaxies, or all spiral galaxies erupt as Seyferts 2 percent of the time. The peculiar heart of our own galaxy suggests that all spirals erupt now and then; we may be living in a galaxy that was a Seyfert only a few hundred million years ago. A second statistic tells us that Seyfert galaxies are three times more common among galaxies interacting with other galaxies. This statistic suggests that galaxy collisions, interactions, and mergers may trigger activity in the cores of galaxies (Figure 14-7).

Notice that some scientific evidence is statistical. For example, we might notice that Seyfert galaxies are three times more likely among interacting galaxies. This evidence is statistical because we can't be certain that any particular interacting galaxy will be a Seyfert. Yet the probability is higher than if it were an isolated galaxy, and that leads us to suspect that interactions between galaxies may trigger the Seyfert phenomenon. Such statistical evidence can tell us something in general about the cause of Seyfert eruptions.

Statistical evidence is common in science, but uncommon in most courts of law. The American legal system is based on the principle of reasonable doubt, so statistical evidence is questionable. For example, plaintiffs have had great trouble suing tobacco companies for causing their cases of lung cancer, even though there is a clear statistical link between smoking and lung cancer. The statistics tell us something in general about smoking but cannot be used to prove that any specific case of lung cancer was caused only by smoking. There are other causes of lung cancer, so there is always a reasonable doubt as to the cause of any specific case.

We can use statistical evidence in science because we do not demand that the statistics demonstrate anything conclusively about any single example. To continue our example of Seyfert galaxies, we don't demand that the statistics predict that any specific interacting galaxy be a Seyfert. Rather, we use the statistical evidence to gain a general insight into the cause of active galactic nuclei. That is, we are trying to understand galaxies as a class, not convict any single galaxy.

Of course, if we surveyed only a few galaxies, our statistics might not be very good, and a critic might be justified in making the common complaint, "Oh, that's only statistics." For example, if we surveyed only four galaxies and found that three had companions, our statistics would not be very reliable. But if we surveyed 1000 galaxies, our statistics could be very good indeed, and our conclusions could be highly significant.

Thus, scientists can use statistical evidence if it passes two tests. It cannot be used to draw conclusions about specific cases, and it must be based on samples large enough that the statistics are significant. With these restrictions, statistical evidence can be a powerful scientific tool.

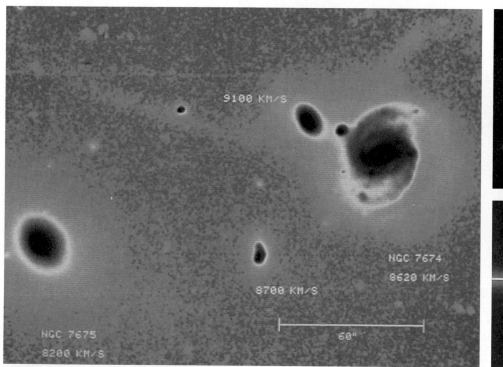

Figure 14-7

(a) Seyfert galaxy NGC7674 is distorted with tails to upper right and upper left (arrows) in this false-color image. Note the companion galaxies. *(John W. Mackenty, Institute for Astronomy, University of Hawaii)* (b) Seyfert galaxy NGC4151 has a brilliant nucleus in visible light images (top). The Space Telescope Imaging Spectrograph reveals Doppler shifts showing that part of the expelled gas has a blue shift (to the right) and is approaching us, and part has a red shift (to the left) and is receding. *(John Hutchings, Dominion Astrophysical Observatory; Bruce Woodgate. GSFC/NASA; Mary Beth Kaiser, Johns Hopkins University; and the STIS Team)*

As we gather more and more evidence that the cores of galaxies can erupt violently, we turn from the spiral Seyferts to **BL Lac objects** (also known as *blazars*), which appear to be associated with elliptical galaxies. These objects are named after the first one discovered—object BL in the constellation Lacerta (the Lizard). They have no absorption or emission lines in their spectra, but when the light of the bright core is blocked, we see the faint spectrum of an elliptical galaxy. Evidently, BL Lac objects are the active cores of elliptical galaxies.

Some galaxies with active nuclei and no radio lobes emit jets. M87, for example, is a giant elliptical galaxy in the nearby Virgo cluster of galaxies. Short-exposure photographs reveal a small, brilliant nucleus and a jet of ionized matter 1800 pc long squirting out of the nucleus (Figure 14-8a). Radio observations of the jet over a few years show that matter in the jet is traveling at about 48 percent the speed of light.

What could produce such violence in the cores of galaxies? Astronomers have found strong evidence that at least some active galaxies contain supermassive black holes at their centers, and the giant elliptical galaxy M87 (see Figure 13-3a) is one of the best examples. Hubble Space Telescope images of the nucleus reveal that the jet springs from a tiny spinning disk of gas (Figure 14-8b), and spectra show that the disk is rotating very fast. To be so small and spin so fast, it must contain a vast amount of matter at its center to hold the disk together. Only about 60 ly (3.8 million AU) from the center, the gas of the disk has an orbital velocity of 750 km/sec. At this speed it would have an orbital period of only about 150,000 years. Kepler's third law tells us the mass inside the orbit is slightly over 2 billion solar masses. Astronomers generally believe that the object is a supermassive black hole. By measuring the velocities of matter very near the centers of galaxies, astronomers have found a number of similar objects in the cores of galaxies.

Figure 14-8
(a) Giant elliptical galaxy M87 (Figure 13-3a) contains mostly cool stars and appears red in this image recorded with the Very Large Telescope (Figure 5-18b). The core of the galaxy is ejecting an 1800-pc-long jet of high-speed gas that is bright at ultraviolet wavelengths (blue in this image). *(European Southern Observatory)* (b) Hubble Space Telescope images show that the core of M87 is located at the center of a small disk of gas spinning at high speed. The rotation of the disk suggests the core contains a supermassive black hole. *(NASA/STScI)*

a

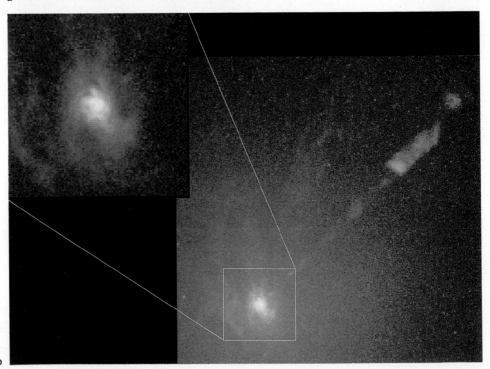

b

An argument can be a shouting match, but another definition of the word is "a discourse intended to persuade." Scientists construct arguments as part of the business of science not because they want to persuade others they are right, but because they want to test their ideas. For example, we can construct a scientific argument to show that the radio lobes that flank some active galaxies have their origin in jets coming from the cores of the galaxies. If we do our best and our argument is not persuasive, we begin to suspect we are on the wrong track. But if our argument seems convincing, we gain confidence that we are beginning to understand radio lobes.

A scientific argument is a logical presentation of evidence and theory with interpretations and explanations that help us understand some aspect of nature. Geologists might construct an argument to explain volcanism on mid-ocean islands, and the argument could include almost anything—maps, for example, or mineral samples compared with seismic evidence and mathematical models of subsurface magma motions. Such an argument might involve the physics of radioactive decay or observations of the shapes of bird beaks on the islands. The scientists would be free to include any evidence or theory that helps persuade, but they must observe one fundamental rule of scientific argument: They must be totally honest. The purpose of a scientific argument is to test our understanding, not to win votes or sell soap. Dishonesty in a scientific argument is self-deluding, and scientists consider such dishonesty the worst possible behavior.

On the other hand, scientists are human, and they often behave no better than other humans. They sometimes defend their own theories and attack opposing theories more vigorously than the evidence warrants. They may unintentionally overlook contradictory evidence or unconsciously misinterpret data. Of course, if a scientist did these things intentionally, we would be shocked because he or she would be violating the most serious rules of honesty in science. In the struggle to understand how nature works, however, it is not surprising to see that human passions rise and that scientific debate becomes heated. That is why science is so exciting—it is a struggle. Nonetheless, no matter how scientists behave in the excitement of a controversy, they all aspire to that ideal of the scientific argument—the cool, logical presentation of evidence that leads to a new understanding of nature.

Much of this book or any other scientific book consists of scientific arguments—logical presentations of evidence. As you read about any science, look for the arguments and practice organizing and presenting them to test your own understanding. The "Critical Inquiry" that ends each chapter section in this book is intended to illustrate how arguments are constructed. Use the same plan when you develop your own arguments.

Even the nearby Andromeda Galaxy and its companion galaxy M32 seem to have supermassive objects in their cores.

These supermassive black holes are not the remains of a single supernova explosion, as described in Chapters 10 and 11, but are the consequence of the accumulation of matter at the center of the galaxy over billions of years. Nevertheless, just as matter flowing into a small black hole can release energy to power an X-ray binary star, matter flowing into a supermassive black hole could release energy to power an active galactic nucleus.

If we combine the double-exhaust model with the proposal that active galactic nuclei harbor massive black holes, we can create a comprehensive model that helps us understand not only what happens inside active galaxies but also what causes galaxies to become active.

The Search for a Unified Model

When a field of research is young, scientists often find many seemingly different phenomena such as double-lobed radio galaxies, Seyfert galaxies, BL Lac objects and so on. As the research matures, the scientists begin seeing similarities and eventually are able to unify the different phenomena as different aspects of a single process. This is the real goal of science, to organize evidence and theory in logical arguments that explain how nature works (Window on Science 14-2). Astronomers studying active galaxies are now struggling to find a unifying model of active galaxy cores.

It seems clear that the cores of active galaxies contain supermassive black holes surrounded by accretion disks that are extremely hot near the black hole but cooler further out. Theory predicts that the central cavity in the disk where the black hole lurks is very narrow but that the disk there is "puffed up" and thick. This means the black hole may be hidden deep inside this narrow central well. The hot inner disk seems to be the source of the jets often seen coming out of active galaxy cores, but the process by which jets are generated is not understood. The outer part of the disk, according to calculations, is a fat, dense doughnut (a torus) of dusty gas (Figure 14-9).

What we see when we view the core of an active galaxy must depend on how this disk is tipped with respect to our line of sight. We should note, at this point, that the accretion disk may be tipped at a steep angle to the plane of a galaxy, so just because we see a galaxy face-on doesn't mean we are looking at the accretion disk face-on. Thus the important factor is not

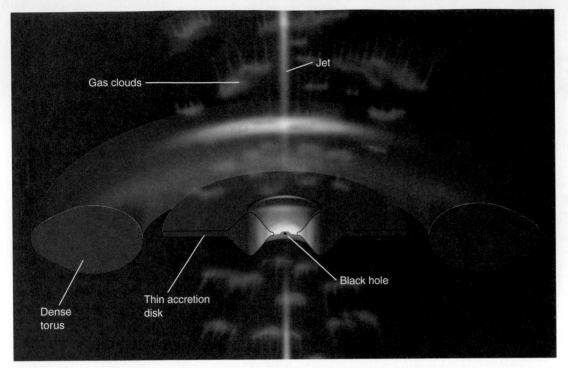

Figure 14-9
An accretion disk around a massive black hole. Matter flowing inward passes first through a large, opaque torus; then into a thinner, hotter disk; and finally into a hot, narrow cavity around the black hole. Gas clouds above and below the disk are excited by radiation from the hot disk. This diagram is not to scale. The central cavity may be only 0.01 in radius, while the outer torus may be 1000 pc in radius.

the inclination of the galaxy, but the inclination of the accretion disk.

BL Lac objects, for example, are believed to be active cores emitting jets that are pointed directly at Earth. We are looking directly into the brilliant jet, and that drowns out the fainter parts of the galaxy. All we see is a brilliant point of light.

If the accretion disk is tipped slightly, we may not be looking directly into the central cavity, but we may be able to see some of the intensely hot gas there. This region emits broad spectral lines because the gas is orbiting at high velocities and the high Doppler shifts smear out the lines. Seyfert 1 galaxies may be explained by this phenomenon.

If we view the accretion disk from the edge, we cannot see the central area at all because the thick dusty torus blocks our view. Instead we see radiation emitted by gas lying above and below the central disk. Because this gas is farther from the center, it orbits more slowly and has smaller Doppler shifts. Thus we see narrow spectral lines. This might account for the Seyfert 2 galaxies.

This unified model is far from complete. The actual structure of accretion disks is poorly understood, and factors other than inclination must also be important. For example, what we see must also depend on the mass of the central black hole, the rate at which matter flows in, and the strength of magnetic fields in the disk. Astronomers are now working to build a unified model that incorporates all of the important factors in a single logical argument to explain the active cores of galaxies.

Galaxy Collisions and Active Cores

We must now consider the central knotty problem of active galaxies. What makes a galaxy erupt? The unified model suggests the core contains a supermassive black hole, and evidence is mounting that most galaxies contain such objects. These are not the black holes produced by dying stars but are the accumulation of mass that probably formed when the galaxy first took shape.

If all galaxies contain supermassive black holes, what triggers an eruption? Observation and theory tell us that mass must flow into the black hole through an accretion disk, and models of colliding galaxies show that during a collision significant amounts of mass are thrown into the center of the galaxies. Thus a collision between galaxies could trigger an eruption by feeding an otherwise dormant black hole. Most galaxies apparently contain supermassive black holes that are inactive because little matter is flowing into them.

There is strong evidence that collisions trigger core eruptions. Seyfert galaxies often have nearby

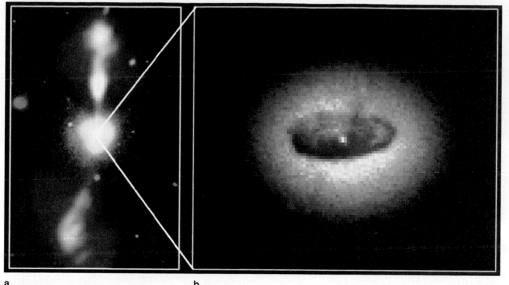

Figure 14-10
(a) Galaxy NGC4261 looks like a distorted fuzzy blob in an Earth-based photograph at visible wavelengths (white), but radio telescopes reveal jets leading into radio lobes (orange). (b) A Hubble Space Telescope image of the core of the galaxy reveals a bright central spot surrounded by a disk with its axis pointing at the radio lobes. *(L. Ferrarese, Johns Hopkins University, and NASA)*

a   b

companions, and active galaxies are often distorted as they would be by an encounter with another galaxy. NGC5128 is thought to be two galaxies in the act of colliding. In fact, high-quality images of this galaxy show that it is surrounded by shells of stars, a common by-product of galaxy mergers (Figure 13-17).

Evidence continues to mount. The Hubble Space Telescope has captured an image of the core of NGC4261, a distorted galaxy flanked by oppositely directed jets (Figure 14-10). The images of the core reveal a small disk of cool gas about 100 pc across spinning on an axis that points toward the radio lobes. Spectra of the disk show that it is rotating so fast that it must contain about 1.2 billion solar masses at its center—evidently a supermassive black hole. Presumably, the cool disk feeds material into a smaller, hot accretion disk near the black hole at the center and thus produces the jets we see.

REVIEW Critical Inquiry

Construct an argument to show that double-lobed radio galaxies and Seyfert galaxies have similar energy sources.

Observations show that double-lobed radio galaxies are produced by two jets flowing outward from a galaxy's core to inflate the radio lobes with hot gas. In many cases, radio maps actually show these jets. Seyfert galaxies do not have radio lobes, but the unified model of active galactic nuclei suggests that Seyferts are produced by matter flowing into an accretion disk around a supermassive black hole. We have seen that such an accretion disk can produce oppositely directed jets as in the case of the X-ray binary star SS433, so a more massive version in the core of a galaxy could presum-

ably produce the jets we see inflating radio lobes. The same geometry, an accretion disk around a supermassive black hole, can also explain the difference between the two types of Seyfert galaxies, so it seems reasonable to suppose that double-lobed radio galaxies and Seyfert galaxies have similar energy sources at their centers.

The preceding paragraph is a quick summary of a logical argument citing evidence and hypotheses to reach a conclusion. Construct an argument to propose that galaxy interactions are necessary to trigger active galactic nuclei.

Observational evidence and basic theory have led us to the unified model of active galactic nuclei. Although the model seems successful, there are still difficulties, and astronomers continue to struggle to understand the details. Nevertheless, the unified model may help us understand the mysterious quasars.

14-2 Quasars

The largest telescopes detect multitudes of faint points of light with peculiar emission spectra, objects called **quasars** (also known as *quasi-stellar objects,* or *QSOs*). Although astronomers now recognize quasars as some of the most distant visible objects in the universe, they were a mystery when they were first identified. The discovery of these objects in the 1960s and the struggle to understand them comprise a classic example of how scientists explore hypotheses, gather evidence, and build confidence in new ways to understand nature. We begin this section by telling that story.

The Discovery of Quasars

In the early 1960s, radio interferometers (see Chapter 5) showed that a number of radio sources are much smaller than normal radio galaxies. Photographs of the location of these radio sources did not reveal a galaxy, not even a faint wisp, but rather a single star-like point of light. The first of these objects so identified was 3C48, and later the source 3C273 was added. Though these objects emitted radio signals like those from radio galaxies, they were obviously not normal radio galaxies. Even the most distant photographable galaxies look fuzzy, but these objects looked like stars. Their spectra, however, were totally unlike stellar spectra, so the objects were called quasi-stellar objects (Figure 14-11).

For a few years, the spectra of quasars were a mystery. A few unidentifiable emission lines were superimposed on a continuous spectrum. But in 1963, Maarten Schmidt at Hale Observatories tried red-shifting the hydrogen Balmer lines to see if they could be made to agree with the lines in 3C273's spectrum. At a red shift of 15.8 percent, three lines clicked into place (Figure 14-12). Other quasar spectra quickly yielded to this approach, revealing even larger red shifts.

The red shift z is the change in wavelength $\Delta\lambda$ divided by the unshifted wavelength λ_0:

$$\text{red shift} = z = \frac{\Delta\lambda}{\lambda_0}$$

Astronomers commonly use the Doppler formula to convert these red shifts into radial velocities. Thus, the large red shifts of the quasars imply large radial velocities.

To understand the significance of these large radial velocities, we must draw on the Hubble law. Recall that the Hubble law states that galaxies have radial velocities proportional to their distances, and thus the dis-

Figure 14-11
Quasars have starlike images clearly different from the images of distant galaxies. The spectra of quasars are unlike the spectra of stars or galaxies. *(C. Steidel, Caltech, NASA)*

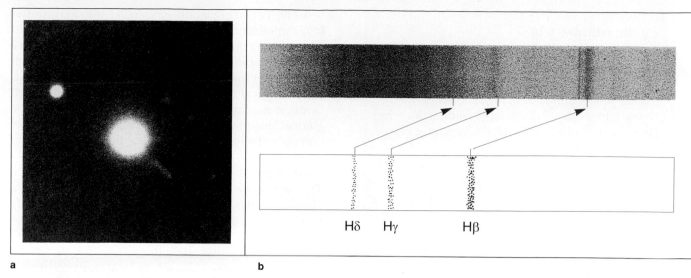

Figure 14-12

(a) This negative image of 3C273 shows the bright quasar at the center surrounded by faint fuzz. Note the jet protruding to lower right. (b) The spectrum of 3C273 (top) contains three hydrogen Balmer lines red-shifted by 15.8 percent. The drawing shows the unshifted positions of the lines.0 *(Courtesy Maarten Schmidt)*

tance to a quasar is equal to its radial velocity divided by the Hubble constant. The large red shifts of the quasars imply that they must be at great distances.

The red shift of 3C273 is 0.158, and the red shift of 3C48 is 0.37. These are large red shifts, but not as large as the largest then known for galaxies, about 1.0. Soon, however, quasars were found with red shifts much larger than that of any known galaxy. Some quasars are evidently so far away that galaxies at those distances are very difficult to detect. Yet the quasars are easily photographed. This leads us to the conclusion that quasars must have 10 to 1000 times the luminosity of a large galaxy. Quasars must be superluminous.

Soon after quasars were discovered, astronomers detected fluctuations in brightness over times as short as a few days. Recall from our discussion of pulsars (Chapter 11) that an object cannot change its brightness appreciably in less time than it takes light to cross its diameter. The rapid fluctuations in quasars showed that they are small objects, not more than a few light-days or light-weeks in diameter.

Thus, by the late 1960s astronomers faced a problem: How could quasars be superluminous but also be very small? Since then, evidence has accumulated that quasars are the active cores of very distant galaxies. The distance to quasars is so great that we need to pay special attention to how astronomers estimate the distance to these objects.

Quasar Distances

Astronomers estimate the distances to quasars by using the Hubble law and dividing their radial veloc-

ity by the Hubble constant (Chapter 13). Thus, knowing the distance to a quasar depends critically on knowing its red shift and thus its radial velocity. Many quasars have red shifts greater than 1, and if we put such large red shifts into the Doppler formula (By the Numbers 6-2), we get radial velocities greater than the velocity of light. This seems unreasonable at first, but if we consider the nature of quasar red shifts, we will see that red shifts greater than 1 do make sense.

First we must note that the Doppler formula given in By the Numbers 6-2 is only an approximation good for low velocities. It is correctly called the classical Doppler formula, and it works fine if the velocity is significantly less than the velocity of light. For very high velocities, however, we need to use the **relativistic red shift** formula derived from Einstein's theory of relativity (By the Numbers 14-1). That formula tells us that very rapidly moving objects will appear to have red shifts greater than 1 even though they are not traveling faster than the velocity of light. Figure 14-13 plots the velocity of an object versus its red shift z, and we see that the true z we observe can become quite large without the object's actually exceeding the velocity of light. Thus, the quasars with red shifts greater than 1 should not give us special concern. They do not violate any rules of physics.

Now that we feel some comfort with quasar red shifts greater than 1, we should note that the red shifts associated with the expansion of the universe are not described precisely by the Doppler effect. In the next chapter, we will discuss the expansion of the universe in detail, and we will conclude that galaxy red shifts are not really Doppler shifts. The exact relation between a quasar's red shift and its expansion velocity is similar to the relativistic red shift formula in that red shifts greater than 1 imply velocities near but not exceeding the velocity of light. But the exact relation depends critically on such parameters as the average density of the universe, parameters that are not well known. Thus, we can't at present convert quasar red shifts into precise distances.

What evidence do we have that quasars really are located at these great distances? A discovery made in 1979 provides a dramatic illustration of the tremendous distances to quasars. The object catalogued as 0957 + 561 lies just a few degrees west of the bowl of the Big Dipper and consists of two quasars separated by only 6 seconds of arc. The spectra of these quasars are identical and even have the same red shift of 1.4136. Quasar spectra are as different as fingerprints, so when two quasars so close together were discovered to have the same spectra and the same red shift, astronomers concluded that they were separate images of the same quasar.

The two images are formed by a **gravitational lens,** an effect first predicted by Einstein in 1936 but never seen before. The gravitational field of a galaxy between us and the distant quasar acts as a lens, bending the light from the quasar and focusing it into multiple images (Figure 14-14a). The lensing galaxy itself is much too far away to be easily visible against the glare of the quasar. However, when astronomers used computer techniques to subtract the upper image of the quasar from the bottom image, the faint image of a giant elliptical galaxy appeared (Figure 14-14b and c). This galaxy is probably the brightest member of a cluster of galaxies. The red shift of the galaxy is only 0.36, much less than that of the quasar, 1.4136. As a result, we conclude that the quasar must lie behind the galaxy and that the galaxy is so far away it is hardly visible even though it is a giant elliptical. Other examples of gravitational lensing of quasars have been found, and they provide very strong evidence that the quasars really are distant objects, as implied by their large red shifts.

For more evidence that quasars are very distant, look again at Figure 14-11. Just above the image of the quasar lies the small, oval image of an elliptical galaxy, which, evidenced by its red shift, is about 7 billion light-years away. The spectrum of the quasar contains absorption lines with the same red shift as the elliptical galaxy. Evidently the lines are produced by the thin gas in the elliptical galaxy. From this we conclude that the quasar, though very bright, is actually further away than the distant elliptical galaxy.

The quasars are very distant and must be superluminous and quite small. To understand quasars, we must examine their observed characteristics and compare that evidence with the unified model of active galaxies.

Evidence of Quasars in Distant Galaxies

Astronomers have now found a few thousand quasars, and the properties of those objects support the theory that they are superluminous objects located in very distant galaxies. In most cases, the quasar is so distant the host galaxy is invisible. The largest quasar red

Figure 14-13
At high velocities, the relativistic red shift must be used in place of the classical approximation. Note that no matter how large z gets, the velocity can never equal that of light.

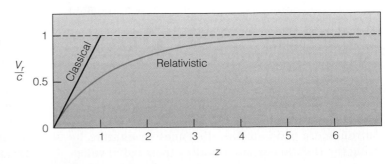

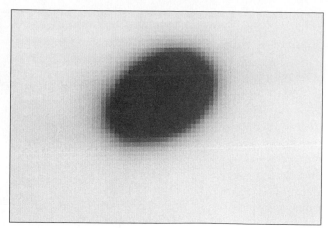

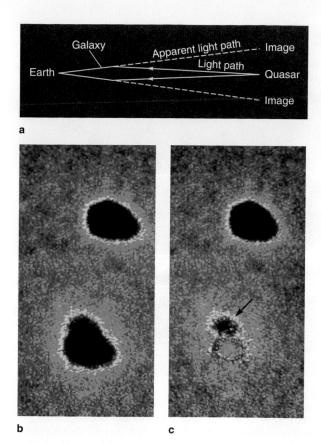

a

b   c

Figure 14-14
The gravitational lens effect. (a) A distant quasar can appear to us as multiple images if its light is deflected and focused by the mass of an intervening galaxy. (b) The two images of quasar 0957 + 561 appear as black blobs in this computer-enhanced false-color image. (c) When the upper image is subtracted from the lower image, we are able to detect the faint image of the giant elliptical galaxy (arrow) whose gravity produces the gravitational lens effect. *(Courtesy Alan Stockton, Institute for Astronomy, University of Hawaii)*

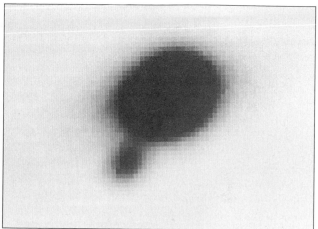

Figure 14-15
Two negative images of the quasar 1059 + 730. The lower image shows an extra point of light—apparently a supernova that exploded in the galaxy associated with the quasar. *(Courtesy Bruce Campbell)*

shifts are about 5, and that means their velocity of recession exceeds 90 percent the velocity of light. These objects are so distant that the look-back times are immense, and we see them as they were when the universe was only about 10 percent of its present age. Thus, when we look at quasars we not only look out into space, but we look back into the past to a time when the universe was younger.

Quasars with lower red shifts are not as distant, and long-exposure photographs of these quasars reveal that they are surrounded by faint fuzzy images. The spectra of these faint images are the same as the spectra of normal spiral and elliptical galaxies with the same red shift as the quasars. Thus, the quasars are evidently objects in galaxies.

Still more evidence appeared when astronomers noticed a new point of light near quasar QSO 1059 + 730. The point of light was a supernova exploding in the galaxy that hosts the quasar (Figure 14-15). Although the galaxy is too distant to detect, the appearance of the supernova confirms that the quasar is located in a galaxy.

Photographs made with the largest telescopes on Earth and with the Hubble Space Telescope provide dramatic evidence of the nature of quasars. For example, quasar 3C273 (see Figure 14-12a) seems to be ejecting a jet much like the jet in M87. Furthermore, images of many quasars reveal that they are not only located inside galaxies, but that the galaxies are distorted or have nearby companions (Figure 14-16). Some quasars even have radio lobes. All of this evidence leads astronomers to believe that quasars are the erupting cores of very distant galaxies triggered into activity by collisions with other galaxies.

A Model Quasar

The most satisfying part of science is combining evidence with theory to understand some aspect of nature. We have learned to interpret quasar red shifts, and we have reviewed the basic characteristics of quasars. Can we now assemble the evidence and theories in a logical argument that describes quasars? That is, can we create a model of a quasar?

Our model quasar is based on the geometry of a rotating accretion disk ejecting jets along its axis of rotation (Figure 14-17). Disks are very common in the universe. We have discovered rotating disks around protostars, neutron stars, and black holes. Our own galaxy has a disk shape, and in a later chapter we will see that our solar system has a disk shape. We have also seen that some disks can produce powerful jets of matter and radiation directed along the axis of rotation. Bipolar flows from protostars are low-energy examples, while the jets expelled by the X-ray binary SS433 are much more powerful. Jets expelled by active galactic nuclei are even more powerful. To create our model quasar, we imagine a galaxy containing this disk-jet structure at its core. Our model is, in fact, just an extension of the unified model of active galactic nuclei discussed earlier in this chapter.

To make the energy typical of a quasar we need a black hole of 10 million to a billion solar masses surrounded by a large, opaque disk. The innermost part of the accretion disk, much less than a light-year in diameter, would be intensely hot. If matter amounting to about 1 solar mass per year flows through the accretion disk into the black hole, the energy released could power a quasar. Exactly how some of this matter and energy are focused into two oppositely directed high-speed jets isn't well understood, but magnetic fields are probably involved. In any case, as matter flows into the black hole, some is ejected in jets and a great deal of energy is radiated away. One astronomer described the situation by commenting, "Black holes are messy eaters."

This model could explain the different kinds of radiation we receive from quasars (see Figure 14-9). A small percentage of quasars are strong radio sources, and the radio radiation may come from synchrotron radiation produced in the high-energy gas and magnetic fields in the jets. Much of the light we see from a quasar is spread in a continuous spectrum, and that is the light that fluctuates quickly. Thus, it must come from a small region. Our model could produce this light in the innermost region surrounding the black hole. Because this is a very small region, roughly the size of our solar system, rapid fluctuations can occur, probably because of random fluctuations in the flow of matter into the accretion disk. The emission lines in quasar spectra don't fluctuate rapidly, and that suggests they are produced in a larger region. In our model, emission lines are emitted by clouds of gas surrounding the core in a region many light-years in diameter and excited by the intense synchrotron radiation streaming out of the central cavity.

Just as in the unified model of active galactic nuclei, the orientation of the accretion disk is important (see Figure 14-9). If the disk faces us so that

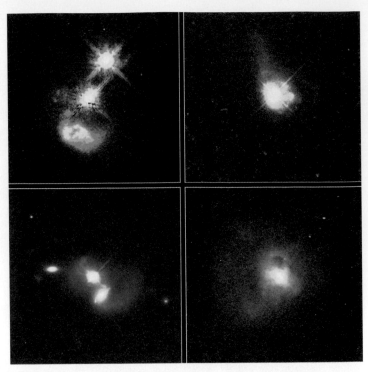

Figure 14-16
The bright object at the center of each of these images is a quasar. Fainter objects near the quasar are galaxies distorted by collisions. Compare the upper left ring-shaped galaxy with Figure 13-20, and compare the tail in the lower left image with The Antennae galaxies in Figure 13-18. *(J. Bahcall, Institute for Advanced Study, Mike Disney, University of Wales, NASA)*

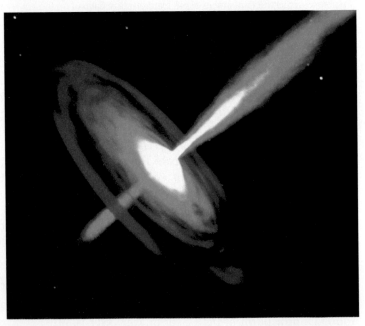

Figure 14-17
A model quasar. In the heart of a galaxy, matter flows into a hot accretion disk around a supermassive black hole. In this artist's conception, the disk ejects powerful jets. Radiation from the accretion disk heats the surrounding gas at the center of the galaxy and turns it into a quasar. *(Dana Berry, Space Telescope Science Institute)*

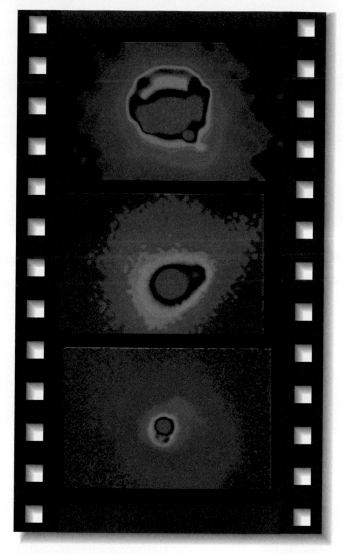

Figure 14-18
These three quasars appear to illustrate three stages in the triggering of a quasar by the tidal interaction between galaxies. The quasar IRAS 00275-2859 (top) is located in a galaxy interacting with a relatively distant companion (not visible in this image) and thus represents an early stage. In the galaxy containing quasar PG1613 (middle), the galaxies are closer together. The nucleus of the companion is visible at the 2 o'clock position, and the outline of the interacting galaxies is distorted. Markarian 231 (bottom) is located in a severely distorted galaxy containing radial streaks and a tail to the lower right. Such features are common when galaxies merge, suggesting that Markarian 231 represents a late stage in the merger of two galaxies. *(NOAO)*

one of its jets points directly toward us, we may see one kind of quasar. But if the disk is tipped slightly so the jet is not pointing straight at us, we may see a slightly different kind of quasar. Astronomers are now using this unified model to sort out the different kinds of quasars and active galaxies so we can understand how they are related. For example, using infrared radiation to penetrate dust, astronomers observed the core of the double-lobed radio galaxy Cygnus A (see Figure 14-2) and found an object much like a quasar.

Astronomers have begun to refer to such objects as buried quasars.

As in the case of active galaxies, many quasars are found in distorted galaxies that are interacting with nearby companions. Such a collision could throw matter into a central black hole and trigger a quasar outburst. Figure 14-18 shows three quasars that appear to illustrate different stages in such an encounter.

It is important to remember that the look-back time to quasars is very large. We see them as they were when the universe was young. In the next chapter, we will see that the universe appears to have begun about 15 billion years ago, and we can suppose that gas collected to form the first galaxies by the time the universe was a few billion years old. If at least some of those galaxies had supermassive black holes at their centers, then collisions between galaxies could have triggered eruptions in their cores that we would see as quasars. Because the universe was so young then, it had not expanded much, and the galaxies were closer together and would have collided more often. Also, theorists suppose that young galaxies would contain more gas because they would not have used the gas up to make stars, and such gas-rich galaxies might erupt more violently during a collision. Nevertheless, the quasar phenomenon must have been rare; at any one time, few galaxies had quasars in their cores.

Quasars are most common with red shifts of about 2, but they become much less common at red shifts above 2.7. The largest-known red shifts are about 5, and such high red-shift quasars are quite rare. This suggests that when we look at quasars with red shifts of 2 or so, we are looking back to an age when galaxies were actively colliding and merging. In that age in the history of the universe, quasars were about 1000 times more common than they are now. When we look at even more distant quasars, we are looking back to an earlier time before most galaxies formed. We see no quasars with really large red shifts because we would be looking back to an age before there were galaxies.

REVIEW Critical Inquiry

Why are most quasars so far away?

To analyze this question we must combine two factors. First, quasars are the active cores of galaxies, and only a small percentage of galaxies contain active cores. In order to sample a large number of galaxies in our search, we must extend our search to great distances. Most of the galaxies lie far away from us because the amount of space we search increases rapidly with distance. Just as most seats in a baseball stadium are far from home plate, most galaxies are far from our Milky

Way Galaxy. Thus, most of the galaxies that might contain quasars lie at great distances.

But a second factor is much more important. The farther we look into space, the farther back in time we look. It seems that there was a time in the distant past when quasars were more common, and thus we see most of those quasars at large look-back times, meaning at large distances. For these two reasons, most quasars lie at great distances.

If our model of quasars is correct, then quasars must be triggered into eruption. What observational evidence supports that argument?

Our study of active galaxies has led us far out in space and back in time to quasars. The light now arriving from the most distant quasars left them when the universe was only a fraction of its present age. The quasars stand at the very threshold of the study of the history of the universe itself, the subject of the next chapter.

The quasars appear to be related objects. Their spectra show emission lines with very large red shifts. The large red shifts are interpreted to mean the quasars are at great distance and thus must be superluminous. Yet, because they fluctuate in brightness over intervals as short as a week or less, quasars must also be very small. The detection of quasars in clusters of galaxies, gravitational lenses, and the galaxylike spectra of quasar fuzz show that quasars are indeed located inside galaxies at great distances.

Growing evidence suggests that quasars are the active cores of very distant galaxies. As in the case of active galaxies, quasars are often found in galaxies that are distorted or interacting with other galaxies. The rapid flow of matter into a central, massive black hole could produce so much energy that the quasars would be visible at great distances.

Because the quasars lie at great distances, we see them as they were long ago when the universe was much younger than it is now. Quasars are most common at a red shift of about 2, which suggests that there was an age when galaxies had formed, were still rich in gas and dust, but were close enough together to collide and merge often and thus produce quasars. Few quasars are found with larger red shifts and no quasars have been found beyond a red shift of about 5 (to date). This suggests that at the largest red shifts we are looking back to a time when the universe had not yet formed galaxies.

Summary

Some galaxies appear to be suffering eruptions in their cores. Such active-core galaxies are powerful sources of energy and appear to be related to the in-fall of matter into central, massive black holes.

Double-lobed radio galaxies emit radio energy from lobes on either side of the galaxy. In some cases, we can see jets leading from the core of the galaxy out into the lobes. The double-exhaust model proposes that the core of the galaxy ejects relativistic particles in two jets that punch into the intergalactic medium and blow up the radio lobes like balloons. The impact of the beams on the intergalactic medium produces the hot spots detected on the outer edges of many radio lobes. We see many active galaxies producing jets, and we can see instances where the jets are blown back by the intergalactic medium to produce head–tail galaxies.

Some galaxies, such as Seyfert galaxies and BL Lac objects, have small, brilliant cores. Giant elliptical galaxies such as M87 have bright cores with jets.

The energy source for active galactic nuclei has been described as the unified model—in which mass flows into an accretion disk around a supermassive black hole at the center of the galaxy. Active galaxies are often distorted by interaction with other galaxies, and such tidal interactions could throw matter into the central region, feed the black hole, and trigger an eruption. What we would see would depend on the inclination of the accretion disk to our line of sight. If the disk were pole-on, we would look directly into the hot cavity and see a brilliant BL Lac object. At higher inclinations, we would see a Seyfert 1 galaxy; and if the disk were edge-on, we would see a Seyfert 2 galaxy.

New Terms

radio galaxy	head–tail radio galaxy
active galaxy	Seyfert galaxy
active galactic nuclei (AGN)	BL Lac object
double-lobed radio galaxy	quasar
hot spot	relativistic red shift
double-exhaust model	gravitational lens

Review Questions

1. What is the difference between the terms *radio galaxy* and *active galaxy*?

2. What evidence do we have that the energy source in a double-lobed radio galaxy lies at the center of the galaxy?

3. How does the peculiar rotation of NGC5128 help us understand the origin of this active galaxy?

4. What statistical evidence suggests that Seyfert galaxies have suffered recent interactions with other galaxies?

5. How does the unified model explain the two kinds of Seyfert galaxies?

6. What observations are necessary to identify the presence of a supermassive black hole at the center of a galaxy?

7. How does the unified model implicate collisions and mergers in triggering active galaxies?

8. Why were quasars first noticed as being peculiar?

9. How do the large red shifts of quasars lead us to conclude they must be very distant?

10. Why do we conclude that quasars are superluminous but must be very small?

11. How do gravitational lenses provide evidence that quasars are distant?

12. What evidence do we have that quasars occur in distant galaxies?

13. How can our model quasar explain the different radiation we receive from quasars?

14. What evidence do we have that quasars must be triggered by collisions and mergers?

15. Why are there few quasars at low red shifts and at high red shifts but many at red shifts of about 2?

Discussion Questions

1. Why do quasars, active galaxies, SS433, and protostars have similar geometry?

2. By custom, astronomers refer to the unified *model* of AGN and not to the unified *hypothesis* or unified *theory*. In your opinion, which of the words seems best?

Problems

1. The total energy stored in a radio lobe is about 10^{53} J. How many solar masses would have to be converted to energy to produce this energy? (*Hints:* Use $E = mc^2$. One solar mass equals 2×10^{30} kg.)

2. If the jet in NGC5128 is traveling at 5000 km/sec and is 40 kpc long, how long will it take for gas to travel from the core of the galaxy to the end of the jet? (*Hint:* 1 pc equals 3×10^{13} km.)

3. Cygnus A is roughly 225 Mpc away, and its jet is about 50 seconds of arc long. What is the length of the jet in parsecs? (*Hint:* See By the Numbers 3-1.)

4. If the average giant elliptical galaxy is 50 kpc in diame-

ter and the visible galaxy associated with 3C449 (see Figure 14-5) is 1 minute of arc in diameter, calculate an estimate of the distance to the galaxy. (*Hint:* See By the Numbers 3-1.)

5. The active core of a galaxy 3.25 Mpc from us is observed to have an angular diameter of 0.0015 seconds of arc. What is its linear diameter? (*Hint:* See By the Numbers 3-1.)

6. If a quasar is 1000 times more luminous than an entire galaxy, what is the absolute magnitude of such a quasar? (*Hints:* The absolute magnitude of a bright galaxy is about -21.)

7. If the quasar in Problem 6 were located at the center of our galaxy, what would its apparent magnitude be? (*Hints:* See By the Numbers 8-2, and ignore dimming by dust clouds.)

8. What is the radial velocity of 3C48 if its red shift is 0.37? (*Hint:* See By the Numbers 14-1.)

9. If the Hubble constant is 70 km/sec/Mpc, how far away is the quasar in Problem 8? (*Hint:* Use the Hubble law.)

10. The hydrogen Balmer line Hβ has a wavelength of 486.1 nm. It is shifted to 563.9 nm in the spectrum of 3C273. What is the red shift of this quasar? (*Hint:* What is $\Delta\lambda$?)

Critical Inquiries for the Web

1. What object currently holds the distinction as the farthest known galaxy? Search the Web for information on this distant object, and find out its red shift and distance. What is the look-back time for this object?

2. Gravitational lenses were first predicted by Einstein in 1936 but were not observed until recently. Search the Web for instances of gravitational lensing of galaxies and quasars. For a particular case, discuss how the lens effect allows astronomers to determine information about the lensing and/or lensed objects that might not have been available without the alignment.

 Go to the Brooks/Cole Astronomy Resource Center (www.brookscole.com/astronomy) for critical thinking exercises, articles, and additional readings from InfoTrac College Edition, Brooks/Cole's online student library.

CHAPTER 15

Cosmology

The Universe, as has been observed

before, is an unsettlingly big place, a

fact which for the sake of a quiet life

most people tend to ignore.

Douglas Adams
*The Restaurant at the End
of the Universe*

Guidepost

This chapter marks a watershed in our study of astronomy. Since Chapter 1, our discussion has focused on learning to understand the universe. Our outward journey has discussed the appearance of the night sky, the birth and death of stars, and the interactions of the galaxies. Now we reach the limit of our journey in space and time, the origin and evolution of the universe as a whole.

The ideas in this chapter are the biggest and the most difficult in all of science. Our imaginations can hardly grasp such ideas as the edge of the universe and the first instant of time. Perhaps it is fitting that the biggest questions are the most challenging.

But this chapter is not an end to our story. Once we complete it, we will have a grasp of the nature of the universe, and we will be ready to focus on our place in that universe—the subject of the rest of this book.

What is the biggest number? A billion? How about a billion billion? That is only 10^{18}. How about 10^{100}? That very large number is called a googol, a number that is at least a billion billion times larger than the total number of atoms in all of the galaxies we can observe. Can you name a number bigger than a googol? Try a billion googols, or better yet, a googol of googols. You can go even bigger than that, 10 raised to a googol. (That's called a googolplex.) No matter how large a number you name, we can name a bigger number. That is what infinity means—big without limit.

If the universe is infinite in size, then you can name any distance you want, and the universe is bigger than that. Try a googol to the googol light-years. If the universe is infinite, there is more universe beyond that distance.

Is the universe infinite or finite? In this chapter, we will try to answer that question and others, not by playing games with big numbers, but through **cosmology,** the study of the universe as a whole.

If the universe is not infinite, then we face the edge–center problem. Suppose that the universe is not infinite, and you journey out to the edge of the universe. What do you see: a wall of cardboard? a great empty space? nothing, not even space? What happens if you try to stick your head out beyond the edge? These are almost nonsensical questions, but they illustrate the problem we have when we try to

think about an edge to the universe. Modern cosmologists believe the universe cannot have an edge. In this chapter, we will see how the universe might be finite but have no edge.

If the universe has no edge, it can have no center. We find the centers of things—galaxies, globular clusters, oceans, pizzas—by referring to their edges. With no edge, there can be no center.

Finally, we must answer a basic question. How old is the universe? If it is not eternal (infinite in age), it must have had a beginning, and we must think about how the universe began and how it will end. Are a beginning and an ending like edges in time?

These are arguably the most challenging questions in human knowledge, and we start our search for answers by trying to understand the structure of the universe. That leads us, as is always the case in astronomy, to compare observations with theories.

15-1 The Structure of the Universe

Modern telescopes reveal a sky filled with clusters of galaxies at great distances, and today's sophisticated instruments can image distant galaxies at many wavelengths. Such observations assure us that the universe is filled with galaxies (Figure 15-1). Yet the most

Figure 15-1
The left image, made by an Earth-based telescope, shows a cluster of galaxies that is 8 billion light-years from Earth. Such clusters are very common. The blue shading shows where orbiting X-ray telescopes detect hot gas filling the cluster. The Hubble Space Telescope image at right shows that the galaxies of this cluster are both spiral and elliptical. *(Megan Donahue, STScI, Isabella Gioia, University of Hawaii, and NASA)*

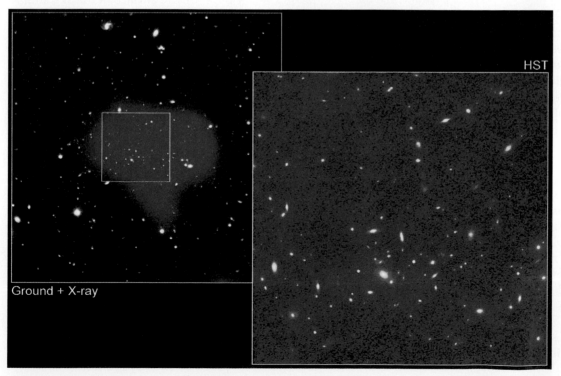

Ground + X-ray

HST

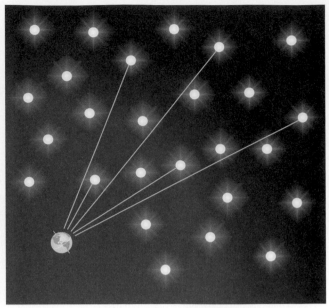

b

Figure 15-2

(a) Every direction we look in a forest eventually reaches a tree trunk, and we cannot see out of the forest. *(Photo courtesy Janet Seeds)* (b) If the universe is infinite and uniformly filled with stars, then any line from Earth should eventually reach the surface of a star. This assumption predicts that the night sky should glow as brightly as the surface of the average star, a puzzle commonly referred to as Olbers' paradox.

important observation in cosmology is a simple fact that you have noticed many times. It gets dark at night.

Olbers' Paradox and the Necessity of a Beginning

We have all noticed that the night sky is dark. However, reasonable assumptions about the geometry of the universe can lead us to the conclusion that the night sky should glow as brightly as a star's surface. This conflict between observation and theory is called **Olbers' paradox** after Heinrich Olbers, a Viennese physician and astronomer who discussed the problem in 1826.

However, Olbers' paradox is not Olbers'. The problem of the dark night sky was first discussed by Thomas Digges in 1576 and was further analyzed by such astronomers as Johannes Kepler in 1610 and Edmond Halley in 1721. Olbers gets the credit through an accident of scholarship on the part of modern cosmologists who did not know of previous discussions. What's more, Olbers' paradox is not a paradox. We will be able to understand why the night sky is dark by revising our assumptions about the nature of the universe.

To begin, let's state the so-called paradox. Suppose the universe is static, infinite, eternal, and uni-formly filled with stars. (The aggregation of stars into galaxies makes no difference to our argument.) If we look in any direction, our line of sight must eventually reach the surface of a star (Figure 15-2). Consequently, every point on the surface of the sky should be as bright as the surface of a star, and it should not get dark at night.

Of course, the most distant stars would be much fainter than the nearer stars, but there would be a greater number of distant stars than nearer stars. The intensity of the light from a star decreases according to the inverse square relation, so distant stars would not contribute much light. However, the farther we look in space, the larger the volume we survey. Thus, the number of stars we see at any given distance increases as the square of the distance. The two effects cancel out, and the stars at any given distance contribute as much total light as the stars at any other distance. Then, given our assumptions, every spot on the sky must be occupied by the surface of a star, and it should not get dark at night.

Imagine the entire sky glowing with the brightness of the surface of the sun. The glare would be overpowering. In fact, the radiation would rapidly heat Earth and all other celestial objects to the average temperature of the surface of the stars, roughly a few thousand degrees. Thus, we can pose Olbers' paradox in another way: Why is the universe so cold?

Olbers assumed that the sky was dark because clouds of matter in space absorb the radiation from distant stars. But this interstellar medium would gradually heat up to the average surface temperature of the stars, and the gas and dust clouds would glow as brightly as the stars.

Today, cosmologists believe they understand why the sky is dark, and the universe is cold. Olbers' paradox makes the incorrect prediction that the night sky should be bright because it is based on an incorrect assumption. The universe is not eternal. That is, it is not infinitely old.

This solution to Olber's Paradox was first suggested by Edgar Allan Poe in 1848. He proposed that the night sky was dark because the universe was not infinitely old but had been created at some time in the past. The more distant stars are so far away that light from them has not reached us yet. That is, if we look far enough, the look-back time is greater than the age of the universe, and we look back to a time before stars began to shine. Thus, the night sky is dark because the universe is not infinitely old.

This is a powerful idea because it clearly illustrates the difference between the universe and the observable universe. The universe is everything that exists, and it could be infinite. But the **observable universe** is the part that we can see. We will learn later that the universe is 10 to 20 billion years old. In that case, the observable universe has a radius of 10 to 20 billion light-years. Do not confuse the observable universe, which is finite, with the universe as a whole, which could be infinite.

One of the assumptions that we made when we described Olbers' paradox was in error. This illustrates the importance of assumptions in cosmology and serves as a warning that our commonsense expectations are not dependable. All of astronomy is reasonably unreasonable — that is, reasonable assumptions often lead to unreasonable results. That is especially true in cosmology, so we must always examine our assumptions with particular care.

Basic Assumptions in Cosmology

Although we could make many assumptions, three are basic — homogeneity, isotropy, and universality.

Homogeneity is the assumption that matter is uniformly spread throughout space. Obviously, this is not true on the small scale, because we can see matter concentrated in planets, stars, and galaxies. Homogeneity refers to the large-scale distribution. If the universe is homogeneous, we should be able to ignore individual galaxies and think of matter as material evenly spread through space. Observation of the distribution of galaxies reveals that the universe is homogeneous only on the largest scales, a subject we will discuss later in this chapter.

Isotropy is the assumption that the universe looks the same in every direction, that it is isotropic. On the small scale this is not true, but if we ignore local variations like galaxies and clusters of galaxies, the universe should look the same in any direction. For example, we should see roughly the same number of galaxies in every direction. Again, observations suggest that the universe is isotropic only on the largest scale.

The most easily overlooked assumption is **universality,** which holds that the physical laws we know on Earth, such as the law of gravity, apply everywhere in the universe. Although this may seem obvious at first, some astronomers challenge universality by pointing out that when we look out in space we look back in time. If the laws of physics change with time, we may see peculiar effects when we look at distant galaxies. For now we will assume that the physical laws observed on Earth apply everywhere in the universe.

The assumptions of homogeneity and isotropy lead to an assumption so fundamental it is called the **cosmological principle.** According to this principle, any observer in any galaxy sees the same general features of the universe. For example, all observers should see the same kinds of galaxies. As in previous assumptions, we ignore local variations and consider only the overall appearance of the universe, so the fact that some observers live in galaxies in clusters and some live in isolated galaxies is only a minor irregularity.

Evolutionary changes are not included in the cosmological principle. If the universe is expanding and the galaxies are evolving, observers living at different times may see galaxies at different stages. The cosmological principle says that once observers correct for evolutionary changes, they should see the same general features.

The cosmological principle is actually an extension of the Copernican principle. Copernicus said that Earth is not in a special place; it is just one of a number of planets orbiting the sun. The cosmological principle says that there are no special places in the universe. Local irregularities aside, one place is just like another. Our location in the universe is typical of all other locations.

If we accept the cosmological principle, then we may not imagine that the universe has an edge or a center. Such locations would not be the same as all other locations.

The first fundamental observation of cosmology, that the night sky is dark, has led us to deep insights about the nature of the universe. The second fundamental observation will be equally revealing.

The Expansion of the Universe

The second fundamental observation of cosmology is that the spectra of galaxies contain red shifts that are proportional to their distances (Figure 15-3). These red shifts are commonly interpreted as Doppler shifts due to the recession of the galaxies, which is why we say the universe is expanding.

Edwin P. Hubble discovered the velocity–distance relationship in 1929 using the spectra of only 46 galaxies (see Figure 13-11). Since then, the Hubble law, as it's known, has been confirmed for large numbers of galaxies out to great distances. This law is clear evidence that the universe is expanding uniformly and has no center of expansion.

To see how the Hubble law implies uniform, centerless expansion, we can think about an analogy (Window on Science 15-1). Imagine that we make a loaf of raisin bread. As the dough rises, the expansion pushes the raisins away from one another. Two raisins that were originally separated by only 1 cm move apart rather slowly, but two raisins that were originally separated by 4 cm of dough are pushed apart more rapidly. The uniform expansion of the dough causes the raisins to move away from each other at velocities proportional to their distances (Figure 15-4). According to the Hubble law, the larger the distance between two galaxies, the faster they recede from each other. This is exactly the result we expect from uniform expansion.

The raisin bread also shows that the expansion has no identifiable center. Bacteria living on one of the raisins would see themselves surrounded by a universe of receding raisins. (Here we must assume that they cannot see to the edge of the loaf—that is, they must not be able to see to the edge of their universe.) It does not matter on which raisin the bacteria live. The bacterial astronomers would measure the distances and velocities of recession and derive a bacterial Hubble law showing that the velocities of recession are proportional to the distances. So long as they cannot see the edge of the loaf, they cannot identify any raisin as the center of the expansion.

Similarly, we see galaxies receding from us, but no galaxy or point in space is the center of the expansion. Any observer in any galaxy would see the same expansion that we see.

Although astronomers and cosmologists commonly refer to these red shifts as Doppler shifts and often speak of the recession of the galaxies, relativity provides a more elegant explanation. Einstein's theory of general relativity explains the expansion of the universe as an expansion of space–time itself. A photon traveling through this space–time is stretched as space–time expands, and the photon arrives with a longer wavelength than it had when it left (Figure 15-5). Photons from distant galaxies travel for a longer time and are stretched more than photons from nearby galaxies. Thus, the expansion of the universe

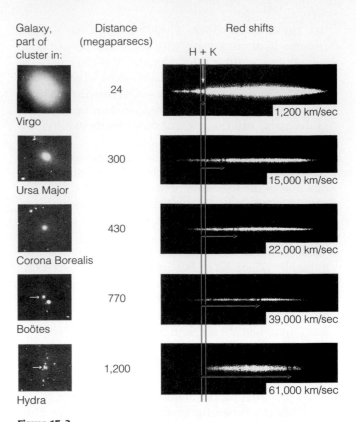

Galaxy, part of cluster in:	Distance (megaparsecs)	Red shifts
Virgo	24	1,200 km/sec
Ursa Major	300	15,000 km/sec
Corona Borealis	430	22,000 km/sec
Boötes	770	39,000 km/sec
Hydra	1,200	61,000 km/sec

Figure 15-3
Galaxies at different distances are shown at left, with their spectra shown at the right. The two vertical blue lines mark the unshifted location of the H and K lines of once-ionized calcium. The expansion of the universe causes a red shift that moves the spectral lines to longer wavelengths, as shown by the red arrows. More distant galaxies have larger red shifts, expressed as radial velocities in kilometers per second. *(Palomar Obs/Caltech)*

is an expansion of the geometry and not just a simple recession of the galaxies.

This explains, by the way, why the equation for the relativistic red shift given in By the Numbers 14-1 is not quite correct for galaxies. That equation comes from the special theory of relativity and expresses the Doppler shift caused by the motion of an object through space. But the galaxies, in receding from one another, are not moving through space any more than the raisins in our raisin bread are swimming through the bread dough. The galaxies are carried away from each other by the uniform expansion of space itself.

We need not worry about the actual form of the red-shift equation; it is different for different assumptions about the nature of space and time. Rather, we must focus on the central concept of modern cosmology, the curvature of space–time.

The Geometry of Space–Time

How can the universe expand if it does not have any extra space to expand into? How can it be finite if it doesn't have an edge? The properties we have ascribed to the universe seem to violate common sense,

"The economy is overheating and it may seize up," an economist might say. Economists like to talk in analogies because economics is often abstract, and one of the best ways to think about abstract problems is to find a more approachable analogy. Rather than discuss the details of the national economy, we might make conclusions about how the economy works by thinking about how a gasoline engine works. Of course, if our analogy is not a good one, it can mislead us.

Much of astronomy is abstract, and cosmology is the most abstract subject in astronomy. Furthermore, cosmology is highly mathematical, and unless we are prepared to learn some of the most difficult mathematics known, we must use analogies, such as raisin bread or an ant on an orange, to discuss the nature of the universe and the curvature of space–time.

Reasoning by analogy is a powerful technique. An analogy can reveal unexpected insights and lead us to further discoveries. But carrying an analogy too far can be misleading. We might compare the human brain to a computer, and that would help us understand how data flow in and are processed and how new data flow out. But our analogy is flawed. Data in computers are stored in specific locations, but memories are stored in the brain in a distributed form. No single brain cell holds a specific memory. If we carry the analogy too far, it can mislead us. Whenever we reason by analogy, we should be alert for potential problems.

As you study any science, be alert for analogies. They are tremendously helpful, but we must always take care not to carry them too far.

a

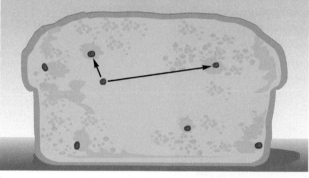
b

Figure 15-4
The uniform expansion of the universe can be represented by raisins in a loaf of raisin bread (a). As the dough rises (b), raisins originally near each other move apart more slowly, and raisins originally farther apart move away from each other more rapidly. A colony of bacteria living on any raisin will find that the velocities of recession of the other raisins are proportional to their distances.

so we must proceed with caution. Our everyday intuition can be misleading. For example, everything we touch and see every day has an edge, so thinking about an edgeless universe can be a challenge. One powerful tool is reasoning by analogy, but we must also use words carefully so they do not mislead us (Window on Science 15-2). With such cautions in mind, we can use general relativity to understand the geometry of space–time.

Einstein's theory of general relativity proposes that the presence of mass can curve space–time and that we sense this curvature as a gravitational field. This theory explains Earth's gravity and that of a black hole, but it also predicts that the entire universe can have a general curvature that determines the geometry and expansion of the universe. Thus, modern cosmological theories are based on general relativity and curved space–time.

To discuss curvature, we can use a two-dimensional analogy for our three-dimensional universe. Suppose an ant lived on an orange and was a perfectly two-dimensional creature. That is, the ant could move in two dimensions over the surface of the orange but could not move in the third dimension perpendicular to the surface of the orange. Furthermore, suppose that light could not travel perpendicular to the surface of the orange. Then the ant would never know there was a third dimension, and it could wander over the surface of the orange thinking it was moving on a horizontal surface that had no edge. If the ant left dirty footprints, however, it might eventually say, "I have visited every square centimeter in my two-dimensional universe, so it must be finite, but I can't find any edge or any center (Figure 15-6). My universe is finite but it is unbounded." Of course, if the ant lived on a sheet of paper it would discover the edges and conclude that it lived in a finite, bounded universe. Our two-dimensional analogy shows that curvature can allow a universe to be finite but have no edge. If we expand our analogy to three dimensions, we see that our universe could be finite in volume if it is curved such that it is unbounded.

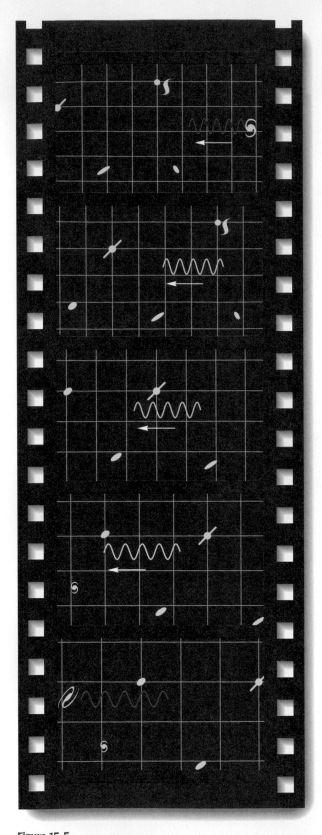

Figure 15-5
General relativity explains that the red shift of light from distant galaxies is produced by the stretching of space–time (shown here as an expanding grid) between the time the photon is emitted (top frame) and the time it arrives at Earth (bottom frame). Notice that the wavelength in the top frame is shorter than in the bottom frame. Photons from distant galaxies spend more time traveling through space and are thus stretched more than photons from nearby galaxies. Consequently, distant galaxies have larger red shifts.

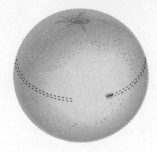

Figure 15-6
An ant confined to the two-dimensional surface of an orange could explore the entire surface without coming to an edge. Were it to leave dirty footprints, it might realize that its two-dimensional universe was finite but unbounded.

To continue our discussion of curved space–time, we can extend our analogy of the ant. There are three possible ways that a universe could be curved. It could be flat (zero curvature), it could be spherical (positive curvature), or it could be saddle-shaped (negative curvature). If an ant lived on a two-dimensional universe, it could detect any curvature if it could draw circles and measure their areas. On the flat surface, a circle would always have an area of exactly πr^2, but on the positively curved surface its area would be less than πr^2, and on the negatively curved surface it would be more (upper part of Figure 15-7). Drawing small circles would not suffice, because the area of small circles would not differ noticeably from πr^2, but if the ant could draw big enough circles, it could actually measure the curvature of its two-dimensional universe.

We are three-dimensional creatures, but our universe might still be curved, and we could measure that curvature by measuring the volume of spheres. If space–time is flat, then no matter how big a sphere we measure, its volume will always be $\frac{4}{3}\pi r^3$. But in positively curved space–time, spheres would have less volume than this; in negatively curved space–time, they would have more. The spheres must be many megaparsecs in radius for the difference to be detectable.

We could measure the volume of such spheres by counting the number of galaxies within a certain distance r from Earth. If galaxies are homogeneously scattered through space, the number within distance r should be proportional to the volume of the sphere. If, as we count to greater and greater distances, we find the number of galaxies increasing proportional to $\frac{4}{3}\pi r^3$, space–time is flat. However, if we find an excess of distant galaxies, space–time is negatively curved, and if we find a deficiency, space–time is positively curved (lower part of Figure 15-7).

By thinking of the two-dimensional analogy shown in Figure 15-7, astronomers usually refer to a universe with positive curvature as a **closed universe.** A zero-curvature universe is said to be a **flat universe,** and a universe with negative curvature is an **open universe.** Notice that flat and open universes are infinite, but closed universes are finite.

There are certain words we should never say, not even as a joke. We should never call our friend "fool," for example, because we might begin to think of our friend as a fool. Words lead thoughts. It works in advertising and politics, and it can work in science, too. Using words carelessly can lead us to think carelessly.

In science, there are certain ways to say things, not because scientists are sticklers for good grammar, but because if we say things wrongly we begin to think things wrongly. For example, a biologist would never let us say a beehive knows it must store food. "No, don't say it that way," the biologist would object. "The hive is just a collection of individual creatures and it can't 'know' anything. Even the individual bees don't really 'know' things. The instinctive behavior of individual bees causes food to accumulate in the hive. That's the way to say it."

All scientists are careful of language because careless words can mislead us.

We would never refer to the center of the universe, for example, but we must also be careful not to say things like "galaxies flying away from the big bang." Those words imply a center to the expansion of the universe, and we know the expansion must be centerless. Rather, we should say "galaxies flying away from each other."

Whatever science you study, notice the customary ways of using words. It is not just a matter of convention. It is a matter of careful thought.

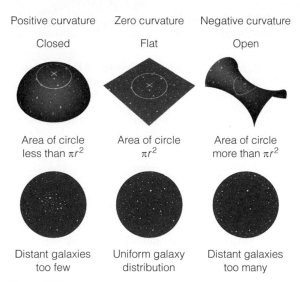

Positive curvature — Closed — Area of circle less than πr^2 — Distant galaxies too few

Zero curvature — Flat — Area of circle πr^2 — Uniform galaxy distribution

Negative curvature — Open — Area of circle more than πr^2 — Distant galaxies too many

Figure 15-7
In two-dimensional space, curvature distorts the area of a circle; in three-dimensional space, it distorts the volume of a sphere. We could measure distortion and detect curvature by counting galaxies.

A closed universe has a finite volume but no edge. Like the ant in Figure 15-6, we might explore our universe and visit every location but never find an edge. Of course, an open universe, being infinite, also has no edge. Thus, we can answer our question about the edge of the universe. Whether it is open or closed, the universe can have no boundary.

We must not try to visualize the expansion of the universe as an outer edge moving into previously unoccupied space. Open, flat, or closed, the universe has no edge, so it does not need additional room to expand. The universe contains all of the volume that exists, and the expansion is a change in the nature of space–time that causes that volume to increase.

REVIEW Critical Inquiry

Why do we say the universe can't have an edge or center?

The quick answer is that an edge or a center would violate the cosmological principle, which says that every place in the universe is similar in its general properties to every other place. Then a place at the edge or center would be different, so there can be no edge and no center. Of course, this isn't critical thinking; it is only an appeal to the arbitrary authority of cosmological principle. What is the real reason we conclude that there can be no edge and no center?

Imagine that the universe had an edge and you went there. What would you see? Of course, you might imagine an edge to the distribution of matter with empty space beyond, but that would not be a real edge. Imagine an edge beyond which there is no space. Could you stick your head beyond the edge and look around? Thinking about an edge to the universe leads us to logical problems we cannot resolve. From this, we conclude that an edge is impossible. Of course, if there is no edge, then there is no center because we define the centers of things by reference to their edges.

We believe the universe has no edge and no center for good logical reasons, not because of some arbitrary rule. But then why do we believe the cosmological principle is correct? What evidence or assumptions support it?

According to the evidence and consistent with general relativity, we live in an expanding universe. Our problem is to review the evidence and understand how an expanding universe could have begun.

15-2 The Big Bang

If the universe were a videotape, we could run it backward and see the universe contracting as the galaxies approached each other and eventually merged into a hot, high-density gas. As we watched, we would see the volume of space decrease and trap the matter and energy in a high-temperature, high-density state, the state from which the universe began expanding—the **big bang.**

The name *big bang* was coined by cosmologist Fred Hoyle as a sardonic comment on the believability of the theory, and it is often the butt of jokes and cartoons (Figure 15-8). At first glance, it is difficult to imagine how the big bang could have occurred, but it is the dominant theory for the origin of the universe. Our challenge is to stretch our imaginations to encompass the beginning of everything.

The Beginning

The big bang occurred long ago, and our instinct is to think of it as a historical event, like the Gettysburg Address—something that ended long ago and happened in a particular place. As we will discover, the big bang isn't over, and it is happening everywhere.

Because light travels at a finite speed, we see distant galaxies, not as they are now, but as they were when the light left them to begin its journey to Earth. The Andromeda Galaxy is about 2 million light-years away, so the look-back time to the Andromeda Galaxy is 2 million years. The look-back time to more distant galaxies is larger. The largest telescopes can detect

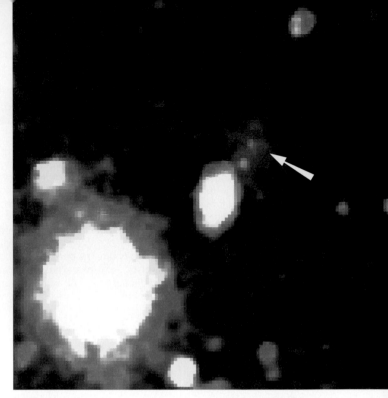

Figure 15-9
One of the most distant galaxies ever photographed (arrow) has a red shift of 4.25. The look-back time to this galaxy is so large we see it as it was when the universe was only 7 to 10 percent of its present age. The galaxy was first mapped by radio astronomers; this image was made using the 10-m Keck Telescope atop Mauna Kea in Hawaii. The two brightest objects in the field of view are less distant galaxies. *(H. Spinrad, A. Dey, and J. Graham, University of California, Keck Observatory)*

galaxies with look-back times that are a large fraction of the age of the universe (Figure 15-9). Thus, the look-back time allows us to see what the universe was like when it was young.

Suppose that we look even farther, past the most distant objects, backward in time to the fiery clouds of matter that filled the universe during the big bang. From these great distances we receive light that was emitted by the hot gas soon after the universe began. That means the big bang is all around us.

Although our imaginations try to visualize the big bang as a localized event, we must keep firmly in mind that the big bang did not occur at a single place but filled the entire volume of the universe. The matter of which we are made was part of that big bang, so we are inside the remains of that event, and the universe continues to expand around us. We cannot point to any particular place and say "The big bang occurred over there." The big bang occurred everywhere, and wherever we look, at great distances we look back to the age when the universe was filled with hot gas (Figure 15-10).

The radiation that comes to us from this great distance has a tremendous red shift. The most distant visible objects are quasars, which have red shifts up

Figure 15-8
The population of Cosmos, Minnesota, is apparently in flux; someone has painted out the numbers on the city limits sign. This sign reminds us of the big bang theory. Although the theory seems at first sight difficult to believe and is sometimes the object of jokes and unwarranted criticism, it is currently the only widely accepted theory for the origin of the universe. The evidence that there was a big bang is overwhelming. *(Richard Fluck)*

a   A region of the universe during the big bang

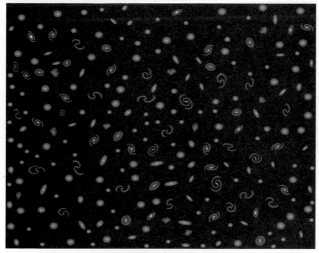

b   A region of the universe now

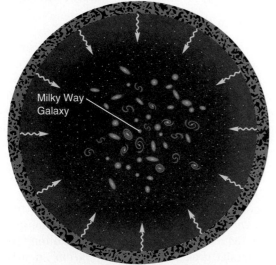

c   The present universe as it appears from our galaxy

Figure 15-10
Three views of a small region of the universe centered on our galaxy. (a) During the big bang, the region is filled with hot gas and radiation. (b) Later, the gas forms galaxies, but we can't see the universe this way because the look-back time distorts what we see. (c) Near us we see galaxies, but farther away we see young galaxies (dots), and at a great distance we see radiation (arrows) coming from the hot clouds of the big bang.

to about 5. In contrast, the radiation from the hot gas of the big bang has a red shift of about 1000. Thus the light emitted by the big bang gases arrives at Earth as infrared radiation and short radio waves. We can't see it with our eyes, but we should be able to detect it with infrared and radio telescopes. Thus, unlike the Gettysburg Address, the big bang isn't over, and we should be able to see it happening if we can detect the radiation it emitted.

The Primordial Background Radiation

If radiation is now arriving from the big bang, then it should be detectable. The story of that discovery begins in the mid-1960s when two Bell Laboratories physicists, Arno Penzias and Robert Wilson, were using a horn antenna to measure the radio brightness of the sky. Their measurements showed a peculiar noise in the system, which they at first attributed to pigeons living inside the antenna (Figure 15-11a). After relocating the birds and cleaning out their droppings, the scientists still detected the noise. Perhaps they would have enjoyed cleaning the antenna more if they had known they would win the 1978 Nobel Prize for physics for the discovery they were about to make.

The explanation for the noise goes back to 1948, when George Gamow predicted that the early big bang would be very hot and would emit copious black body radiation. A year later, physicists Ralph Alpher and Robert Herman pointed out that the large red shift of the big bang gas clouds would lengthen the wavelengths of the radiation and make the hot gas clouds seem very cold. There was no way to detect this radiation until the mid-1960s, when Robert Dicke at Princeton concluded the radiation should be just strong enough to detect with newly developed techniques. Dicke and his team began building a receiver.

When Penzias and Wilson heard of Dicke's work, they recognized the noise they had detected as radiation from the big bang, the **primordial background radiation.**

The background radiation was measured at many wavelengths as astronomers tried to confirm that it did indeed follow a black body curve. Some measurements could be made from the ground, but balloon and rocket observations were critical and difficult. A few disturbing observations suggested that the radiation departed inexplicably from a black body curve and thus might not come from the big bang.

In November 1989, the orbiting Cosmic Background Explorer (COBE) began taking the most accurate measurements of the background radiation ever made. Within 2 months, the verdict was in (Figure 15-11b). To an accuracy of better than 1 percent, the radiation follows a black body curve with a temperature of 2.735 ± 0.06 K—in good agreement with accepted theory.

a

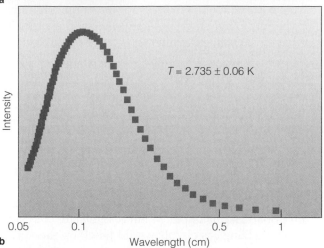

$T = 2.735 \pm 0.06$ K

Intensity

0.05   0.1   0.5   1

b   Wavelength (cm)

Figure 15-11

(a) Robert Wilson (left) and Arno Penzias pose before the horn antenna with which they discovered the primordial background radiation. *(AT&T Archives)* (b) The primordial background radiation from the big bang is observed in the infrared and radio part of the spectrum. Until recently, the critical measurements near the peak of the curve had to be made from balloons and rockets, but the COBE satellite observed from orbit. These COBE data show that the radiation fits a black body curve with a temperature of 2.735 K.

It may seem strange that the hot gas of the big bang seems to have a temperature of only 2.7K, but recall the tremendous red shift. We see light that has a red shift of about 1000—that is, the wavelengths of the photons are about 1000 times longer than when they were emitted. The gas clouds that emitted the photons had a temperature of about 3000 K, and they emitted black body radiation with a λ_{max} of about 1000 nm (see By the Numbers 6-1). Although this is in the near infrared, the gas would also have emitted enough visible light to glow orange-red. But the red shift has made the wavelengths about 1000 times longer, so λ_{max} is about 1 million nm (equivalent to 1 mm). Thus, the hot gas of the big bang seems to be 1000 times cooler, about 2.7 K.

Although the radiation is almost perfectly isotropic, precise observations reveal small departures from complete isotropy. The background radiation has slightly shorter wavelengths—it seems hotter—in one direction and has slightly longer wavelengths—it seems cooler—in the opposite direction. This difference is caused by the motion of our Milky Way Galaxy along with the Local Group at about 540 km/sec toward the massive Virgo cluster. This motion causes a slight blue shift in the direction of motion and that makes the background radiation appear slightly hotter. In the opposite direction there is a red shift, and the radiation looks cooler. (Smaller irregularities in the background radiation will be discussed later in this chapter.)

The greatest importance of the primordial background radiation lies in its widely accepted interpretation as radiation from the big bang. If that interpretation is correct, the radiation is evidence that a big bang really did occur. In fact, the evidence is so strong that an alternative theory of the universe was abandoned. The **steady state theory,** popular with some astronomers during the 1950s and 1960s, held that the universe was eternal and unchanging. If the universe had no beginning, then it could never have had a big bang. According to the theory, new matter continuously appeared from nothing to maintain the density of the expanding, steady-state universe. The discovery of the primordial background radiation in 1965 led astronomers to abandon the steady state theory within a few years. We know there was a big bang because, using radio and infrared telescopes, we can see it happening.

A History of the Big Bang

Modern cosmologists have been able to reconstruct the history of the early universe to reveal how energy and matter interacted as the universe began. As we review that history, remember that the big bang did not occur in a specific place. The big bang filled the entire volume of the universe from the first moment.

We cannot begin our history at time zero, because we do not understand the physics of matter and energy under such extreme conditions, but we can come close. If we could visit the universe when it was very young, only 10 millionths of a second old, for instance, we would find it filled with high-energy photons having a temperature well over 1 trillion (10^{12}) K and a density greater than 5×10^{13} g/cm^3, nearly the density of an atomic nucleus. When we say the photons had a given temperature, we mean that the photons were the same as black body radiation emitted by an object of that temperature. Thus, the photons in the early universe were gamma rays of very short wavelength and therefore very high energy. When we say that the radiation had a certain density, we refer to Einstein's equation $E = m_0 c^2$. We can express a given amount of energy in the form of radiation as if it were matter of a given density.

If a photon has enough energy, it can decay and convert its energy into a pair of particles—a particle of normal matter and a particle of **antimatter.** When an antimatter particle meets its matching particle of normal matter—when an antiproton meets a normal proton, for example—the two particles annihilate each other and convert their mass into energy in the form of gamma rays. In the early universe, the photons had enough energy to produce proton–antiproton pairs; but when the particles collided, they converted their mass back into energy. Thus, the early universe was a dynamic soup of energy flickering from photons into particles and back again.

While all of this went on, the universe was expanding, and the wavelengths of the photons were lengthened by the expansion. This lowered the energy of the gamma rays, and the universe cooled. By the time the universe was 0.0001 second old, its temperature had fallen to 10^{12} K. By this time, the average energy of the gamma rays had fallen below the energy equivalent to the mass of a proton or a neutron, so the gamma rays could no longer produce such heavy particles. The particles combined with their antiparticles and quickly converted most of the mass into photons.

It would seem from this that all the protons and neutrons should have been annihilated with their antiparticles, but for quantum-mechanical reasons a small excess of normal particles existed. For every billion protons annihilated by antiprotons, one survived with no antiparticles to destroy it. Thus, we live in a world of normal matter, and antimatter is very rare.

Although the gamma rays did not have enough energy to produce protons and neutrons, they could produce electron–positron pairs, which are about 1800 times less massive than protons and neutrons. This continued until the universe was about 4 seconds old, at which time the expansion had cooled the gamma rays to the point where they could no longer create electron–positron pairs. Thus, the protons, neutrons, and electrons of which our universe is made were produced during the first 4 seconds of its history.

This soup of hot gas and radiation continued to cool and eventually began to form atomic nuclei. High-energy gamma rays can break up a nucleus, so the formation of such nuclei could not occur until the universe had cooled somewhat. By the time the universe was about 2 minutes old, protons and neutrons could link to form deuterium, the nucleus of a heavy hydrogen atom; and, by the end of the next minute, further reactions began converting deuterium into helium. But no heavier atoms could be built because no stable nuclei exist with atomic weights of 5 or 8 (in

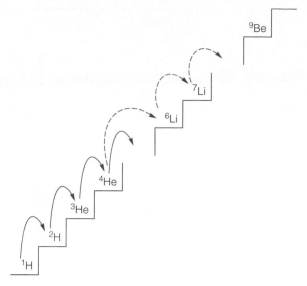

Figure 15-12
Cosmic element building. During the first few minutes of the big bang, temperatures and densities were high, and nuclear reactions built heavier elements. Because there are no stable nuclei with atomic weights of 5 or 8, the process built very few atoms heavier than helium.

units of the hydrogen atom). Cosmic element building during the big bang had to proceed rapidly, step by step, like someone hopping up a flight of stairs (Figure 15-12). The lack of stable nuclei at atomic weights of 5 and 8 meant there were missing steps in the stairway, and the step-by-step reactions could not jump over these gaps.

By the time the universe was 30 minutes old, it had cooled sufficiently that nuclear reactions had stopped. About 25 percent of the mass was helium nuclei, and the rest was in the form of protons—hydrogen nuclei. This is the cosmic abundance we see today in the oldest stars. (The heavier elements, remember, were built by nucleosynthesis inside many generations of massive stars.) The cosmic abundance of helium was fixed during the first minutes of the universe.

For the first million years, the universe was dominated by radiation. The gamma rays interacted continuously with the matter, and they cooled together as the universe expanded. The gas was ionized because it was too hot for the nuclei to capture electrons to form neutral atoms, and the free electrons made the gas very opaque. A photon could not travel very far before it collided with an electron and was deflected. Thus, radiation and matter were locked together.

When the universe reached an age of about 1 million years, it was cool enough for nuclei and electrons to form neutral atoms. The free electrons were captured by atomic nuclei, the gas became transparent, and the radiation was free to travel through the universe. The temperature at the time of this **recombination** was

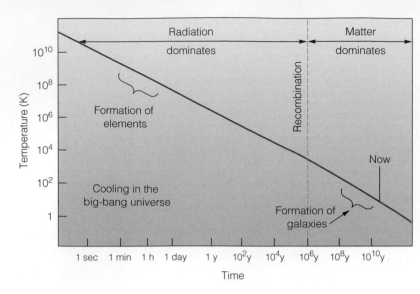

Figure 15-13
During the first few minutes of the big bang, some hydrogen became deuterium and helium and a few heavier atoms. Later, after recombination, when radiation was no longer dominant, galaxies formed, and nuclear reactions inside stars made the rest of the chemical elements.

about 3000 K. We see these photons arriving now as primordial background radiation.

After recombination, the universe was no longer dominated by radiation. Instead, matter was free to move under the influence of gravity, so we say the universe after recombination is dominated by matter (Figure 15-13). How the matter cooled and collected into clouds and eventually gave birth to galaxies in clusters and superclusters is not well understood. We will return to this problem at the end of this chapter.

For now, we must turn our attention away from the first moments of the big bang and instead consider its final moments—the ultimate fate of the universe.

The End of the Universe: A Question of Density

Will the universe ever end? Will it go on expanding forever with stars burning out, galaxies exhausting their star-forming gas and becoming cold, and dark systems expanding forever through an endless, dark universe? How else could the universe end?

Earlier, we decided that the universe could be either open, flat, or closed. The geometry of the universe determines how it will end, so we must consider these three possible universes.

The general curvature of the entire universe is determined by its density. If the average density of the universe is equal to the **critical density,** or 4×10^{-30} g/cm^3, space–time will be flat. If the average density of the universe is less than the critical density, the universe is negatively curved and open. If the average density is greater than the critical density, the universe is positively curved and closed.

We decided earlier that an open universe and a flat universe are both infinite. Both of these universes will expand forever. The gravitational field of the material in the universe, present as a curvature of space–time, will cause the expansion to slow; but, if the universe is open, it will never come to a stop (Figure 15-14). If the universe is flat, it will barely slow to a stop after an infinite time. Thus if the universe is open or flat, it will expand forever, and the galaxies will eventually become black, cold, solitary islands in a universe of darkness.

If the universe is closed, however, its fate will be quite different. In a closed universe, the gravitational field, present as curved space–time, is sufficient to slow the expansion to a stop and make the universe contract. Eventually, the contraction will compress all matter and energy back into the high-energy, high-density state from which the universe began. This end to the universe, a big bang in reverse, has been called the big crunch. Nothing in the universe could avoid being destroyed by this "crunch."

Some theorists have suggested that the big crunch will spring back to produce a new big bang and a new expanding universe. This theory, called the **oscillating universe theory,** predicts that the universe undergoes alternate stages of expansion and contraction. Theoretical work, however, suggests that successive bounces of an oscillating universe would be smaller until the oscillation ran down, and the theory is no longer taken seriously.

Whether the universe is open, closed, or flat depends on its density, but it is quite difficult to measure the density of the universe. We could count galaxies in a given volume, multiply by the average mass of a galaxy, and divide by the volume; but we are not sure of the average mass of a galaxy, and many small galaxies are too faint to see even if they are nearby. We would also have to include the mass equivalent to the energy in the universe, because mass and energy are related. The best attempts yield a density of about 5×10^{-31} g/cm^3, about 10 percent of the mass needed to close the universe.

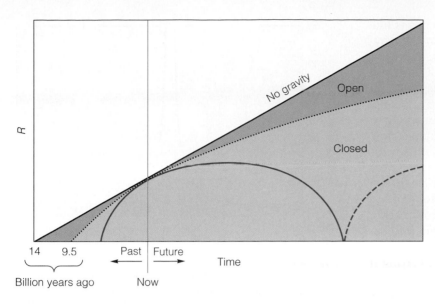

Figure 15-14
The expansion of the universe as a function of time. Models of the universe can be represented as curves in a graph of *R*, a measure of the size of the universe, and time. Open-universe models expand without end, and the corresponding curves fall in the region shaded orange. Closed models expand and then contract back to a high-density state (solid curve) from which they might expand again. Curves representing closed models fall in the region shaded blue. The dotted line represents a flat universe, the dividing line between open and closed models. Note that the estimated age of the universe depends on the rate at which the expansion is slowing down. (This figure assumes $H = 70$ km/sec/Mpc.)

Before we jump to any conclusions about the density of the universe, we need to consider dark matter. Modern astronomers have found extensive evidence that much of the matter in the universe is invisible.

Dark Matter in Cosmology

There is more to the universe than meets the eye. In Chapter 13, we discovered that much of the mass of the universe was invisible, the dark matter. Galaxies have invisible, massive coronas, and clusters of galaxies contain more mass than the sum of the visible galaxies. This is not a small correction. Apparently 90 to 99 percent of the mass of the universe is dark matter. This is a big effect. To correct for the dark matter, we would not just add a small percentage to the visible matter. We have to multiply by a *factor* between 10 and 100.

If we have any doubt how insubstantial the visible universe is, we have only to look at gravitational lensing of distant galaxies. In Chapter 14 we saw how light from distant quasars could be bent by the gravitational field of an intervening galaxy to create false images of the quasar. Using the largest telescopes in the world, astronomers are finding that light from the most distant galaxies can be bent by the mass of a nearby cluster of galaxies. The distant galaxies are extremely faint and blue, perhaps because of rapid star formation. As the light from these faint galaxies passes through a nearby cluster of galaxies, the images of the distant galaxies can be distorted into short arcs like blue rainbows (Figure 15-15). From such arcs, astronomers can calculate the mass of the cluster of galaxies. For instance, the arcs in Figure 15-15 show that over 90 percent of the mass in the cluster is dark matter.

Figure 15-15
Galaxy cluster Abell 2218 (orange images) distorts the images of background galaxies to produce short blue arcs. This gravitational lens effect outlines regions of high mass concentration. *(Courtesy Tony Tyson, AT&T Bell Labs)*

Looking at the universe of visible galaxies is like looking at a ham sandwich and seeing only the mayonnaise. Most of the universe is invisible, and most astronomers now believe that invisible matter is not the normal matter of which you and the stars are made. During the first few minutes of the big bang, nuclear reactions converted some protons into helium and a small amount of other heavier elements. How much of these elements was created depends critically on the density of the material, so astronomers have tried to measure the abundance of these elements. Deuterium, for example, is an isotope of hydrogen in which the nucleus contains a proton and a neutron. This element is so easily converted into helium that none can be produced in stars. In fact, stars destroy what deuterium they have by converting it into helium. Nevertheless, deuterium lines do appear in spectra of gas clouds in space, and astronomers using the giant Keck telescopes have observed gas clouds near quasars at large look-back times. At such large look-back times, we can see how much deuterium existed when the universe was much younger —before stars could destroy significant amounts of deuterium. The gas clouds contain roughly 25 deuterium atoms for every million hydrogen atoms, and that tells us something about the big bang.

The amount of deuterium in the universe depends critically on the density of normal matter during the big bang. If there were a lot of normal particles like protons and neutrons, then the big bang would have destroyed most of the deuterium. The observed amount of deuterium seems to tell us that the normal matter in the big bang wasn't very dense—only about 5 percent of the mass needed to make a closed universe. But many observations show that dark matter is abundant, so modern astronomers believe the dark matter can't be normal matter.

Theorists have proposed that the dark matter might be made up of exotic particles that don't often interact with normal matter. Axions and photinos, for example, have never been detected in the laboratory but, if they exist, they could be part of the dark matter. Data from a large particle detector hint that WIMPs, Weakly Interacting Massive Particles, may indeed exist. How much they contribute to the dark matter will be revealed once physics confirm their existence and determine how common they are.

More promising are neutrinos. Observations in 1998 suggest that neutrinos oscillate, and to do that, they must have some mass. Even if that mass is very small, there are over 10^8 neutrinos in the universe for every particle of normal matter, so if neutrinos prove to have even a small mass, they could be an important part of the dark matter.

A contrasting theory holds that each galaxy is orbited by large numbers of very-low-mass stars or planets too faint to see. These **MACHOs,** for Massive

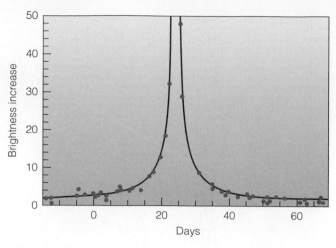

Figure 15-16
Measurements of the brightness of a distant star (dots) increased and then decreased exactly as predicted by general relativity (curve) for a star's passage behind a gravitational lens caused by a MACHO. In this example, the star brightened by a factor of over 40.

Compact Halo Objects, can be detected if they pass between our telescopes on Earth and a distant star. The gravitational field of the MACHO acts as a gravitational lens, focusing the light of the star and making it grow brighter for a period of a few weeks. Massive searches have detected such events (Figure 15-16), but not in the large numbers expected. It does not appear that MACHOs can explain the dark matter.

Although there is good evidence that dark matter exists, it has proven very difficult to identify. If it is as abundant as cosmologists think, it could dominate the evolution of the universe and the formation and evolution of galaxies. Thus, dark matter is one of astronomy's most important problems.

REVIEW Critical Inquiry

How do we know there was a big bang?

The primordial background radiation consists of photons emitted by the hot gas of the big bang, so when we detect those photons, we are "seeing" the big bang. Of course, all scientific evidence must be interpreted, so we must understand how the big bang could produce isotropic radiation before we can accept the background radiation as evidence. First, we must recognize that the big bang event filled all of the universe with hot, dense gas. The big bang didn't happen in a single place; it happened everywhere. At recombination, the expansion of the universe reached the point where the matter became transparent and the radiation was free to travel through space. Today we see that radiation from the age of recombination arriving from all over the sky. It is isotropic because we are part of the big bang event, and as we look out into space to great distance, we look back in time and see the hot gas in any direction we look. We can't see the radiation as light because of the large red shift that has lengthened the wavelengths by a factor of 1000 or so, but we can detect the radiation as infrared and radio photons.

With this interpretation, the primordial background radiation is powerful evidence that there was a big bang. That tells us how the universe began. How can the density of the universe tell us how it will end?

Whether the universe is open or closed, the big bang theory has been phenomenally successful in explaining the origin and evolution of the universe. At present, no other theory is a serious contender. Yet there are important questions to be resolved about the big bang theory.

15-3 Refining the Big Bang: Theory and Observation

Many problems still puzzle cosmologists, but the four we examine here will illustrate some important trends in future research.

The Quantum Universe

A theory developed in 1980 that combined general relativity and quantum mechanics may be able to tell us how the big bang began and why the universe is in the particular state that we observe. Some theorists claim that the new theory will explain why there was a big bang in the first place. To introduce the new theory, we can consider two unsolved problems in the current big bang theory.

One of the problems is called the **flatness problem.** The universe seems to be balanced near the boundary between an open and a closed universe. That is, it seems nearly flat. Given the vast range of possibilities, from zero to infinite, it seems peculiar that the density of the universe is within a factor of 10 of the critical density that would make it flat. If dark matter is as common as it seems, the density may be even closer than a factor of 10.

Even a small departure from critical density when the universe was young would be magnified by subsequent expansion. To be so near critical density now, the density of the universe during its first moments must have been within 1 part in 10^{49} of the critical density. So the flatness problem is: Why is the universe so nearly flat?

Another problem with the big bang theory is the isotropy of the primordial background radiation.

When we correct for the motion of our galaxy, we see the same background radiation in all directions to at least 1 part in 1000. Yet when we look at background radiation coming from two points in the sky separated by more than a degree, we look at two parts of the big bang that were not causally connected when the radiation was emitted. That is, when recombination occurred and the gas of the big bang became transparent to the radiation, the universe was not old enough for any signal to have traveled from one of these regions to the other. Thus, the two spots we look at did not have time to exchange heat and even out their temperatures. Then how did every part of the entire big bang universe get to be so nearly the same temperature by the time of recombination? This is called the **horizon problem** because the two spots are said to lie beyond their respective light–travel horizons.

The key to these two problems and to others involving subatomic physics may lie with a theory called the **inflationary universe** because it predicts a sudden expansion when the universe was very young, an expansion even more extreme than that predicted by the big bang theory.

To understand the inflationary universe, we must consider recent attempts to unify the four forces of nature. Recall from Chapter 9 that physicists know of only four forces—gravity, the electromagnetic force, the strong nuclear force, and the weak nuclear force.

For many years, theorists have tried to unify these forces; that is, they have tried to describe the forces with a single mathematical law. A century ago, James Clerk Maxwell showed that the electric force and the magnetic force were really the same effect, and we now count them as a single electromagnetic force. In the 1960s, theorists succeeded in unifying the electromagnetic force and the weak force in what they called the electroweak force, effective only for processes at very high energy. At lower energies, the electromagnetic force and the weak force behave differently. Now theorists have found ways of unifying the electroweak force and the strong force at even higher energies. These new theories are called **grand unified theories,** or **GUTs** (Figure 15-17).

Studies of GUTs suggest that the universe expanded and cooled until about 10^{-35} second after the big bang, when it became so cool that the forces of nature began to separate. This released tremendous energy, which suddenly inflated the universe by a factor between 10^{20} and 10^{30}. At that time the part of the universe that we can see now, the entire observable universe, was no larger than the volume of an atom, but it suddenly inflated to the volume of a cherry pit and then continued its slower expansion to its present extent.

That sudden inflation can solve the flatness problem and the horizon problem. The sudden inflation of the universe would have forced whatever curvature it

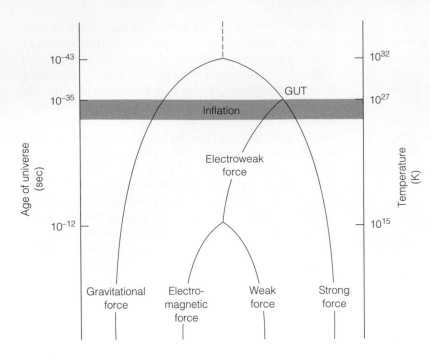

Figure 15-17
When the universe was very young and hot (top), the four forces of nature were indistinguishable. As the universe began to expand and cool, the forces separated and triggered a sudden inflation in the size of the universe.

had toward zero, just as inflating a balloon makes a small spot on its surface flatter. Thus, we now see the universe nearly flat because of that sudden inflation long ago. In addition, because the part of the universe we can see was once no larger in volume than an atom, it had plenty of time to equalize its temperature before the inflation occurred. Now we see the same temperature for the background radiation in all directions.

The inflationary theory predicts that the universe is almost perfectly flat. That is, the true density must equal the critical density. A theory can never be used as evidence, but the beauty of the inflationary theory has given many cosmologists confidence that the universe must be flat and that the dark matter must make up the difference between the matter we can see and the critical density.

The newest observations, however, may be telling us the universe is not flat, so we must now turn from theory to evidence. We begin by trying to find the age of the universe.

The Age of the Universe

In 1929, Edwin Hubble announced that the universe was expanding and reported the rate of expansion, the number now known as the Hubble constant (H). Within a few years, astronomers had proposed the idea that the universe had a violent beginning and had used H to extrapolate backward to find the age of the universe. But the value of H that Hubble reported, 530 km/sec/Mpc, led them to conclude that the universe was only half as old as Earth. Something was wrong.

In the decades that followed, astronomers figured out that Hubble's distances to galaxies were systemat-

ically too small, and thus his value of H was too large (Figure 16-11). Modern measurements of H yield an age of the universe that is at least twice the age of Earth, but controversy still rages.

The calculation of the age of the universe from its expansion is really very simple. We make such calculations whenever we take a drive and divide distance by speed. We say, "It is 100 miles to the city and I can average 50 miles an hour, so I can get there in 2 hours." We know the distance to a galaxy and its radial velocity, so we can divide distance by velocity to find out how long the expansion of the universe has taken to separate the galaxies to their present distance. That is the age of the universe.

The Hubble constant makes this calculation easy. We need to divide distance by velocity, and the Hubble constant is just velocity divided by distance. The reciprocal of the Hubble constant ($\frac{1}{H}$) must be proportional to the age of the universe. The only problem is the units. Astronomers always express H as a velocity in kilometers per second divided by a distance in megaparsecs. To make the division give us an age, we must convert megaparsecs to kilometers, and then the distances will cancel out and we will have an age in seconds. To get years, we must divide by the number of seconds in a year. If we make these simple changes in units, the age of the universe in years is approximately 10^{12} divided by H in its normal units, km/sec/Mpc.

$$T \approx \frac{1}{H} \times 10^{12} \text{ years}$$

This is known as the **Hubble time.** For example, if H is 70 km/sec/Mpc, then the age of the universe can be no older than $10^{12}/70$, which equals 14 billion years.

When we calculate the Hubble time, we are assuming the galaxies have receded at a constant velocity since the big bang, but we know that gravity slows the expansion of the universe. Then the galaxies must have traveled faster in the past, and the Hubble time is just an upper limit. The universe is no older than one Hubble time, but it could be younger. To know how much younger, we must know the density of the universe. If the density exactly equals the critical density (if the universe is flat), then the true age equals two-thirds of a Hubble time ($\frac{2}{3} \frac{1}{H}$). If the density of the universe is less than the critical density, the universe is older; and if the density is higher, the universe is younger.

Astronomers have measured the distances to galaxies in different ways and obtained different values for H. The lowest are about 50 km/sec/Mpc and the highest about 100 km/sec/Mpc. This generated a heated controversy, because a value of H higher than about 80 predicts an age for the universe that is younger that the age astronomers derive for the globular clusters. It is obviously impossible for objects to be older than the universe.

Two spacecraft have helped solve the problem. High-precision parallaxes from the Hipparcos satellite have recalibrated the H-R diagrams of the globular clusters, and astronomers now estimate that the globular clusters are about 11.5 billion years old, give or take a billion years. Hipparcos data have also allowed the recalibration of the period–luminosity diagram for Cepheid variable stars. This has slightly increased the distance to galaxies and decreased the observed value of H.

The Hubble Space Telescope has allowed astronomers to find more accurate distances to galaxies. Cepheids can now be detected in galaxies as far away as the Virgo cluster of galaxies. Also, astronomers have been able to calibrate the brightness of supernovae caused by the collapse of a white dwarf. These are called Type Ia supernovae. Because all of these supernovae have about the same maximum luminosity, they can be used to find the distances to galaxies and thus the value of H.

The most recent values of H fall between 60 and 70 km/sec/Mpc. This implies a Hubble time between 16 and 14 billion years, just large enough to permit the formation of the galaxies and the globular clusters.

The controversy over the Hubble constant seems to be easing as better observations are made, but a new controversy is brewing about the actual expansion of the universe.

The Cosmological Constant

When Einstein published his theory of general relativity in 1916, he recognized that his equations describing space and time implied that space had to expand or contract. Thinking such motion unreasonable, he introduced a constant called the **cosmological constant.** The constant plays the mathematical role of a force of repulsion that balances the gravitation in the universe so it does not have to expand or collapse. When, in 1929, Edwin Hubble announced that the universe was expanding, Einstein said introducing the cosmological constant was his biggest blunder. Modern astronomers aren't so sure.

The ages of the oldest star clusters and the age of the universe derived from H are uncomfortably close. Since 1929, astronomers have generally believed that the cosmological constant was zero and so has no effect, but the age conflict could be resolved if the cosmological constant was not zero. Then the repulsion represented by the cosmological constant would allow the universe to expand slower at first and gradually accelerate to its present rate. If this is true, then the universe could be older than implied by H.

Two international teams of astronomers working independently announced in 1998 that their data show that the expansion of the universe is accelerating, a result that requires that the cosmological constant be nonzero. Both teams studied Type Ia supernovae to calibrate them as distance indicators. They then used that type supernovae seen in very distant galaxies to measure the distance to the galaxies. The astronomers expected to measure the Hubble constant and detect the rate at which the expansion is slowed by gravity. Instead, both teams got the same amazing result. The expansion is accelerating.

If the expansion of the universe is being accelerated, then the cosmological constant tells us that empty space contains tremendous energy. Because energy and matter are equivalent, the energy of empty space acts as a mass, and that mass interests astronomers. Although the inflationary theory predicts that the universe must be exactly flat, astronomers find that normal matter makes up only a few percent of the matter needed to make the universe flat. Dark matter may make up roughly a quarter of the necessary matter, and the mass equivalent of the cosmological constant may make up the rest. Thus the universe may indeed be flat, although the expansion is accelerating.

For 80 years, Einstein's cosmological constant was out of fashion, but it seems to have returned dramatically. More observations and more theoretical work are needed to help us understand this new cosmology. Nevertheless, this is not the only excitement in cosmology. While some astronomers struggle to understand the behavior of space and time, others are trying to understand how the hot gas of the big bang made galaxies.

Figure 15-18
The distribution of brighter galaxies in the sky reveals the great Virgo Cluster (center), containing over 1000 galaxies only about 17 Mpc away. Other clusters fill the sky, such as the more distant Coma cluster just above the Virgo cluster in this diagram. The Virgo cluster is linked with others to form the Local Supercluster.

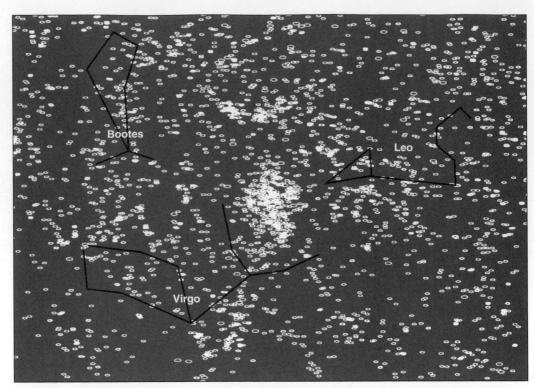

Figure 15-19
By measuring the red shifts and positions of thousands of galaxies, astronomers have been able to create this map of the location of galaxies in two slices of the universe looking north and south out of our Milky Way Galaxy. Filaments, walls, and voids are clearly evident, with the largest of the features, the Great Wall, stretching from left to right across the northern slice, while the Southern Wall reaches diagonally across the southern slice.
(Margaret J. Geller, John P. Huchra, Luis A. N. da Costa, and Emilio E. Falco, Smithsonian Astrophysical Observatory ©1994)

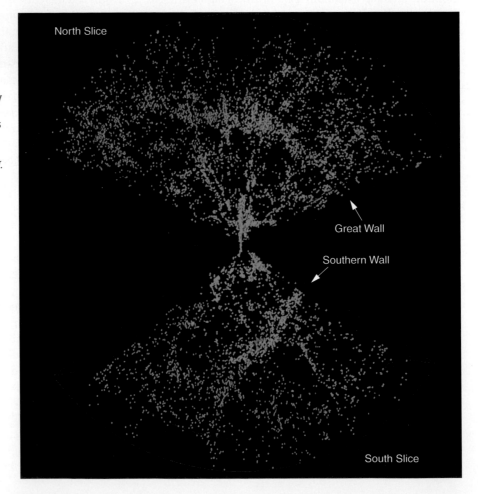

The Origin of Structure

When astronomers study the distribution of galaxies, they find them in clusters and in clusters of clusters called **superclusters** (Figure 15-18). The Local Supercluster, in which we live, is a roughly disk-shaped swarm of galaxy clusters 50 to 75 Mpc in diameter. Furthermore, the superclusters are distributed in long, narrow filaments and thin walls that outline great voids that are nearly empty of galaxies (Figure 15-19). Astronomers refer to all of this as structure, and they wonder how the galaxies, clusters, superclusters, filaments, walls, and voids formed from the hot gas of the big bang.

This structure is a problem for cosmologists because the primordial background radiation is extremely uniform, and thus the gas of the big bang must have been extremely uniform at the time of recombination. If the universe was so uniform when it was young, how did it get to be so lumpy now?

The problem of the origin of structure has developed slowly since the 1960s, when the background radiation was first discovered. Then, the most distant galaxies visible had red shifts no greater than 1. These galaxies had look-back times no greater than 4 billion years, so it was not difficult to imagine how the uniform gas of the big bang could eventually form galaxies. That is, there seemed to be plenty of time between recombination and the origin of galaxies.

Through the 1970s and 1980s, however, that age of mystery narrowed. Astronomers recognized quasars as the active nuclei of galaxies much farther away than 4 billion light years. Brilliant starburst galaxies have been found at great distances. At present, the most distant galaxies have look-back times that are 93 percent the age of the universe. Thus, the age of mystery when the great clouds of galaxies formed must be less than 7 percent of the age of the universe.

How did all of this structure originate? The universe had to be very uniform during recombination and then clump quickly to form the structure we see. Cosmologists now work on a two-pronged solution: dark matter provides the gravitation, and galaxy seeds provide nuclei around which structure grows.

We have seen clear evidence that dark matter must exist, and cosmologists are now trying to build models that show how the gravitation of dark matter could cause the matter of the early universe to clump quickly into large structures. One kind of dark matter is called hot dark matter. If it exists, it would consist of particles that travel at nearly the speed of light, such as neutrinos with nonzero masses. Cold dark matter, on the other hand, would consist of slow-moving particles such as WIMPs, axions, and similar as-yet-undetected particles. Cold dark matter, because its particles travel more slowly, seems better able to clump together quickly (Figure 15-20).

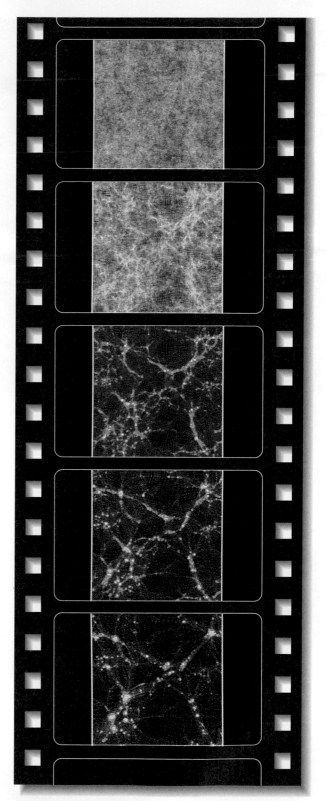

Figure 15-20

A computer simulation of the evolution of a small region of a universe containing dark matter shows small irregularities growing into filaments, walls, and voids. The area visible in each frame remains the same as the universe expands. Galaxy formation presumably occurs at about the third frame. Although this model spans the history of the universe from soon after the big bang to the present, the time steps between frames are not uniform. The first two frames show the young universe, and the last three frames show more recent times. *(Courtesy James Gelb and Edmund Bertschinger, MIT)*

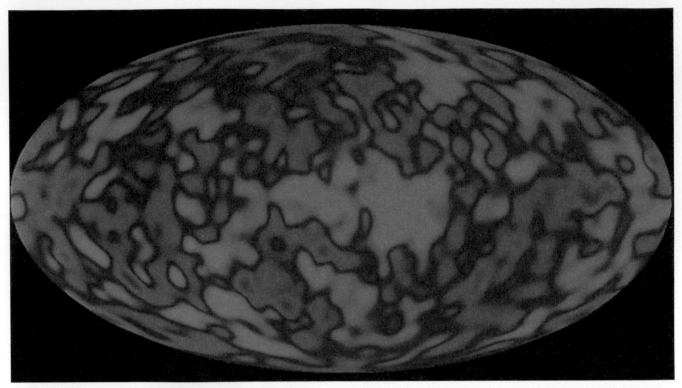

Figure 15-21
Over 70 million measurements by the COBE satellite went into this all-sky map of the primordial background radiation. Patchy areas are irregularities in the background radiation that suggest the gas of the big bang was not perfectly uniform. The gravitational influence of dark matter could have forced these denser regions to grow into the clouds of galaxies we see today. *(NASA/Goddard Space Flight Center)*

Something must trigger the clumping of matter to make a galaxy, and these clumping centers have been called **galaxy seeds.** Two theories have been proposed to account for them. In the inflationary model of the universe, the sudden inflation could have magnified tiny quantum fluctuations into huge density differences that could have attracted matter to begin forming galaxies.

The second theory proposes that the galaxy seeds are defects in space–time that develop as the forces of nature separate from one another in what cosmologists call a phase transition. We might compare such defects with the flaws we see in ice cubes. When the water freezes, beginning at the outer edges, different parts of the ice cube have different crystal orientations; and as the water freezes inward, these different regions meet and form defects that we see inside the ice cube as lines and sheets. Defects in space–time would possess strong gravitational fields and could act as galaxy seeds around which dark matter clumps to create galaxies, clusters, filaments, walls—all of the structure we see.

One of the theories can be tested against observation. In 1992, the Cosmic Backgroud Explorer (COBE)

satellite carefully mapped the background radiation over the entire sky and detected tiny fluctuations from place to place amounting to only 0.0006 percent (Figure 15-21). Presumably these irregularities in the early universe grew to become the structure we see. Theoretical studies show that the observed fluctuations cannot give rise to the observed structure if the original galaxy seeds were space–time defects. This dramatic result seems to spell the end for the defect theory. Cosmologists have not yet found a way to test the alternative theory.

The structure we see today in the great clouds of galaxies had its beginning in the first moments of the big bang. How the galaxies grew, merged, and evolved is being revealed by new photos (Figure 13-21), but the origin of the structure itself remains a challenging puzzle.

The four subjects we have reviewed here—the inflationary universe, the age of the universe, the cosmological constant, and the origin of structure—in no way invalidate the big bang theory. They are refinements that extend our understanding of how our universe began and has evolved from a first moment we call the big bang.

REVIEW Critical Inquiry

Why do we have to understand dark matter to understand the origin of galaxies?

The big bang created a universe filled with gas too hot to collect and form galaxies, so astronomers believe the galaxies began to form roughly a billion years after the beginning when the expansion of the universe had cooled the gas enough. Somehow the gas must have collected to form galaxies, and inflation could have created seeds around which gas collected to form galaxies in clusters, superclusters, filaments, and walls. Just the visible matter in the universe, however, does not appear to have enough gravity to have pulled itself together quickly, so the added gravity of the dark matter may have been important in aiding the first contractions. Once galaxies and galaxy clusters formed, the presence of dark matter stabilized them as the galaxies continued to collide, merge, and evolve. One of the unknown factors in this story is the age of the universe. The current best estimates of the Hubble constant seem to make the universe too young for the matter to have contracted to form the structure we see.

Cosmology is definitely a work in progress; but astronomers have made dramatic progress in understanding the history of the universe, and they have done it by comparing observation with theory. As an example, how would you compare observation with the inflationary theory to explain the horizon problem?

Although we have traced the origin of the universe, the origin of the elements, and the birth and death of stars, we have left out one important class of objects—planets. How do Earth and the other planets fit into this grand scheme of origins? That is the subject of the next unit.

Summary

The fact that the night sky is dark shows that the universe is not infinitely old. If it were infinite in extent and age, then every spot on the sky would glow as brightly as the surface of a star. This problem, known incorrectly as Olbers' paradox, illustrates how important assumptions are in cosmology.

The basic assumptions of cosmology are homogeneity, isotropy, and universality. Homogeneity says that matter is spread uniformly through the universe. Isotropy says the universe looks the same in any direction. Both deal only with general features. Universality assumes that the laws of physics known on Earth apply everywhere. In addition, the cosmological principle asserts that the universe looks the same from any location.

The Hubble law implies that the universe is expanding. Tracing this expansion backward in time, we come to an initial high-density state commonly called the big bang. From the Hubble constant, we can conclude that the expansion began 10 to 15 billion years ago.

The universe can be infinite or finite, but it cannot have an edge or a center. Such regions would violate the cosmological principle. Instead, we assume that the universe occupies curved space–time and thus could be finite but unbounded. Depending on the average density of the universe, it could be open, flat, or closed. Measurements of the density of the universe are uncertain because of the presence of dark matter, which we cannot detect easily.

During the first few minutes of the big bang, about 25 percent of the matter became helium, and the rest remained hydrogen. Very few heavy atoms were made. As the matter expanded, instabilities caused the formation of clusters of galaxies, which are still receding from each other.

Whether the universe expands forever or slows to a stop and begins to contract depends on the amount of matter in the universe. If the average density is greater than the critical density of 4×10^{-30} g/cm^3, it will provide enough gravity to slow the expansion and force the universe to collapse. The collapse will smash all matter back to a high density from which a new big bang may emerge. Such a universe is termed closed because the curvature of space is positive. If the average density is less than 4×10^{-30} g/cm^3, gravity will be unable to stop the expansion, and it will continue forever. Such a universe is termed open because the space curvature is negative.

The inflationary universe theory combines quantum mechanics, general relativity, and cosmology. It predicts that the universe underwent a sudden inflation when it was only 10^{-35} seconds old. This seems to explain why the universe is so flat and why the primordial background radiation is so isotropic.

We can estimate an upper limit to the age of the universe from the Hubble constant, but we need to know the average density of the universe to find a true age. Also, the value of the Hubble constant is not well known. It is believed to be between 60 and 70 km/sec/Mpc. Such a value of H makes the universe just slightly older than the oldest globular clusters. This has led some astronomers to reconsider the cosmological constant, a constant in Einstein's theory of relativity that represents a repulsive force. Observations of Type Ia supernovae in distant galaxies have revealed that the expansion of the universe is accelerating, and that requires that the cosmological constant be nonzero. If so, the universe may be older than implied by the value of H.

The primordial background radiation is highly isotropic, but small irregularities in the radiation imply that the gas of the big bang was not perfectly uniform. These irregularities may be tiny quantum mechanical fluctuations magnified by inflation. Theorists believe that such galaxy seeds, aided by the gravitational influence of dark matter, could grow quickly after the big bang into the structure we see in the distribution of the galaxy.

New Terms

cosmology

Olbers' paradox

observable universe

homogeneity

isotropy

universality

cosmological principle

closed universe

flat universe

open universe

big bang

primordial background radiation

steady state theory

antimatter

recombination

critical density

oscillating universe theory

MACHOs

flatness problem

horizon problem

inflationary universe

grand unified theories (GUTs)

Hubble time

cosmological constant

superclusters

galaxy seeds

Review Questions

1. Would the night sky be dark if the universe were only 1 billion years old and was contracting instead of expanding? Explain your answer.

2. How is the cosmological principle related to the Copernican principle?

3. How can we be located at the center of the observable universe if we say the universe has no center?

4. Why can't an open universe have a center? Why can't a closed universe have a center?

5. What evidence do we have that there was a big bang?

6. Why do astronomers say the hot gases of the big bang have a temperature of 2.7 K?

7. Why couldn't atomic nuclei exist when the universe was younger than 3 minutes?

8. Why do we conclude that elements heavier than helium were made in generations of stars and not during the big bang?

9. Why is it difficult to determine the present density of the universe?

10. Why does the age of the universe depend on the density?

11. How does the inflationary universe theory resolve the flatness problem? the horizon problem?

12. If the Hubble constant was really 100 km/sec/Mpc, then much of what we understand about the evolution of stars and star clusters would have to be wrong. Explain why.

13. Why do we conclude that the universe must have been very uniform during its first million years?

14. How is the presence of structure in the early universe a problem for cosmologists?

15. What is the difference between hot dark matter and cold dark matter? What difference does it make to cosmology?

Discussion Questions

1. Do you think Copernicus would have accepted the cosmological principle? Why or why not?

2. Suppose a civilization lives in a galaxy that we see as a very distant quasar. What would the universe look like to them?

Problems

1. Use the data in Figure 15-3 to plot a velocity–distance diagram, find H, and determine the approximate age of the universe.

2. If a galaxy is 8 Mpc away from us and recedes at 456 km/sec, how old is the universe, assuming that gravity is not slowing the expansion? How old is the universe if it is flat?

3. If the temperature of the big bang had been 10^6 K at the time of recombination, what maximum wavelength would the primordial background radiation have as seen from Earth?

4. If the average distance between galaxies is 2 Mpc and the average mass of a galaxy is 10^{11} solar masses, what is the average density of the universe? (*Hints:* The volume of a sphere is $\frac{4}{3}\pi r^3$. The mass of the sun is 2×10^{33} g.)

5. Figure 15-14 is based on an assumed Hubble constant of 50 km/sec/Mpc. How would you change the diagram to fit a Hubble constant of 70 km/sec/Mpc?

6. Hubble's first estimate of the Hubble constant was 530 km/sec/Mpc. If his distances were too small by a factor of 7, what answer should he have obtained?

7. What is the maximum age of the universe predicted by Hubble's first estimate of the Hubble constant?

8. If the value of the Hubble constant were found to be 60 km/sec/Mpc, how old would the universe be if it were not slowed by gravity? if it were flat?

Critical Inquiries for the Web

1. Will the universe go on expanding forever? Search the Web for information on recent investigations that shed light on the question of the density of matter in the uni-

verse. What predictions do these studies make about the fate of the universe? What kinds of observations were necessary to make these predictions?

2. The steady state theory was once a rival cosmology of the big bang. Search for Web sites that provide information on steady state cosmology. (Be careful to locate legitimate sites that discuss the theory, rather than sites where individuals use steady state ideas as part of non-scientific arguments on cosmology.) What were the key predictions of steady state cosmology? How has recent evidence led to its decline?

Go to the Brooks/Cole Astronomy Resource Center (www.brookscole.com/astronomy) for critical thinking exercises, articles, and additional readings from InfoTrac College Edition, Brooks/Cole's online student library.

CHAPTER 16

The Origin of the Solar System

What place is this?

Where are we now?

Carl Sandburg
Grass

Guidepost

The preceding 15 chapters have described the appearance, origin, structure, and evolution of the physical universe, but they have largely neglected one important class of objects—planets. In this chapter, we can look back on what we have learned and find our place in the universe. We live on a planet. What does that mean? Where do we fit?

Each time we have studied a new object, we have asked how it formed and how it evolved to its present state. We have done that with stars and galaxies and the universe, so it is appropriate to begin our discussion of the solar system by considering its origin.

Another reason for discussing the origin of the solar system here is to give ourselves a framework into which we can fit the planets as we discuss them in detail in the chapters that follow. Without a theoretical framework, science is nothing but a jumble of facts. With a good framework in hand, we will be ready to make sense of the solar system through the next four chapters.

Microscopic creatures live in the roots of your eyelashes. Don't worry. Everyone has them, and they are harmless.* They hatch, fight for survival, mate, lay eggs, and die in the tiny spaces around the roots of our eyelashes without doing us any harm. Some live in renowned places—the eyelashes of a glamorous movie star—but the tiny beasts are not self-aware; they never stop to say, "Where are we?" Humans are more intelligent; we have the ability and the responsibility to wonder where we are in the universe and how we came here.

In this chapter, we begin exploring our solar system: the sun and its family of planets. We must avoid, however, the great pitfall of science. We must avoid memorizing facts instad of understanding nature. Jupiter is 317.83 times more massive than Earth. That is a fact and it is boring. The excitement of science comes from understanding nature. We want to understand how the sun and planets formed, and that will tell us *why* Jupiter is so massive.

Facts fit together to build understanding, so we begin our study of the solar system by developing a unifying hypothesis for the origin of the sun and planets. As we learn facts about the solar system we can use them to test and improve our hypothesis, and that will increase our understanding of nature. A unifying hypothesis is like a basket into which we can put facts. The facts are easier to remember because they make sense, and the combination of hypothesis and supporting facts represents real knowledge—real understanding—and not just memory.

We have no choice; we must understand our solar system out of self-preservation. We live on a planet with 5 billion other humans, and we are altering our world's environment in unplanned and unknown ways. We could make Earth uninhabitable. The deadly deserts of Mars and sulfuric acid fogs of Venus may help us understand our own more comfortable world.

We must also study planets as astronomers. It seems likely that roughly half of all stars have planetary systems, so there are more planets in the universe than stars. Also, the chemical reactions that give rise to life seem to need the moderate conditions of the surface of planets. To search for life in the universe, we must understand planets.

Above all, we study the solar system because it is our home in the universe. Because we are an intelligent species, we have the right and the responsibility to wonder what we are. Our kind have inhabited this solar system for at least a million years, but only within our lifetime have we begun to understand what a solar system is. Like sleeping passengers on a train, we waken, look out at the passing scenery, and mutter: "What place is this? Where are we now?"

Demodex folliculorum has been found in 97 percent of individuals and is a characteristic of healthy skin.

16-1 The Great Chain of Origins

We are linked through a great chain of beginnings that leads backward through time to the first instant when the universe began 10 to 20 billion years ago. The gradual discovery of the links in that chain is one of the most exciting adventures of the human intellect. In earlier chapters, we discussed some of that story: the origin of the universe in the big bang, the formation of galaxies, the origin of stars, and the growth of the chemical elements. Here we will explore further to consider the formation of planets.

A Review of the Origin of Matter

Look at your thumb. The matter in your thumb came into existence within minutes of the beginning of the universe. Astronomers have strong evidence that the universe began in an event called the big bang (Chapter 15), and by the time the universe was 3 minutes old, the protons, neutrons, and electrons in your thumb had come into existence. You are made of very old matter.

Although those particles formed quickly, they were not linked together as they are today. Most of the matter was hydrogen and about 25 percent was helium. Very few heavier atoms were made in the big bang. Although helium is very rare in our bodies, your thumb contains many of those ancient hydrogen atoms unchanged since the universe began.

Within a billion years or so after the big bang, matter began to collect to form galaxies containing billions of stars. Astronomers understand that nuclear reactions inside stars combine low-mass atoms such as hydrogen to make heavier atoms (Chapter 9). Generation after generation of stars cooked the original particles, linking them together to build atoms such as carbon, nitrogen, and oxygen (Chapter 10). Look at your thumb. Even the calcium in the bone and the iron in the blood was assembled inside stars.

The atoms heavier than iron were created by rapid nuclear reactions that can only occur when a massive star explodes (Chapter 10). Gold and silver are rare in our bodies, but iodine is critical in your thyroid gland, and there are no doubt a few of those iodine atoms circulating through your thumb at this very moment thanks to the violent deaths of massive stars.

Our galaxy contains about 100 billion stars of which our sun is one. It formed from a cloud of gas and dust about 5 billion years ago, and the atoms in your thumb were part of that cloud. How the sun took shape, how the cloud gave birth to the planets, how those atoms in your thumb found their way onto Earth and into you is the story of this chapter. As we expore the origin of our solar system, keep in mind the great

Many theories in science can be classified as either evolutionary, in that they involve gradual processes, or catastrophic, in that they depend on specific, unlikely events. Scientists have generally preferred evolutionary theories. Nevertheless, catastrophic events do occur.

Something in people prefers catastrophic theories, perhaps because we like to see spectacular violence from a safe distance, which may explain the success of movies that include lots of car crashes and explosions. Also, cataclysmic theories resonate with Old Testament accounts of catastrophic events and special acts of creation. Thus, we have an understandable interest in catastrophic theories.

Nevertheless, most scientific theories are evolutionary. Such theories do not depend on unlikely events or special acts of biblical creation, and thus they are open to scientific investigation. For example, geologists much prefer theories of mountain building that are evolutionary, with the mountains being pushed up slowly as centuries pass. All the evidence of erosion and the folding of rock layers show that the process is gradual. Because most such natural processes are evolutionary, scientists sometimes find it difficult to accept any theory that depends on catastrophic events.

We will see in this and later chapters that catastrophes do occur. Earth, for example, is bombarded by debris from space, and some of those impacts are very large. As you study astronomy or any other natural science, notice that most theories are evolutionary but that we must allow for the possibility of unpredictable catastrophic events.

---

chain of origins that created the atoms. As the geologist Preston Cloud remarked, "Stars have died that we might live."

The Origin of Planets

Over roughly the last two centuries, astronomers have proposed two kinds of theories for the origin of the planets. Catastrophic theories proposed that the planets formed from some improbable cataclysm such as the collision of the sun and another star. Gradualistic theories proposed that the planets formed naturally with the sun. Since about the time of World War II, evidence has accumulated to support the gradualistic theories. In fact, nearly all astronomers now believe that planets form naturally as a by-product of star formation (Window on Science 16-1).

In earlier chapters, we saw how stars can form from the gravitational contraction of gas and dust clouds. In at least some cases, that contraction must be triggered by compression of the gas cloud, perhaps by a nearby supernova explosion.

As stars form in these contracting clouds, they remain surrounded by cocoons of dust and gas, and the rotation of the cloud causes that dust and gas to form a spinning disk around the protostar. When the center of the star grows hot enough to ignite nuclear reactions, its surface quickly heats up, becomes more luminous, and blows away the gas and dust cocoon.

We have clear evidence that disks of gas and dust around young stars are common. Infrared observations of T Tauri stars, for instance, show that some are surrounded by gas clouds rich in dust, and spectra show that these stars are blowing away their nebulae at speeds up to 200 km/sec. The presence of bipolar

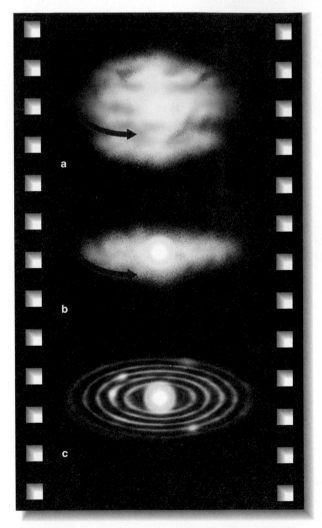

Figure 16-1
(a) Because the solar nebula was rotating, (b) it contracted into a disk, and (c) the planets formed with orbits lying in nearly the same plane.

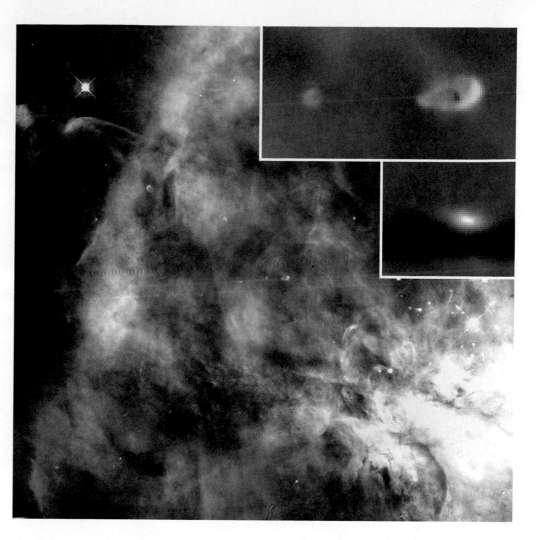

Figure 16-2
This Hubble Space Telescope image of the central part of the Orion Nebula reveals that many young stars there are surrounded by disks of gas and dust (insets). Most of the disks glow brightly because they are illuminated by the hottest stars in the nebula, but a few, such as the dark disk seen nearly edge-on (lower inset), are silhouetted against the bright nebula. Note that the central star in the silhouetted nebula is visible by light scattered by dust above and below the nebula. This particular disk is the largest detected, with a diameter about 17 times that of our solar system. *(C. R. O'Dell, Rice University, NASA; lower inset, Mark Mc-Caughrean, Max-Planck Institute for Astronomy, C. R. O'Dell, and NASA)*

flows from young stars confirms that such systems contain gas and dust distributed in a disk-shaped cloud (see Chapter 9).

Our own planetary system probably formed in such a disk-shaped cloud around the sun. When the sun became luminous enough, the remaining gas and dust were blown away into space, leaving the planets orbiting the sun. This is known as the **solar nebula theory** because the planets form from the nebula around the proto-sun (Figure 16-1). If the theory is right, then planets do form as a by-product of star formation, and thus most stars should have planetary systems.

Planets Orbiting Other Stars

If the solar nebula theory is correct, then planets should be very common, and we could test the theory by looking for planets orbiting other stars, objects astronomers call **extra-solar planets.** Unfortunately, detecting a planet orbiting another star is a bit like detecting a bug crawling on the bulb in a distant searchlight. Nevertheless, astronomers have found evidence that other planetary systems exist.

Both visual and radio observations have found dense disks orbiting young stars. For example, at least 50 percent of the stars in the Orion nebula are encircled by dense disks of gas and dust (Figure 16-2). Radio astronomers have found dense disks of gas and dust around stars like HL Tauri (itself a T Tauri star). These appear to be the kind of disks in which planets form.

Infrared observations have identified very cold, very low-density disks of dust around stars such as Beta Pictoris. The Beta Pictoris disk is about 100 times the diameter of our solar system and has an inner hole around the star that may be a place where planets have formed (Figure 16-3). These tenuous disks of cold dust are believed to be debris released from comets or produced by collisions among small bodies such as asteroids and comets. The outer part of our own solar system contains this kind of cold dust. Thus the cold dust disks are not evidence of active planet formation but are evidence that planetary systems have already formed.

In the middle 1990s, astronomers obtained further evidence of extra-solar planets. Although such a planet is too faint to be seen directly, it can be detected

Beta Pictoris

Size of Pluto's orbit

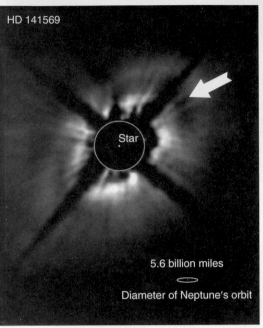

HD 141569

Star

5.6 billion miles

Diameter of Neptune's orbit

HR 4796A

Star

5.6 billion miles

Diameter of Neptune's orbit

Figure 16-3
The star Beta Pictoris (top) is hidden behind the black bar at center, and the dust disk around it is seen edge-on extending to right and left. *(Al Schultz, CSC/STScI, Sally Heap, GSFC, and NASA)* In the lower two images the stars are hidden behind the round masks with dust disks extending outward. The gap in the ring around HD141569 (arrow) may have been cleared by a planet. The narrowness of the ring around HR4796A suggests it is confined by planets. *(Alycia Weinberger, Eric Becklin, UCLA, Glenn Schneider, University of Arizona, and NASA)*

from the motion of the star it orbits. The first planet detected this way orbits the star 51 Pegasi. As the star and the planet orbit each other, they revolve around the center of mass of the system. The planet, having the lower mass, does most of the moving, but the star does move slightly around the center of mass, and that motion is detectable as Doppler shifts in the star's spectrum (Figure 16-4). From the motion of the star and estimates of its mass, astronomers can deduce that the planet has half the mass of Jupiter and orbits only 0.05 AU from the star. Presumably the star has other planets at greater distances, but they are harder to detect.

Astronomers greeted the announcement of a planet orbiting 51 Pegasi with cheers and skepticism (Window on Science 16-2), but more such planets were found. Roughly two dozen are known, including at least three orbiting the star Upsilon Andromidae—the first evidence astronomers have of a planetary system.

The planets discovered so far tend to be massive and have short periods because lower-mass planets or planets with larger orbits and longer periods are harder to detect. Nevertheless, the discovery of planets orbiting other stars gives astronomers added confidence in the solar nebula theory. The theory predicts

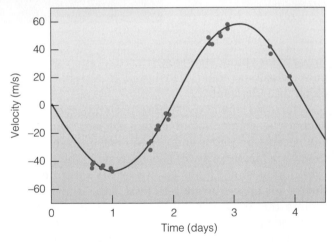

Figure 16-4
The velocity of the sunlike star 51 Pegasi can be deduced from its spectrum. The star moves with a period of 4.2 days and an amplitude of only 50 meters per second. This wobble in the star is apparently produced by a planet, the first planet found to orbit a normal star other than our sun. *(Adapted from data by Geoff Marcy and Paul Butler)*

"Scientists are just a bunch of skeptics who don't believe in anything." That is a common complaint about scientists, but it misinterprets the fundamental characteristic of the scientist. Yes, scientists treat new discoveries with skepticism, but scientists do hold strong beliefs about how nature works, and their skepticism is the tool they use to test their beliefs. Ultimately, scientists try to tell stories about how nature works, and if those stories are to be correct, then every idea must be tested over and over.

When the discovery of a planet orbiting the star 51 Pegasi was first announced, astronomers were skeptical—not because they didn't expect other stars to have planets and not because they thought the observations were wrong, but because that is the way science works.

Every observation is tested, and every discovery must be confirmed. Only an idea that survives many tests begins to be accepted as a scientific truth, what a scientist would call a law of nature.

This makes scientists seem like irritable skeptics to nonscientists, but among scientists it is not bad manners to ask, "Really, how do you know that?" or "Why do you think that?" or "What evidence do you have?"

The goal of science is to tell stories about how nature works, stories that are sometimes called theories or hypotheses. A story by John Steinbeck can be a brilliant work of art invented to help us understand some aspect of human nature, and another story, such as a TV script, can be little more than entertaining chewing gum for the mind. But a scientist tries to

tell a story that is different in one critical way—it is the truth. To create such stories about nature, scientists cannot be artistic and create facts to fit their story, and they cannot invent a plot just to be entertaining. Every single link in the scientist's story must be based on evidence and clear, logical steps. Of course, evidence can be misunderstood, and logical slipups happen all too often. To test every aspect of the story, the scientist must be continuously skeptical. That is the only way to discover those scientific stories that help us understand how nature works.

Skepticism is not a refusal to hold beliefs. Scientists often believe sincerely in their theories and hypotheses once they have been tested over and over. Rather, skepticism is the tool scientists use to find those natural principles worthy of belief.

---

that planets are common, and astronomers are finding them orbiting many stars.

REVIEW Critical Inquiry

Why does the solar nebula theory imply planets are common?

Often, the implications of a theory are more important in its analysis than its own conjecture about nature. If the solar nebula theory is correct, the planets of our solar system formed from the disk of gas and dust that surrounded the sun as it condensed from the interstellar medium. If that is true, then it should be a common process. Most stars should form with disks of gas and dust around them, and so we would expect them to have planets.

Without knowing any more about the theory, the implication that planets are common suggests a way we could test the theory. If we look at a large number of stars and find lots of planets, the solar nebula theory is confirmed. But if we search and find few planets, it is contradicted. What observational evidence do we have so far on this issue?

So far only a few other stars are known to have planets. We can expect more such worlds to be discovered, and that raises the question of just how planets form. To answer that question, we begin by surveying our own solar system to discover the distinguish-

ing characteristics of planetary systems. Those are the characteristics that our theory must eventually explain.

16-2 A Survey of the Solar System

To test our theories, we must search the present solar system for evidence of its past. In this section, we survey the solar system and compile a list of its most significant characteristics, potential clues to how it formed.

We begin with the most general view of the solar system. It is, in fact, almost entirely empty space (Figure 16-5). Imagine that we reduce the solar system until Earth is the size of a grain of table salt, about 0.3 mm (0.01 in.) in diameter. The moon is a speck of pepper about 1 cm (0.4 in.) away, and the sun is the size of a small plum 4 m (13 ft) from Earth. Mercury, Venus, and Mars are grains of salt. Jupiter is an apple seed 20 m (66 ft) from the sun, and Saturn is a smaller seed over 36 m (120 ft) away. Uranus and Neptune are slightly larger than average salt grains, and Pluto, the farthest planet (on average), is a speck of pepper over 150 m (500 ft) from the central plum. We would need a powerful microscope to detect the asteroids. These tiny specks of matter scattered around the sun are the remains of the solar nebula.

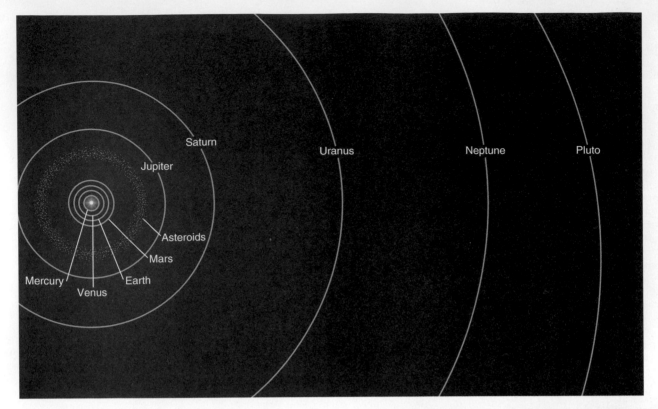

Figure 16-5
The solar system. The radii of the orbits are drawn to scale here. Note the eccentricity of Pluto's orbit. At this scale, only the sun would be visible. The planets are too small to be seen. Asteroids are discussed later in the chapter.

Figure 16-6 ▶
The relative sizes of the planets compared with the disk of the sun illustrate the two kinds of planets. The four terrestrial planets are small, and the Jovian planets are large. Pluto is not included in the classification scheme.

Revolution and Rotation[*]

The planets revolve around the sun in orbits that lie close to a common plane. The orbit of Mercury, the planet closest to the sun, is tipped 7° to Earth's orbit, and Pluto's orbit is tipped 17.2°. The rest of the planets' orbital planes are inclined by no more than 3.4°. Thus, the solar system is basically disk-shaped.

The rotation of the sun and planets on their axes also seems related to this disk shape. The sun rotates with its equator inclined only 7.25° to Earth's orbit, and most of the other planets' equators are tipped less than 30°. The rotations of Venus, Uranus, and Pluto are peculiar, however. Venus rotates backward compared with the other planets, and both Uranus and Pluto rotate on their sides (with their equators almost

[*]Recall from Chapter 2 that English distinguishes between the words *revolve* and *rotate*. A planet revolves around the sun but rotates on its axis.

Figure 16-7
Mercury is about 40 percent larger in diameter than Earth's moon, shown here to the same scale. Like the moon, Mercury lacks an atmosphere and is heavily cratered by the impact of large meteorites. *(Mercury, NASA, moon, Lick Observatory photograph)*

perpendicular to their orbits). We will discuss these planets in detail in Chapters 17 and 18, but later in this chapter we will try to understand how they could have acquired their peculiar rotations.

Apparently, the preferred direction of motion in the solar system—counterclockwise as seen from the north—is also related to its disk shape. All the planets revolve counterclockwise around the sun, and with the exception of Venus, Uranus, and Pluto they rotate counterclockwise on their axes. Thus, with only a few exceptions, most of which are understood, revolution and rotation in the solar system follow a disk theme.

Two Kinds of Planets

Perhaps the most important clue we have to the origin of the solar system comes from the division of the planets into two categories: terrestrial, or Earthlike, and Jovian, or Jupiterlike (Figure 16-6). The **terrestrial planets** are small, dense, rocky worlds with less atmosphere than the Jovian planets. The terrestrial planets—Mercury, Venus, Earth, and Mars—lie in the inner solar system. By contrast, the **Jovian plan-**

ets are large, gaseous, low-density worlds. The Jovian planets—Jupiter, Saturn, Uranus, and Neptune—lie in the outer solar system beyond the asteroids. Note that Pluto does not fit either category very well. It is small like the terrestrial planets but lies far from the sun and has a low density like the Jovian planets.

Although *terrestrial* means "Earthlike," Mercury is more like a moon. It is a small world, only 40 percent larger than Earth's moon, and thus it has held no atmosphere. Its gray, rocky surface is covered with thousands of overlapping craters, which, like the craters on Earth's moon, are the scars left behind by the impact of large meteorites (Figure 16-7). Such impact craters are very common in the solar system.

Venus is nearly as large as Earth, and its atmosphere is thick with heavy clouds that hide its surface (Figure 16-8). Radio signals can penetrate these clouds and bounce back from its rocky surface to give us a radar image of its terrain. These radar maps reveal that, although Venus is nearly the same size as Earth, it totally lacks Earth's oceans of water. Venus is a deadly, dry, desert world marked by impact craters, volcanoes, lava flows, and geological faults (Figure 16-9). In contrast, the atmosphere on Mars is thin, and we can see

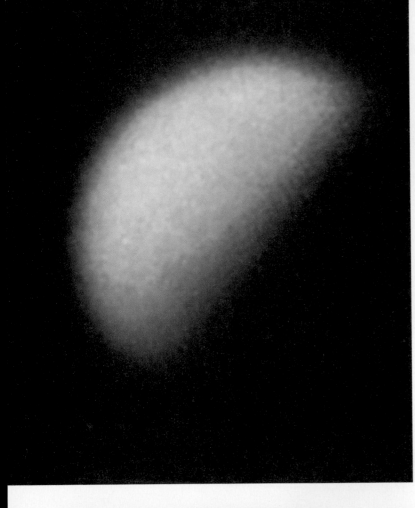

its surface clearly from space probes. These photos show that Mars, about half the diameter of Earth, is marked by impact craters and volcanoes.

Impact craters seem typical of all but the youngest surfaces in our solar system. The terrestrial planets, including Earth, are marked by such craters, and spacecraft have found craters on the moons of Mars, Jupiter, Saturn, Uranus, and Neptune. (There are no craters on the planets Jupiter, Saturn, Uranus, or Neptune because, as we will see, those planets do not have solid surfaces.) This suggests that craters are a characteristic of every object in the solar system with a surface capable of retaining such features.

The terrestrial planets have densities ranging from about 3 to 5 g/cm³. Typically, they contain cores of iron and nickel surrounded by a mantle of dense rock. In contrast, the Jovian planets are rich in hydrogen and helium, so their average density is low—less than 1.75 g/cm³. Saturn's very low density, 0.7 g/cm³, is less than the density of water, so the planet would float if we could find a bathtub big enough. Nevertheless, all the Jovian planets are massive. Jupiter is over

Figure 16-8
Venus is covered by a thick layer of clouds that never allows us to see its surface. In this photo taken through an Earth-based telescope, the planet is visible at quarter phase. *(NASA)*

Figure 16-9
Earth, Venus, and Mars to scale. Earth (left) is mostly a water planet with large masses of exposed rocky surface and an atmosphere of changing cloud patterns. The surface of Venus is perpetually hidden below thick clouds, so we must depend on radar images to reveal its surface (right), a dry desert marked by impact craters, volcanism, lava flows, and faults. Mars (center) is a smaller world with impact craters and old volcanoes on its rocky surface. *(NASA)*

318 times as massive as Earth, and Saturn is about 95 Earth masses. Uranus and Neptune are smaller—15 and 17 Earth masses, respectively.

Although photographs of the Jovian planets reveal swirling cloud patterns (Figure 16-10), their atmospheres are not very deep compared to their radii. In fact, if these so-called gas giants were shrunk to a few centimeters in diameter, their gaseous hydrogen and helium atmospheres would be no deeper than the fuzz on a badly worn tennis ball.

Mathematical models predict that the interiors of Jupiter and Saturn are mostly hydrogen in two different states (Figure 16-11). Beneath the cloud belts of these planets are layers where pressure forces the hydrogen into a liquid state. Even deeper, the pressure is so high that the hydrogen atoms can no longer hold their electrons. With the electrons free to move about, the material becomes an excellent electrical conductor and is known as **liquid metallic hydrogen.** The centers of Jupiter and Saturn are believed to be hot cores

Jupiter

Great Red Spot

Uranus

Neptune

Saturn

Figure 16-10
Of the Jovian planets (reproduced here to scale), Jupiter and Saturn are the largest. Jupiter is marked by dramatic cloud belts and the Great Red Spot, while Saturn has a bright ring system. Uranus looks featureless, but faint clouds are visible in computer-enhanced photos. The dark blue color of Neptune and the lighter blue of Uranus are caused by atmospheric methane, which absorbs red photons. (*Adapted from NASA photographs*)

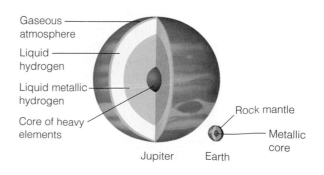

Gaseous atmosphere

Liquid hydrogen

Liquid metallic hydrogen

Core of heavy elements

Rock mantle

Metallic core

Jupiter   Earth

Figure 16-11 ▶
Cross sections of Earth and Jupiter illustrate the differences between the interiors of terrestrial and Jovian planets. Earth contains a metallic core with a rock mantle. Jupiter, 11.18 times larger in diameter, is composed mostly of liquid hydrogen. A core composed of heavy elements such as iron, nickel, and silicon lies at the center.

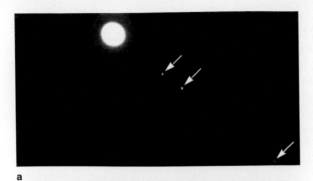

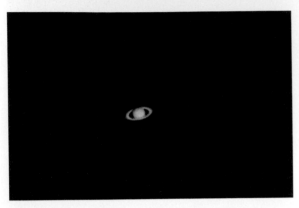

a

b

Figure 16-12
Jupiter and Saturn through a small telescope. (a) Jupiter has at least 16 satellites, but only the four Galilean moons are visible in small telescopes. In this photo, the planet was overexposed to show three of the moons. (b) The icy rings of Saturn are so reflective that they are visible through even a small telescope. *(Grundy Observatory)*

of heavy elements slightly larger than Earth. These are often erroneously called "rocky cores"; the heat and pressure would not allow rock as it is known on Earth to exist.

Uranus and Neptune, being smaller, have less pressure inside and consequently cannot force hydrogen into the liquid metallic state. Rather, model calculations suggest they contain heavy-element cores surrounded by deep oceans of liquid water with dissolved minerals. Above that is a very dense, deep atmosphere made mostly of hydrogen gas. Although Uranus and Neptune have low densities, hydrogen-rich compositions, and atmospheric features similar to those of Jupiter and Saturn, the two smaller Jovian worlds are clearly different kinds of planets.

Another characteristic of the Jovian planets is their large satellite systems. Jupiter has at least 16 known moons; 4 of them, the **Galilean satellites** (named after Galileo, who discovered them in 1610), are visible through small telescopes or even binoculars (Figure 16-12a). Saturn, Uranus, and Neptune also have systems of moons.

All of the Jovian planets have ring systems. Saturn has bright rings made of ice particles (Figure 16-

12b), and Jupiter has a thin ring of dark, rocky dust. Uranus and Neptune also have thin rings composed of myriads of small particles.

We will examine the satellites and rings of the Jovian planets in later chapters. Here it is sufficient to note that the Jovian planets have large numbers of satellites and that all four have ring systems. In contrast, the terrestrial planets have few satellites and no rings at all. These observations will help us understand how planets were formed in the solar nebula. For now, however, we turn our attention to the smaller members of our solar system.

Space Debris

The sun and planets are not the only remains of the solar nebula. The solar system is littered with three kinds of space debris: asteroids, comets, and meteoroids. Although these objects represent a tiny fraction of the mass of the system, they are a rich source of information about the origin of the planets.

The **asteroids,** sometimes called minor planets, are small rocky worlds, most of which orbit the sun in a belt between the orbits of Mars and Jupiter. Roughly 20,000 asteroids have been identified, of which about 1000 follow orbits that bring them into the inner solar system, where they can occasionally collide with a planet. Earth has been struck many times in its history. Some asteroids are located in Jupiter's orbit, some have been found beyond the orbit of Saturn, and a growing number are being discovered among and even beyond the outermost planets.

About 200 asteroids are more than 100 km (60 mi) in diameter, and over 2000 are more than 10 km (6 mi) in diameter. There are probably 500,000 that are larger than 1 km (0.6 mi) and billions that are smaller. Because even the largest are only a few hundred kilometers in diameter, Earth-based telescopes can detect no details on their surfaces.

We do, however, have clear evidence that asteroids are irregularly shaped, cratered worlds. The Galileo spacecraft, on its way to Jupiter, flew past two asteroids, Gaspra and Ida, and photographed their surfaces. Both asteroids are irregular lumps of rock pocked by impact craters (Figure 16-13). Furthermore, the Hubble Space Telescope has provided images of some of the larger asteroids and confirms that their surfaces are irregular. We will discuss these observations in detail in Chapter 19, but in our quick survey of the solar system we can note that all evidence suggests the asteroids have suffered many impacts from collisions with other asteroids.

Older theories proposed that the asteroids are the remains of a planet that broke up, but modern astronomers recognize the asteroids as the debris left over by a planet that failed to form at a distance of 2.8

AU from the sun. When we discuss the formation of planets later in this chapter, we must be prepared to explain why material in the solar nebula failed to form a planet at a distance of 2.8 AU.

In contrast to asteroids, the brightest **comets** are impressively beautiful objects (Figure 16-14a). Most comets are faint, however, and are difficult to locate even at their brightest. A comet may take months to sweep through the inner solar system, during which time it appears as a glowing head with an extended tail of gas and dust.

According to the **dirty snowball model** of comets, the tail, which can be greater than 1 AU in length for a bright comet, is produced by the nucleus, a ball of dirty ices (mainly water and carbon dioxide) only a few dozen kilometers in diameter. When the nucleus enters the inner solar system, the sun's radiation begins to vaporize the ices, releasing gas and dust. The pressure of sunlight and the solar wind pushes the gas and dust away, forming a long tail. The motion of the nucleus along its orbit, the pressure of sunlight, and the outward flow of the solar wind can create comet tails that are long and straight or gently curved, but in either case the tails of comets always point generally away from the sun (Figure 16-14b).

Figure 16-13
The asteroid Gaspra as photographed by the Galileo spacecraft. The asteroid is 20 by 12 by 11 km and rotates once every 7.04 hours. Its irregular shape and numerous craters testify that it has suffered many impacts. The color here is exaggerated to reveal subtle color variation. Gaspra would look gray to the human eye. *(NASA/Galileo Imaging Team)*

a

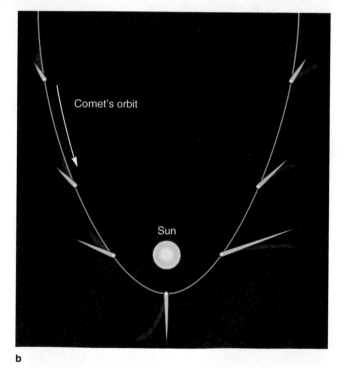

b

Figure 16-14
(a) A comet may remain visible in the evening or morning sky for weeks as it moves through the inner solar system. Comet West was in the sky during March 1976. *(Celestron International)* (b) A comet in a long, elliptical orbit becomes visible when the sun's heat vaporizes its ices and pushes the gas and dust away in a tail.

Figure 16-15
A meteor (right edge of photo) is the streak of glowing gases produced by a bit of material falling into Earth's atmosphere. Friction with the air vaporizes the material about 80 km (50 mi) above Earth's surface. *(Daniel Good)*

The nuclei of comets are icy bodies left over from the origin of the planets. Thus, we must conclude that at least some parts of the solar nebula were rich in ices. We will see later in this chapter how important ices were in the formation of the Jovian planets, and we will discuss comets in more detail in Chapter 19.

Unlike the stately comets, **meteors** flash across the sky in momentary streaks of light (Figure 16-15). They are commonly called "shooting stars." Of course, they are not stars but small bits of rock and metal falling into Earth's atmosphere and bursting into incandescent vapor about 80 km (50 mi) above the ground because of friction with the air. This vapor condenses to form dust, which settles slowly to Earth, adding about 40,000 tons per year to the planet's mass.

Technically, the word *meteor* refers to the streak of light in the sky. In space, before its fiery plunge, the object is called a **meteoroid,** and any part of it that survives its fiery passage to Earth's surface is called a **meteorite.** Most meteoroids are specks of dust, grains of sand, or tiny pebbles. Almost all the meteors we see in the sky are produced by meteoroids that weigh less than 1 g. Only rarely is one massive enough and strong enough to survive its plunge and reach Earth's surface.

Thousands of meteorites have been found, and we will discuss their particular forms in Chapter 19. We mention meteorites here for one specific clue they can give us concerning the solar nebula: Meteorites can tell us the age of the solar system.

The Age of the Solar System

If the solar nebula theory is correct, the planets should be about the same age as the sun. The most accurate way to find the age of a celestial body is to bring a sample into the laboratory and determine its age by analyzing the radioactive elements it contains.

When a rock solidifies, it incorporates known percentages of the chemical elements. A few of these elements are radioactive and can decay into another element, called the daughter element. For example, the isotope of uranium ^{238}U can decay into an isotope of lead, ^{206}Pb. The **half-life** of a radioactive element is the time it takes for half of the atoms to decay. The half-life of ^{238}U is 4.5 billion years. Thus, the abundance of a radioactive element gradually decreases as it decays (Figure 16-16), and the abundance of the daughter element gradually increases. If we know the abundance of the elements in the original rock, we can measure the present abundance and find the age of the rock. For example, if we studied a rock and found that only 50 percent of the ^{238}U remained and the rest had become ^{206}Pb, we would conclude that one half-life must have passed and the rock was 4.5 billion years old.

Uranium isn't the only radioactive element used in radioactive dating. Potassium (^{40}K) decays with a half-life of 1.3 billion years to form calcium (^{40}Ca) and argon (^{40}Ar). Rubidium (^{87}R) decays to strontium (^{87}Sr) with a half-life of 47 billion years. Any of these elements can be used as a radioactive clock to find the age of mineral samples.

Of course, to find a radioactive age, we need a sample in the laboratory, and the only celestial bod-

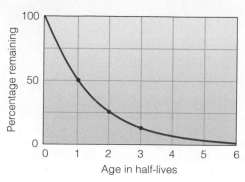

Figure 16-16
The number of nuclei remaining in a radioactive sample decreases such that 50 percent are left after one half-life, 25 percent after two half-lives, 12.5 percent after three half-lives, and so on.

ies from which we have samples are Earth, the moon, Mars, and meteorites.

The oldest Earth rocks so far discovered and dated are about 3.9 billion years old. That does not mean that Earth formed 3.9 billion years ago. The surface of Earth is active, and the crust is continually destroyed and reformed from material welling up from beneath the crust (see Chapter 17). Thus, the age of these oldest rocks tells us only that Earth is *at least* 3.9 billion years old.

One of the most exciting goals of the Apollo lunar landings was bringing lunar rocks back to Earth's laboratories, where they could be dated. Because the moon's surface is not being recycled like Earth's, some parts of it might have survived unaltered since early in the history of the solar system. Dating the rocks showed the oldest to be 4.48 billion years old. Thus, the solar system must be *at least* 4.48 billion years old.

Although no one has yet been to Mars, a few meteorites found on Earth have been identified by their chemical composition as having come from Mars. Most of these have ages of only 1.3 billion years, but one has an age of 4.5 billion years. Thus, Mars must be at least that old.

Another important source for determining the age of the solar system is meteorites. Radioactive dating of meteorites yields a range of ages, with the oldest about 4.6 billion years old. This figure is widely accepted as the age of the solar system.

One last celestial body deserves mention: the sun. Astronomers estimate the age of the sun to be about 5 billion years, but this is not a radioactive date because they cannot obtain a sample of solar material. Instead, they estimate the sun's age from the radioactive ages of Earth, the moon, and meteorites. Computer models of the sun give only approximate ages, but they generally agree with the age of Earth.

Apparently, all the bodies of the solar system formed at about the same time some 4.6 billion years

ago. This completes our survey of the solar system. We have found five significant characteristics (Table 16-1). We must now try to understand how the formation of the planets in the solar nebula explains these significant characteristics.

REVIEW Critical Inquiry

In what ways is the solar system a disk?

First, the general shape of the solar system is that of a disk. The planets follow orbits that lie in nearly the same plane. The orbit of Mercury is inclined 7° to the plane of Earth's orbit, and the orbit of Pluto is inclined a bit more than 17°. Thus, the planets follow orbits confined to a thin disk with the sun at its center.

Second, the motions of the sun and planets also follow this disk theme. The sun and most of the planets rotate in the same direction, counterclockwise as seen from the north, with their equators near the plane of the solar system. Also, all of the planets revolve around the sun in that same direction. Thus, our solar system seems to prefer motion in the same direction, which further reflects a disk theme.

One of the basic characteristics of our solar system is its disk shape, but another dramatic characteristic is the division of the planets into two groups. What are the distinguishing differences between the terrestrial and Jovian planets?

We have completed our survey of the solar system and gathered the preliminary evidence we can use to evaluate the solar nebula hypothesis. Now it is time to ask how the planets formed from the solar nebula and took on the characteristics we see today.

Scientists often face problems in which they must reconstruct the past. Some of these problems are obvious, such as those faced by an archaeologist excavating the ruins of a burial tomb, but other such problems are less obvious. In each case, success requires the interplay of hypotheses and evidence to re-create a past that no longer exists.

The reconstruction of the past is obvious when we use the chemical abundance of stars to reconstruct the story of the formation of our galaxy, but a biologist studying a centipede is also reconstructing the past. How did this creature come to have a segmented body with so many legs? How did it develop the metabolism that allows it to move quickly and hunt prey? Although the problem might at first seem to be one of mere anatomy, the scientist must reconstruct an environment that no longer exists.

The astronomer's problem is not just to understand what the planets are like, but to understand how they got that way. That means planetary astronomers must look at the evidence they can see today and reconstruct a history of the solar system, a past that is quite different from the present. If we had a time machine, it would be a fantastic adventure to go back and watch the planets form. Time machines may be impossible, but we can use the grand interplay of evidence and theory, the distinguishing characteristic of science, to journey back billions of years and reconstruct a past that no longer exists.

16-3 The Story of Planet Building

The challenge for modern planetary astronomers is to compare the characteristics of the solar system with the solar nebula theory and tell the story of how the planets formed. Thus they must re-create the solar system's early past (Window on Science 16-3).

The Chemical Composition of the Solar Nebula

Everything we know about the solar system and star formation suggests that the solar nebula was a fragment of an interstellar gas cloud. Such a cloud would have been mostly hydrogen with some helium and tiny traces of the heavier elements.

This is precisely what we see in the composition of the sun (see Table 6-2). Analysis of the solar spectrum shows that the sun is mostly hydrogen, with 25 percent of its mass being helium and only about 2 percent being heavier elements. Of course, nuclear reactions have fused some hydrogen into helium, but this happens in the sun's core and has not affected its surface composition. Thus, the composition revealed in its spectrum is essentially the composition of the gases from which it formed.

This must have been the composition of the solar nebula, and we can see that composition reflected in the chemical compositions of the planets. One factor, however, can dramatically alter the composition of a planet. We must recognize that planets grow by two processes. Planets begin growing by the sticking to-gether of solid bits of matter. Only when a planet has grown to a mass of about 15 Earth masses does it have enough gravitation to begin capturing gas directly from the solar nebula in a process called **gravitational collapse.** The Jovian planets began forming by the aggregation of bits of rock and ice. Once they accumulated enough mass, they began growing by gravitational collapse. That is, they could capture large amounts of gas, mostly hydrogen and helium, directly from the solar nebula. Jupiter and Saturn grew rapidly and captured the most hydrogen and helium, while Uranus and Neptune grew more slowly and never became massive enough to capture gas as rapidly as Jupiter and Saturn. Thus, the Jovian worlds grew to be low-density, hydrogen-rich planets, with Jupiter and Saturn the largest and least dense.

The terrestrial planets, in contrast, contain very little hydrogen and helium. These small planets have such low masses they were unable to keep the hydrogen and helium from leaking into space. Indeed, because of their small masses, they were probably unable to capture very much of these lightweight gases from the solar nebula. The terrestrial planets are dense worlds because they are composed of the heavier elements from the solar nebula.

Although we can see how the chemical composition of the solar nebula is reflected in the present composition of the sun and planets, a very important question remains: How did gas and dust in the solar nebula come together to form the solid matter of the planets? We must answer that question in two stages. First, we must understand how gas and dust formed billions of small solid particles, and then we must explain how these particles built the planets.

The Condensation of Solids

The key to understanding the process that converted the nebular gas into solid matter is the variation in density among solar system objects. We have already noted that the four inner planets are high-density, terrestrial bodies, whereas the outermost planets are low-density, giant planets (except for Pluto, which is low in density but not a giant). This division is due to the different ways gases condensed into solids in the inner and outer regions of the solar nebula.

Even among the four terrestrial planets, we find a pattern of subtle differences in density. Merely listing the observed densities of the terrestrial planets does not reveal the pattern, because Earth and Venus, being more massive, have stronger gravity and have squeezed their interiors to higher densities. We must look at the **uncompressed densities**—the densities the planets would have if their gravity did not compress them. These densities (Table 16-2) show that the closer a planet is to the sun, the higher its uncompressed density.

This density variation probably originated when the solar system first formed solid grains. The kind of matter that condensed in a particular region would depend on the temperature of the gas there. In the inner regions, the temperature may have been 1500 K or so. The only materials that could form grains at this temperature are compounds with high melting points, such as metal oxides and pure metals, which are very dense. Farther out in the nebula it was cooler, and silicates (rocky material) could condense. These are less dense than metal oxides and metals. In the cold outer regions, ices of water, methane, and ammonia could condense. These are low-density materials.

The sequence in which the different materials condense from the gas as we move away from the sun is called the **condensation sequence** (Table 16-3). It suggests that the planets, forming at different distances from the sun, accumulated from different kinds of materials. Thus, the inner planets formed from high-density metal oxides and metals, and the outer planets formed from low-density ices.

We must also remember that the solar nebula did not remain the same temperature throughout the formation of the planets but may have grown progressively cooler. Thus, a particular region of the nebula may have begun by producing solid particles of metals and metal oxides but, after cooling, began producing particles of silicates. Allowing for the cooling of the nebula makes our theory much more complex, but it also makes the processes by which the planets formed much more understandable.

The Formation of Planetesimals

In the development of a planet, three groups of processes operate. First, grains of solid matter grow larger, eventually reaching diameters ranging from a few centimeters to kilometers. The larger of these objects, called **planetesimals,** are believed to be the bodies that the second group of processes collects into planets. Finally, a third set of processes clears away the solar nebula. The study of planet building is the study of these three groups of processes.

According to the solar nebula theory, planetary development in the solar nebula began with the growth of dust grains. These specks of matter, whatever their composition, grew from microscopic size by two processes: condensation and accretion.

A particle grows by **condensation** when it adds matter one atom at a time from a surrounding gas. Thus, snowflakes grow by condensation in Earth's atmosphere. In the solar nebula, dust grains were continuously bombarded by atoms of gas, and some of these stuck to the grains. A microscopic grain capturing a gas atom increases its mass by a much larger fraction than a gigantic boulder capturing a single atom. Thus, condensation can increase the mass of a

Table 16-2 Observed and Uncompressed Densities

Planet	Observed Density (g/cm³)	Uncompressed Density (g/cm³)
Mercury	5.44	5.4
Venus	5.24	4.2
Earth	5.50	4.2
Mars	3.94	3.3
(Moon)	3.36	3.35

Table 16-3 The Condensation Sequence

Temperature (K)	Condensate	Planet (Estimated Temperature of Formation; K)
1500	Metal oxides	Mercury (1400)
1300	Metallic iron and nickel	
1200	Silicates	
1000	Feldspars	Venus (900)
680	Troilite (FeS)	Earth (600)
		Mars (450)
175	H_2O ice	Jovian (175)
150	Ammonia–water ice	
120	Methane–water ice	
65	Argon–neon ice	Pluto (65)

small grain rapidly, but as the grain grows larger, condensation becomes less effective.

The second process is **accretion,** the sticking together of solid particles. In building a snowman, we roll a ball of snow across the snowy ground so that it grows by accretion. In the solar nebula, the dust grains were, on the average, no more than a few centimeters apart, so they collided frequently. Their mutual gravitation was too small to hold them to each other, but other effects may have helped. Static electricity generated by their passage through the gas could have held them together, as could compounds of carbon that might have formed a sticky surface on the grains. Ice grains might have stuck together better than some other types. Of course, some collisions might have broken up clumps of grains; on the whole, however, accretion must have increased grain size. If it had not, the planets would not have formed.

There is no clear distinction between a very large grain and a very small planetesimal, but we can consider an object a planetesimal when its diameter becomes a kilometer or so. Objects this size and larger were subject to new processes that tended to concentrate them. One important effect may have been that the growing planetesimals collapsed into the plane of the solar nebula. Dust grains could not fall into the plane because the turbulent motions of the gas kept them stirred up, but the larger objects had more mass, and the gas motions could not have prevented them from settling into the plane of the spinning nebula. This would have concentrated the solid particles into a thin plane about 0.01 AU thick and would have made further planetary growth more rapid.

This collapse of the planetesimals into the plane is analogous to the flattening of a forming galaxy. However, an entirely new process may have become important once the plane of planetesimals formed. Computer models show that the rotating disk of particles should have been gravitationally unstable and would have broken up into small clouds (Figure 16-17). This would further concentrate the planetesimals and help them coalesce into objects up to 100 km (60 mi) in diameter. Thus, the theory predicts that the nebula became filled with trillions of solid particles ranging in size from pebbles to tiny planets. As the largest began to exceed 100 km in diameter, new processes began to alter them, and a new stage in planet building began, the growth of protoplanets.

The Growth of Protoplanets

The coalescing of planetesimals eventually formed **protoplanets,** massive objects destined to become planets. As these larger bodies grew, new processes began making them grow faster and altered their physical structure.

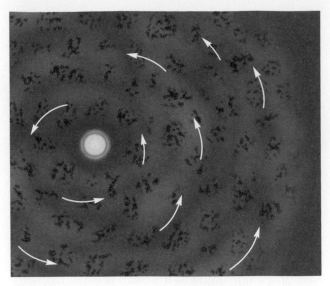

Figure 16-17
Gravitational instabilities in the rotating disk of planetesimals may have forced them to collect in clumps, accelerating their growth.

If planetesimals collided at orbital velocities, it is unlikely that they would have stuck together. The average orbital velocity in the solar system is about 30 km/sec (67,000 mph). Head-on collisions at this velocity would have pulverized the material. However, the planetesimals were moving in the same direction in the nebular plane and thus avoided head-on collisions. Instead, they merely rubbed shoulders at low relative velocities. Such collisions would have been more likely to fuse them than to shatter them.

In addition, some adhesive effects probably helped. Sticky coatings and electrostatic charges on the surfaces of the smaller planetesimals probably aided formation of larger bodies. Collisions would have fragmented some of the surface rock, but if the planetesimals were large enough, their gravity would have held on to some fragments, forming a layer of soil composed entirely of crushed rock. Such a layer on the larger planetesimals may have been effective in trapping smaller bodies.

The largest planetesimals would grow the fastest because they had the strongest gravitational field. Not only could they hold on to a cushioning layer to trap fragments, but their stronger gravity could also attract additional material. These planetesimals probably grew quickly to protoplanetary dimensions, sweeping up more and more material. When massive enough, they trapped some of the original nebular gas to form primitive atmospheres. At some point, they crossed the boundary between planetesimals and protoplanets.

To trace the growth of a protoplanet, we can think of the formation of Earth. In its simplest form, our theory of protoplanet growth supposes that all the planetesimals had about the same chemical composition.

The planetesimals accumulated gradually to form a planet-size ball of material that was of homogeneous composition throughout. Once the planet formed, heat began to accumulate in its interior from the decay of short-lived radioactive elements, and this heat eventually melted the planet and allowed it to differentiate. **Differentiation** is the separation of material according to density. When the planet melted, the heavy metals such as iron and nickel settled to the core, while the lighter silicates floated to the surface to form a low-density crust. The story of planet formation from planetesimals of similar composition is shown in the left half of Figure 16-18.

In this simple form, the theory predicts that Earth's present atmosphere was not its first. The first atmosphere consisted of gases trapped from the solar nebula—mostly hydrogen and helium. They were later driven off by the heat, aided perhaps by outbursts from the infant sun, and new gases that baked from the rocks formed a secondary atmosphere. This creation of a planetary atmosphere from the planet's interior is called **outgassing.**

We can improve this simple theory of planet formation in two ways. First, it seems likely that the solar nebula cooled during the formation of the planets, so they did not accumulate from planetesimals of common composition. As planet building began, the first particles to condense in the inner solar system were metals and metal oxides, so the protoplanets may have begun by accreting metallic cores. Later, as the nebula cooled, more silicates could form, and the protoplanets added silicate mantles. A second improvement in the theory proposes that the planets grew so rapidly that the heat released by the in-falling particles, the **heat of formation,** did not have time to escape. This heat rapidly accumulated and melted the protoplanets as they formed. If this is true, then the planets would have differentiated as they formed. This improved story of planet formation is shown in the right half of Figure 16-18.

Our improved theory suggests that our present atmosphere was not produced entirely by outgassing. If Earth formed in a molten state, then there was never a time when it had a primitive atmosphere of hydrogen and helium accumulated from the solar nebula. Rather, the gases of the atmosphere were released by the molten rock as the protoplanet grew. Those gases would not have included much water, however, so some astronomers now think that Earth's water and much of its present atmosphere accumulated late in the formation of the planets as Earth swept up volatile-rich planetesimals forming in the cooling solar nebula. Such icy planetesimals may have formed in the outer parts of the solar nebula and been scattered by encounters with the Jovian planets in a bombardment of comets.

Figure 16-18
At left, a planet grows slowly by the accretion of planetesimals of similar composition. After the protoplanet has formed, the accumulation of heat from radioactive decay in its interior melts it and allows it to differentiate. An improved theory, right, proposes that the solar nebula cooled as the planets formed. The first particles to accumulate were metals, but later particles were silicates. Also, it seems likely that the planets formed fast enough to prevent the loss of the heat of formation, which melted the protoplanets and allowed them to differentiate as they formed.

The Jovian planets grew to their present size in about 10 million years. How do astronomers know this? The T Tauri stars that have gas disks are all younger than 10 million years; older T Tauri stars have blown their gas disks away. The Jovian planets grew mainly from the gas in the solar nebula, so they must have finished forming by the time the sun blew its gas away. The terrestrial planets probably took longer to form because they did not grow from the gas in the nebula but had to accrete from planetesimals. Nevertheless, the terrestrial planets must have been nearly complete by the time the sun reached the main sequence at an age of about 30 million years. Models suggest that the planets were fully formed by about 100 million years.

The details of the formation of the planets are still being unraveled, but it seems clear that the solar nebula hypothesis can explain the formation of planets around the forming sun. But can it explain the distinguishing characteristics of the solar system that we compiled earlier in this chapter?

Explaining the Characteristics of the Solar System

Table 16-1 contains the list of distinguishing characteristics of the solar system. Any theory of the origin of the solar system should explain these characteristics.

The disk shape of the solar system is inherited from the solar nebula. The sun and planets revolve and rotate in the same direction because they formed from the same rotating gas cloud. The orbits of the planets lie in the same plane because the rotating solar nebula collapsed into a disk, and the planets formed in that disk.

The solar nebula hypothesis is evolutionary in that it calls on continuing processes to gradually build the planets. To explain the rotation of Venus, Uranus, and Pluto, however, we must introduce catastrophic events. Apparently, these planets acquired their peculiar rotations when they were struck by very large planetesimals late in their formation. Such impacts may be unusual, but some planets probably experienced them, and an off-center impact by a planetesimal as large as Mars might have made Venus rotate backward. A similar event might have forced Uranus to rotate on its side. In Chapter 18, we will discuss Pluto, and we will see that such a catastrophic event may have been involved in its formation as well.

The second item in Table 16-1, the division of the planets into terrestrial and Jovian worlds, can be understood through the condensation sequence. The terrestrial planets formed in the inner part of the solar nebula, where the temperature was high and most compounds remained gaseous. Only compounds such as the metals and silicates could condense to form solid particles. Planets must begin forming by the

sticking together of solid particles because a small blob of matter does not have enough gravitation to capture gas. Thus, the planets that began growing in the inner solar system had to form mainly from metals and silicates. They became the small, dense terrestrial planets.

In contrast, the Jovian planets formed in the outer solar nebula, where the lower temperature allowed the gas to form large amounts of ices, perhaps three times more ices than silicates. Thus, the Jovian planets grew rapidly and became massive.

The heat of formation (the energy released by infalling matter) was tremendous for these massive planets. Jupiter must have grown hot enough to glow with a luminosity of about 1 percent that of the present sun. However, because it never got hot enough to generate nuclear energy as a star would, it never generated its own energy. Jupiter is still hot inside. In fact, both Jupiter and Saturn radiate more heat than they absorb from the sun, so they are evidently still cooling.

Although planets must begin growing by the accretion of solid particles, once they become a few times more massive than Earth they have enough gravity to capture gas directly from the solar nebula. Thus, the Jovian planets grew by attracting vast amounts of nebular gas and grew rich in light gases such as hydrogen and helium. The terrestrial planets could not do this because they never became massive enough and because the gas in the inner nebula was hotter and more difficult to capture.

When planets were first discovered orbiting other stars, some astronomers wondered if the solar nebula theory was quite right. The newly discovered planets are massive, so they are presumably Jovian, but they are close to their stars where a terrestrial planet should be. Also, some of the new planets follow elliptical orbits. Models of planet formation resolved the problem. If a Jovian planet forms in a disk rich in planetesimals, it may sweep up mass so fast it migrates inward toward its star. Also, interactions among massive planets could throw a planet into an elliptical orbit. Thus the newly discovered planets do not pose a problem for the solar nebula explanation of terrestrial and Jovian planets.

A glance at the solar system suggests that we should expect to find a planet between Mars and Jupiter at the present location of the asteroid belt. Evidently, we have asteroids there and not a planet, because Jupiter grew into such a massive planet it was able to gravitationally disturb the motion of nearby planetesimals. Thus, the bodies that should have formed a planet just inward in the solar system from Jupiter were broken up, thrown into the sun, or ejected from the solar system. The asteroids we see today are the last remains of those objects.

The large satellite systems of the Jovian worlds may contain two kinds of moons. Some moons may

a

b

Figure 16-19
Every old, solid surface in the solar system is scarred by craters. (a) Earth's moon is scarred by craters ranging from basins hundreds of kilometers in diameter down to microscopic pits. (b) The surface of Mercury as photographed by a passing spacecraft shows vast numbers of overlapping craters. *(NASA)*

have formed in orbit around the forming planet in a miniature of the solar nebula. But some of the smaller moons may be captured planetesimals and asteroids. The large masses of the Jovian planets would have made it easier for them to capture satellites.

In Table 16-1, we noted that all four Jovian worlds have ring systems, and we can understand this by considering the large mass of these worlds and their remote location in the solar system. A large mass makes it easier for a planet to hold onto orbiting ring particles; and being farther from the sun, the ring particles are not as easily swept away by the pressure of sunlight and the solar wind. It is hardly surprising, then, that the terrestrial planets, low-mass worlds located near the sun, have no planetary rings.

The last entry in Table 16-1 is the common ages of solar system bodies, and the solar nebula hypothesis has no difficulty explaining that characteristic. If the hypothesis is correct, then the planets formed at the same time as the sun and thus should have roughly the same age. Although we have mineral samples for radioactive dating only from Earth, the moon, Mars, and meteorites, so far the ages seem to agree.

Our general understanding of the origin of the solar system also explains the origin of asteroids, meteors, and comets. They appear to be the last of the debris left behind by the solar nebula. These objects are such important sources of information about the history of our solar system that we will discuss them in detail in Chapter 19. But for now we ask: What happened to the solar nebula?

Clearing the Nebula

The planets in our solar system formed in a disk of gas, dust, and planetesimals around the sun 4.6 billion years ago. That solar nebula vanished while the sun was still young, and the clearing of the nebula ended planet building.

Four effects helped clear the nebula. The most important was **radiation pressure.** When the sun became a luminous object, light streaming from its surface pushed against the particles of the solar nebula. Large bits of matter such as planetesimals and planets were not affected, but low-mass specks of dust and individual gas atoms were pushed outward and eventually driven from the system.

The second effect that helped clear the nebula was the solar wind, the flow of ionized hydrogen and other atoms away from the sun's upper atmosphere. This flow is a steady breeze that rushes past Earth at about 400 km/sec (250 mi/sec). When the sun was young, it may have had an even stronger solar wind, and irregular fluctuations in its luminosity, like those observed in T Tauri stars, may have produced surges in the wind that helped push dust and gas out of the nebula.

The third effect for clearing the nebula was the sweeping up of space debris by the planets. All of the old, solid surfaces in the solar system are heavily cratered by meteorite impacts (Figure 16-19). Earth's moon, Mercury, Venus, Mars, and most of the moons in the solar system are covered with craters. A few of these craters have been formed recently by the steady rain of meteorites that falls on all the planets in the solar system, but most of the craters we see appear to have been formed roughly 4 billion years ago in what is called the **heavy bombardment,** as the last of the debris in the solar nebula was swept up by the planets. We will find more evidence of this bombardment in later chapters. Thus, many of the last remaining solid objects in the solar nebula were swept up by the planets soon after they formed.

The fourth effect was the ejection of material from the solar system by close encounters with planets. If a small object such as a planetesimal passes close to a planet, it can gain energy from the planet's gravitational field and be thrown out of the solar system. Ejection is most probable in encounters with massive planets, so the Jovian planets were probably very efficient at ejecting the icy planetesimals that formed in their region of the nebula.

Together, these four effects cleared away the solar nebula. This may have taken considerable time, but eventually the solar system was relatively clear, the planets could no longer gain mass, and planet building ended.

REVIEW Critical Inquiry

Why are there two kinds of planets in our solar system?

Planets begin forming from solid bits of matter, not from gas. Consequently, the kind of planet that forms at a given distance from the sun depends on the kind of compounds that can condense out of the gas to form solid particles. In the inner parts of the solar nebula, the temperature was so high that most of the gas could not condense to form solids. Only metals and silicates could form solid grains, and thus the innermost planets grew from this dense material. Much of the mass of the solar nebula consisted of hydrogen, helium, water vapor, and other gases, and they were present in the inner solar nebula but couldn't form solid grains. The small terrestrial planets couldn't grow from these gases, so the terrestrial planets are small and dense.

In the outer solar nebula, the composition of the gas was the same; but it was cold enough for water vapor to condense to form ice grains, and because hydrogen and oxygen are so abundant, there was lots of ice available. Thus, the outer planets grew from solid bits of metal and silicate combined with large amounts of ice. The outer planets grew so rapidly that they became big enough to capture gas directly, and thus they became the hydrogen- and helium-rich Jovian worlds.

The condensation sequence combined with the solar nebula hypothesis gives us a way to understand the formation of the planets. Can we extend our insight to other stars? What would a planetary system be like if it formed around a star that was slightly hotter than the sun?

The great chain of origins leads from the first instant of the big bang through the birth of galaxies, the formation of stars, the origin of our solar system, and finally to us. One of the most elegant and complex links in that chain is planet building. As we explore the solar system in detail in the following chapters, we must stay alert for further clues to the birth of the planets.

Summary

The solar nebula hypothesis proposes that the solar system began as a contracting cloud of gas and dust that flattened into a rotating disk. The center of this cloud eventually became the sun, and the planets formed in the disk of the nebula.

Observational evidence gives astronomers confidence in this theory. Disks of gas and dust have been found around many young stars, so astronomers suspect that planetary systems are common. Planets orbiting other stars are too faint and too close to their star to image directly, but astronomers have found a number of these planets by observing the motion of the star as the star and planet revolve around their center of mass.

The solar nebula theory explains many of the characteristic properties of the solar system. For example, the solar system has a disk shape. The orbits of the planets lie in nearly the same plane, and they all revolve around the sun in the same direction, counterclockwise as seen from the north. This is also true of many of the satellites of the planets. With only three exceptions, the planets rotate counterclockwise around axes roughly perpendicular to the plane of the solar system. This disk shape and the motion of the planets appear to have originated in the disk-shaped solar nebula.

Another striking feature of the solar system is the division of the planets into two families. The terrestrial planets, which are small and dense, lie in the inner part of the system. The Jovian planets are large, low-density worlds in the outer part of the system. In general, the closer a planet lies to the sun, the higher its uncompressed density.

The solar system is now filled with smaller bodies such as asteroids, comets, and meteors. The asteroids are small, rocky worlds, most of which orbit the sun between Jupiter and Mars. They appear to be material left over from the formation of the solar system.

Another important characteristic of the solar system bodies is their similar ages. Radioactive dating tells us that Earth, the moon, Mars, and meteorites are no older than 4.6 billion years. Thus, it seems our solar system took shape about 4.6 billion years ago.

According to the condensation sequence, the inner part of the nebula was so hot that only high-density minerals could form solid grains. The outer regions, being cooler, condensed to form icy material of lower density. The planets grew from these solid materials, with the denser planets forming in the inner part of the nebula and the lower-density Jovian planets forming farther from the sun.

Planet building began as dust grains grew by condensation and accretion into planetesimals ranging from a kilometer to hundreds of kilometers in diameter. These planetesimals settled into a thin plane around the sun and accumulated into larger bodies, the largest of which grew the fastest and eventually became protoplanets.

Once a planet had formed from a large number of planetesimals, heat from radioactive decay could have melted it and allowed it to differentiate into a dense metallic core and a lower-density silicate crust. In fact, it is possible that the solar nebula cooled as the protoplanets grew so that the first planetesimals were metallic and later additions were silicate. It is also likely that the planets grew rapidly enough that the heat of formation released by the in-falling material melted the planets and allowed them to differentiate as they formed.

The Jovian planets probably grew rapidly from icy materials and became massive enough to attract and hold vast amounts of nebular gas. Heat of formation raised their temperatures very high when they were young, and Jupiter and Saturn still radiate more heat than they absorb from the sun.

Once the sun became a luminous object, it cleared the nebula as its light and solar wind pushed material out of the system. The planets helped by absorbing some planetesimals and ejecting others from the system. Once the solar system was clear of debris, planet building ended.

New Terms

solar nebula theory	gravitational collapse
extra-solar planet	uncompressed density
terrestrial planets	condensation sequence
Jovian planets	planetesimal
liquid metallic hydrogen	condensation
Galilean satellites	accretion
asteroid	protoplanet
comet	differentiation
dirty snowball model	outgassing
meteor	heat of formation
meteoroid	radiation pressure
meteorite	heavy bombardment
half-life	

Review Questions

1. What produced the helium now present in the sun's atmosphere? in Jupiter's atmosphere? in the sun's core?

2. What produced the iron in Earth's core and the heavier elements like gold and silver in Earth's crust?

3. What evidence do we have that disks of gas and dust are common around young stars?

4. According to the solar nebula theory, why is the sun's equator nearly in the plane of Earth's orbit?

5. Why does the solar nebula theory predict that planetary systems are common?

6. Why do we think the solar system formed about 4.6 billion years ago?

7. If you visited another planetary system, would you be surprised to find planets older than Earth? Why or why not?

8. Why is almost every solid surface in our solar system scarred by craters?

9. What is the difference between condensation and accretion?

10. Why don't terrestrial planets have rings and large satellite systems like the Jovian planets?

11. How does the solar nebula theory help us understand the composition and location of asteroids and comets?

12. How does the solar nebula theory explain the dramatic density difference between the terrestrial and Jovian planets?

13. If you visited some other planetary system in the act of building planets, would you expect to see the condensation sequence at work, or was it unique to our solar system?

14. Why do we expect to find that planets are differentiated?

15. What processes cleared the nebula away and ended planet building?

Discussion Questions

1. In your opinion, if planets formed with one of the first stars to form in our galaxy, how would those planets differ from the planets in our solar system? (*Hint:* To what population would the star belong?)

2. If the solar nebula hypothesis is correct, then there are probably more planets in the universe than stars. Do you agree? Why or why not?

Problems

1. The nearest star is about 4.2 ly away. If you looked back at the solar system from that distance, what would the maximum angular separation be between Jupiter and the sun? (*Hint:* 1 ly equals 63,000 AU.)

2. Venus is the brightest planet in our sky at a time when it is about 1 AU from Earth. How many times fainter would it look from the nearest star 4.2 ly away? (*Hints:* 1 ly equals 63,000 AU. Remember the inverse square law.)

3. What is the smallest-diameter crater you can identify in Figure 16-7? (*Hint:* See the Appendix A Table A-13, Properties of the Planets, to find the diameter of Mercury in kilometers.)

4. A sample of a meteorite has been analyzed, and the result shows that out of every 1000 nuclei of ^{40}K originally in the meteorite, only 125 have not decayed. How old is the meteorite? (*Hint:* See Figure 16-16.)

5. In Table 16-2, which object's density differs least from its uncompressed density? Why?

6. What composition might we expect for a planet that formed in a region of the solar nebula where the temperature was about 100 K?

7. Suppose that Earth grew to its present size in 10 million years through the accretion of particles averaging 100 g each. On the average, how many particles did Earth capture per second? (*Hint:* See Appendix A, Table A-13, to find Earth's mass.)

8. If you stood on Earth during its formation as described in Problem 7 and watched a region covering 100 m^2, how many impacts would you expect to see in an hour? (*Hints:* Assume that Earth had its present radius. The surface area of a sphere is $4\pi r^2$.)

9. The velocity of the solar wind is roughly 400 km/sec. How long does it take to travel from the sun to Pluto?

Critical Inquiries for the Web

1. How does our solar system compare with the others that have been found? Search the Internet for sites that give information about planetary systems around other stars. What kinds of planets have been detected by these searches so far? Discuss the selection effects (see Window on Science 13-2) that must be considered when interpreting these data.

2. How is radioactive dating carried out on meteorites and rocks from surfaces of various bodies in the solar system? Look for Web sites on the details of radioactive dating, and summarize the methods used to uncover the abundances of radioactive elements in a particular sample. (*Hint:* Try looking for information on how a particular meteorite—for example, the Martian meteorite ALH84001—was studied, what age range was determined, and what radioactive elements were used to arrive at the age.)

Go to the Brooks/Cole Astronomy Resource Center (www.brookscole.com/astronomy) for critical thinking exercises, articles, and additional readings from InfoTrac College Edition, Brooks/Cole's online student library.

CHAPTER 17

The Earthlike Planets

That's one small step for a man . . .

one giant leap for mankind.

Neil Armstrong
On the moon

Beautiful, beautiful.

Magnificent desolation.

Edwin E. Aldrin, Jr.
On the moon

Guidepost

The preceding chapter explained how planetary systems form, and now we are ready to begin examining our planetary system to see what further evidence we can find to help us understand how Earth and the other planets formed and evolved.

This chapter seems at first glance to cover an odd mixture of planets, but they all share similar processes with Earth. Not only does Earth help us understand these Earthlike planets, but they will help us understand our own planet.

The principles we learn from these worlds will continue to help us in the next chapter. There, we will discuss the Jovian planets, which are certainly un-Earthlike, but we will also explore the Jovian moons. Those moons experience geological processes that we will recognize from our study of the Earthlike worlds.

Our understanding of planets will carry us through the last chapter of this book where we think about life on other worlds. Understanding life requires that we understand the planets where it flourishes.

The first two people to visit Earth's moon responded in dramatically different ways. Neil Armstrong responded to the significance of the first human step on the surface of another world, but Buzz Aldrin responded to the moon itself. It *is* desolate, and it *is* magnificent. But it is not unusual. Most planets in the universe probably look like Earth's moon, and astronauts may someday walk on such worlds and compare them with our moon.

The comparison of one planet with another is called **comparative planetology,** and it is one of the best ways to analyze the worlds in our solar system. We will learn much more by comparing planets than we would by studying them individually.

The basis for this comparison is Earth. We know a great deal about Earth, and it is our home, so it is natural for us to compare other worlds with our own. But in addition, our planet is a planet of extremes. Its interior is molten and generates a magnetic field. Its crust is active, with moving sections that push against each other and trigger earthquakes, volcanoes, and mountain building. Almost 75 percent of Earth's surface is covered by a layer of liquid water with an average depth of 3 km. Even Earth's atmosphere is extreme. Processes have altered Earth's air from its natural composition to a highly unusual, oxygen-rich sea of gas. Once we understand Earth's complex properties, we should find it easier to understand other planets that resemble Earth.

17-1 The Early History of Earth

If Earth is to be the benchmark for our study of other worlds, then we must examine our planet in detail. We will consider its general history, interior, crust, and atmosphere.

Four Stages of Planetary Development

As Earth formed in the inner solar nebula, it passed through four developmental stages (Figure 17-1), stages that other worlds also experienced to varying degrees.

The first stage of planetary evolution is *differentiation,* the separation of material according to density. Earth now has a dense core and a lower-density crust, so we know it differentiated. Some of that differentiation may have occurred very early as the heat of formation released by in-falling matter melted the growing Earth. Some of the differentiation, however, may have occurred later as radioactive decay released heat and further melted Earth, allowing the denser metals to sink to the core.

The second stage, *cratering,* could not begin until a solid surface formed. The heavy bombardment of the early solar system cratered Earth just as it did our moon. As the debris in the solar nebula cleared away,

Figure 17-1
The four stages of planetary development. First: Differentiation into core and crust as dense material sinks to the center and low-density material floats to the surface. Second: Cratering of the newly formed crust by the heavy bombardment of meteorites in the young, debris-filled solar nebula. Third: Flooding of lowlands by lava from the interior and later by water from the cooling atmosphere. Fourth: Slow surface evolution due to geological activity and erosion.

the rate of cratering impacts fell rapidly to its present low rate.

The third stage, *flooding,* began as the decay of radioactive elements heated Earth's interior. Lava welled up through fissures in the crust and flooded the deeper basins. Later, as the atmosphere cooled, water fell as rain and flooded the basins to form the first oceans. Note that on Earth flooding includes both lava and water.

The fourth stage, *slow surface evolution,* has continued for at least the past 3.5 billion years or more. Earth's surface is constantly changing as sections of crust slide over each other, push up mountains, and shift continents. In addition, moving air and water erode the surface and wear away geological features. Almost all trace of the first billion years of Earth's geology has been destroyed by the active crust and erosion.

Terrestrial planets pass through these four stages, but differences in mass, temperature, and composition emphasize some stages over others and produce surprisingly different worlds.

Earth's Interior

We expect that Earth differentiated, but what evidence do we have to tell us what the interior of Earth is really like?

Earth's mass divided by its volume tells us its average density, about 5.50 g/cm³. (See Data File Two.) But the density of Earth's rocky crust is only about half that. Clearly, a large part of Earth's interior must be made of material denser than rock.

Although the deepest wells extend only a few kilometers down, Earth scientists have clear proof that Earth did differentiate. Each time an earthquake occurs, seismic waves travel through the interior and register on seismographs all over the world. Analysis of these waves shows that Earth's interior is divided into a metallic core, a dense rocky mantle, and a thin, low-density crust.

The core has a density of 14 g/cm³, denser than lead, and is believed to be composed of iron and nickel at a temperature of roughly 6000 K. In other words, the core of Earth is as hot as the surface of the sun, but the high pressure keeps the metal a solid near the middle of the core and a liquid in its outer parts. Two kinds of seismic waves tell us the outer core is liquid. The **P waves** travel like sound waves, and they can penetrate a liquid; but **S waves** travel as a side-to-side vibration of particles like the jiggle in a bowl of jelly. *S* waves can't travel through a liquid, so Earth scientists can deduce the size of the liquid core by observing where *S* waves get through and where they don't (Figure 17-2). The outer boundary of the core lies just over halfway to the surface.

Earth's magnetism gives us further clues about the core. Convection currents stir the molten liquid; and, because it is a very good conductor of electricity and is rotating as Earth rotates, it generates a magnetic field through the dynamo effect—the same process that creates the sun's magnetic field (Chapter 7). From traces of magnetic field retained by rocks that formed long ago, geologists conclude that Earth's magnetic field reverses itself every million years or so. Periodic

Earth Data File Two

A shaded relief map of Earth with the oceans removed reveals the geological features on the continents and under the oceans. *(National Geophysical Data Center)*

Average distance from the sun	1.00 AU (1.495979 × 10⁸ km)
Eccentricity of orbit	0.0167
Maximum distance from the sun	1.0167 AU (1.5210 × 10⁸ km)
Minimum distance from the sun	0.9833 AU (1.4710 × 10⁸ km)
Inclination of orbit to ecliptic	0°
Average orbital velocity	29.79 km/sec
Orbital period	1.00 y (365.25 days)
Period of rotation (with respect to the sun)	24ʰ00ᵐ00ˢ
Inclination of equator to orbit	23°27′
Equatorial diameter	12,756 km
Mass	5.976 × 10²⁴ kg
Average density	5.497 g/cm³ (4.2 g/cm³ uncompressed)
Surface gravity	1.0 Earth gravity
Escape velocity	11.2 km/sec
Surface temperature	−50° to 50°C (−60° to 120°F)
Average albedo*	0.39
Oblateness	0.0034

*Albedo is the fraction of light striking an object that the object reflects. The albedo of a perfect reflector is 1.

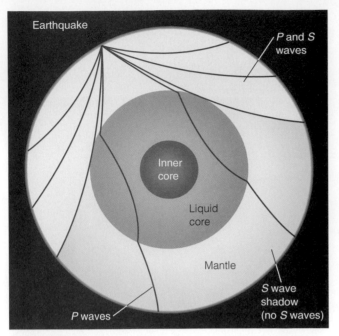

Figure 17-2
P and *S* waves give us clues to Earth's interior. That no direct *S* waves reach the far side shows that Earth's core is liquid. The size of the *S* wave shadow tells us the size of the outer core.

reversals, though poorly understood, are a characteristic of the dynamo effect. Thus, the presence of a magnetic field is evidence that part of Earth's core is a liquid metal.

The **mantle** is the layer of dense rock and metal oxides that lies between the molten core and the solid crust. The mantle material is a **plastic,** a material with the properties of a solid but capable of flowing under pressure. Asphalt used in paving roads is a common plastic; it shatters if struck with a sledgehammer, but it bends under the weight of a truck. Just below Earth's crust, where the pressure is less than at great depths, the mantle is most plastic.

The Earth's rocky crust is made up of low-density rocks and floats on the denser mantle. The crust is thickest under the contintents, up to 60 km thick, and thinnest under the oceans where it is only about 10 km thick.

While Earth's hot interior seems far from our everyday experience, it is in fact responsible for most of the scenery we see on the surface.

Plate Tectonics

Earth's surface is active in that it is constantly renewed as large sections of crust drift about. Geologists refer to this activity as **plate tectonics.** *Plate* refers to the sections of crust, and *tectonics* comes from the Greek word for "builder." Plate tectonics is the builder of Earth's surface. It pushes up mountains, destroys old crust, and creates new crust.

The plates move because the mantle is highly plastic just below the crust. Convection currents rise through the mantle and help break the crust into segments, which then slide on the plastic mantle at speeds of a few centimeters per year. The recognition of the importance of plate tectonics is one of the great scientific revolutions of the 20th century.

Two features of the ocean floor give us dramatic evidence of plate tectonics. In many places, the ocean floor is marked by a long underwater mountain range called a **midocean rise,** which is split by a **midocean rift** where the ocean crust is pulling apart and lava flows are welling up to form new ocean crust. (You may have heard this called *sea floor spreading.*) In other places, sections of ocean crust are pushing below continents into **subduction zones** where they build mountains and trigger volcanism (Figure 17-3).

The evidence is clear that midocean rises are young. The rocks near a midocean rift are geologically young **basalt,** rock formed by solidified lava, and the farther basalt is from the rift the older it is. Also, the ocean floor nearest a rift has accumulated less sediment than the ocean floor further away. Further evidence lies in the magnetic properties of the ocean floor. As molten rock wells up along a midocean rift and hardens, it traps a record of the Earth's magnetic field. As the ocean crust spreads and Earth's magnetic field periodically reverses, the newly created basalt records the changing direction of Earth's magnetic field in a series of bands that run parallel to the rift and mirror each other on opposite sides of the rift. This was first detected along the midocean rift that runs from the South Atlantic to north of Iceland (Figure 17-4).

Like a conveyor belt, ocean crust is created in midocean rifts, spreads outward, and disappears down subduction zones to be remelted by the heat of Earth's mantle. Where ocean crust slides downward under a continent, it crumples the crust at the edge of the con-

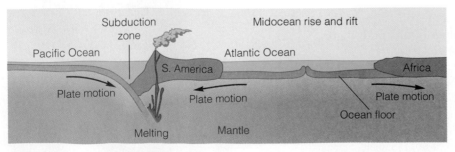

Figure 17-3
A cross section of Earth's crust shows the thick crust of the continents and the thinner crust of the sea floors. The motion of crustal plates allows rising magma to build new crust at the midocean rift, and old sea floor descends under the continents in subduction zones where it triggers mountain building and volcanism.

Figure 17-4
A color-coded relief map of Earth's continents and ocean floors reveals midocean rises such as that running the length of the Atlantic Ocean. The ocean floor spreads apart along the rift as the crustal plates drift apart. Magma rising into the rift hardens and records the alternate reversals of Earth's magnetic field as parallel bands of fossil magnetism in the seafloor (right). Other rises mark the Pacific and Indian oceans. (Compare with Figure 17-6.) *(National Geophysical Data Center)*

tinent to build mountains, and the melting of the ocean crust releases low-density magma that rises to the surface to create volcanoes. The coastal volcanoes (including Mount Saint Helens) extending from northern California into Alaska are caused by a tectonic plate plunging below the continent.

The seafloor does not always slip below a continent. The floor of the Atlantic Ocean is locked to North and South America and is pushing the continents westward. Tracing this continental drift backward in time, Earth scientists find that North and South America were in contact with Europe and Africa only 200 million years ago. At that point, the land masses on Earth were joined in one great continent, Pangaea (Figure 17-5). About 200 million years ago, Pangaea broke into two land masses, Laurasia in the north and Gondwanaland in the south. These land masses further fragmented and drifted apart to form the continents we know.

Continental drift continues today, driven by moving plates. Continents can split apart in a long, straight, deep depression called a **rift valley**. Africa has split from Arabia, opening a rift valley now filled by the Red Sea. In contrast, continents can collide. The collision of Africa with the east coast of North America, roughly 250 million years ago, folded the crust into the Ap-

palachian Mountains, just as the collision of India with southern Asia is now building the Himalaya Mountains. These **folded mountain chains** are typical of plate collisions. Sections of crust can slip past each other, as in the case of the Pacific plate carrying part of southern California northward along the San Andreas fault and generating frequent earthquakes.

Plotting the location of earthquakes and volcanism on a world map reveals the edges of the moving plates (Figure 17-6). The Pacific plate is bounded by a ring of Earthquake zones known as the "ring of fire." Shallow earthquakes and undersea volcanism also occur along the midocean rises, where new crust forms and the seafloor spreads.

Because Earth's crust is active, all geological features gradually change. The oldest existing portions of the crust, including parts of Canada and parts of South Africa and Australia, are only about 3.9 billion years old. The constant churning of Earth's surface has wiped away all record of older crust.

Earth's Atmosphere

Three questions confront us when we consider Earth's atmosphere: How did it form? How has it evolved? How

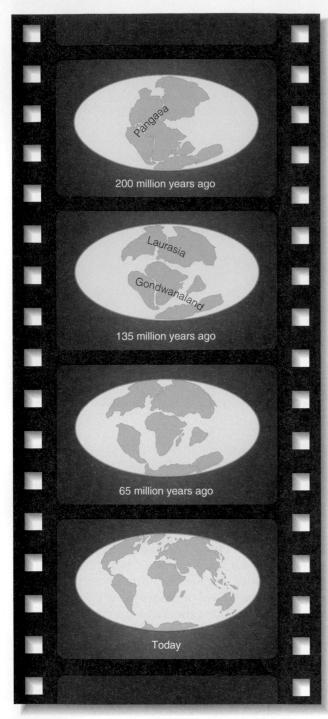

Figure 17-5
Continental drift has broken the original land mass Pangaea first into Laurasia and Gondwanaland and eventually into the continents we know. *(Adapted from R. S. Dietz and J. C. Holden,* Scientific American *223, April 1970)*

are we changing it? Answering these questions will help us understand other planets as well as our own.

Our planet's first atmosphere, its **primeval atmosphere,** was once thought to contain gases from the solar nebula, such as hydrogen, methane, and ammonia. Modern studies, however, tell us that the planets formed hot, so gases such as carbon dioxide, nitrogen,

and water vapor would have been cooked out of the rock and metal as Earth grew. In addition, the final stages of planet building may have seen planets accreting planetesimals rich in easily vaporized materials such as water, ammonia, and carbon dioxide. Thus the primeval atmosphere must have been rich in carbon dioxide, nitrogen, and water vapor.

Once Earth cooled enough, oceans began to form, and the carbon dioxide began to dissolve in the water. Carbon dioxide is highly soluble in water—which explains the easy manufacture of carbonated beverages. As the oceans removed carbon dioxide from the atmosphere, it reacted with dissolved compounds in the ocean water to form silicon dioxide, limestone, and other mineral sediments. Thus, the oceans transferred the carbon dioxide from the atmosphere to the seafloor and left air rich in nitrogen.

This removal of carbon dioxide is critical to us because an atmosphere rich in carbon dioxide can trap heat by a process called the **greenhouse effect.** When sunlight shines through the glass roof of a greenhouse, it heats the benches and plants inside. The warmed interior radiates heat in the form of infrared radiation, which can't get out through the glass. Heat is trapped in the greenhouse, and the temperature climbs until the glass itself grows warm enough to radiate heat away as fast as sunlight enters (Figure 17-7a). (Of course, a greenhouse also retains its heat because the walls prevent the warm air from mixing with the cooler air outside.) With a planet, sunlight can enter the atmosphere and warm the surface, but carbon dioxide, being opaque to infrared, prevents the heat from escaping. Thus, carbon dioxide in a planetary atmosphere can trap energy and warm the surface (Figure 17-7b).

For roughly 4 billion years, Earth's oceans and plant life have been absorbing carbon dioxide and burying it in the form of carbonates such as limestone and in carbon-rich deposits of coal, oil, and natural gas. But in the last century or so, human civilization has begun digging up those fuels, burning them for energy, and releasing the carbon back into the atmosphere as carbon dioxide. This process is steadily increasing the carbon dioxide concentration in our atmosphere and warming our climate (Figure 17-7c). Although the temperature increase is very small now, it is a serious issue for the future. A temperature rise of only a few degrees would begin melting the polar caps and would cause a rise in sea levels of as much as 50 meters. Florida, for example, would vanish. When we discuss Venus later in this chapter, we will see an extreme greenhouse effect.

Earth's early atmosphere may have contained some methane (CH_4) and ammonia (NH_3), but ultraviolet photons would have penetrated deep into the atmosphere and broken those molecules up to form carbon, nitrogen, and hydrogen. The hydrogen would escape to space; the carbon would form more carbon

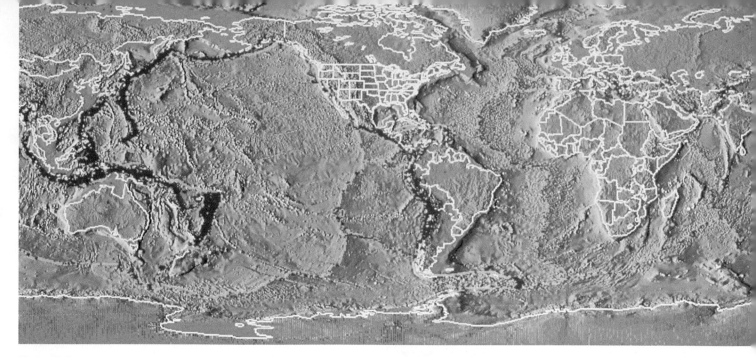

Figure 17-6
Earthquakes, plotted as red dots, outline the edges of Earth's crustal plates (yellow). Volcanism follows the same pattern, and much of the Pacific is ringed by volcanic peaks that extend from Chile through Central America, Alaska, and south through Japan—the "ring of fire." The earthquakes and volcanism in Hawaii are produced where a hot spot in the mantle penetrates upward through the middle of a plate. *(National Geophysical Data Center)*

Figure 17-7 ►
The greenhouse effect. (a) Short-wavelength light can enter a greenhouse and heat its contents, but the longer-wavelength infrared radiated by the warm interior cannot get out. (b) The same process can heat a planet's surface if its atmosphere contains CO_2. (c) The concentration of CO_2 in Earth's atmosphere as measured in Antarctic ice cores has remained roughly constant until the Industrial Revolution. *(Adapted from a figure by Etheridge, Steele, Langenfelds, Francey, Barnola, and Morgan.)*

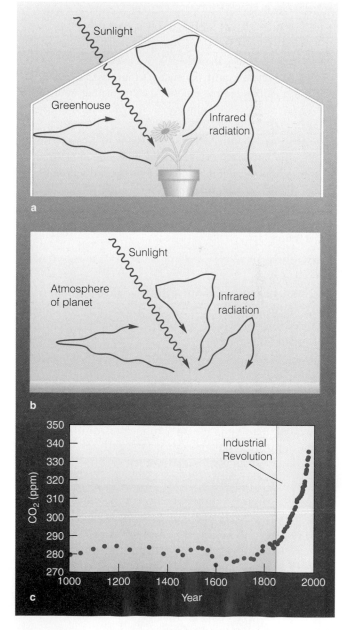

dioxide; and the nitrogen would be left in the atmosphere. This no longer happens because the atmosphere now contains free oxygen, which forms a layer of ozone (O_3) at altitudes of 15 to 30 km.

The ozone layer now protects the lower atmosphere from high-energy photons, but when Earth was young, its atmosphere had no free oxygen. Oxygen is very reactive and quickly forms oxides in the soil, so plant life is needed to keep a steady supply of oxygen in the atmosphere. Although life seems to have begun on Earth within a billion years of its formation, life did not generate free oxygen until the onset of photosynthesis about 3.3 billion years ago. Photosynthesis makes energy for the plant by absorbing carbon dioxide and releasing free oxygen. Apparently, the development of large, shallow seas along the continental margins half a billion years ago allowed ocean plants to manufacture oxygen faster than chemical reactions could remove it. Atmospheric oxygen then increased rapidly, and it is still increasing at the rate of about 1 percent every 36 million years.

The ozone layer exists because our atmosphere contains free oxygen, and it protects us from harmful ultraviolet photons; but certain compounds called chlorofluorocarbons (CFCs), which are used in refrigeration and industry, can destroy ozone when they leak into the atmosphere. Since the late 1970s, the ozone concentration has been falling, and the intensity of harmful ultraviolet radiation at Earth's surface has been observed increasing year by year. While this poses an immediate problem for public health on Earth, it is also of interest astronomically. When we discuss Mars, we will see the effects of an atmosphere without ozone.

REVIEW Critical Inquiry

What evidence do we have that Earth has a molten metallic core?

In science, the critical analysis of ideas must eventually return to evidence. In this case, the evidence is indirect because we can never visit Earth's core. Seismic waves from distant earthquakes pass through Earth, but a certain kind of wave, the S waves, cannot pass through the core. Because the S waves cannot propagate through a liquid, we suspect that Earth's core is a liquid. Earth's magnetic field gives us further evidence of a metallic core. The theory for the generation of magnetic fields, the dynamo effect, requires a liquid, convective, conducting core. If the core were not a liquid, it would not be able to generate a magnetic field. Thus, we have two different kinds of evidence to tell us that our planet has a liquid core.

Every theory must be supported by evidence. What evidence do we have to support the theory of plate tectonics?

As we learn more about other worlds and their atmospheres, we also learn more about our own. Ironically, the next world we will discuss has no atmosphere at all.

17-2 The Moon

The Earth's moon is the only other world that humans have visited, and although it is small and airless, it has a fascinating geological history.

Lunar Geology

The man in the moon is formed by the smooth, dark lunar lowlands. In contrast, the lunar highlands are lighter in color and heavily battered by uncountable craters, one on top of another. Even with the naked eye, we can see these regions on Earth's moon, but in

a photograph they are dramatic (Data File Three). The key to understanding Earth's moon lies in understanding why the lowlands, on average 3 km lower than the highlands, are so free of craters.

The lunar craters are impact features formed when meteorites struck the lunar surface at high velocity (Figure 17-8). These craters range from hundreds of miles in diameter down to microscopic pits. From studies of Earth's meteorite craters, we know that such impacts are rare now, so we must conclude that most craters of Earth's moon date from early in its history during the heavy bombardment at the end of planet building. Although no one has seen a new crater appear on Earth's moon, meteorites continue to fall, and craters must continue to form. This bombardment of the lunar surface has covered the surface with a layer of crushed rock called **ejecta,** and the youngest craters are bright and marked by radial **rays** of fresh ejecta.

Twelve Apollo astronauts visited both the highlands and the lowlands on Earth's moon between 1969 and 1972 (Figure 17-9). Most of the rocks they found were basalts typical of hardened lava, and some were **vesicular basalt,** which contains holes formed by bubbles in the molten rock. These bubbles form when the rock flows out onto the surface, and the lower pressure allows gases dissolved in the rock to form bubbles. The same thing happens when we open a bottle of carbonated beverage and bubbles form. The presence of vesicular basalts shows that much of the surface of Earth's moon has been covered by successive lava flows, and the flat plains of the lunar lowlands, known as **maria** (singular **mare**), Latin for "seas," are actually ancient lava flows. The highlands, in contrast, are rich in **anorthosite,** a light-colored rock of low density that would be among the first to solidify and float to the top of molten rock. Many of the rocks all over Earth's moon are **breccias,** rocks made up of fragments of broken rock cemented together under pressure. The breccias tell us how extensively the lunar surface has been shattered by meteorites. Nowhere did the astronauts find what they could call bedrock; the entire surface of Earth's moon is fractured by meteorite impacts.

As the astronauts bobbed across the lunar surface, their boots kicked up the powdery dust that covers everything on Earth's moon. This lunar dust is produced by the continuous bombardment of the lunar surface by micrometeorites, which slowly grind exposed rocks into fine gray grit about the consistency of talcum powder.

The color of moon rocks is a dark gray, but Earth's moon looks quite bright in the night sky. In fact, the **albedo** of Earth's moon, the fraction of the light that it reflects, is only 0.06. In other words, it reflects only 6 percent of the light that hits it. Earth, thanks mostly to its bright clouds, has an average albedo of 0.39. Thus, our moon looks bright in the night sky only in contrast. It is in reality, a cold, airless, gray world.

Figure 17-8
The formation of a crater. An approaching meteorite (a) strikes the lunar surface (b) and penetrates deeply. The meteorite vaporizes, and shock waves fracture the rock and blast the surface away (c) to form a circular crater. (d) Ejected material forms a raised crater rim. The rebound of the rock can form a raised central peak, and slumping along the crater wall can form terraces. The rock below the crater is severely fractured.

The History of Earth's Moon

The four-stage history of Earth's moon is dominated by a single fact — it is small, only one-fourth the diameter of Earth. The escape velocity is low, so it has been unable to hold any atmosphere, and it cooled rapidly as its internal heat flowed outward into space. As we will see when we study other worlds, a planet's

The Moon Data File Three

The lunar lowlands (the maria) are dark and smooth, while the highlands are bright and cratered. *(Lick Observatory photograph)*

Average distance from Earth	384,400 km (center to center)
Eccentricity of orbit	0.055
Maximum distance from Earth	405,500 km
Minimum distance from Earth	363,300 km
Inclination of orbit to ecliptic	5°9′
Average orbital velocity	1.022 km/sec
Orbital period (sidereal)	27.321661 days
Orbital period (synodic)	29.5305882 days
Inclination of equator to orbit	6°41′
Equatorial diameter	3476 km
Mass	7.35×10^{22} kg (0.0123 $M_\oplus$)
Average density	3.36 g/cm³ (3.35 g/cm³ uncompressed)
Surface gravity	0.167 Earth gravity
Escape velocity	2.38 km/sec (0.21 $V_\oplus$)
Surface temperature	−170° to 130°C (−274° to 266°F)
Average albedo	0.07

a

b

Figure 17-9

(a) Edwin E. Aldrin, Jr., one of the three Apollo 11 astronauts, stands on the dusty surface of Mare Tranquillitatis in the lunar lowlands, site of the first lunar landing. The flat lava plain is almost featureless, and the horizon is straight. (b) In the lunar highlands at Taurus–Littrow, the surface is mountainous and irregular. Note the horizon. Harrison Schmitt, a scientist-astronaut on Apollo 17, is shown passing huge boulders. *(NASA)*

geology is largely driven by energy in the form of heat flowing outward from its interior (Window on Science 17-1). Small worlds have less heat and lose it more rapidly, so Earth's moon's small size has been critical in its history.

During the first stage of planet formation, Earth's moon formed and differentiated. The Apollo moon rocks show that our moon formed in a molten state, allowing the denser materials it contained to sink to form a small core. Earth's moon has a low density and is poor in iron, so the core is not large and must contain little iron. The lowest-density minerals rose to form a low-density anorthosite crust. Radioactive ages of moon rocks tell us that the surface solidified 4.6 to 4.1 billion years ago.

The second stage, cratering, began as soon as the crust solidified, and the older highlands show that the cratering was intense during the first 0.5 billion years—during the heavy bombardment at the end of planet building. Earth's moon's anorthosite crust was shattered, and the largest impacts formed giant crater basins hundreds of kilometers in diameter (Figure 17-10). The basin that became Mare Imbrium, for instance, was blasted out by the impact of an object about the size of Rhode Island. This Imbrium event occurred about 4 billion years ago and blanketed 16 percent of our moon with ejecta. Between 4.1 and 3.9 billion years ago, the cratering rate fell rapidly to the current rate, roughly 1000 times less than during the early heavy bombardment.

The tremendous impacts that formed the lunar basins cracked the anorthosite crust as deep as a few kilometers and led to the third stage—flooding. Though Earth's moon cooled rapidly after its formation, some process, perhaps radioactive decay, heated the subsurface material; and part of it melted, producing lava that followed the cracks up into the giant basins (Figure 17-11a). The basins were flooded by successive lava flows of dark basalts from 3.8 to 3.2 billion years ago, thus forming the maria.

Studies of our moon show that its crust is thinner on the side toward Earth, perhaps due to tidal effects. Consequently, while lava flooded the basins on the Earthward side, it was unable rise through the thicker crust to flood the lowlands on the far side. The largest impact basin in the solar system is the South Pole–Aitkin Basin (Figure 17-11b). It is about 2500 kilometers (1500 miles) in diameter and as deep at 13 kilometers (8 miles) in places, but flooding has never filled it with smooth lava flows.

The fourth stage, slow surface evolution, is limited because our moon lacks water and has cooled rapidly. Flooding on Earth included water, but our moon has never had an atmosphere and thus has never had liquid water. With no air and no water, erosion is limited to the constant bombardment of micrometeorites and rare larger impacts. Indeed, a few

One of the best ways to think about a scientific problem is to follow the energy. According to the principle of cause and effect, every effect must have a cause, and every cause must involve energy (Window on Science 4–3). Energy moves from regions of high concentration to regions of low concentration and, in doing so, produces changes. For example, we may burn coal to make steam in a power plant, and the steam passes through a turbine and then escapes into the air. In flowing from the burning coal to the atmosphere, the heat spins the turbine and makes electricity.

It is surprising how commonly scientists use energy as a key to understanding nature. A biologist might ask where certain birds get the energy to fly thousands of miles, and an economist might ask where the economic energy to support the creation of new investments comes from. Energy is everywhere, and when it moves, whether it is birds, money, or molten magma, it causes change. Energy is the cause in cause and effect.

In earlier chapters, we used the flow of energy from the inside of a star to its surface to understand how stars, including the sun, work. We saw that the outward flow of energy supports the star against its own weight, drives convection currents that produce magnetic fields, and causes surface activity such as spots, prominences, and flares. We were able to understand stars because we could follow the flow of energy outward from their interiors.

We can think of a planet by following the energy. The heat in the interior of a planet may be left over from the formation of the planet, or it may be heat generated by radioactive decay; but it must flow outward toward the cooler surface, where it is radiated into space. In flowing outward, the heat can cause convection currents in the mantle, magnetic fields, plate motions, quakes, faults, volcanism, mountain building, and more.

When you think about any world, be it a small asteroid or a giant planet, think of it as a source of heat that flows outward through the planet's surface into space. If we can follow that energy flow, we can understand a great deal about the world. A planetary astronomer once said, "The most interesting thing about any planet is how its heat gets out."

Figure 17–10 ▶

The evolution of the lunar surface is illustrated in this artist's conception of the formation of Mare Imbrium. A major impact (a) excavated a large basin, which was repeatedly flooded by lava (b). Since the end of the lava flows, minor cratering has created the surface we see (c). *(Courtesy Don Davis)*

meteorites found on Earth have been identified as moon rocks ejected from our moon by major impacts within the last few million years. As our moon lost its internal heat, volcanism died down; and Earth's moon became geologically dead. Its crust never divided into moving plates—we find no folded mountain chains— and it is now a one-plate planet, frozen between stages 3 and 4.

Although we can tell the story of the lunar surface, we have neglected one important point. We have not said how Earth's moon formed. We will explore that idea in the next section.

The Origin of Earth's Moon

Over the last two centuries, astronomers have developed three different hypotheses for the origin of Earth's moon, but these traditional ideas have failed to survive comparison with the evidence. A theory proposed in the 1980s may hold the answer. We begin by testing the three unsuccessful theories against the evidence.

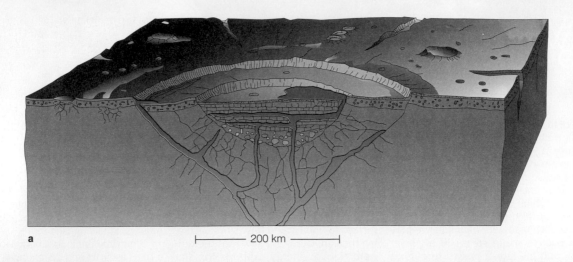

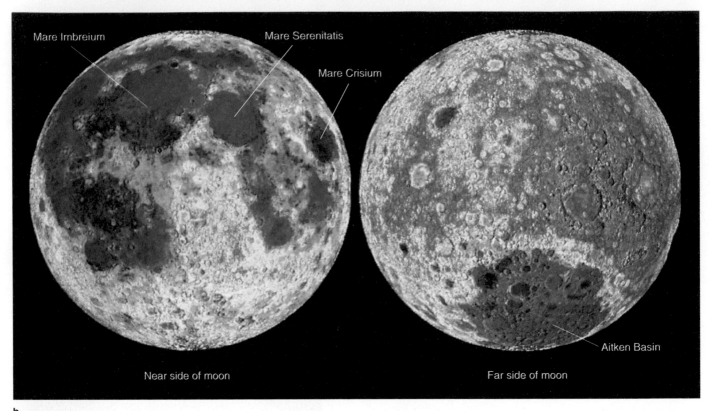

a

|—— 200 km ——|

Mare Imbreium

Mare Serenitatis

Mare Crisium

Aitken Basin

Near side of moon

Far side of moon

b

Figure 17-11

(a) Major impacts broke the lunar crust and produced large basins. Lava flooded up through the fractures and flooded the basins to produce maria. *(Adapted from a diagram by William Hartmann)* (b) In these images color marks elevation, with red the highest regions and purple the lowest. Maria on the near side are generally circular, low, lava-filled basins. Note the circular outline of the labeled maria. The crust on the far side is thicker, and there is less flooding. Even the Aitken Basin contains little lava flooding. *(NASA/Clementine)*

The **fission hypothesis** proposed that our moon broke from a rapidly spinning proto-Earth. If this happened after the proto-Earth differentiated, our moon would have formed from iron-poor material. However, moon rocks differ chemically from those of Earth. Also, if the proto-Earth had spun fast enough to break up, the Earth–moon system should now contain much more angular momentum than it does.

The **condensation hypothesis** suggested that Earth and its moon condensed from the same cloud of matter in the solar nebula. This idea doesn't work, however, because Earth and its moon have different densities and compositions. The moon, for example, is very poor in volatiles—materials such as water, sodium, and calcium, which are easily vaporized.

The **capture hypothesis** suggested that our moon formed elsewhere in the solar nebula and was later captured by Earth. However, if our moon passed near enough to Earth to be captured, Earth's gravity would have ripped it to fragments.

Until the mid-1980s, astronomers had no acceptable hypothesis for the origin of Earth's moon, but at that point a hybrid theory offered hope. The **large-impact hypothesis** supposes that our moon formed when a planetesimal at least as large as Mars smashed into the proto-Earth and ejected debris into an orbit, where it formed our moon (Figure 17-12).

This would explain a number of phenomena. If the collision occurred off center, it would have spun the Earth–moon system rapidly and would thus explain the present angular momentum. If the proto-Earth and impactor had already differentiated, the ejected material would have been mostly iron-poor mantle and crust. Models show the iron core of the impactor would have fallen into Earth within 4 hours, leaving our moon to form from iron-poor rock. Also, the material would have remained in a disk around Earth long enough to have lost the volatile materials that our moon lacks.

The large-impact hypothesis survives comparison with the known evidence and is now a widely considered hypothesis.

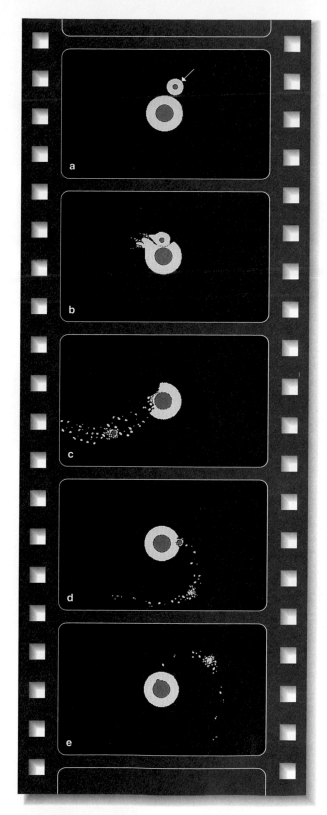

Figure 17-12
A Mars-sized planetesimal strikes the young, molten Earth, blasting iron-poor mantle material into orbit. In this simulation, the iron core of the planetesimal falls back into Earth within 4 hours (d), and the moon begins to form within about 10 hours (e).

REVIEW Critical Inquiry

Why are the maria nearly free of craters?

Timing is everything, someone has said, and in this case our analysis must carefully consider the sequence of events. Earth's moon's anorthosite crust formed and was heavily cratered before about 4 billion years ago. The impacts of the heavy bombardment cratered the entire lunar surface and created some very large crater basins. Later, after the end of that heavy cratering, lava welled up and filled the lowlands of the largest crater basins with basalt to form the maria. Any craters in the maria were covered over, and there were few impacts to form new craters. Thus, the maria are nearly free of craters; but the ancient highlands are heavily cratered.

Explain how timing is important in explaining the formation of an iron-poor moon in the large-impact hypothesis.

The evolution of Earth's moon has been restricted by its small size. Another world in our solar system has also suffered from its diminutive size. Mercury, like Earth's moon, is a small, rocky world, cratered by impacts and flooded by ancient lava flows that are driven by its internal heat.

Mercury   **Data File Four**

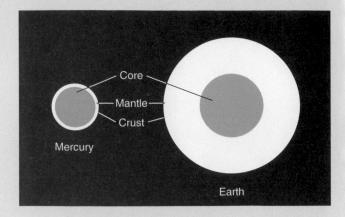

In proportion to its size, Mercury has a very large metallic core.

Average distance from the sun	0.387 AU (5.79×10^7 km)
Eccentricity of orbit	0.2056
Maximum distance from the sun	0.467 AU (6.97×10^7 km)
Minimum distance from the sun	0.306 AU (4.59×10^7 km)
Inclination of orbit to ecliptic	7°00′16″
Average orbital velocity	47.9 km/sec
Orbital period	0.24085 y (87.969 days)
Period of rotation	58.646 days (direct)
Inclination of equator to orbit	0°
Equatorial diameter	4878 km (0.382 $D_\oplus$)
Mass	3.31×10^{23} kg (0.0558 $M_\oplus$)
Average density	5.44 g/cm³ (5.4 g/cm³ uncompressed)
Surface gravity	0.38 Earth gravity
Escape velocity	4.3 km/sec (0.38$V_\oplus$)
Surface temperature	−170°C to 430°C (−280°F to 800°F)
Average albedo	0.1
Oblateness	0

17-3 Mercury

Mercury is intermediate in size between Earth and its moon (Data File Four and Figure 16-7). Like Earth's moon, Mercury cooled too quickly to develop plate tectonics. Thus, it is a cratered, dead world.

Mariner 10 at Mercury

Because Mercury's orbit keeps it near the sun, its surface temperature in direct sunlight can reach 430°C (800°F), hot enough to melt lead. So close to the sun, Mercury is difficult to observe from Earth, and little was known about it until 1974–75, when the Mariner spacecraft flew past Mercury three times and revealed a planet whose surface is heavily cratered, much like that of Earth's moon (Figure 17-13). Analysis shows that large areas have been flooded by lava and then cratered.

The largest impact feature on Mercury is the Caloris Basin, a ringed area 1300 km (800 miles) in diameter (Figure 17-14a), which looks much like Mare Orientale, the ringed basin on Earth's moon. Both features consist of concentric rings of cliffs formed by a large impact.

Figure 17-13
Photomosaic of Mercury made by the Mariner 10 spacecraft. Caloris Basin is in shadow at left. Almost no surface detail is visible from Earth (inset). *(NASA; inset courtesy Lowell Observatory)*

Like any technical subject, science includes a mass of details, facts, figures, measurements, and observations. It is easy to be overwhelmed by the flood of details, but one of the most important characteristics of science comes to our rescue. The goal of science is not to discover more details, but to explain the details with a unifying hypothesis or theory. A good theory is like a basket that makes it easier for us to carry a large assortment of details.

This is true of all the sciences. When a psychologist begins studying the way the human eye and brain respond to moving points of light, the data are a sea of detailed measurements and observations. Once the psychologist forms a hypothesis about the way the eye and brain interact, the details fall into place as parts of a log-

ical story. If we understand the hypothesis, the details all fit together and make sense, and thus we can remember the details without blindly memorizing tables of facts and figures. The goal of science is understanding, not memorization.

Scientists are in the storytelling business. The stories are often called hypotheses or theories, but they are really just stories to explain how nature works. The difference between scientific stories and works of fiction lies in the use of facts. Scientific stories are constructed to fit all known facts and are then tested over and over against new facts obtained by observation and experiment.

When we try to tell the story of each planet in our solar system, we pull together all the hypotheses and theories

and try to make them into a logical history of how the planet got to be the way it is. Of course, our stories will be incomplete because astronomers don't understand all the factors in planetary astronomy. Nevertheless, our story of each planet will draw together the known facts and details and attempt to make them a logical whole.

Memorizing a list of facts can give us a false feeling of security, just as when we memorize the names of things without understanding them (Window on Science 2-2). Rather than memorizing facts, we should search for the unifying hypothesis that pulls the details together into a single story. Our goal in studying science is understanding nature, not just remembering facts.

Figure 17-14
(a) Caloris Basin on Mercury lies partly in darkness to the left in this Mariner 10 photo. It is a vast impact basin, 1300 km in diameter, filled with lava flows and ringed by concentric mountain ranges up to 3 km high. (b) Discovery Rupes, a lobate scarp, cuts through craters that must predate the formation of the lobate scarps. *(NASA)*

a

b

Though Mercury looks moonlike, it does have one characteristic feature Earth's moon lacks. Mariner 10 photos revealed long curving ridges called **lobate scarps** up to 3 km (1.9 miles) high and 500 km (310 miles) long (Figure 17-14b). The scarps even cut through craters, indicating that they formed after most of the heavy bombardment.

Mercury is quite dense, and models tell us it must have a large metallic core (Data File Four). In fact, the metallic core occupies about 70 percent of the radius of the planet. In a sense, Mercury is a metal planet with a thin rock mantle.

A History of Mercury

The accumulated facts about Mercury don't really help us understand the planet until we have a unifying hypothesis. Like a story, it must make sense and bring the known facts together in a logical argument that explains how Mercury got to be the way it is (Window on Science 17-2).

Mercury is small, and that fact has determined much of its history. Not only is it too small to retain an atmosphere, but it has also lost much of its internal heat and thus is not geologically active.

In the first stage, Mercury differentiated to form a metallic core and a rocky mantle. Mariner 10 discovered a magnetic field about 10^{-4} times that of Earth—further evidence of a metallic core. In Chapter 16, we saw that the condensation sequence could explain the abundance of metals in Mercury, but detailed calculations show that Mercury contains even more iron than we would expect. Drawing on the large-impact hypothesis for the origin of Earth's moon, astronomers have proposed that Mercury suffered a major impact soon after it differentiated, an impact so large it shattered the rocky mantle and drove much of it away. The remaining iron and rock then re-formed the present Mercury with its unusually large iron core. Such catastrophic events are rare in nature, but they do occur, so astronomers must be prepared to consider such hypotheses. (See Window on Science 16-1.)

In the second and third stages of planet formation, cratering battered the crust, and lava flows welled up to fill the lowlands, just as they did on our moon. As the small world lost internal heat, its large metal core contracted and its crust was compressed, breaking to form the lobate scarps much as the peel of a drying apple wrinkles.

Lacking an atmosphere to erode it, Mercury has changed little since the last lava hardened, and it is now a one-plate planet much like Earth's moon.

REVIEW Critical Inquiry

Why don't Earth and Earth's moon have lobate scarps?

At first glance, we might suppose that any world with a metallic interior should have lobate scarps, but other factors are also important. Earth has a fairly large metallic core; but, being a large world, it has not cooled very much, so it presumably hasn't shrunk much. Of course, the geological activity on Earth's surface would erase such scarps if they did form. Earth's moon is not geologically active, but it does not contain a significant metallic core. Although Earth's moon has lost much of its internal heat, its interior is mostly rock and didn't shrink as much as metal would have.

We can, in a general way, understand lobate scarps; but not all materials contract when they cool and solidify. Water, for instance, expands when it freezes. What kinds of features would you expect to find on small, cold, icy worlds that once had interiors of liquid water? Such icy worlds exist as Earth's moons of the outer planets, and we will explore them in Chapter 18.

Mercury, like Earth's moon, is now an inactive world. To find planets where the heat flowing outward from the interior still drives geological activity, we must look at larger worlds such as Venus and Mars.

17-4 Venus

We might expect Venus to be much like Earth. It is 95 percent Earth's diameter (Data File Five), has a similar average density, and is only 30 percent closer to the sun (see Figure 16-9). Unfortunately, the surface of Venus is perpetually hidden below thick clouds (Figure 17-15a), and only in the last few decades have astronomers discovered that Venus is a deadly hot desert world of volcanoes, lava flows, and impact craters lying at the bottom of a deep ocean of hot gases.

The Atmosphere of Venus

In composition, temperature, and density, the atmosphere of Venus is more Hades than Heaven. The air is unbreathable, very hot, and almost 100 times denser than Earth's air. How do we know? Because U.S. and Soviet space probes have descended into the atmosphere and in some cases landed on the surface.

In composition, the atmosphere of Venus is roughly 96 percent carbon dioxide. The rest is mostly nitrogen, with some argon, sulfur dioxide, and small amounts of sulfuric acid, hydrochloric acid, and hydrofluoric acid. There is only a tiny amount of water vapor. On the whole, the composition is deadly unpleasant, and spectra show that the impenetrable clouds that hide the surface are made up of droplets of sulfuric acid and microscopic crystals of sulfur (Figure 17-15b).

This unbreathable atmosphere is 90 times denser than Earth's atmosphere. The air we breathe is 1000 times less dense than water, but on Venus the air is only 10 times less dense than water. If we could survive the unpleasant composition and intense heat, we could strap wings on our arms and fly.

The surface temperature on Venus (Data File Five) is hot enough to melt lead, and we can understand that because the thick atmosphere creates a severe greenhouse effect. Sunlight filters down through the clouds and warms the surface, but heat cannot escape because the carbon dioxide gas is opaque to infrared. Traces of sulfur dioxide and water vapor also help trap the infrared, but it is the overwhelming abundance of carbon dioxide that makes the greenhouse effect on Venus much more important than on Earth.

The Surface of Venus

Although the thick clouds on Venus are opaque to visible light, they are transparent to radio waves, so astronomers have been able to map Venus using radar. As early as 1965, Earth-based radio telescopes made low-resolution maps, but later both U.S. and Soviet spacecraft orbited Venus and mapped its surface by radar. Maps made in the early 1990s by the

Magellan spacecraft reveal objects as small as 100 m in diameter.

The radar maps show that Venus is dramatically different from Earth. Roughly 65 percent of Venus is covered by low rolling plains, 24 percent is highlands, and the rest is very high volcanic uplands. Names of features on Venus are all female, with three exceptions—Maxwell, Alpha Regio, and Beta Regio. The highland region Ishtar Terra, for example, is named after Ishtar, the Babylonian goddess of love (Figure 17-16a).

The highlands resemble continents on Earth, but they are geologically quite different. Ishtar Terra, for example, is about the size of Australia. At its eastern edge, the mountain called Maxwell Montes thrusts up 12 km (Everest is 8.8 km high) with the impact crater Cleopatra on its lower slopes (Figure 17-16b). Bounded by mountain ranges in the north and west, the center of Ishtar Terra is occupied by Lakshmi Planum, a great plateau about 4 km above the surrounding plains. The collapsed calderas Colette and Sacajawea suggest that Lakshmi Planum is a great lava plain. The mountains bounding Ishtar Terra, including Maxwell, resemble folded mountain ranges, and that suggests that limited horizontal motion in the crust as well as volcanism has helped form the highlands.

Volcanism is clearly important on Venus. Many volcanic peaks rise high above the surface, and lava flows spread outward (Figure 17-17). These are **shield volcanoes** much like those in Hawaii. They are formed by highly fluid lava and have shallow slopes. Long, narrow lava channels meander for thousands of kilometers across the surface. At a length of 6800 km, Baltis Vallis is the longest known channel in the solar system. The length of these channels testifies that the lava was highly fluid and the environment was hot enough to prevent the lava from cooling rapidly. Although no eruption has been detected in progress, it seems safe to assume that some volcanoes on Venus are currently active.

Smaller volcanoes, numerous faults, and sunken regions produced when magma below the surface drained away are further evidence of volcanism. Related features, the **coronae,** are circular bulges up to 2100 km in diameter bordered by fractures, volcanoes, and lava flows (Figure 17-18a on page 356). These appear to be caused by rising convection currents of molten magma below the crust.

Radar images show that Venus is marked by numerous craters (Figure 17-18b). The atmosphere protects the surface from smaller meteorites that would produce craters smaller than 3 km in diameter. Larger meteorites penetrate the atmosphere and have formed about 10 percent as many craters on Venus as on Earth's moon's maria. The number of craters show that the crust is not as ancient as the lunar maria, but not as young as Earth's active surface. The average age

Venus Data File Five

This radar image of Venus strips the cloudy atmosphere away to reveal a world marked by volcanoes and lava flows (bright regions). *(NASA)*

Average distance from the sun	0.7233 AU (1.082×10^8 km)
Eccentricity of orbit	0.0068
Maximum distance from the sun	0.7282 AU (1.089×10^8 km)
Minimum distance from the sun	0.7184 AU (1.075×10^8 km)
Inclination of orbit to ecliptic	3°23′40″
Average orbital velocity	35.03 km/sec
Orbital period	0.61515 y (224.68 days)
Period of rotation	243.01 days (retrograde)
Inclination of equator to orbit	177°
Equatorial diameter	12,104 km (0.95 $D_\oplus$)
Mass	4.870×10^{24} kg (0.815 $M_\oplus$)
Average density	5.24 g/cm^3 (4.2 g/cm^3 uncompressed)
Surface gravity	0.903 Earth gravity
Escape velocity	10.3 km/sec (0.92$V_\oplus$)
Surface temperature	745 K (472°C, or 882°F)
Albedo (cloud tops)	0.76
Oblateness	0

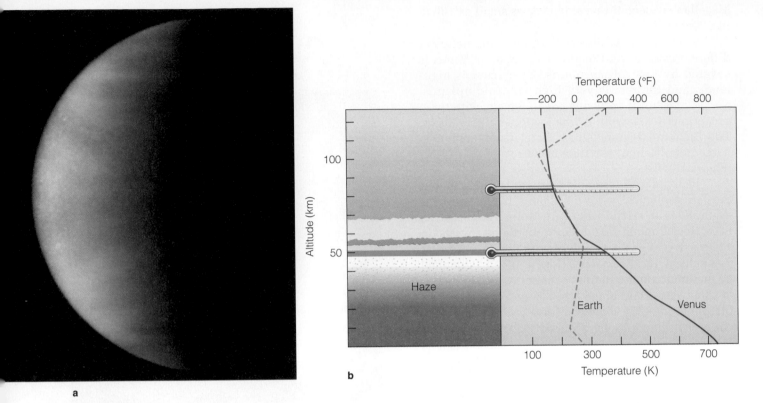

Figure 17-15
(a) The clouds of Venus are composed not of water droplets but of sulfuric acid droplets. This ultraviolet image made with the Hubble Space Telescope shows typical weather patterns in the clouds, but no part of the surface is visible. *(L. Esposito, University of Colorado, Boulder, and NASA)* (b) The four principal cloud layers in the atmosphere of Venus are shown at left. Thermometers inserted into the atmosphere at different levels would register the temperatures indicated by the red line in the graph at the right. The blue dashed line shows the temperatures in Earth's atmosphere for comparison. Note the difference between the surface temperature of Earth and that of Venus.

of the surface of Venus is roughly 300 million years. Clearly, plate tectonics cannot be renewing the surface as rapidly as it does on Earth.

No astronaut has ever stood on Venus, but both U.S. and Soviet spacecraft have landed on the surface and survived the heat and pressure for a few hours. A few of those spacecraft have analyzed the rock and snapped a few photographs (Figure 17-19). The surface rocks on Venus appear to be basalts much like those in Earth's ocean floors. This confirms our suspicion that volcanism is important. Photographs radioed back to Earth reveal a rocky world bathed in the orange glow of sunlight filtering through the thick clouds. In fact, Magellan radar images were given an arbitrary orange tint to mimic this lighting effect. In reality, rocks on Venus appear to be dark gray, typical of basalts.

Although Venus is Earth's twin in size, its surface geology is dramatically different.

The History of Venus

To tell the story of Venus we must draw together all the evidence and find hypotheses to explain two things: the thick carbon dioxide atmosphere and the peculiar geology.

Calculations show that Venus and Earth have outgassed about the same amount of carbon dioxide, but Earth's oceans have dissolved it and converted it to sediments such as limestone. If all of Earth's carbon were dug up and converted back to carbon dioxide, our atmosphere would be about as dense as the air on Venus. This suggests that the main difference between Earth and Venus is the lack of water on Venus. Venus may have had small oceans when it was young; but, being closer to the sun, it was warmer, and the carbon dioxide in the atmosphere created a greenhouse effect that made the planet even warmer. That process could have dried up any oceans that did exist and reduced the ability of the planet to purge its atmosphere of carbon dioxide. As more carbon dioxide was outgassed, the greenhouse effect grew even more severe. Thus, Venus seems to have been trapped in a runaway greenhouse effect.

The intense heat at the surface may have affected the geology of Venus by making the crust drier and stronger so that it was unable to break into moving plates as on Earth. There is no sign of plate tectonics on Venus, but rather evidence that convection currents below the crust are deforming the crust to create coronae and push up mountains such as Maxwell.

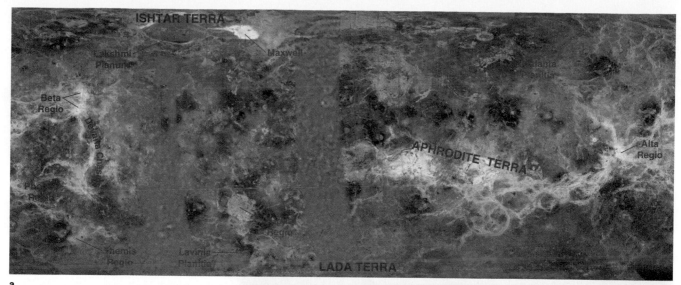

a

Figure 17-16

(a) Magellan radar images have been combined to form this map of Venus. North and south polar regions are not shown. The highlands consist mainly of Ishtar Terra, Aphrodite Terra, and Lada Terra. *(NASA)* (b) The largest feature in Ishtar Terra is Lakshmi Planum, a volcanic plain produced by lava flooding from vents such as Colette and Sacajawea. Here, violet is the smoothest terrain, green rougher, and orange the roughest. *(USGS)*

b

Figure 17-17

Below the clouds. A computer-generated image based on Magellan radar data shows volcanoes Gula Mons (left) and Sif Mons (right) and associated lava flows. The vertical scale has been stretched by a factor of about 20. Volcanoes on Venus are shallow-sloped shield volcanoes (inset). *(NASA)*

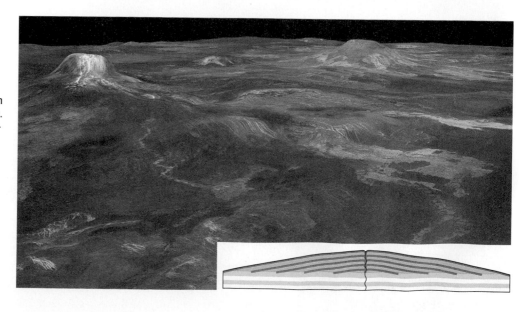

Detailed measurements of the strength of gravity over the mountains on Venus show that some must be held up not by deep roots like mountains on Earth but by rising currents of magma. Other mountains, like those around Ishtar Terra, appear to be folded mountains caused by limited horizontal motions in the crust, driven perhaps by convection in the mantle.

The surface of Venus contains only about 900 craters, far fewer than our moon but far more than Earth. This tells astronomers that the surface of Venus has been totally replaced within the last 300 to 500 million years, perhaps by a planetwide outburst of volcanism. Models of the climate on Venus suggest that major volcanism could increase the greenhouse effect and drive the surface temperature up by as much as 100°C. This could soften the crust and increase the extent of volcanism. These resurfacing episodes may occur periodically on Venus, or the

a

Figure 17-18
(a) This Magellan radar image shows coronae Bahet (left) and Onatah (right). Concentric and radial faults suggest the coronae are scars left behind by rising currents of molten rock below the crust. The vertical black bar is data lost in transmission from the spacecraft. (b) Impact crater Howe in the foreground of this Magellan radar image is 37 km in diameter. Craters in the background are 47 km and 63 km in diameter. *(NASA)*

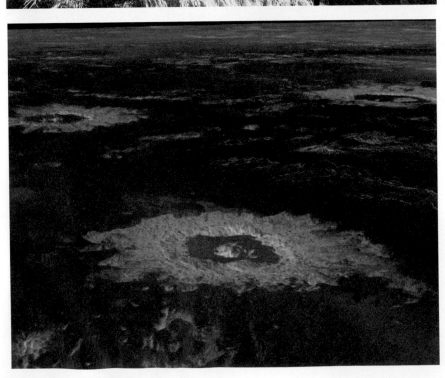

b

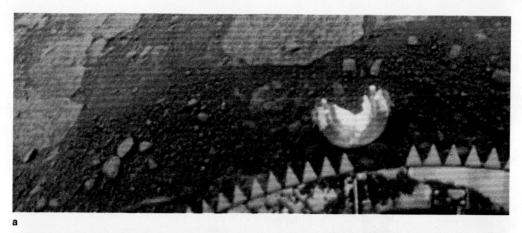

a

b

Figure 17-19
The color of the surface of Venus appears orange in the original Venera 13 photograph (a), but analysis shows that the color is produced by the thick atmosphere reddening sunlight into a day-long sunset glow. The corrected image (b) reveals slabs of gray rock and dark soil typical of iron oxides at high temperatures. An instrument cover and the base of the spacecraft are visible at lower right. *(C. M. Pieters through the Brown/Vernadsky Institute to Institute Agreement and the U.S.S.R. Academy of Science)*

planet may have had a geology more like Earth's until the surface was flooded by lava. In either case, Earth's neighboring planet, its apparent twin, is in fact astonishingly unearthly.

REVIEW Critical Inquiry

What evidence do we have that Venus does not have plate tectonics?

When we talked about Earth, we identified certain geological features that are characteristic of plate tectonics: rift valleys, subduction zones, and folded mountain chains. We don't see these features on Venus even in the highest-resolution radar maps made by the Magellan spacecraft. There are no rift valleys or subduction zones outlining plates, and there are no major chains of folded mountains. Rather, we see evidence of volcanism and lava flows, and the small regions of folded mountains suggest only limited horizontal motion in the crust. Rather than being dominated by the motion of rigid crustal plates, Venus may have a crust made more flexible by the heat, a crust dominated by rising plumes of molten rock that strain the crust to produce coronae and that break through to form volcanoes and lava flows.

At first glance, comparative planetology suggested that Earth and Venus should be sister worlds, but it seems they can be no more than distant cousins. We can blame the thick atmosphere of Venus for altering its geology, but that raises the question: Why isn't Earth's atmosphere similar?

Venus is certainly an unearthly world, but the familiar principles of comparative planetology help us understand it. As we turn our attention to Mars, we will see the same principles in action, but we will discover that Mars is a world that lacks what Venus has in abundance—internal heat and air.

Mars Data File Six

This Hubble Space Telescope view of Mars shows the northern polar cap at the top, impact craters, and dark surface features. The spot near the left edge is the volcano Ascraeus Mons, 25 km (16 mi) high, poking up through morning clouds of water-ice crystals in the thin CO_2 atmosphere. *(Philip James, University of Toledo; Steven Lee, University of Colorado, Boulder; and NASA)*

Average distance from the sun	1.5237 AU (2.279×10^8 km)
Eccentricity of orbit	0.0934
Maximum distance from the sun	1.6660 AU (2.492×10^8 km)
Minimum distance from the sun	1.3814 AU (2.066×10^8 km)
Inclination of orbit to ecliptic	1°51'09"
Average orbital velocity	24.13 km/sec
Orbital period	1.8808 y (686.95 days)
Period of rotation	$24^h37^m22.6^s$
Inclination of equator to orbit	23°59'
Equatorial diameter	6796 km (0.53 $D_{\oplus}$)
Mass	0.6424×10^{24} kg (0.1075 $M_{\oplus}$)
Average density	3.94 g/cm³ (3.3 g/cm³ uncompressed)
Surface gravity	0.379 Earth gravity
Escape velocity	5.0 km/sec (0.45$V_{\oplus}$)
Surface temperature	−140° to 20°C (−220° to 68°F)
Average albedo	0.16
Oblateness	0.009

17-5 Mars

Mercury and Earth's moon are small. Venus and Earth are, for terrestrial planets, large. But Mars, with a diameter 53 percent that of Earth's, is intermediate (Data File Six and Figure 16-9). In some ways, however, Mars is much like Earth. It rotates on its axis in 24 hours 37 minutes, and its year is 1.88 Earth years long. Because its axis of rotation is tipped, it has seasons as summer and winter come and go. Nevertheless, Mars is a forbidding world, with an ancient surface marked by craters and volcanoes and a thin, cold atmosphere.

The Atmosphere of Mars

The Martian air contains 95 percent carbon dioxide, 3 percent nitrogen, and 2 percent argon. In composition, that is much like the air on Venus, but the Martian atmosphere is very thin, less than 1 percent as dense as Earth's atmosphere. It is also never warmer than an autumn afternoon and can become as cold as −140°C (−220°F).

There is very little water in the Martian atmosphere, and the polar caps are composed of frozen carbon dioxide (dry ice) with an unknown amount of frozen water trapped below. As summer comes to a hemisphere, the carbon dioxide in its polar cap turns from solid to vapor and adds carbon dioxide to the atmosphere, while winter in the opposite hemisphere is freezing carbon dioxide out of the atmosphere and adding it to the polar cap.

Liquid water cannot exist on the surface of Mars because the air presure is too low. Any liquid water would immediately boil away; and if we stepped out of a spaceship on Mars without our space suits, our body heat would make our blood boil. Whatever water is present on Mars must be frozen below the polar caps or frozen as permafrost in the soil.

Although the present atmosphere of Mars is very thin, we will see evidence that the climate once permitted liquid water to flow over the surface, so we must conclude that Mars once had a thicker atmosphere. As a terrestrial planet, it should have outgassed significant amounts of carbon dioxide, nitrogen, and water vapor; but because it was small, it could not hold onto its gases. The escape velocity on Mars is only 5 km/sec, less than half of Earth's, so it was easier for rapidly moving gas molecules to escape into space. Another factor is the temperature of a planet. If Mars had been colder, the gas molecules in its atmosphere would have been traveling more slowly and would not have escaped as easily (Figure 17-20). Finally, with no free oxygen, Mars had no ozone layer to protect its atmosphere from ultraviolet radiation, which can break molecules up into smaller

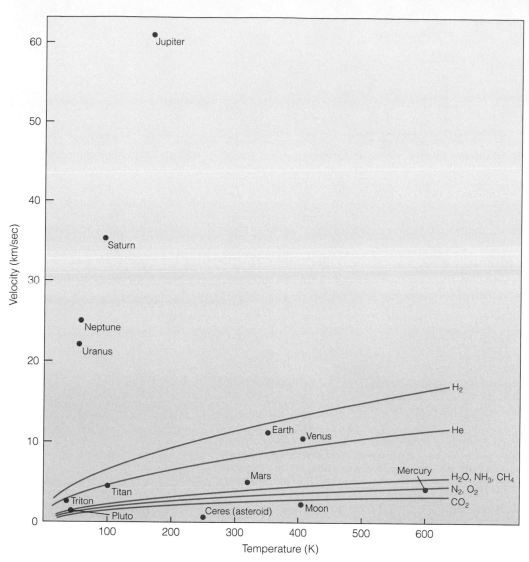

Figure 17-20
The loss of planetary gases. Dots represent the escape velocity and temperature of various solar system bodies. The lines represent the typical highest velocities of molecules of various masses. The Jovian planets have high escape velocities and can hold on to even the lowest-mass molecules. Mars can hold only the more massive molecules, and the moon has such a low escape velocity that even the most massive molecules can escape.

fragments—and these escape more easily. Water, for example, can be broken up into hydrogen and oxygen. Thus, Mars was large enough to have a substantial atmosphere when it was young and may have had water falling as rain and collecting in rivers and lakes. But it gradually lost its atmosphere and is now a cold world where liquid water is impossible.

The Geology of Mars

Spacecraft orbiting Mars have sent back photographs of its surface, and three spacecraft have landed. The two Viking probes landed in 1976, and the Pathfinder probe carried a robotic explorer to the surface in 1997. All of this data tells us that the surface of Mars is an old, dusty volcanic surface with hints that water was once present (Figure 17-21).

The northern half of Mars contains giant volcanoes, and lava flows have covered over the older cratered surface. The average age of the northern hemisphere is about a billion years. The southern hemisphere, however, contains many craters and is 2 to 3 billion years old (Figure 17-22). The reason for this asymmetry is unknown.

Volcanoes on Mars are shield volcanoes, and the largest, Olympus Mons, is 600 km (370 miles) in diameter at its base and 25 km (16 miles) high. The largest volcano on Earth is Mauna Loa in Hawaii, rising only 10 km (6 miles) above its base on the ocean floor. Mauna Loa is so heavy that it has sunk into Earth's crust, producing an undersea moat around its base, but Olympus Mons, 2.5 times higher, has no moat and is supported entirely by the Martian crust (Figure 17-23). Evidently, the crust of Mars is much thicker and stronger than Earth's.

Olympus Mons is only one of a number of volcanoes that lie in the Tharsis region (see Figure 17-22), a bulge rising about 10 km (6 miles) above the surrounding surface. A similar uplifted volcanic plain, the Elysium region, is more heavily cratered and eroded and so appears to be older than the Tharsis bulge. The Elysium region was volcanically active about 1.5 billion

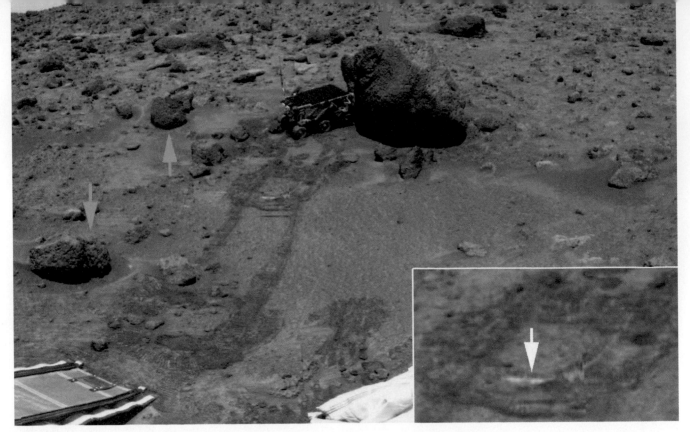

Figure 17-21
Pathfinder's rover, Sojourner, rolled down the ramp at lower left and tested the composition of rocks such as Barnacle Bill (blue arrow at left), Cradle (blue arrow at top), and Yogi (red arrow). The soil is red because it is rich in iron oxides. White material exposed in Sojourner's tracks (inset) may be ancient deposits made by water. Tests of the rocks and soil suggest the surface is volcanic lavas fragmented over long periods of time by meteorite impacts. *(NASA)*

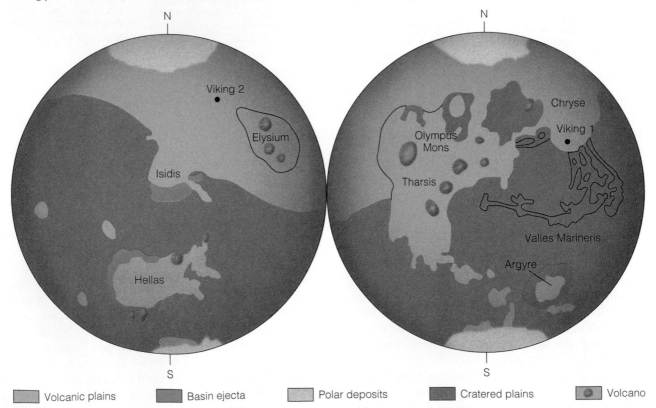

Figure 17-22
A geological map of Mars. Most of the southern hemisphere is old cratered terrain, whereas most of the northern hemisphere is younger volcanic plains. Note the volcanic areas, Tharsis and Elysium, and the large basins, Hellas and Argyre.

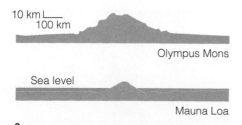

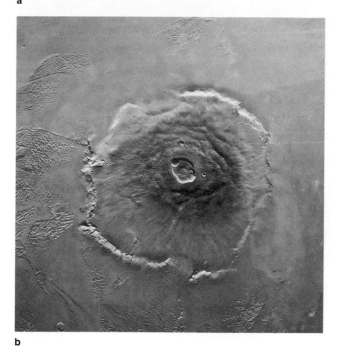

b

Figure 17-23
(a) Olympus Mons, the tremendous shield volcano on Mars, is much larger than Mauna Loa, the largest volcano on Earth. Mauna Loa has sunk into the crust, producing a moat around its base. (b) The caldera of Olympus Mons shows evidence of repeated eruptions. *(NASA)*

Figure 17-24
This mosaic of photographs taken by the orbiting Viking spacecraft shows the reddish surface of Mars marked by circular impact craters and three large, dark volcanoes (left). Valles Marineris (arrows) stretches across the planet for a distance equal to that from New York to Los Angeles. *(NASA/USGS, courtesy Alfred S. McEwen)*

years ago, while the Tharsis volcanoes may have been active as recently as 0.2 billion years ago.

When the crust of a planet is strained, it may break, producing faults and rift valleys. Near the Tharsis region is a great valley, Valles Marineris, named after the Mariner spacecraft that first photographed it (Figure 17-24). The valley is apparently a block of crust that has dropped downward along parallel faults. Erosion and landslides have further modified the valley into a great canyon nearly 4000 km (2500 miles) long, stretching almost 19 percent of the way around the planet. It is as deep as 6 km (4 miles) and as wide as 200 km (120 miles).

Where Valles Marineris begins, near the Tharsis region, it is marked by numerous faults, suggesting that the valley is related to the uplifted volcanic plain (see Figure 17-22). The number of craters in the valley indicates that it is 1 to 2 billion years old, placing its origin sometime before the end of volcanism in the Tharsis region. Some astronomers suggest that the

valley is the begining of a rift valley, but it does not connect to outline an entire plate.

Does Mars show signs of plate tectonics? That is a key question. To attempt an answer, we must first study the history of water on Mars.

Water on Mars

Spacecraft visiting Mars have photographed two kinds of erosional features caused by running water. Although the features are old and liquid water no longer exists on Mars, the signs of erosion suggest that Mars was once a more hospitable world.

Many parts of Mars are marked by erosional features that appear to have been cut by massive floods of moving water. For example, Figure 17-25a shows the Kasei Valles region on Mars where floods about a billion years ago stripped away layers or rock to expose an ancient crater (Figure 17-25b). Some of these eroded areas are so large they must have been formed by floods 10,000 times greater than the flow of the Mississippi over a period of hours or days. One way to account for these features is permafrost, water frozen in

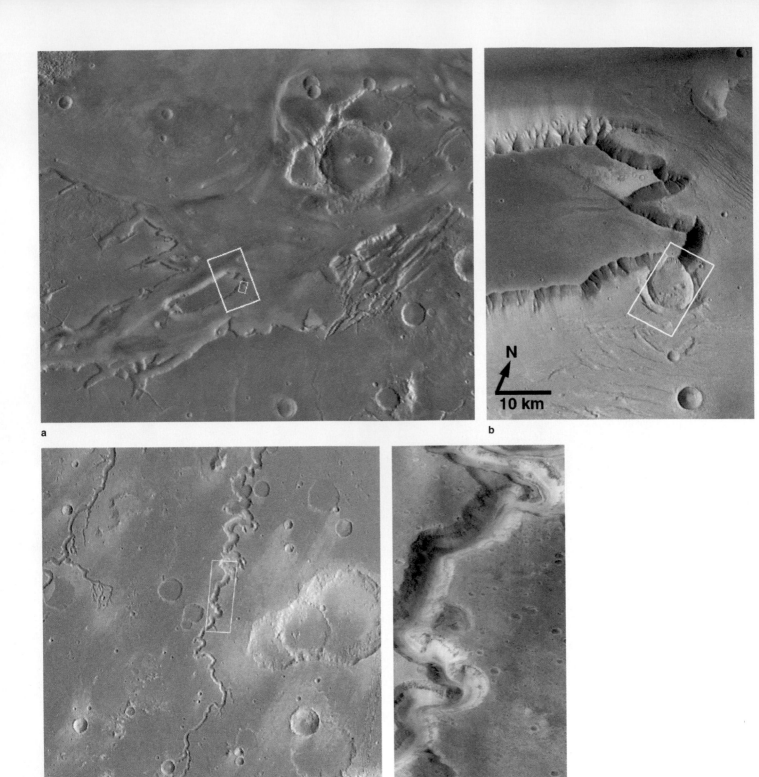

Figure 17-25
(a) The elongated flow features in this image were produced by one or more massive floods that carried away the upper layers of rock and exposed the ancient crater shown in the white box in part (b). (c) The steady flow of streams on ancient Mars is suggested by meandering canyons such as the two shown here. (d) A detailed image of the central canyon recorded by the Mars Global Surveyor spacecraft reveals a narrow streambed at the bottom. *(Malin Space Science Systems/NASA)*

the soil. Volcanic heating, a large meteorite impact, or internal pressure might have melted the ice to produce sudden floods. Photos of Mars do show jumbled valleys typical of the withdrawal of subsurface water and craters surrounded by "splashes" of ejecta, suggesting that the soil was water-laden at the time of impact.

Meandering drainage valleys winding downhill through nearly flat terrain (Figure 17-25c and d) are a second sign of erosion, suggesting rain or snowfall and long-term erosion rather than sudden floods. The thin atmosphere that Mars has now would not permit liquid water to form these features, but the number of craters in the flow features tells us they formed between 1 and 3 billion years ago, with the flood features being the youngest. This suggests that the atmosphere of Mars was denser in the past.

Evidence of climatic variation is visible near the polar caps. The terrain near the polar caps is made up of layers of sediments, each roughly 10 meters thick. These layers are apparently built up as dust accumulates on the polar cap in the winter and is left behind as the carbon dioxide vaporizes each spring. At 1 mm of dust per year, this process would take about 10,000 years to build up a single layer. The variation in the layering shows that some periodic change in climate occurs, perhaps due to orbital changes that affect the frequency and intensity of the planetwide dust storms as well as the extent of the winter polar caps (Figure 17-26). Recall that Earth's climate may go through cycles caused by variations in its orbit and rotation (Chapter 3).

One astonishing bit of evidence of water on Mars arrived on Earth in meteorites. Astronomers have identified a dozen meteorites as rocks that were originally on Mars and were blasted into space by large meteorite impacts on Mars. The meteorites are basalts, typical of lava flows, but chemical analysis shows that the basalt solidified from magma that contains about 1.4 percent water. If all of the ancient lava flows on Mars contained this much water, then Mars must have had significant amounts of water when it was young. Furthermore, one of the meteorites contains carbonate deposits that could have been formed only if liquid water was present in the rock sometime after it was formed, further evidence that the climate on Mars has changed. (This meteorite also contains

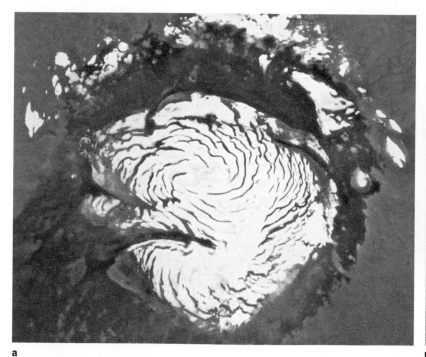

a

b

Figure 17-26
(a) The north polar cap of Mars is made of many regions of ice separated by narrow valleys free of ice. *(NASA)* (b) An image from the Mars Global Surveyor spacecraft in one of the valleys of the polar cap reveals superimposed layers of sediment as thin as 10 m (33 feet). *(Malin Space Science Systems/NASA)* (c) In places, different sets of layers are superimposed, suggesting climate change. *(Adapted from a diagram by J. A. Cutts, K. R. Balasius, G. A. Briggs, M. H. Carr, R. Greeley, and H. Masursky)*

c

Geologists are fond of saying "The present is the key to the past." By that they mean that we can learn about the history of Earth by looking at the present condition of Earth's surface. The position and composition of various rock layers in the Grand Canyon, for example, tell us that the western United States was once at the floor of an ocean. This principle of geology was astonishing when it was first formulated in the late 1700s, and it continues to be relevant today as we try to understand other worlds, such as Venus and Mars.

In the late 18th century, naturalists first recognized that the present gave them clues to the history of Earth. This was astonishing because most people assumed that Earth had no history. That is, they assumed either that Earth had been created in its present state as described in the Old Testament or that Earth was eternal. In either case, people commonly assumed that the hills and mountains they saw around them had always existed more or less as they were. The 18th-century naturalists began to see evidence that the hills and mountains were not eternal but were the result of past processes and were slowly changing. That gave birth to the idea that Earth had a history.

As the naturalists of the 18th century made the first attempts to thoughtfully and logically explain the nature of Earth by looking at the evidence, they were inventing modern geology as a way of understanding Earth. What Copernicus, Kepler, and Newton did for the heavens in the 1600s, the first geologists did for Earth beginning in the late 1700s. Of course, the invention of geology as the study of Earth led directly to our modern attempts to understand the geology of other worlds.

Geologists and astronomers share a common goal: They are attempting to reconstruct the past (Window on Science 16-3). Whether we study Earth, Venus, or Mars, we are looking at the present evidence and trying to reconstruct the past history of the planet by drawing on observations and logic to test each step in the story. How did Venus get to be covered with lava, and how did Mars lose its atmosphere? The final goal of planetary astronomy is to draw together all of the available evidence (the present) to tell the story (the past) of how the planet got to be the way it is. Those first geologists of the late 1700s would be fascinated by the stories planetary astronomers tell today.

features that may be traces of ancient bacterial life on Mars, a result we will discuss in Chapter 20.)

The evidence tells us that Mars once had significant amounts of water on its surface, but today we see a deadly dry desert world. Clearly, the history of Mars is a complex story.

The History of Mars

Did Mars ever have plate tectonics? Where did the water go? These fundamental questions challenge us to assemble the evidence and hypotheses for Mars and tell the story of its evolution (Window on Science 17-3).

The four-stage history of Mars is a case of arrested development. The planet began by differentiating into a crust and core. Studies of its rotation reveal that it has a dense core. The Mars Global Surveyor spacecraft detected no planetwide magnetic field, but it did find 8 regions of the crust with fields a bit over 1 percent as strong as Earth's. Apparently, the young Mars had a molten iron core and generated a magnetic field, which became frozen into regions of the crust. The core must have cooled quickly and shut off the dynamo effect, making the planetwide field. Nevertheless, the magnetic regions of the crust remain behind like fossils.

The crust of Mars is now quite thick, as shown by the mass of Olympus Mons, but it was thinner in the past. Cratering may have broken or at least weakened the crust, triggering lava flows that flooded some basins. Why most of the flows occurred in the northern hemisphere is unknown. Mantle convection may have pushed up the Tharsis and Elysium volcanic regions and broken the crust to form Valles Marineris, but moving crustal plates seem never to have formed. We see no folded mountain ranges on Mars and no sign of plate boundaries. The small planet cooled rapidly, and the crust grew thick and immobile.

The large size of volcanoes on Mars is evidence that the crust does not move. On Earth, volcanoes like those that formed the Hawaiian Islands occur over rising currents of hot material in the mantle. Because the plate moves, the hot material heats the crust in a string of locations and forms a chain of volcanoes instead of a single large feature. The Hawaiian Islands are merely the most recent of a series of volcanic islands called the Hawaiian–Emperor island chain (see Figure 17-4), which stretches nearly 3800 km (2300 miles) across the Pacific Ocean floor. A lack of plate motion on Mars would have allowed a rising current of magma to heat the crust repeatedly in the same place and build a very large volcanic cone.

The fourth stage in the history of Mars has been one of slow decline. Volcanic activity probably reached a maximum about 1 billion years ago, and since then the planet has cooled and the crust has thickened.

Much of the atmosphere has been lost to space, and the climate is now too cold and the air pressure too low for liquid water. What water remains must be locked up in the polar caps or in permafrost.

The Moons of Mars

Unlike Mercury or Venus, Mars has moons. Small and irregular in shape, Phobos (28 × 23 × 20 km in diameter) and Deimos (16 × 12 × 10 km) may be captured asteroids, or they may have formed with Mars.

The photographs reveal a unique set of narrow, parallel grooves on Phobos. Averaging 150 m (500 ft) wide and 25 m (80 ft) deep, the grooves run from Stickney, the largest crater, to an oddly featureless region on the opposite side of the satellite. One theory suggests that the grooves are deep fractures produced by the impact that formed the crater (Figure 17-27a).

The Mars Global Surveyor spacecraft reveals more about the dust. Observations made with the spacecraft's infrared spectrometer show that the moon's surface cools quickly from −4°C to −112°C (from 25°F to −170°F) as it passes from sunlight into darkness. Solid rock would retain heat and cool more slowly, so the dust must be at least a meter deep and very fine. In photos made by the spacecraft camera, the dust blankets the terrain, but the photos also show boulders a few meters in diameter thought to be ejected by impacts (Figure 17-27b).

Deimos not only has no grooves, but it also looks smoother because of a thicker layer of dust on its surface. This material partially fills craters and covers

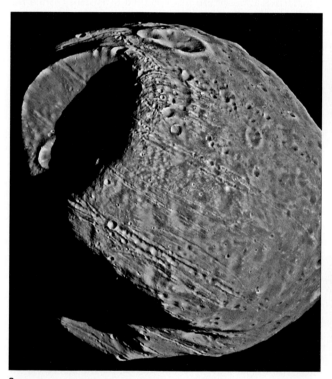

a

b

c

Figure 17-27
The moons of Mars are too small to pull themselves into spherical shape. (a) Phobos has an impact crater 10 km in diameter at one end with groves radiating away. *(Damon Simonelli and Joseph Ververka, Cornell University/NASA)* (b) A photo from the Mars Global Surveyor spacecraft shows features as small as 8 meters in diameter. *(Malin Space Science Systems/NASA)* (c) Deimos, smaller than Phobos, looks more uniform because of dusty soil covering the smaller features. *(NASA)*

minor surface irregularities (Figure 17-27c). It seems certain that Deimos experienced collisions in its past, so fractures may be hidden below the debris.

The debris on the surfaces of the moons raises an interesting question: How can the weak gravity of small bodies hold any fragments from meteorite impacts? The escape velocity on Phobos is only about 12 m/sec (40 ft/sec). An athletic astronaut who could jump 2 m (6 ft) high on Earth could jump 2.8 km (1.7 miles) on Phobos. Certainly most of the fragments from an impact should escape, but the slowest particles could fall back in the weak gravity and accumulate on the surface.

Because Deimos is smaller than Phobos, its escape velocity is smaller, so it seems surprising that it has more debris on its surface. This may be related to Phobos's orbit close to Mars. The Martian gravity is almost strong enough to pull loose material off of Phobos's surface, so Earth's moon may be able to retain less of its cratering debris.

REVIEW Critical Inquiry

Why would we be surprised to find volcanism on Phobos or Deimos?

In discussing Earth's moon, Mercury, Venus, Earth, and Mars, we have seen illustrations of the principle that the larger a world is, the more slowly it loses its internal heat. It is the flow of that heat from the interior through the surface into space that drives geological activity such as volcanism and plate motion. A small world, like Earth's moon, cools quickly and remains geologically active for a shorter time than does a larger world like Earth. Phobos and Deimos are not just small; they are tiny. However they formed, any interior heat would have leaked away very quickly; with no energy flowing outward, there can be no volcanism.

Some futurists suggest that our first human missions to Mars will be not to land on the surface but to build a colony on Phobos or Deimos. These plans speculate that there may be water deep inside Earth's moons that colonists could use. What would happen to water released in the sunlight on the surface of such small worlds?

Venus is nearly the same size as Earth, but it is unlikely that humans will ever colonize its surface. The smaller and colder Mars, however, may someday be home to large colonies. In the next chapter, we turn our attention to Jupiter and Saturn, giant worlds so totally unlike Earth that we will need entirely new principles of comparative planetology to describe them.

Summary

All terrestrial planets pass through a four-stage history: (1) differentiation; (2) cratering; (3) flooding of the crater basins by lava, water, or both; and (4) slow surface evolution. The importance of each of these processes in the evolution of a planet depends on the planet's mass and temperature.

Earth is the largest terrestrial planet in our solar system, and it has passed through all four stages. Studies show that Earth has differentiated into a metallic core and a silicate crust. Currents in the molten portion of the core produce Earth's magnetic field.

Because Earth is still partially molten, its surface is active. Plate tectonics refers to the motion of large crustal sections. As new crust appears in the midocean rifts, it pushes the massive plates apart and destroys old crust where plates slide over one another.

Earth's atmosphere has changed drastically since its formation. The first atmosphere was rich in carbon dioxide. As the carbon dioxide dissolved into the oceans and was incorporated in the bottom sediment, the atmosphere was left rich in nitrogen. Later, plant life produced oxygen.

When Earth's moon formed, it differentiated into a dense core and a low-density crust of anorthosite. Cratering broke up this crust, and lava flows of dark basalt filled the lowlands, producing the maria. Cratering has pulverized the surface of the highlands and the lowlands, covering the lunar surface with ejecta.

Mercury, like Earth's moon, has no atmosphere and is covered by craters; but unlike Earth's moon's, its surface is marked by lobate scarps. The long, curved cliffs suggest that, after formation, Mercury cooled and contracted, and its crust wrinkled.

The surface of Venus is hidden below a dense, cloudy atmosphere of carbon dioxide. The thick atmosphere creates a strong greenhouse effect, and the surface temperature is high enough to melt lead. Radar maps reveal that its geology is dominated by rising currents of molten rock below the surface. Volcanism has flooded much of the surface with lava flows. There is no evidence that plate tectonics is important on Venus.

Because Mars is small, it has cooled relatively rapidly and is no longer active. Nevertheless, its surface contains lava plains, volcanoes, and a long feature, possibly a rift valley. These suggest that the surface was active in the past.

Venus and Mars illustrate two principles of planetary atmospheres. A thick atmosphere that is opaque to infrared radiation can trap heat via the greenhouse effect and produce a high surface temperature. Earth avoided this by dissolving its CO_2 in the oceans, but Venus was too warm for liquid water to form. The second principle is that the density and composition of a planet's atmosphere depend on the mass and temperature of the planet. A warm, low-mass planet has difficulty retaining an atmosphere. Also, since light gases leak away most easily, a planet may be able to retain heavier gases like CO_2, while H_2 and He leak into space.

New Terms

comparative planetology
P waves
S waves
mantle
plastic
plate tectonics
midocean rise
midocean rift
subduction zone
basalt
rift valley
folded mountain chain
primeval atmosphere
greenhouse effect

ejecta
ray
vesicular basalt
mare (maria)
anorthosite
breccia
albedo
fission hypothesis
condensation hypothesis
capture hypothesis
large-impact hypothesis
lobate scarp
shield volcano
coronae

Review Questions

1. What are the four stages in the development of a terrestrial planet?

2. Why do we expect planets to have differentiated?

3. How does plate tectonics create and destroy Earth's crust?

4. Why do we suspect that Earth's primeval atmosphere was rich in carbon dioxide?

5. Why doesn't Earth have as many craters as its moon or Venus?

6. What kind of erosion is now active on Earth's moon?

7. Discuss the evidence and hypotheses concerning the origin of Earth's moon.

8. Why doesn't Earth or its moon have lobate scarps?

9. How did Earth avoid the greenhouse effect that made Venus so hot?

10. What evidence do we have that plate tectonics does not occur on Venus? on Mars?

11. What evidence suggests that Venus has been resurfaced within the last 300 million years?

12. Why is the atmosphere of Venus rich in carbon dioxide? Why is the atmosphere of Mars rich in carbon dioxide?

13. What evidence do we have that the climate on Mars has changed?

14. Why do we conclude that the crust on Mars must be thick?

15. What evidence do we have that there has been liquid water on Mars?

Discussion Questions

1. If we visited a planet in another solar system and discovered oxygen in its atmosphere, what might we guess about its surface?

2. If liquid water is rare on the surface of planets, then most terrestrial planets must have CO_2-rich atmospheres. Why?

Problems

1. If the Atlantic seafloor is spreading at 30 mm/year and is now 6400 km wide, how long ago were the continents in contact?

2. Why do small planets cool faster than large planets? (*Hint:* Compare surface area to volume.)

3. The smallest detail visible through Earth-based telescopes is about 1 second of arc in diameter. What size is this on Earth's moon? (*Hint:* See By the Numbers 3-1.)

4. Midocean rifts and the trenches where the seafloor slips downward are 1 km or less wide. Could Earth-based telescopes resolve such features on Earth's moon? Why are we sure such features are not present on Earth's moon?

5. How long would it take radio signals to travel from Earth to Venus and back if Venus were at its farthest point from Earth? Why are such observations impractical?

6. Repeat Problem 5 for Mercury.

7. Imagine that we have sent a spacecraft to land on Mercury, and it has transmitted radio signals to us at a wavelength of 10 cm. If we see Mercury at its greatest angular distance west of the sun, to what wavelength must we tune our radio telescope to detect the signals? (*Hints:* See Data File 4 to find Mercury's orbital velocity, and then see By the Numbers 6-2.)

8. The smallest feature visible through an Earth-based telescope has an angular diameter of about 1 second of arc. If a crater on Mars is just visible when Mars is nearest to Earth, what is its linear diameter? (*Hint:* See By the Numbers 3-1.)

9. What is the maximum angular diameter of Phobos as seen from Earth? What surface features should we be able to see using Earth-based telescopes? (*Hint:* See By the Numbers 3-1.)

10. Phobos orbits Mars at a distance of 9380 km from the center of the planet and has a period of 0.3189 days. Calculate the mass of Mars. (*Hints:* Convert to AU and years, and then see By the Numbers 4-1.)

Critical Inquiries for the Web

1. Who decides how planetary features are named? Surface features on Venus are (mostly) named after female figures from history and mythology, while figures from the arts and music are used to name features on Mercury. Look for information on planetary nomenclature, and summarize the way different types of features on Venus are assigned names.

2. "Martians" have fascinated humans for the last century or more. There are many online sources that chronicle the representation of life on Mars throughout history and in literature. Read about the Martians as represented by a particular literary work or nonfiction account, and discuss to what extent it is (or is not) based on realistic views of the nature of Mars both in terms of our current understanding and the views of that period.

3. What would it be like to walk on the lunar surface? Apollo astronauts visited six different locations on our moon, exploring a variety of lunar terrain. Describe the horizons and general relief of the landing locations of the different missions by exploring Web sites that provide lunar surface photography from the missions. What differences do you see between images from landings in highlands and maria?

Go to the Brooks/Cole Astronomy Resource Center (www.brookscole.com/astronomy) for critical thinking exercises, articles, and additional readings from InfoTrac College Edition, Brooks/Cole's online student library.

There wasn't a breath

in that land of death . . .

Robert Service
The Cremation of Sam McGee

CHAPTER 18

Worlds of the Outer Solar System

Guidepost

As we begin this chapter, we leave behind the psychological security of planetary surfaces. We can imagine standing on a terrestrial world, but the Jovian worlds have no surfaces. Thus, we face a new challenge—to use comparative planetology to study worlds so unearthly we cannot imagine being there.

One reason we find Earth's moon and Mars of interest is that we might go there someday. Humans may become the first Martians. But the worlds of the outer solar system seem much less useful, and that gives us a chance to think about the cultural value of science.

Among the outer worlds, we see strong evidence that the smaller bodies that fall through the solar system impact planets and satellites. The next chapter will allow us to study these small bodies in detail and will give us new evidence that our solar system formed from a solar nebula.

369

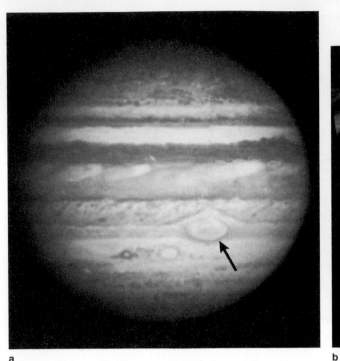

a   b

Figure 18-1
(a) Jupiter's atmosphere is marked by complex circulation patterns, including the Great Red Spot (arrow).
The planet has a large system of moons, including two in this image, and a faint ring, not visible here.
(b) In its longest dimension, the Great Red Spot is about twice the diameter of Earth. This infrared image
from the Galileo spacecraft shows lower clouds as blue, higher thin hazes as red, and high, thick clouds
as white. The Red Spot is a region of rising gases forced to rotate like a giant hurricane by Jupiter's rapid
rotation. *(NASA)*

The sulfuric acid fogs of Venus seem totally alien, but compared with the planets of the outer solar system, Venus is a tropical oasis. Of the five planets beyond the asteroid belt, four have no solid surface, and one, Pluto, is so far from the sun that most of its atmosphere lies frozen on its surface. As we begin our study of these strange worlds, we will discover new principles of comparative planetology.

The four Jovian planets—Jupiter, Saturn, Uranus, and Neptune—are strikingly different from the terrestrial planets. Their thick atmospheres of hydrogen, helium, methane, and ammonia are filled with clouds of ammonia, methane, and water ice crystals; and their interiors are mostly liquid. They seem to be worlds dominated by cloud patterns (Figure 18-1).

The Jovian worlds can be studied from Earth, but much of what we know has been radioed back to Earth from space probes. Pioneer 10 and Pioneer 11 explored Jupiter and Saturn in the mid-1970s, but their instruments were not highly sensitive. The Voyager 1 and Voyager 2 spacecraft were launched in 1977 on a mission to visit all four Jovian worlds and concluded when Voyager 2 flew past Neptune in 1989. In December 1995, the Galileo spacecraft reached Jupiter. While an instrumented probe dropped into Jupiter's atmosphere, the main Galileo spacecraft went into orbit to begin a long-term study of the planet and its moons. Throughout this discussion of the Jovian worlds, you will find images and data returned by these robotic explorers (Window on Science 18-1).

Today's astronomers have a tremendous library of photos, measurements, and facts about the Jovian planets; and although no spacecraft has visited Pluto, Earth's astronomers are beginning to understand its role. Our task is to discover the relationships that explain how these worlds got to be the way they are. We begin with the closest and largest Jovian planet, named after the ruler of the Roman gods.

18-1 Jupiter

Jupiter is the most massive of the Jovian planets, containing 71 percent of all the planetary matter in our solar system. This high mass accentuates some processes that are less obvious or nearly absent on the other Jovian worlds. Just as we used Earth as the basis for comparison in our study of the terrestrial planets, we will examine Jupiter in detail so we can use it as a standard in our comparative study of the other Jovian planets.

Sending spacecraft to another world is expensive, and it may seem pointless when that world seems totally hostile to human life. What practical value is there in sending a space probe to Jupiter? To resolve that question, we need to consider the distinction between science, technology, and engineering.

Science is nothing more than the logical study of nature, and the goal of science is understanding. Although much scientific knowledge proves to have tremendous practical value, the only goal of science is a better understanding of how nature works. Technology, in contrast, is the practical application of scientific knowledge to solve a specific problem. People trying to find a faster way to paint automobiles might use all the tools and techniques of science, but if their goal has some practical outcome, we should more properly call it technology rather than science. Engineering is the most practical form of technology. An engineer is likely to use well-understood technology to find a practical solution to a problem.

Of course, there are situations in which science and technology blur together. For example, humanity has a practical and tragic need to solve the AIDS problem; we need a cure and a method of prevention. Unfortunately, we don't understand the AIDS virus itself or viruses in general well enough to design a simple solution, and thus much of the work involves going back to basic science and trying to better understand how viruses interact with the human body. Is this technology or science? It is hard to decide.

We might describe science that has no known practical value as basic science or basic research. Our exploration of worlds such as Jupiter would be called basic science, and it is easy to argue that basic science is not worth the effort and expense because it has no known practical use. Of course, the problem is that we have no way of knowing what knowledge will be of use until we acquire that knowledge. In the middle of the 19th century, Queen Victoria is supposed to have asked physicist Michael Faraday

what good his experiments with electricity and magnetism were. He answered, "Madam, what good is a baby?" Of course, Faraday's experiments were the beginning of the electronic age. Many of the practical uses of scientific knowledge that fill our world—transistors, vaccines, plastics—began as basic research. Basic scientific research provides the raw materials that technology and engineering use to solve problems.

Basic scientific research has yet one more important use that is so valuable it seems an insult to refer to it as merely practical. Science is the study of nature, and as we learn more about how nature works, we learn more about what our existence in this universe means for us. The seemingly impractical knowledge we gain from space probes to other worlds tells us about our planet and our own role in the scheme of nature. Science tells us where we are and what we are, and that knowledge is beyond value.

The Interior

Density is mass divided by volume, and the density of Jupiter is only a third greater than that of water (Data File Seven). For comparison, Earth is over 4 times more dense than Jupiter. Theoretical models of Jupiter based on this density conclude that it is composed mostly of hydrogen and helium. In the interior, the pressure compresses the hydrogen into a liquid. At the center, a so-called rocky core contains heavier elements, such as iron, nickel, silicon, and so on. With a temperature 4 times hotter than the surface of the sun and a pressure 50 million times Earth's air pressure at sea level, this material is unlike any rock on Earth. The term *rocky core* refers to the chemical composition and not to the consistency of the material.

Careful measurements of the heat flowing out of Jupiter reveal that it emits about twice as much energy as it absorbs from the sun. This energy appears to be heat left over from the formation of the planet. In Chapter 16 we saw that Jupiter should have grown very hot when it formed, and some of this heat remains trapped in its interior.

Although Jupiter is mostly liquid hydrogen, there is no ocean surface on which we might imagine sail-

ing a Jovian boat. The base of the atmosphere is so hot and the pressure is so high that there is no real distinction between liquid and gas, and if we jumped into Jupiter carrying a rubber raft we would be disappointed. As we fell deeper and deeper through the atmosphere, we would see the gas density increasing around us until we were sinking through a liquid, but we would never splash into a liquid surface.

Deep inside Jupiter the pressure is so high that liquid hydrogen is not stable, and it is compressed to form liquid metallic hydrogen—a material that is a very good conductor of electricity. This vast globe of conducting liquid, stirred by convection currents and spun by the planet's rapid rotation, drives the dynamo effect and generates a powerful magnetic field. Jupiter's field is over 10 times stronger than Earth's.

One consequence of this magnetic field is aurora. Just as in the case of Earth, interactions between Jupiter's magnetic field and the solar wind generate powerful electric currents that flow down into the atmosphere in two rings around the planet's magnetic poles. These are visible at ultraviolet wavelengths as rings of aurora larger in diameter than Earth (Figure 18-2a).

Jupiter Data File Seven

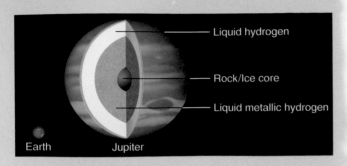

A theoretical model of the interior of Jupiter. Note the oblateness of the planet.

Average distance from the sun	5.2028 AU (7.783×10^8 km)
Eccentricity of orbit	0.0484
Maximum distance from the sun	5.455 AU (8.160×10^8 km)
Minimum distance from the sun	4.951 AU (7.406×10^8 km)
Inclination of orbit to ecliptic	1°18′29″
Average orbital velocity	13.06 km/sec
Orbital period	11.867 y (4334.3 days)
Period of rotation	$9^h55^m30^s$
Inclination of equator to orbit	3°5′
Equatorial diameter	142,900 km (11.20 $D_\oplus$)
Mass	1.899×10^{27} kg (317.83 $M_\oplus$)
Average density	1.34 g/cm³
Gravity at base of clouds	2.54 Earth gravities
Escape velocity	61 km/sec (5.4 $V_\oplus$)
Temperature at cloud tops	−130°C (−200°F)
Albedo	0.51
Oblateness	0.0637

A planet's magnetic field deflects the solar wind and dominates a volume of space around the planet called the **magnetosphere.** Jupiter's magnetosphere is 100 times larger than Earth's (Figure 18-2b). If we could see it in the sky, it would be 6 times larger than the full moon.

The magnetic field around Jupiter traps charged particles from the solar wind in large, doughnut-shaped radiation belts that surround the planet just as Earth is surrounded by its Van Allen radiation belts. Because Jupiter's magnetic field is stronger, its trapped radiation is a billion times more intense than Earth's. The spacecraft that have flown through these regions received over 400 times the radiation that would have been lethal for a human.

However unearthly Jupiter may be, it obeys the same rules of nature that govern Earth. Thus, it has a magnetic field and aurora just as Earth does. Jupiter's atmosphere, however, seems even more unearthly than its interior.

Jupiter's Atmosphere

The atmosphere on Jupiter is about 78 percent hydrogen, and the rest is mostly helium with small traces of water vapor, methane, ammonia, and similar molecules. The clouds we see are layers of ammonia and ammonia hydrosulfide crystals. Through gaps in these clouds, astronomers can see deeper, warmer layers of water-droplet clouds (Figure 18-3). Below that the atmosphere merges with the liquid interior.

Jupiter's atmosphere is dominated by **belt–zone circulation,** which circles the planet parallel to its equator (Figure 18-1a). Even a small telescope can reveal the dark belts and lighter zones. The colors of the belts and zones are believed to arise from molecules in the clouds produced by sunlight or by lightning interacting with ammonia and other compounds.

Jupiter's belt–zone circulation is related to the high- and low-pressure areas we see on Earth's weather maps. Zones are high-pressure regions of rising gas, and belts are low-pressure regions where gas sinks.

On Earth, the temperature difference between the equator and poles drives a wavelike wind pattern that organizes such high- and low-pressure regions into cyclonic circulations familiar from weather maps (Figure 18-4a). On Jupiter, the equator and poles appear to be about the same temperature, perhaps because of heat rising from the interior. Thus, there is no temperature difference to drive the wave circulation, and Jupiter's rapid rotation draws the high- and low-pressure regions into bands that circle the planet. On Earth, high- and low-pressure regions are bounded by winds induced by the wave circulation. On Jupiter, the same circulation appears as high-speed winds that blow around the planet at the boundaries of the belts

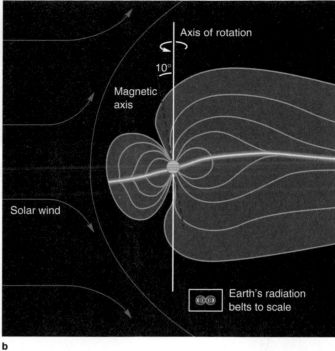

a

b

Figure 18-2
(a) Aurora on Jupiter is confined to rings around the north magnetic pole and the south magnetic pole, as shown in these ultraviolet images. Earthly auroras follow the same pattern. The small comet-shaped spots are caused by powerful electrical currents flowing from Jupiter's moon Io. *(John Clarke, University of Michigan, and NASA)* (b) Jupiter's large conducting core and rapid rotation create a powerful magnetic field that holds back the solar wind and dominates a region called the magnetosphere. High-energy particles trapped in the magnetic field form giant radiation belts.

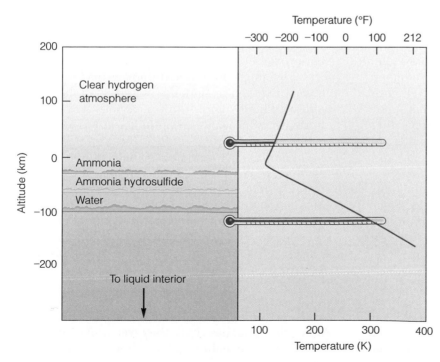

Figure 18-3
The three principal cloud layers on Jupiter are shown at left. Thermometers inserted into the Jovian atmosphere would register temperatures as in the graph at right. Note that the upper atmosphere is clear, cold hydrogen, but that the temperature rises rapidly below the cloud belts.

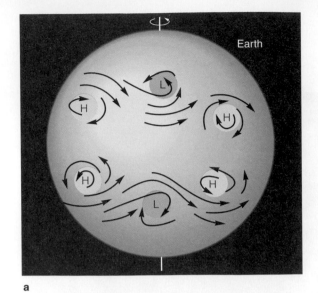

a

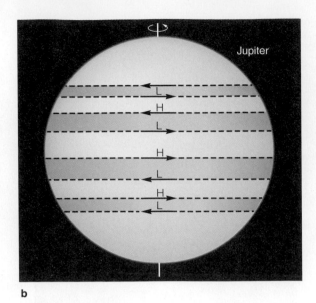

b

c

Altitude

North   Equator

Figure 18-4

(a) In Earth's atmosphere, a wave circulation separates low- and high-pressure regions. (b) On Jupiter, low- and high-pressure regions take the form of dark belts of sinking gas and bright zones of rising gas. (c) Zones are bright because the rising gas forms clouds high in the atmosphere, where sunlight is strong. On both worlds, winds circulate in the same way—counterclockwise around low-pressure regions in the northern hemisphere and clockwise in the southern hemisphere.

and zones (Figure 18-4b and 4c). Near the poles, the belt–zone pattern is not stable, and the circulation is more turbulent.

Mixed among the belts and zones are light and dark spots, a few larger than Earth. The largest is the Great Red Spot, a reddish oval that has been in one of the southern zones for at least 300 years (see Figure 18-1). Data from the Voyager and Galileo spacecraft show that the Great Red Spot is a vast circulating storm where winds between adjacent belt and zone meet. Smaller spots (visible in Figure 18-1a below and to the left of the Great Red Spot) are also long-lived storms. Two white spots that have been observed for 58 years collided and merged into a single larger storm in 1998, so we can conclude that although weather patterns on Jupiter are long lived, they are not entirely permanent.

Astronomers around the world watched the atmosphere of Jupiter in the summer of 1994 as fragments of a comet slammed into the planet at high velocity. Over a period of 6 days, 20-some objects 0.5 km or so in diameter hit with the energy of millions of megatons of TNT, each impact creating a fireball of

hot gases and leaving behind dark smudges that remained visible for months afterward (Figure 18-5). Planetary scientists continue trying to model the complex physics of the impacts, but two things are becoming clear. First, the violent impacts had little effect on the long-term circulation patterns in Jupiter's atmosphere. And second, the fragments of the comet were very fragile and didn't penetrate down to the cloud layers before they exploded. In the long run, such impacts on Jupiter probably occur roughly once every century or two. Major impacts on Earth occur less often because Earth is smaller, but they are inevitable. All of the planets in the solar system are occasionally struck by these objects.

One of the interesting features of Jupiter does not lie within its cloudy atmosphere, but orbits above its equator. Jupiter, like the other three Jovian planets, has a ring.

Jupiter's Rings

Although Jupiter has rings, they were not discovered from Earth. The main ring was photographed by the

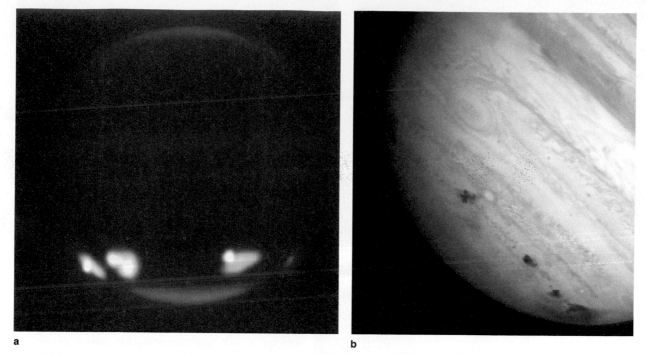

Figure 18-5

(a) Over 6 days in July 1994, fragments of a disrupted comet struck Jupiter and produced violent fireballs as they exploded in the atmosphere. This infrared photograph shows the glowing sites of impacts against the darker globe of Jupiter. *(University of Hawaii)* (b) At visible wavelengths, the impact sites show as dark smudges that were slowly dispersed by circulation in the atmosphere. The largest scar at extreme right is the size of Earth. *(NASA)*

Voyager 1 spacecraft in 1979. The ring is less than 30 km thick and extends out to 1.81 planetary radii. Less than 100 times less opaque than Saturn's rings, the main ring around Jupiter is composed of reddish dust. In 1998, the Galileo spacecraft was able to confirm the existence of even more tenuous rings, the **gossamer rings,** extending over twice as far from Jupiter.

We have strong evidence that Jupiter's rings are formed of very small particles. The photo in Figure 18-6a was taken by the Galileo spacecraft looking back with the sun hidden behind Jupiter. The ring, illuminated from behind, glows brightly in what astronomers call **forward scattering.** This effect can only occur when the dust particles are very small. Objects the size of snowflakes, baseballs, or even larger do not scatter light forward. Thus we can be sure that Jupiter rings are made up of very small dust specks.

The main ring lies inside the **Roche limit,** the distance inside which a moon cannot hold itself together by its gravity alone. It would be pulled apart by the planet's tidal forces. Any moon that orbits inside the Roche limit must be made of physically strong rock. (For a moon and planet of similar density, the Roche limit is about 2.44 times the radius of the planet.)

Because the rings lie inside the Roche limit, we might suspect the rings are material that could not form a moon so close to Jupiter, but the rings can't be very old. The pressure of sunlight and magnetic forces rapidly push such small particles into Jupiter or out

into space. Also, the radiation near Jupiter grinds the particles down to nothing in a century or so. Thus, it is clear the rings must be constantly supplied with new particles.

Observations with the Galileo spacecraft show that the main ring is densest at its outer edge where the small moon Adrastea orbits, and that another small moon, Metis, orbits inside the main ring. Evidently, meteorite impacts on the moons blast dust into space, and tidal forces inside the Roche limit prevent the moons from pulling the dust back. Electromagnetic effects in the main ring cause some particles to rise above and below the ring to form a faint halo (Figure 18-6b). The gossamer rings are most dense at the orbits of two other small moons, Amalthea and Thebe, again giving us evidence that the dust is being blasted into space by impacts on the inner moons.

Jupiter's Family of Satellites

Jupiter has well over a dozen moons, but most are small, only a few tens of kilometers in diameter, and some of these have retrograde (backward) orbits. This evidence suggests that many of Jupiter's smaller moons are captured asteroids.

In contrast, the four largest moons are called Galilean moons (Figure 18-7) after their discoverer, Galileo. These four moons are clearly related to each other and probably formed with Jupiter. We know

a

b

Figure 18-6
(a) The main ring of Jupiter, illuminated from behind, glows brightly in this photo made by the Galileo spacecraft from within Jupiter's shadow. (b) Digital enhancement and false color reveal the halo of ring particles that extends above and below the main ring. The halo is just visible in part a. *(NASA)*

these worlds surprisingly well because the two Voyager spacecraft and the Galileo spacecraft have studied them in detail.

The outermost Galilean moon, Callisto, is half again as big as Earth's moon, but it has a low density of only 1.79 g/cm³. This means it must consist roughly of 50:50 rock and ice. Observations of its gravitational field by the Galileo spacecraft reveal that it does have a dense core and a lower-density exterior, so it presumably differentiated. Also, it interacts with Jupiter's magnetic field in such a way that astronomers suspect it has a mineral-rich ocean of liquid

Figure 18-7
The Galilean moons of Jupiter with Earth's moon at center for scale. The Galilean moons, clockwise from upper left, are Io, Europa, Callisto, and Ganymede. *(NASA and Lick Observatory)*

a

b

Figure 18-8

(a) Callisto's surface is heavily cratered, dirty ice, where the youngest craters look bright because they have dug into cleaner ice. The bull's-eye impact feature is called Valhalla and is roughly 3000 km in diameter. (b) A Galileo image of Callisto's surface shows the dirty icy crust and impact features. *(NASA)*

water 100 km below its icy crust. Photos of its surface, however, show thousands of impact craters with little sign of geological activity (Figure 18-8).

Next inward from Callisto is Ganymede, with a density of 1.9 g/cm³, suggestive of a mix of rock and ice. The Galileo spacecraft has detected a weak magnetic field, which suggests a differentiated metallic core. The icy crust is marked by old, cratered, dark areas, and younger, brighter regions of **grooved terrain** believed to be systems of faults in the brittle crust (Figure 18-9a). Sets of groves overlap other sets of grooves, and that suggests extended episodes of geological activity (Figure 18-9b).

The density of the next moon inward, Europa, is 3.03 g/cm³, high enough to suggest the moon is mostly rock with a thin icy crust. The visible surface is very

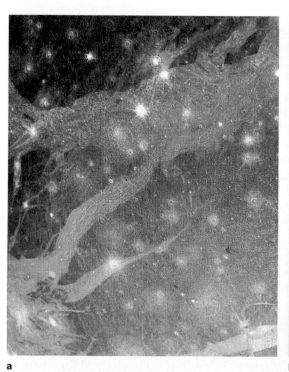

a

b

Figure 18-9

(a) The icy surface of Ganymede is divided into old, dark, cratered terrain and bands of bright grooves (see Figure 18-7). In this Voyager image, the dark terrain is marked by a few bright craters where more recent impacts have broken through the dirty surface ice to clean ice below. (b) A Galileo spacecraft image shows how a system of icy grooves covered an older region at bottom while an even more recent system of grooves at top covered the previous grooves. The age of the grooved terrain is uncertain. *(NASA)*

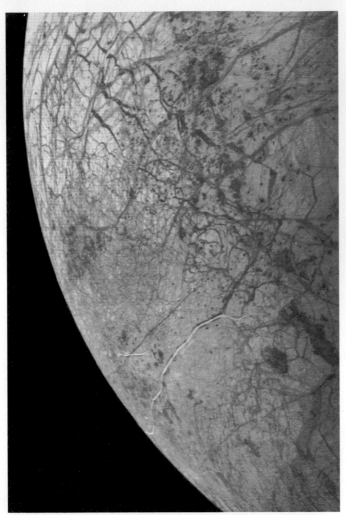

a

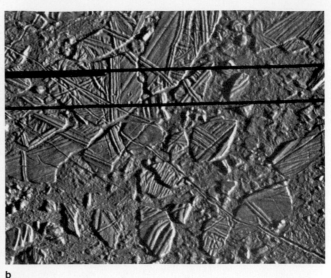

b

Figure 18-10

(a) Long lines on the surface of Europa appear to be cracks in its icy crust where water from below has refrozen. Note the bright surface and lack of craters—signs of youth. (b) This image from the Galileo spacecraft shows ice blocks on Europa that have moved like a disturbed puzzle. This resembles icebergs floating in pack ice on Earth. *(NASA)*

clean ice, contains very few craters, and has long scars suggestive of cracks in the icy crust (Figure 18-10). Evidently the icy crust we see overlays an ocean of liquid water or watery slush. The surface is estimated to be only about 10 million years old—very young for a geological surface. Thus Europa is apparently still an active world.

Io, the innermost of the four moons, is also the most active. Its density of 3.55 g/cm³, combined with gravity measurements by the Galileo spacecraft, shows that it has a large metallic core and a rocky, sulfur-rich crust. Spectra show no trace of water. Spacecraft photos reveal active volcanoes venting gases and sulfur-rich ash high above the surface (Figure 18-11). A new crater on Io would be buried under the sulfur ash at the rate of a few millimeters per year. That burying explains why photos show no impact craters on Io.

The activity we see in the Galilean moons must be driven by energy flowing outward. The violently active volcanism of Io is apparently caused by **tidal heating.** Io is too small to have remained hot from its formation, but its orbit is slightly elliptical; and as it moves closer to and then farther from Jupiter, the

planetary gravitational field flexes the moon with tides, and friction heats its interior. That heat flowing outward causes the volcanism. Europa is not as active as Io, but it too must have a heat source, presumably tidal heating. Ganymede is no longer active, but when it was younger, it had internal heat that broke the crust to produce the grooved terrain. Callisto, the most distant of the four moons, appears never to have been geologically active.

Tidal heating has probably affected the Galilean moons, but we must also suspect they formed together in a disk-shaped nebula around the proto-Jupiter. Heat from Jupiter would have heated the inner part of that disk, and the outer parts would have remained colder. Thus, we can understand the densities and compositions of the four Galilean moons by the condensation sequence (Chapter 16) acting in a miniature solar nebula around Jupiter.

A History of Jupiter

Can we put all of the evidence together and tell the story of Jupiter? Creating such a logical argument of

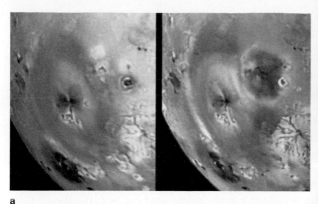

a

Figure 18-11
False color images of Io. (a) These two photos recorded by the Galileo spacecraft show that a new volcanic vent became active on Io during the summer of 1997. The larger feature is the ejecta around the older volcano Pele, but the new dark ejecta sheet is Pillan Patera covering a region about as big as Arizona. (b) A computer-enhanced image of Io reveals the volcano Loki spewing ash high above the surface where it falls back to blanket the surface. *(NASA)*

b

evidence and hypotheses is the ultimate goal of planetary astronomy.

Jupiter formed far enough from the sun to incorporate large numbers of icy planetesimals, and it must have grown rapidly. Once it was a dozen times more massive than Earth, it could grow by gravitational collapse (Chapter 16), the capture of gas directly from the solar nebula. Thus, it grew rich in hydrogen and helium from the solar nebula. Its present composition is quite sunlike and resembles the composition of the solar nebula. The location of Jupiter's point in Figure 17-20 shows that its gravity is strong enough to hold on to all of its gases.

The large family of moons may be mostly captured asteroids, and Jupiter may still encounter a wandering asteroid or comet now and then. Some of these are deflected, some captured into orbit, and some, like the comet that struck Jupiter in 1994, actually fall into the planet. Dust blasted off of the inner moons by micrometeorites settles into the equatorial plane to form Jupiter's rings.

The four Galilean moons seem to have formed in a disk of gas and dust around the forming planet. They have different compositions much like planets in our solar system, but they have also been affected to different degrees by tidal heating.

REVIEW **Critical Inquiry**

Why is Jupiter so big?

We can analyze this question by constructing a logical argument that relates the formation of Jupiter to the solar nebula theory. Jupiter is rich in hydrogen and helium, but Earth is relatively poor in these elements. While the solar nebula existed, Earth grew by the accretion of solid, rocky planetesimals, but it never became massive enough to capture gas directly from the solar nebula. That is, it never grew by gravitational collapse. Jupiter, however, grew so rapidly from icy planetesimals in the outer solar nebula that it was eventually able to grow by gravitational collapse. By the time the solar nebula cleared away and ended planet building, Jupiter had captured large amounts of hydrogen and helium and was quite massive.

Often the present nature of a world can be traced back to the way it formed. Can you create a logical argument to explain the nature of the Galilean moons?

We have studied Jupiter in detail because it is the basis of comparison for the other Jovian worlds. Now we can turn our attention to the most beautiful Jovian world, Saturn.

Saturn Data File Eight

Saturn's rings of ice particles are bright and beautiful, but its belt–zone circulation is subdued. A moon and its shadow lie just above center. *(NASA)*

Average distance from the sun	9.5388 AU (14.27 × 10^8 km)
Eccentricity of orbit	0.0560
Maximum distance from the sun	10.07 AU (15.07 × 10^8 km)
Minimum distance from the sun	9.005 AU (13.47 × 10^8 km)
Inclination of orbit to ecliptic	2°29'17"
Average orbital velocity	9.64 km/sec
Orbital period	29.461 y (10,760 days)
Period of rotation	10^{h}39^{m}25^s
Inclination of equator to orbit	26°24'
Equatorial diameter	129,660 km (9.42 D$_\oplus$)
Mass	5.69 × 10^{26} kg (95.147 M$_\oplus$)
Average density	0.69 g/cm^3
Gravity at base of clouds	1.16 Earth gravities
Escape velocity	35.6 km/sec (3.2 V$_\oplus$)
Temperature at cloud tops	−180°C (−292°F)
Albedo	0.61
Oblateness	0.102

18-2 Saturn

Saturn has played second fiddle to its own rings since Galileo first viewed it through a telescope in 1610. The rings are dramatic, strikingly beautiful, and easily seen through even a small telescope; but Saturn itself, only slightly smaller than Jupiter (Data File Eight), is a fascinating planet.

Although Saturn lies roughly 10 AU from the sun, we know a surprising amount about it. The beautiful rings are easily visible through the telescopes of modern amateur astronomers, and large Earth-based telescopes have explored the planet's atmosphere, rings, and moons. The two Voyager spacecraft flew past Saturn in 1979, transmitting back to Earth detailed measurements and images, and the Cassini spacecraft now on its way to Saturn will arrive in 2004 (Window on Science 18-2). That mission will tell us even more about the Saturn system.

Saturn the Planet

Seen from Earth, Saturn shows only faint evidence of belt–zone circulation, but Voyager and Hubble Space Telescope photos show that belts and zones are present and that the associated winds blow up to three times faster than on Jupiter. The belts and zones on Saturn are less visible because they occur deeper in the cold atmosphere below a layer of methane haze (Figure 18-12).

Saturn is less dense than water (it would float), and that suggests that it is, like Jupiter, rich in hydrogen and helium. In fact, its density is so low that it must have a relatively small core of heavy elements.

The shape of a Jovian planet can tell us about the interior. All of the Jovian planets, being mostly liquid and rotating rapidly, are slightly flattened. A planet's **oblateness** is the fraction by which its equatorial diameter exceeds its polar diameter. As photographs show, Saturn is the most oblate of the planets, and that evidence tells us that it is mostly liquid. A world with a large rocky core and mantle would not be flattened much by rotation, but an all-liquid planet would flatten significantly. Thus the oblateness of a Jovian planet, combined with its average density, can help astronomers model the interior.

Models of Saturn predict that it must have a small core of heavy elements and less liquid metallic hydrogen than Jupiter because Saturn's internal pressure is lower. Perhaps this is why Saturn's magnetic field is 20 times weaker than Jupiter's. Like Jupiter, Saturn has a hot interior and radiates more energy than it receives from the sun (Figure 18-13).

Saturn may look a bit bland, but it makes up for that in the splendor of its rings.

Science is an expensive enterprise, and that raises the question of payment. Some science has direct technological applications, but some basic science is of no immediate practical value. Who pays the bill?

In Window on Science 18-1, we saw that technology is of immediate practical use; thus, business funds much of this kind of scientific research. Auto manufacturers need inexpensive, durable, quick-drying paint for their cars, and they find it worth the cost to hire chemists to study the way paint dries. Many industries have large research budgets, and some industries, such as pharmaceutical manufacturers, depend exclusively on scientific research to discover and develop new products.

If a field of research has immediate potential to help society, it is likely that government will supply funds. Much of the research on public health is funded by government institutions such as the National Institutes of Health and the National Science Foundation. The practical benefit of finding new ways to prevent disease, for example, is well worth the tax dollars.

Basic science, however, has no immediate practical use. That doesn't mean it is useless, but it does mean that the practical-minded stockholders of a company will not approve major investments in such research. Digging up dinosaur bones, for instance, is very poorly funded because no industry can make a profit from the discovery of a new dinosaur. Astronomy is another field of science that has few direct applications, and thus very little astronomical research is funded by industry.

The value of basic research is twofold. Discoveries that have no known practical use today may be critically important years from now. Thus, society needs to continue basic research to protect its own future. But basic research, such as studying Saturn's rings or digging up dinosaur bones, is also of cultural value because it tells us what we are. Each of us benefits in intangible ways from such research, and thus society needs to fund basic research for the same reason it funds art galleries and national parks—to make our lives richer and more fulfilling.

Because there is no immediate financial return from this kind of research, it falls to government institutions and private foundations to pay the bill. The Keck Foundation has built two giant telescopes with no expectation of financial return, and the National Science Foundation has funded thousands of astronomy research projects for the benefit of society. Debates rage as to how much money is enough and how much is too much, but, ultimately, funding basic scientific research is a public responsibility that society must balance against other needs. There isn't anyone else to pick up the tab.

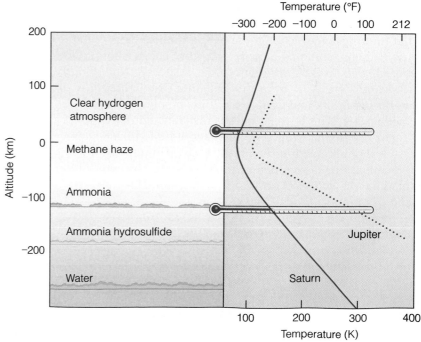

Figure 18-12

Because Saturn is farther from the sun, its atmosphere is colder than Jupiter's (dotted line). The cloud layers on Saturn form at the same temperature as do the cloud layers on Jupiter, but that puts them deeper in Saturn's hazy atmosphere, where they are not as easy to see as the clouds on Jupiter. (See Figure 18-3.)

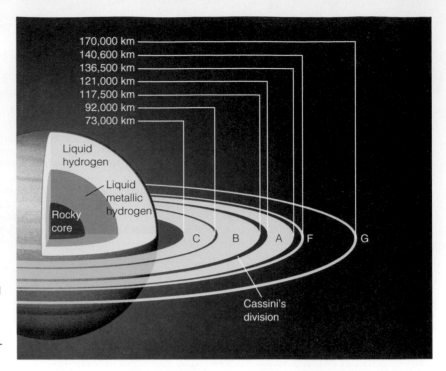

Figure 18-13
Models of the interior of Saturn suggest that it has a core of heavy elements that is smaller in proportion to that in Jupiter and that it contains less liquid metallic hydrogen because the pressure inside Saturn is lower. The major rings of Saturn are identified by letters of the alphabet. The gap called Cassini's division is a bit over one-fourth the diameter of Earth.

The Rings of Saturn

Saturn's rings contain billions of icy particles ranging from specks of dust to rare chunks the size of houses. These particles are confined in a thin layer perhaps only meters thick in the equatorial plane of the planet. A few gaps, such as Cassini's division, are visible from Earth (Figure 18-13), but the true complexity of the rings was not evident until the Voyager 1 spacecraft flew past Saturn in 1980. The Voyager images showed that the rings were made up of over a thousand narrow ringlets, some as narrow as 2 km. There were even ringlets in Cassini's division.

From Earth, astronomers had identified three main rings, A, B, and C, and had hints of a few other ring features. The Voyager images revealed a dusty E ring extending far from the planet and a threadlike F ring lying just outside the bright A ring (Figure 18-14). This F ring is confined in a narrow strand by two small moons that orbit just inside and just outside the ring. As a particle drifts away from the F ring, the moons shepherd it back into the ring, and thus the moons are known as **shepherd satellites.** A similar G ring and the sharp outer edge of the A ring are believed to be confined by shepherd satellites.

Some gaps in the rings seem to be caused by resonances with satellites. A particle in Cassini's division, for example, orbits Saturn twice in the time the moon Mimas takes to orbit once. Thus, on every other orbit, the ring particle overtakes Mimas at the same place in the particle's orbit, and the gravitational tugs of Mimas gradually pull the particle into a slightly elliptical orbit. Such an orbit is dangerous because the particle crosses the paths of millions of other particles.

Our unfortunate particle is almost certain to suffer a collision and have its orbit altered dramatically. Thus, particles cannot remain in orbit at a resonance with a moon, and gaps in the rings mark such spots.

But not all gaps are caused by resonances. Encke's division in the A ring is caused by a small satellite that keeps the gap clear of particles, and the hundreds of other ringlets cannot all be produced by resonances. Small moons in the rings and waves, such as spiral density waves, moving through the rings are believed to explain many of the gaps.

Saturn's icy rings can't be debris left over from the formation of the planet. The planet would have formed hot and would have vaporized ices and driven away the gas. Also, micrometeorites would tend to darken the icy particles over periods of 100 million years. These facts suggest that the material in Saturn's rings is debris from collisions between the heads of comets and icy moons. Such collisions must occur every few tens of millions of years and occasionally supply the rings with fresh ice. Saturn's moons certainly show signs of impacts.

The Moons of Saturn

Saturn has almost two dozen moons, many of which are small and all of which contain mixtures of ice and rock. We know these moons because the Voyager spacecraft flew past Saturn; another spacecraft, Cassini, will orbit Saturn in 2004 and study the moons as well as the planet and its rings.

The largest of Saturn's moons, Titan (Figure 18-15), is a few percent larger than the planet Mercury.

Figure 18-14
The rings of Saturn are made up of ice particles orbiting in the planet's equatorial plane. The transparency of the rings is evident in this image, where they cross in front of the planet at left. Generally pale and colorless, the rings are marked by many gaps and ringlets that are produced by dynamic effects. Note the tenuous C ring near the planet and the narrow F ring beyond the outer edge of the main rings. *(NASA)*

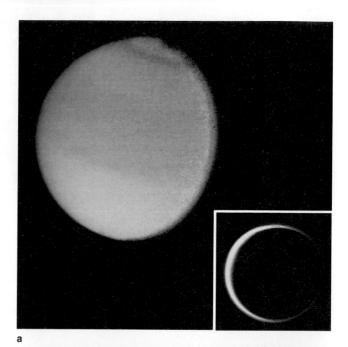

a

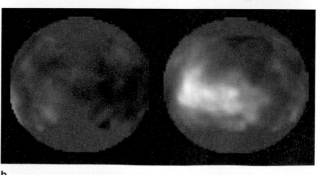

b

Its density suggests that it must contain a rocky core under a thick mantle of ice, but its surface is not visible through its cloudy atmosphere. We might not expect such a small world to have an atmosphere, but Titan is so cold its gas molecules do not travel fast enough to escape (see Figure 17-20). The gas is about 90 percent nitrogen with nearly 10 percent argon and a small amount of methane. Sunlight acting on the methane and nitrogen produces the thick orange smog of organic molecules. Organic molecules are common in living things on Earth but do not have to be derived from living things. One chemist defines an organic molecule as "any molecule with a carbon backbone." The organic smog particles produced in Titan's atmosphere drift down and collect in an organic goo on the moon's surface.

Organic goo on the surface of Titan sounds exciting because it could harbor primitive life, but the surface of Titan is very cold at −179°C (−290°F). Also, models of the atmosphere suggest that the surface may be partly covered by oceans of liquid methane and ethane. Oddly, Titan is cooled by an inverse greenhouse effect as the organic haze blocks much of the incoming sunlight but allows infrared radiation to escape to space. Clearly, Titan is a peculiar world.

In 2004, the Cassini spacecraft will drop an instrumented probe into the atmosphere of Titan before the spacecraft goes into orbit around Saturn. Cassini's cameras are sensitive at a number of different wavelengths and will be able to image the surface of Titan through the smog. We will soon know much more about this mysterious world.

Figure 18-15
(a) The surface of Titan is hidden by its smoggy atmosphere (with a circulation pattern near the north pole). The extent of the atmosphere is visible in photos of the night side (inset). (b) Hubble Space Telescope can detect infrared radiation from the surface. Here images show the side of Titan facing Saturn (left) and the side facing forward in its orbit (right). The surface is not a uniform ethane and methane ocean but consists of at least some solid surface with differing geology. *(NASA)*

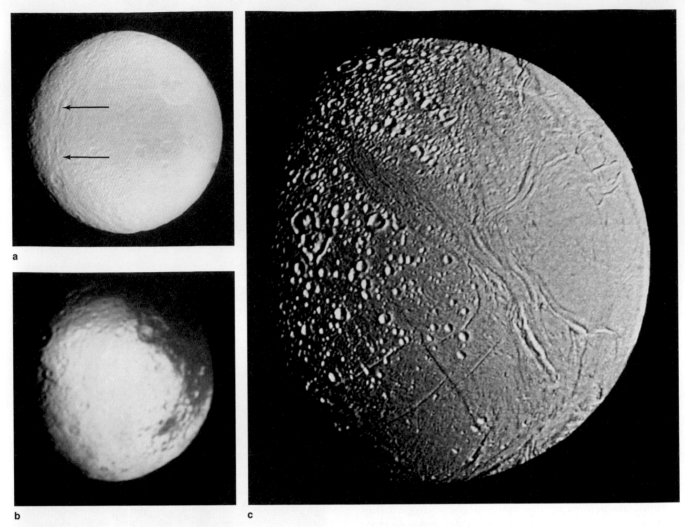

Figure 18-16
(a) Heavily cratered Tethys appears to be a dead world, although it is marked by a deep valley (arrows).
(b) The leading side of Iapetus (facing upper right) is much darker than the trailing side. (c) Some portions
of Enceladus have fewer craters than the rest. These areas apparently have been resurfaced recently,
which suggests that Enceladus is an active moon. *(NASA)*

The remaining moons of Saturn are small, icy, have no atmospheres, and are heavily cratered. Most have ancient surfaces. Tethys, for example, is less than a third the diameter of Earth's moon, and its heavily cratered crust seems quite old (Figure 18-16a). A valley 3 km deep trails three-fourths of the way around the satellite. Such cracks are found on a few other satellites and appear to have formed long ago, when the interiors of the icy moons froze and expanded.

The moon Enceladus, a bit smaller than Tethys, nevertheless shows signs of recent activity (Figure 18-16c). Some parts of its surface contain 1000 times fewer craters than other regions, showing that these lightly cratered regions must be younger. At some point in its history, this moon had internal heat that produced geological activity and resurfaced some areas of its crust. The energy source is unknown, but tidal heating is likely.

Like nearly all moons in the solar system, Saturn's moons are tidally locked to their planet, rotating to keep the same side facing the planet. The leading side of these moons, the side facing forward in the orbit, is sometimes modified by debris. Iapetus, for example, has a cratered trailing side about as dark as dingy snow, but its leading side is as dark as fresh asphalt (Figure 18-16b). One hypothesis is that the dark material is carbon-rich dust from meteorite impacts on the next moon out, Phoebe.

The History of Saturn

Can we put all of the evidence and hypotheses together and describe the history of Saturn? Doing so is a real test of our understanding.

Saturn formed in the outer solar nebula, where ice particles were stable and may have contained more trapped gases. The protoplanet grew rapidly and became massive enough to attract hydrogen and helium by gravitational collapse. The heavier elements sank to the middle to form a small core, and the hydrogen formed a liquid mantle containing liquid metallic hydrogen. The outward flow of heat from the interior is believed to drive convection inside the planet that produces its magnetic field. Because Saturn is smaller than Jupiter, the internal pressure is less, the planet contains less liquid metallic hydrogen, and its magnetic field is weaker.

The rings can't be primordial. That is, they can't be material left over from the formation of the planet. Such ices would have been vaporized and driven away by the heat of the protoplanet. Rather, we can suppose that the ring material is debris from occasional collisions between comets and Saturn's icy moons. Such giant impacts are not unheard of in the solar system. Jupiter was hit by a comet in 1994.

Some of Saturn's moons are probably captured asteroids that wandered too close, but some of the moons probably formed with Saturn. Many have ancient surfaces. The giant moon Titan may have formed with Saturn, or it may be a very large icy planetesimal captured into orbit around Saturn. We will see more evidence for this capture hypothesis when we explore farther from the sun.

REVIEW Critical Inquiry

Why do the belts and zones on Saturn look so indistinct?

In a Jovian planet, the light-colored zones form in high-pressure regions where rising gas cools and condenses to form icy crystals of ammonia, which we see as bright clouds. Saturn is twice as far from the sun as Jupiter, so sunlight is weaker and the atmosphere is colder. The rising gas currents don't have to rise as high to reach temperatures cold enough to form clouds. Because the clouds form deeper in the hazy atmosphere, they are not as brightly illuminated by sunlight and look dimmer. A layer of methane haze above the clouds makes the belts and zones look even less distinct.

We have used some simple physics to construct a logical argument that explains the hazy cloud features on Saturn. Can you explain why Saturn's rings have gaps and ringlets?

In any contest, Saturn would win the ribbon for most beautiful planet. But beyond Saturn's orbit are three more planets that could compete for the ribbon for most striking personality.

18-3 Uranus

Uranus is a Jovian planet, but it is quite different from Jupiter and Saturn. Its atmosphere, interior, rings, and satellites are profitable subjects for comparative planetology.

Uranus the Planet

Uranus is only a third the diameter of Jupiter, only a twentieth as massive, and, being four times farther from the sun, its atmosphere is over 100 degrees colder than Jupiter's (Data File Nine).

Because Uranus is smaller than Jupiter, its internal pressure is lower, and it does not contain liquid metallic hydrogen. Models of Uranus based in part on its density and oblateness suggest that it has a small core of heavy elements and a deep mantle of partly frozen water containing rocky material and dissolved ammonia and methane. Circulation in this electrically conducting mantle may generate the planet's peculiar magnetic field, which is highly inclined to its axis of rotation. Above this mantle lies the deep hydrogen and helium of the planet's atmosphere.

Uranus rotates on its side, with its equator inclined 98° to its orbit. With an orbital period of 84 years, each of its four seasons lasts 21 years and is extreme with the sun passing near each of its celestial poles (Figure 18-17). This peculiar rotation may have been produced when Uranus collided with a very large planetesimal late in its formation. When Voyager 2 flew past in 1986, the planet's south pole was pointed almost directly at the sun.

Voyager 2 photos show a nearly featureless blue-green world (Figure 18-18a). The atmosphere is mostly hydrogen and helium (12 percent), but traces of methane absorb red light and thus make the atmosphere look blue. There is no belt–zone circulation visible in the Voyager photographs, although extreme computer enhancement does reveal a few clouds and bands around the south pole.

Over a decade after the Voyager 2 photographs, the seasons on Uranus have changed. Spring has come to the northern hemisphere and fall to the southern hemisphere. The Hubble Space Telescope has imaged the planet in the near infrared and detected a number of bright cloud features in its atmosphere (Figure 18-18b). Thus we may be watching seasonal changes in the atmosphere.

Atmospheric circulation on a Jovian world may be driven by heat flowing out of the interior, but Uranus has little heat flow. It has cooled and now radiates about the same amount of energy that it receives from the sun. This may account for its limited atmospheric activity.

Uranus Data File Nine

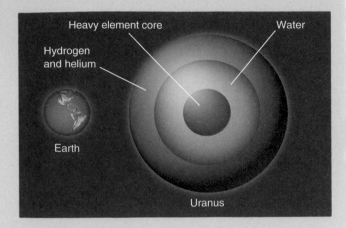

About four times the diameter of Earth, Uranus is believed to contain a liquid interior and dense core.

Average distance from the sun	19.18 AU (28.69 × 10⁸ km)
Eccentricity of orbit	0.0461
Maximum distance from the sun	20.1 AU (30.0 × 10⁸ km)
Minimum distance from the sun	18.3 AU (27.4 × 10⁸ km)
Inclination of orbit to ecliptic	0°46′23″
Average orbital velocity	6.81 km/sec
Orbital period	84.013 y (30,685 days)
Period of rotation	17^h14^m
Inclination of equator to orbit	97°55′
Equatorial diameter	51,118 km (4.01 $D_\oplus$)
Mass	8.69 × 10²⁵ kg (14.54 $M_\oplus$)
Average density	1.29 g/cm³
Gravity	0.919 Earth gravity
Escape velocity	22 km/sec (1.96 $V_\oplus$)
Temperature above cloud tops	−220°C (−364°F)
Albedo	0.35

The Rings of Uranus

The rings of Uranus were discovered in 1977 when astronomers watched the planet occult—cross in front of—a distant star. Although the rings are not easily visible from ground-based telescopes, the astronomers detected them when the star dimmed momentarily as each ring passed in front of it. Voyager 2 captured images of the rings in detail when it visited Uranus in 1986 (Figure 18-19). The Hubble Space Telescope can image the rings, and they are shown in Figure 18-18b where they are enhanced to make them visible along with the planet.

The rings of Uranus are tenuous and dark. Although they are 100 to 1000 times more substantial than Jupiter's ring, they total less mass than the thin material in Cassini's division—the largest "gap" in Saturn's rings. In addition, the rings of Uranus are composed of very dark material as black as lumps of coal.

Eleven narrow rings have been found. The widest, the ε ring, is slightly elliptical and varies from 12 to 60 km in width. Planetary rings should gradually spread into thin sheets, so the narrowness of these rings suggests that they are confined by shepherd moons. Voyager 2 photos show two moons less than 25 km in diameter shepherding the ε ring (Figure 18-19b), and the other rings are presumably confined by moons too small to detect.

The black color of the ring particles seems to be caused by the radiation trapped in Uranus's magnetosphere. The radiation can convert traces of methane ice into black organic polymers. The 10 smallest moons are also blackened by this radiation.

When Voyager 2 looked back through the rings from their dark side, it saw very little forward-scattered light, and this means the rings contain little dust. Particles must be boulder- to house-sized. Collisions between ring particles must occasionally chip off dust particles, but the tenuous high atmosphere of Uranus slows these particles, and they fall into the planet.

The rings of Uranus could not have survived from the formation of the planets. They must have been produced within the last few million years by debris that got trapped in the spaces between the moons' orbits. Astronomers now think that debris may be produced when comets strike the larger satellites.

The Uranian Moons

Until recently, astronomers could see only five moons orbiting Uranus. Voyager 2 discovered ten small moons in 1986, and two more were located in photos taken from Earth in 1997. Each of the Jovian planets probably has more small, undiscovered moons.

The five major satellites were photographed by Voyager 2, and we can analyze their surface features. Oberon and Titania, the two outer moons, are about

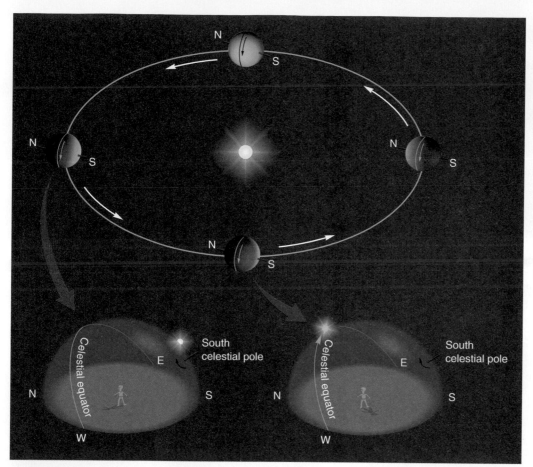

Figure 18-17
Uranus rotates on an axis that is tipped 97.9° from the perpendicular to its orbit, so its seasons are extreme. When one of its poles is pointed nearly at the sun (a solstice), a citizen of Uranus would see the sun near a celestial pole, and it would never rise or set. As it orbits the sun, the planet maintains the direction of its axis in space, and thus the sun moves from pole to pole. At the time of an equinox on Uranus, the sun would be on the celestial equator and would rise and set with each rotation of the planet. Compare with Figure 3-5.

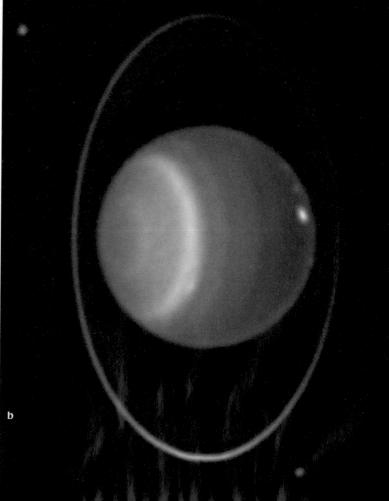

Figure 18-18
(a) The planet Uranus shows almost no atmospheric features in this visible-light image made by Voyager 2 in 1986 when it was high summer in the planet's southern hemisphere. *(NASA)* (b) In the late 1990s, the seasons on Uranus had changed, and this near infrared image reveals banding and cloud features in the atmosphere. The rings have been enhanced to make them visible in this image. *(Erich Karkoschka, University of Arizona and NASA)*

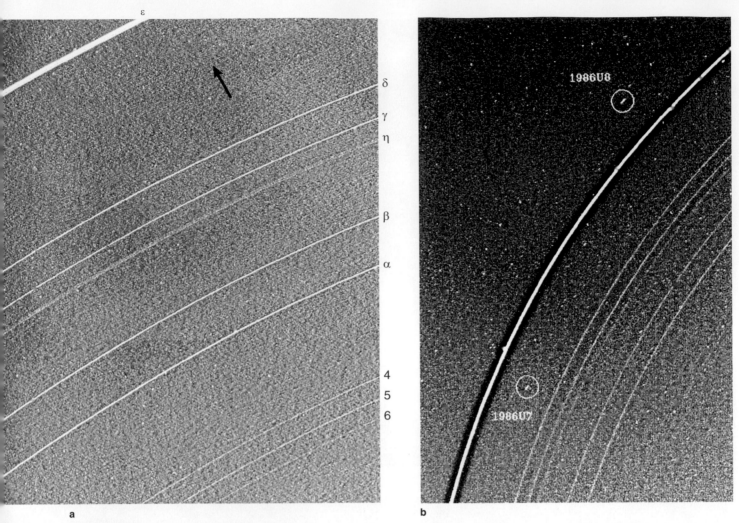

ε

δ
γ
η

β

α

4
5
6

a

b

1986U8

1986U7

Figure 18-19

(a) The narrow rings of Uranus are identified by Greek letter or by number. Voyager images reveal another faint ring (arrow) halfway from the ε ring to the next innermost (δ) ring. Tip this page and view the faint ring along its length. (b) Two shepherd satellites confine the ε ring into a narrow band. Because the satellites have slightly elliptical orbits, the ring is elliptical and slightly wider where it is farther from the planet. *(NASA)*

45 percent the diameter of Earth's moon, and they have old, cratered surfaces with faults hundreds of kilometers long. Some regions of these icy moons have been resurfaced by liquid water "lava" that covered old craters and froze. In contrast, Umbrial, the next moon inward, is about a third the diameter of Earth's moon and shows no sign of activity on its ancient, cratered surface. Ariel, the fourth moon inward, is slightly smaller than Umbrial, but its cratered crust is marked by broad, smooth-floored valleys that may have been cut by flowing ice (Figure 18-20a).

Miranda, the innermost moon, is only 14 percent the diameter of Earth's moon, but its surface is marked by oval patterns of grooves called **ovoids** (Figure 18-20b). These are believed to have been caused by internal heat that produced convection in the icy mantle. Rising currents of ice have deformed the crust and

created the ovoids. From craters on the ovoids, astronomers conclude that the entire surface is old and the moon is no longer active. Perhaps it was warmed by tidal heating long ago.

A History of Uranus

The challenge of comparative planetology is to tell the story of a planet, and Uranus is the most difficult planet to explain. It is not only far away and hard to study, but it is peculiar in a number of ways.

It must have formed from the outer solar nebula, but it must have grown more slowly than Jupiter or Saturn, perhaps because the nebula was less dense beyond Saturn. Uranus captured some gas from the nebula, but it did not grow as rich in hydrogen as Jupiter or Saturn. It is now a mostly liquid world.

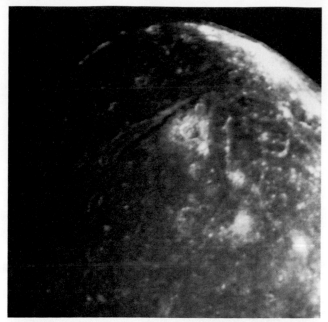

a

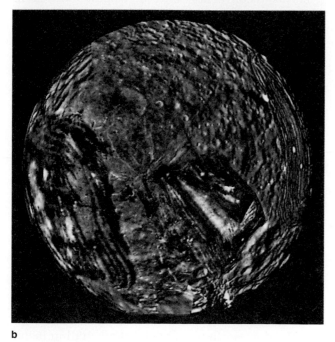

b

Figure 18-20
Geological activity on moons of Uranus. (a) Ariel has an old cratered surface, but some regions are marked by broad, shallow valleys. These are evidence of past activity. (b) The face of Miranda is marred by ovoids, which are believed to have formed when internal heating caused slow convection in the ice of the moon's mantle. Note the 5-km-high cliff at the lower edge of the moon.

Perhaps as it was forming it was struck by a large planetesimal and given its highly inclined rotation. These large impacts seem to have been common when the solar system was young. We have seen evidence for giant impacts on Earth's moon, Mercury, Mars, and the satellites of Jupiter and Saturn. Also, we suspect that impacts between the icy heads of comets and moons may produce the debris that replenishes planetary rings.

The highly inclined magnetic field of Uranus may be produced by convection in its electrically conducting mantle. With very little heat flowing out of the interior, this convection must be limited. The lack of interior heat in Uranus is not well understood, but some astronomers suspect that the impact that gave Uranus its odd rotation also stirred it so thoroughly that it lost much of its heat.

REVIEW Critical Inquiry

Why are the rings of Uranus so narrow?

Unlike the rings of Jupiter and Saturn, the rings of Uranus are quite narrow, like hoops of wire. We would expect collisions among ring particles to gradually spread the rings out into thin sheets, so something must be confining the narrow rings. In fact, two small moons have been found orbiting just inside and outside the ε ring. If a ring particle drifts away from the ring, the corresponding moon's gravity will boost it back into the ring. Moons too small to detect are thought to shepherd the other rings. Thus, the rings of Uranus resemble Saturn's narrow F ring.

This explains the narrowness of the rings, but isn't it odd that moons just happen to be in the right place to keep the rings narrow? How did that happen?

Uranus seems to be a bland, featureless planet, but it hides some peculiar secrets. Its neighbor, Neptune, is also a peculiar world.

18-4 Neptune

Through a small telescope, Neptune looks like a small blue dot with no visible cloud features. In 1989, Voyager 2 flew past and revealed some of Neptune's secrets.

Planet Neptune

Only 4 percent smaller in diameter than Uranus, Neptune has a similar interior. A small core of heavy elements lies within a slushy mantle of water below a hydrogen-rich atmosphere (Data File Ten). Yet Neptune looks quite different; it is dramatically blue and has active cloud formations.

The dark-blue tint to the atmosphere is understandable because Neptune contains 3 percent methane compared to only 2 percent in Uranus. The methane absorbs red photons better than blue, giving Neptune a blue tint. Also, its atmosphere scatters

Neptune Data File Ten

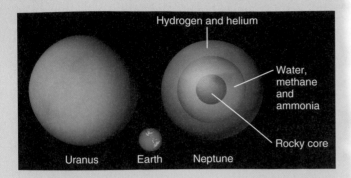

Neptune is a bit smaller than Uranus and four times larger in diameter than Earth. Its interior contains a rocky core with a deep mantle of slushy ices mixed with rock.

Average distance from the sun	30.0611 AU (44.971 × 10⁸ km)
Eccentricity of orbit	0.0100
Maximum distance from the sun	30.4 AU (45.4 × 10⁸ km)
Minimum distance from the sun	29.8 AU (44.52 × 10⁸ km)
Inclination of orbit to ecliptic	1°46′27″
Average orbital velocity	5.43 km/sec
Orbital period	164.793 y (60,189 days)
Period of rotation	$16^h 3^m$
Inclination of equator to orbit	28°48′
Equatorial diameter	49,500 km (3.93 $D_\oplus$)
Mass	1.030×10^{26} kg (17.23 $M_\oplus$)
Average density	1.66 g/cm³
Gravity	1.19 Earth gravities
Escape velocity	25 km/sec (2.2 $V_\oplus$)
Temperature at cloud tops	−216°C (−357°F)
Albedo	0.35
Oblateness	0.027

shorter-wavelength (blue) photons better than red photons, just as does Earth's atmosphere, and that fact helps make Neptune look blue. Finally, Uranus and Neptune are quite cold, and their temperatures may prevent the formation of the molecules that give the clouds their colors on Jupiter and Saturn.

Atmospheric circulation on Neptune is much more dramatic than on Uranus. When Voyager 2 flew by Neptune in 1989, the largest feature was the Great Dark Spot (Figure 18-21a). Roughly the size of Earth, the spot seemed to be an atmospheric circulation much like Jupiter's Great Red Spot. Smaller spots were visible in Neptune's atmosphere, and photos showed they were circulating like hurricanes. More recently, the Hubble Space Telescope has photographed Neptune and found that the Great Dark Spot is gone and new cloud formations have appeared (Figure 18-21b). Evidently, the weather on Neptune is changeable.

The atmospheric activity on Neptune is apparently driven by heat flowing from the interior. The heat causes convection in the atmosphere, which the rapid rotation of the planet converts into high-speed winds, high-level white clouds of methane ice crystals, and rotating storms that we see as spots. Neptune has more activity than Uranus because it has more heat flowing out of its interior.

Like Uranus, Neptune has a highly inclined magnetic field that must be linked to circulation in the interior. In both cases, we suspect that ammonia dissolved in the liquid water mantle makes the mantle a good electrical conductor and that convection in the water, coupled with the rotation of the planet, drives the dynamo effect and generates the magnetic field.

Neptune's Rings

Voyager 2 photos revealed that Neptune has two narrow rings with radii of about 2 and 2.5 planetary radii with a broad, faint ring between (Figure 18-22a). Small moons shepherd the narrow rings and keep the ring particles from spreading.

The narrow rings are brightest when illuminated from behind—forward scattering—so they must contain large amounts of dust. The broader rings, however, seem to contain less dust. Because the ring particles are not very reflective, they are probably rocky rather than icy.

The outer narrow ring contains short arcs where the ring is denser. The ring arcs are difficult to explain because collisions among particles should smooth out the rings in relatively short times. However, a mathematical model of the rings shows that the small, inner moon Galatea creates waves in the ring, waves that confine the particles in certain small arcs. Furthermore, the model predicts the location of other fainter arcs, and computer-enhanced images confirm their

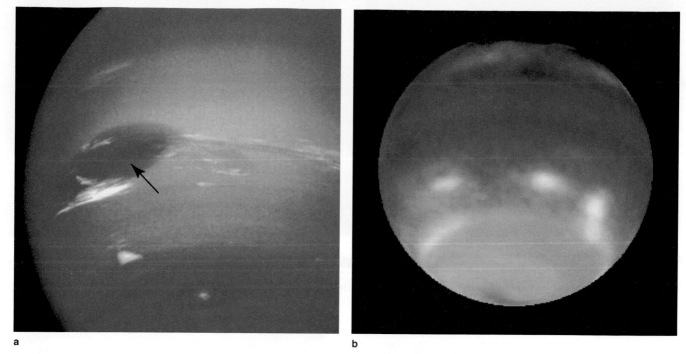

a

b

Figure 18-21

(a) A 1989 Voyager 2 image of Neptune revealed an anticyclonic disturbance about the size of Earth, which was dubbed The Great Dark Spot (arrow). High-altitude winds were deflected around the spot and formed high clouds of methane-ice crystals. *(NASA)* (b) A 1998 Hubble Space Telescope image of Neptune shows how cloud patterns have changed since Voyager 2 flew past in 1989. The Great Dark Spot has disappeared, but new spots have appeared. *(Lawrence A. Sromovsky, University of Wisconsin-Madison, and NASA)*

a

b

Figure 18-22

(a) The brilliant disk of Neptune is hidden behind the central black bar in this Voyager 2 photo of the rings. Although the rings look uniform here, they actually contain short arcs where the ring particles are denser. *(NASA)* (b) This computer-enhanced image shows the arcs in Neptune's rings. A mathematical model reveals that the arcs are created by the gravitational influence of Neptune's moon Galatea. Arrows point to fainter arcs whose existence was predicted by the model. *(NASA image courtesy of Carolyn C. Porco)*

existence (Figure 18-22b) and thus give us added confidence in the model.

The Moons of Neptune

Two moons are visible from Earth, and both have peculiar orbits. Nereid is a small moon in a large, elliptical orbit taking nearly an Earth year to circle Neptune once. Triton orbits Neptune in only 5.877 days, but it orbits backward—clockwise as seen from the north. These odd orbits suggest that the system was disturbed long ago in an interaction with some other body, such as a massive planetesimal.

Voyager 2 photographs revealed six more moons orbiting among the rings. They are all small, dark, rocky bodies. No additional moons were found beyond Triton, and some astronomers suggest that Triton in its retrograde orbit would have consumed such moons.

Although Triton is only 78 percent the radius of Earth's moon, it has an atmosphere of nitrogen and methane about 10^5 times less dense than Earth's. Triton can keep this gas because it is so far from the sun it is very cold—only 37 K ($-393°F$) (see Figure 17-20). In fact, a significant part of Triton is ice, and deposits of nitrogen frost are visible at the southern pole (Figure 18-23a), which has been turned toward sunlight for the last 30 years. This nitrogen appears to be vaporizing in the sunlight and is probably refreezing in the darkness at the north pole.

Many features on Triton suggest it has had an active past. It has few craters on its surface, but it does have long faults that appear to have formed when the icy crust broke. Some approximately round basins about 400 km in diameter appear to have been flooded time after time by liquids from the interior (Figure 18-23b). Most exciting of all are the dark smudges visible in the southern polar cap (Figure 18-23a). Analysis of the photos reveals that these are deposits produced when liquid nitrogen in the crust, warmed by the sun, erupts through vents and spews up to 8 km high into the atmosphere. Methane in the gas is converted by sunlight into dark deposits that fall to the surface, leaving the black smudges. These geysers make Triton one of only three worlds in our solar system observed to be active—Earth, Io, and Triton.

By counting craters on Triton, astronomers conclude that the surface has been active as recently as a million years ago and may still be active. The energy source for this volcanism could come from radioactive decay. The moon is two-thirds rock, and although such a small world would not be able to generate sufficient radioactive decay to keep molten rock flowing to its surface, frigid Triton may suffer from water–ammonia volcanism. A mixture of water and ammonia could melt at very low temperatures and resurface parts of the moon.

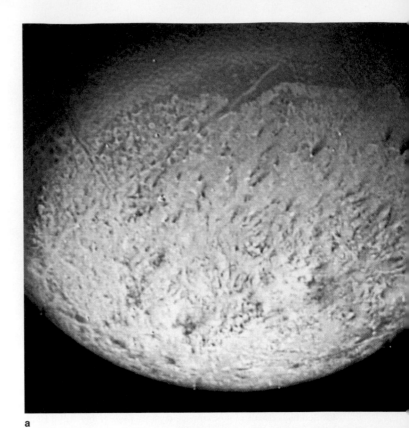

a

b

Figure 18-23
(a) Triton's southern polar cap is formed of nitrogen frost. Note the absence of craters and the dark smudges caused by nitrogen geysers. (b) Such round basins on Triton appear to have been repeatedly flooded by liquid from the interior. *(NASA)*

The History of Neptune

Our history of Neptune is quite short because many of its characteristics are understandable in terms of comparative planetology. On the other hand, its peculiarities are very difficult to understand because we have little data.

Neptune must have formed much as Uranus did, growing slowly in the outer solar nebula and never becoming massive enough to trap large amounts of hydrogen and helium. It developed a core of heavy elements, a mantle of dirty water, and a deep hydrogen-rich atmosphere. It has an internal heat source provided, perhaps, by radioactive decay. The heat flowing to the surface generates its magnetic field and creates atmospheric circulation.

The moons of Neptune suggest some cataclysmic encounter long ago that put Nereid into a long-period elliptical orbit and put Triton into a retrograde orbit. We have seen evidence of major impacts throughout the solar system, so such interactions may have been fairly common. Certainly, impacts on the satellites could provide the debris that we find trapped among the smaller moons to form the rings.

R E V I E W   Critical Inquiry

Why is Neptune blue?

To analyze this question, we must be careful not to be misled by the words we use. When we look *at* something, we really turn our eyes toward it and receive light *from* the object. The light we receive from Neptune is sunlight that is scattered from various layers of Neptune and journeys to our eyes. Because sunlight contains a distribution of photons of all visible wavelengths, it looks white to us; but sunlight entering Neptune's atmosphere must pass through hydrogen gas that contains a small amount of methane, which is a good absorber of longer wavelengths. As a result, red photons are more likely to be absorbed than blue photons, and that makes the light bluer. Furthermore, when the light is scattered in deeper layers, the shorter-wavelength photons are most likely to be scattered, and thus the light that finally emerges from the atmosphere and reaches our telescopes is poor in longer wavelengths. It looks blue.

This discussion shows how a careful, step-by-step analysis of a natural process can help us better understand how nature works. For example, can you explain why the clouds on Neptune look white?

The Voyager 2 spacecraft left the solar system after visiting Neptune. It will not pass near any other worlds unless, millions of years from now, it enters some distant solar system. Nevertheless, we have one more planet to analyze.

18-5 Pluto

Pluto is a small icy world clearly different from the Jovian worlds yet also different from the terrestrial planets of rock and ice. Since its discovery in 1930, it has been a mystery on the edge of the solar system.

Pluto as a Planet

Pluto is very difficult to observe from Earth. Only a bit larger than 0.1 second of arc in diameter, the little planet is only 65 percent the diameter of Earth's moon and shows little surface detail. The best photos by the Hubble Space Telescope reveal large areas of light and dark terrain (Data File Eleven), promising that Pluto will reveal itself to be an interesting world when spacecraft finally visit it.

Most planetary orbits in our solar system are nearly circular; but Pluto's is quite elliptical (see Figure 1-7). On the average, it is the most distant planet, but it can come closer to the sun than Neptune. In fact, from January 21, 1979, to March 14, 1999, Pluto was closer to the sun than Neptune is. The planets will never collide, however, because Pluto's orbit is inclined 17°.

Orbiting so far from the sun, Pluto is cold enough to freeze most compounds we think of as gases, and spectroscopic observations have found evidence of solid nitrogen ice on its surface with traces of methane ice. It also has a thick atmosphere of gaseous methane mixed with nitrogen and smaller amounts of carbon monoxide. There is no direct evidence of water ice on Pluto, but most astronomers believe there must be water frozen into the crust.

In 1978, Pluto's moon Charon was discovered in a highly inclined orbit with a period of 6.387 days and an average distance of about 19,640 km (Figure 18-24). The moon was named after the mythological ferryman who transports souls across the river Styx into the underworld.

From the size of the orbit and its period, astronomers could calculate the mass of Pluto, about 0.002 Earth masses. Given its diameter of about 2300 km, this places its density at roughly 2 g/cm^3, typical for a mixture of rock and ices. Charon appears to be about half the size of Pluto and less dense at 1.8 g/cm^3.

The History of Pluto

Is Pluto really a planet? It is so dramatically different from its Jovian neighbors that it doesn't seem to fit into the solar system, and some astronomers have argued that it isn't a planet at all. In fact, this depends almost entirely on our arbitrary definition of the word "planet." A better question is, "How did Pluto reach its present state?"

Pluto Data File Eleven

The Hubble Space Telescope is just able to reveal bright and dark regions on Pluto. Some of the bright regions may be deposits of bright methane ice. *(Alan Stern, Southwest Research Institute, Marc Buie, Lowell Observatory, NASA, ESA)*

Average distance from the sun	39.44 AU (59.00 × 10⁸ km)
Eccentricity of orbit	0.2484
Maximum distance from the sun	49.24 AU (73.66 × 10⁸ km)
Minimum distance from the sun	29.64 AU (44.34 × 10⁸ km)
Inclination of orbit to ecliptic	17°9′3″
Average orbital velocity	4.73 km/sec
Orbital period	247.7 y (90,465 days)
Period of rotation	$6^d9^h21^m$
Inclination of equator to orbit	122°
Equatorial diameter	2300 km (0.19 $D_\oplus$)
Mass	1.2 × 10²² kg (0.002 $M_\oplus$)
Average density	2.0 g/cm³
Surface gravity	0.06 Earth gravity
Escape velocity	1.2 km/sec (0.11 $V_\oplus$)
Surface temperature	−230°C (−382°F)
Albedo	0.5

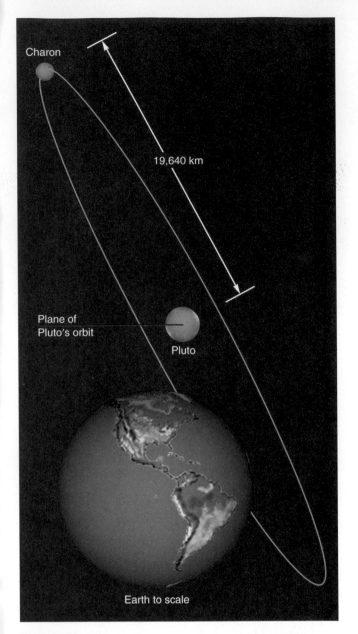

Figure 18-24
The circular orbit of Charon is seen here at an angle. The orbit is only a few times bigger than Earth and is tipped 118° to the plane of Pluto's orbit.

An old theory proposed that Pluto was an escaped moon of Neptune. Pluto is much like Neptune's moon Triton. It is roughly the same density, and it is only 85 percent the diameter of Triton; it could easily fit in with the icy moons of the Jovian planets. The disturbed orbits of Triton and Nereid could be the result of a complex interaction with a passing massive planetesimal in which Pluto might have been ejected from an orbit around Neptune. This theory has been abandoned, however, because Pluto has a moon, and it doesn't seem possible for it to have kept a moon if it was once a moon itself. Also, a newer interpretation seems more probable.

Modern astronomers have begun thinking of Pluto as the largest of a family of small icy bodies that were once common in the outer solar system. Searches for other icy worlds beyond the orbit of Neptune have turned up some 60 icy bodies with hints that there are a vast number of more distant bodies. (We will discuss this belt of icy planetesimals in detail in Chapter 19.) If this is true, then Pluto is just a large, icy planetesimal that never became part of a larger planet.

The moon Charon is presumably another of these icy worlds. A collision long ago between Charon and Pluto could have tilted the axis of rotation of Pluto and trapped Charon in an elliptical orbit around Pluto. Tidal forces between Pluto and Charon could then have forced the orbit of Charon to become circular as it is today. If that happened, then we might hope to see dramatic evidence of tidal heating in Pluto and in Charon when spacecraft finally reach Pluto.

At present no spacecraft is heading for Pluto. Plans have been proposed, but no funds have been set aside. It may be a long time before we see the surface of these cold worlds, but our study of comparative planetology helps our imagination go where our cameras cannot.

REVIEW Critical Inquiry

What evidence do we have that cataclysmic impacts have occurred in our solar system?

We would expect objects that formed together from the solar nebula to rotate and revolve in the plane of the solar system, so the high inclination of Pluto and the orbit of Charon suggest a past collision or, at least, a close gravitational interaction with a passing body. And the Pluto–Charon system is not the only evidence for major impacts in the past. The peculiar orbits of Neptune's moons Triton and Nereid, the fractured and cratered state of the Jovian satellites, and the peculiar rotation of Uranus all hint that impacts and encounters with large planetesimals or the heads of comets have been important in the history of these worlds. Furthermore, the existence of planetary rings suggests that impacts have scattered small particles and replenished the ring systems. Even in the inner solar system, we see the backward rotation of Venus, the density of Mercury, and the formation of Earth's moon as possible consequences of major impacts when the solar system was young.

Certainly, we do not imagine that planets bounce around like billiard balls, but we must also recognize that the planets are not totally isolated now and may have been victims (or beneficiaries) of large impacts as they formed long ago. For example, how does the size and composition of Pluto and Charon compared with the moons of the outer Jovian worlds suggest an unexpected population of icy worlds in the outer solar system?

Like clues at the scene of a crime, the evidence we find in the outer solar system alerts us that all of the planets in our solar system have been and still are sitting ducks for impacts by the smaller objects that hurtle along their orbits around the sun. What are these objects and what can they tell us about the origin and evolution of our solar system? We explore that in the next chapter.

Summary

The Jovian planets are so massive that they have been able to capture and retain large amounts of hydrogen and helium. This excess of light elements gives the Jovian planets densities much lower than those of the terrestrial planets. In their interiors, Jupiter and Saturn have such high pressures that the hydrogen becomes a liquid, and deep inside, it becomes liquid metallic hydrogen, a very good electrical conductor. Circulation in this rotating fluid drives the dynamo effect and produces the planets' magnetic fields, which trap particles from the solar wind to form radiation belts.

The atmosphere of Jupiter is marked by high- and low-pressure regions in the form of belts and zones that encircle the planet. Chemical reactions in the clouds are believed to produce compounds that color the belts. The Great Red Spot on Jupiter is believed to be a cyclonic disturbance much like a hurricane on Earth.

Jupiter's ring is a tenuous sheet of dust particles orbiting in the planet's equatorial plane. Because certain processes rapidly remove dust from the ring, astronomers believe it must be continuously resupplied by new dust chipped from small satellites.

Jupiter's smaller moons may be captured asteroids, but its four largest, the Galilean moons, clearly formed with the planet. The outermost moon, Callisto, is an icy, dead world, and the next moon inward, Ganymede, shows grooves produced by ancient breaks in its icy crust. Europa is less rich in ice, but it does have an icy crust free of craters with a liquid-water ocean underneath. Thus, it must be geologically active. Io is very dry and has active volcanoes powered by tidal heating.

Saturn's cold atmosphere has belt–zone circulation, thought it is partially hidden below methane haze. Saturn's rings are made of icy particles shepherded by small moons and broken into many ringlets by the presence of small moons within the rings and effects such as spiral density waves. The rings cannot be as old as the planet, so they must be accumulations of icy debris from impacts on Saturn's moons. Most of Saturn's moons are small, icy, dead worlds, but at least one, Enceladus, shows signs of an active surface.

Uranus and Neptune are smaller than Jupiter and Saturn and thus have lower internal pressures. They contain no liquid hydrogen, but they do have mantles of dirty water in which the dynamo effect produces highly inclined magnetic fields.

Uranus rotates on its side, perhaps because it collided with a large planetesimal late in its formation. Its atmosphere was bland when Voyager 2 flew past, but Hubble

Space Telescope photos show that the changing seasons have produced cloud patterns. Uranus's narrow rings of dark boulders appear to be confined by shepherd satellites. Although it has many small moons, only five are large enough to be easily visible from Earth. Voyager 2 images revealed traces of past impacts. The innermost moon, Miranda, is marked by ovoids that suggest past convection in its icy mantle and crust.

Heat is flowing outward from the interior of Neptune, and that drives strong weather patterns in its atmosphere which produce hurricanelike storms and high methane clouds. Neptune's rings are thin hoops with the outermost ring containing denser arcs caused by a resonance with one of its smaller moons. Triton, its largest moon, is an icy world where methane-water volcanism may still be active.

Pluto and its moon Charon are small and made of a mixture of ice and rock. The origin of Pluto and Charon is unknown, but they are similar in many ways to Triton, and astronomers now speculate that such icy bodies may once have been common in the outer solar system.

New Terms

magnetosphere	grooved terrain
belt–zone circulation	tidal heating
gossamer rings	oblateness
forward scattering	shepherd satellite
Roche limit	ovoid

Review Questions

1. Why is Jupiter so much richer in hydrogen and helium than Earth?

2. How can Jupiter have a liquid interior and not have a liquid surface?

3. How does the dynamo effect account for the magnetic fields of Jupiter and Saturn?

4. Why are the belts and zones on Saturn less distinct than those on Jupiter?

5. Why do we conclude that neither Jupiter's ring nor Saturn's ring can be left over from the formation of the planets?

6. How can a moon produce a gap in a planetary ring system?

7. Explain how geological activity on Jupiter's moons varies with distance from the planet.

8. What makes Saturn's F ring and the rings of Uranus and Neptune so narrow?

9. Why is the atmospheric activity of Uranus less than that of Saturn and Neptune?

10. Why do we suspect that Enceladus has been geologically active more recently than some other moons?

11. What are the seasons on Uranus like?

12. Why are Uranus and Neptune blue?

13. What evidence do we have that Triton has been geologically active recently?

14. How do astronomers account for the origin of Pluto?

15. What evidence do we have that catastrophic impacts have occurred in the solar system's past?

Discussion Questions

1. Some astronomers argue that Jupiter and Saturn are unusual, while other astronomers argue that all solar systems should contain one or two such giant planets. What do you think? Support your argument with evidence.

2. Why don't the terrestrial planets have rings? If you were to search for a ring among the terrestrial planets, where would you look first?

Problems

1. What is the maximum angular diameter of Jupiter as seen from Earth? Repeat this calculation for Saturn and Pluto. (*Hints:* See Data Files 7, 8, and 11, and also By the Numbers 3-1.)

2. What is the angular diameter of Jupiter as seen from Callisto? From Io? (*Hint:* See By the Numbers 3-1.)

3. Measure the photograph in Data File 8 and calculate the oblateness of Saturn.

4. If we observe light reflected from Saturn's rings, we should see a red shift at one edge of the rings and a blue shift at the other edge. If we observe a spectral line and see a difference in wavelength of 0.056 nm, and the unshifted wavelength (observed in the laboratory) is 500 nm, what is the orbital velocity of particles at the outer edge of the rings? (*Hint:* See By the Numbers 6-2.)

5. One way to recognize a distant planet is by its motion along its orbit. If Uranus circles the sun in 84 years, how many seconds of arc will it move in 24 hours? (*Hint:* Ignore the motion of Earth.)

6. If the ε ring is 50 km wide and the orbital velocity of Uranus is 6.81 km/sec, how long a blink should we expect to see when the ring crosses in front of the star?

7. What is the angular diameter of Pluto as seen from the surface of Charon? (*Hint:* See Figure 18-24.)

8. If Pluto has a surface temperature of 50 K, at what wavelength will it radiate the most energy? (*Hint:* See By the Numbers 6-1.)

9. How long did it take radio commands to travel from Earth to Voyager 2 as it passed Neptune?

10. Use the orbital radius and orbital period of Charon to calculate the mass of the Pluto–Charon system. (*Hints:* Express the orbital radius in AU and the period in years. Then see By the Numbers 8-4.)

Critical Inquiries for the Web

1. If you lived on the surface of Pluto and looked into the sky to observe Charon, what phases would you see? (*Hint:* Be sure to consider your location on the planet when answering this question.)

2. What factors caused Voyager 2 to see a bland atmosphere when it encountered Uranus in 1986? Given these circumstances, would images of the Uranian atmosphere taken by a space probe arriving at Uranus in 2006 be similar to those taken in 1986, or would there be significant differences?

3. Imagine what you'd think if you had been the first person ever to see Saturn through a telescope. When Galileo first observed Saturn in 1610, he did not recognize that it was a ringed planet. It was many years later before the strange apparition of Saturn was finally attributed to a ring structure. Search the Web for information on historical observations of Saturn, summarize the observations of Galileo and others, and determine who was first to recognize what he saw as a ring. Why do you suppose it took so long to understand that Saturn is a ringed planet?

Go to the Brooks/Cole Astronomy Resource Center (www.brookscole.com/astronomy) for critical thinking exercises, articles, and additional readings from InfoTrac College Edition, Brooks/Cole's online student library.

CHAPTER 19

Meteorites, Asteroids, and Comets

Guidepost

In Chapter 16 we began our study of planetary astronomy by asking how our solar system formed. In the two chapters that followed, we surveyed the planets, but we found only limited insight into the origin of the solar system. The planets are big, and they have evolved as heat has flowed out of their interiors. In this chapter we have our best look at unevolved matter left over from the solar nebula. These small bodies are, in fact, the last re- mains of the nebula that give birth to the planets.

This chapter is unique in that it covers small bodies. In past chapters we have used the principles of com- parative planetology to study large objects—the planets. In this chapter, we see that some principles also apply to the smaller bodies, but we also see that we need some new tools to think about the tiniest worlds in the solar system.

Figure 19-1
Comet Hyakutake swept through the inner solar system in 1996 and was dramatic in the northern sky. Seen here from Kitt Peak National Observatory, the comet passed close to the north celestial pole (behind the observatory dome in this photo). Notice the Big Dipper below and to the left of the head of the comet. *(Courtesy Tod Lauer)*

In 1910, Comet Halley was spectacular, and on the night of May 19, Earth actually passed through the tail of the comet—and millions of people panicked. The spectrographic discovery of cyanide gas in the tails of comets led many to believe that life on Earth would end. Householders in Chicago stuffed rags around doors and windows to keep out the gas, and bottled oxygen was sold out. Con artists in Texas sold comet pills and inhalers to ward off the noxious fumes. An Oklahoma newspaper reported (in what was apparently a hoax) that a religious sect tried to sacrifice a virgin to the comet.

Throughout history, bright comets have been seen as portents of doom. Even in our own time, the appearance of bright comets has generated predictions of the end of the world. Comet Kohoutek in 1973 and Comet Halley in 1986 caused concern among the superstitious. A bright comet moving slowly through the night sky is so out of the ordinary (Figure 19-1) that we should not be surprised if it generates some instinctive alarm.

In fact, comets are graceful and beautiful visitors to our skies. Astronomers think of comets as messengers from the age of planet building. By studying comets, we learn about the conditions in the solar nebula from which planets formed. But comets are only the icy remains of the solar nebula; the asteroids are the rocky debris left over from planet building. Because we cannot easily visit comets and asteroids, we begin our discussion with the fragments of those bodies that fall into our atmosphere—the meteorites.

19-1 Meteorites

We discussed meteorites in Chapter 16 when we discussed the age of the solar system. There we saw that the solar system is filled with small particles called meteoroids, which can fall into Earth's atmosphere at speeds of 10 to 40 km/sec. Friction with the air heats the meteoroids to glowing and we see them vaporize as meteors streaking across the night sky. If a meteoroid is big enough and strong enough, it can survive its plunge through the atmosphere and reach Earth's surface. Once the object strikes Earth's surface, we refer to it as a meteorite. The largest of these objects

a   b

Figure 19-2
(a) An iron meteorite sliced open, polished, and etched with acid reveals a Widmanstätten pattern. The detection of these bands conclusively identifies an object as an iron meteorite. (b) A stony meteorite sliced open and polished reveals small, spherical inclusions called chondrules. *(North Museum, Franklin and Marshall College)*

can blast out craters on Earth's surface (see Figure 17-8), but such impacts are rare. The vast majority of meteorites are too small to form craters. We find these meteorites all over Earth, and their value is in what they can tell us about the origin of the planets.

Inside Meteorites

Meteorites can be divided into two broad categories: *Iron* meteorites are solid chunks of iron and nickel. *Stony* meteorites are silicate masses that resemble Earth rocks.

When iron meteorites are sliced open, polished, and etched with nitric acid, they reveal regular bands called **Widmanstätten patterns** (Figure 19-2a). The patterns arise from crystals of nickel-iron alloys that have grown very large, indicating that the meteorite cooled no faster than a few degrees per million years. Explaining how iron meteorites could have cooled so slowly will be a major step in analyzing their history.

In contrast to iron meteorites, most stony meteorites appear never to have been heated hot enough to melt. In fact, we can classify stony meteorites into three types, according to the degree to which they have been heated. **Chondrites** are stony meteorites that contain **chondrules,** rounded bits of glassy rock ranging in size from microscopic to as big as a pea (Figure 19-2b). The origin of chondrules is unknown, but they appear to be very old; and because melting would have destroyed them, their presence in chondrites indicates that these meteorites never melted.

Nevertheless, chondrites have been heated slightly. The rock in which the chondrules are imbed-

ded is also old, but it contains no volatiles. The condensing solar nebula should have incorporated carbon compounds and water into the forming solids, but no such material is present in the chondrites. They have been heated enough to drive off these volatile compounds but not enough to melt the chondrules.

The **carbonaceous chondrites** contain both chondrules and volatile compounds. Had they been heated, even slightly, the volatiles would have evaporated. Thus the carbonaceous chondrites are the least altered remains of the solar nebula.

Stony meteorites of the third type are called **achondrites** because they contain no chondrules. They also lack volatiles and appear to have been subjected to intense heat that melted chondrules and drove off volatiles, leaving behind rock with compositions similar to Earth's lavas.

The Origin of Meteors and Meteorites

The mineral composition of the meteorites gives us a clue to the origins of these objects. Somehow they have survived almost unchanged since the age of planet building in the solar nebula.

We can find evidence of the origin of meteors through one of the most pleasant observations in astronomy. We can observe a **meteor shower,** a display of meteors that are clearly related in a common origin. For example, the Perseid meteor shower occurs each year in August (Table 19-1), and on that night we might see as many as 60 meteors an hour if we stretched out on a lawn chair at a dark site and watched the sky. Like many natural phenomena, a meteor shower is more en-

We can enjoy a meteor shower as Mother Nature's fireworks, but our enjoyment is much greater once we begin to understand what causes meteors and why meteors in a shower follow a pattern. When we know that meteor showers help us understand the origin of our world, the evening display of shooting stars is even more exciting. Science typically increases our enjoyment of the natural world by revealing the significance of things we might otherwise enjoy only in a casual way.

Everyone likes flowers, for example. We enjoy them in a casual way, admiring their colors and fragrances, their graceful petals and massed blooms. But botanists know that evolution has carefully designed flowers to attract insects and spread pollen. The bright colors attract insects, and the shapes of the flowers provide little runways so the insect will find it easy to land. Some color patterns even guide the insect in like landing lights at an airport, and many flowers, such as orchids and snapdragons, force the insect to crawl inside in just the right way to exchange pollen and fertilize the flower. The nectar is nothing more than bug bait. Once we begin to understand what flowers are for, a visit to a garden becomes not only an adventure of color and fragrance, but also an adventure in understanding the beauty of design and function, an adventure in meaning as well as an adventure of the senses.

And our understanding of a natural phenomenon helps us understand and enjoy related phenomena. For example, some flowers attract flies for pollination, and such blossoms smell like rotting meat. A few flowers depend on bats, and they open their blossoms at night. Many plants, such as pine trees, depend on the wind to spread their pollen, and those plants do not have colorful flowers at all; but flowers that depend on hummingbirds have long trumpet-shaped blossoms that just fit the hummingbirds' beaks.

The more science tells us about nature, the more enjoyable the natural world is. We can enjoy a meteor shower as nothing more than a visual treat, but the more we know about it, the more interesting and exciting it becomes. The natural world is filled with meaning, and science, as a way of discovering and understanding that meaning, gives us new opportunities to enjoy the world around us.

---

joyable when we know more about it (Window on Science 19-1).

On any clear, moonless night of the year we might see 5 to 15 meteors per hour, but these are not related to each other. During a meteor shower, we see meteors that are related to each other in that they seem to come from a single spot on the sky. The Perseid shower, for example, appears to come from a spot in the constellation Perseus. These showers are seen when Earth passes near the orbit of a comet; the meteors in meteor showers must be produced by dust and debris released by the icy head of the comet. The meteors appear to come from a single place in the sky because they are produced by particles traveling along parallel paths through space. Like railroad tracks extending from a point on the horizon, the meteors appear to approach from a point in space (Figure 19-3). The orbits of comets are filled with such debris. For example, the telescope aboard the Infrared Astronomy Satellite detected the dusty orbits of a number of comets glowing in the far infrared because of the sun-warmed dust scattered along the orbits.

Studies of meteors show that most of the meteors we see on any given night (whether or not there is a shower) are produced by tiny bits of debris from comets. These specks of matter are so small and so weak they vaporize completely in the atmosphere and never reach the ground, but from their motions astronomers can deduce their orbits, and those orbits match the orbits of comets. Thus nearly all of the meteors come from comets.

Meteorites that reach Earth's surface are structurally much stronger than comet dust. They appear to have had a more complex history, and the different kinds of meteorites suggest that they are fragments of larger bodies that grew hot enough to melt and

Table 19-1 Meteor Showers

Shower	Dates	Hourly Rate	Radiant* R. A.	Dec.	Associated Comet
Quadrantids	Jan. 2–4	30	15^h24^m	50°	
Lyrids	April 20–22	8	18^h4^m	33°	1861 I
η Aquarids	May 2–7	10	22^h24^m	0°	Halley?
δ Aquarids	July 26–31	15	22^h36^m	−10°	
Perseids	Aug. 10–14	40	3^h4^m	58°	1982 III
Orionids	Oct. 18–23	15	6^h20^m	15°	Halley?
Taurids	Nov. 1–7	8	3^h40^m	17°	Encke
Leonids	Nov. 14–19	6	10^h12^m	22°	1866 I Temp
Geminids	Dec. 10–13	50	7^h28^m	32°	

*R. A. and Dec give the celestial coordinates (right ascension and declination) of the radiant of each shower.

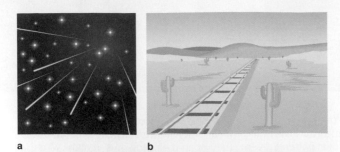

a   b

Orbit of comet

Earth

Sun

c

Figure 19-3
(a) Meteors in a meteor shower enter Earth's atmosphere along parallel paths, but perspective makes them appear to diverge from a radiant point in the sky. (b) Similarly, parallel railroad tracks appear to diverge from a point on the horizon. (c) Meteors in a shower are debris left behind as a comet's icy nucleus vaporizes. The rocky and metallic bits of matter spread along the comet's orbit. If Earth passes through such material, we see a meteor shower.

differentiate to form a molten iron core and a rocky mantle and crust (Figure 19-4).

If an object were 100 km in diameter, then the outer layers of rock would insulate the iron core. It would lose heat slowly, and the iron would cool so gradually that large crystals could grow. A sample of such material sliced, polished, and etched would show Widmanstätten patterns. Thus the material in iron meteorites could have formed in the cores of such bodies.

Collisions could break up such an object, and fragments from the core would look like the iron meteorites. Fragments of the object's mantle, which was hot enough to melt chondrules, would make achondrites. Also, some achondrites appear to be pieces of lavas that flowed out onto the surfaces of larger bodies. Fragments from a crust that has never been hot

Figure 19-4
If a planetesimal were large enough, it could melt and differentiate (a) into an iron-rich core surrounded by layers of different silicate compositions (b). Impacts (c, d, e) could break up these layers and produce various kinds of meteorites. Presumably, these remaining planetesimals are the asteroids. *(Adapted from a diagram by C. R. Chapman)*

enough to melt the chondrules might look very much like the meteorites called chondrites.

The carbonaceous chondrites may have formed in bodies that were smaller and cooler. Planetesimals farther from the sun would have been cooler and could have retained volatiles more easily. They could not have been large, or heating and differentiation would have altered them.

These theories trace the origin of meteorites to planetesimal-like parent bodies, but they produce a mystery. The small meteorites in the solar system cannot be fragments of the planetesimals that formed the planets because small meteorites would have been swept up by the planets in only a billion years or less. They could not have survived for 4.6 billion years.

In fact, when astronomers study the orbits of objects seen to fall to Earth as meteorites, those orbits lead back into the asteroid belt. Thus we have good evidence to believe that the meteorites now in museums all over the world must have broken off asteroids within the last billion years. Although nearly all meteors are pieces of comets, the meteorites are pieces of asteroids.

REVIEW Critical Inquiry

How can we say that meteors come from comets, but meteorites come from asteroids?

A selection effect can affect what we notice when we observe nature, and a very strong selection effect prevents us from finding meteorites that originated in comets. Cometary particles are physically weak, and they vaporize in our atmosphere easily. Thus, very few ever reach the ground, and we are unlikely to find them. Furthermore, even if a cometary particle reached the ground, it would be so fragile that it would weather away rapidly, and, again, we would be unlikely to find it. Asteroidal particles, however, are made from rock and metal and so are stronger. They are more likely to survive their plunge through the atmosphere and more likely to survive erosion on the ground. Meteors from the asteroid belt are rare. Almost all of the meteors we see come from comets, but not a single meteorite is known to be cometary.

The meteorites are valuable because they give us hints about the process of planet building in the solar nebula. What evidence do we find in the meteorites that tells us they were once part of larger bodies broken up by impacts?

The vast majority of meteorites must have originated in the asteroids. Our next goal is to try to understand the nature and origin of these tiny rocky worlds.

19-2 Asteroids

Space pirates lurk in the asteroid belt in old-time pulp science fiction, but astronomers have found that there isn't much in the asteroid belt for a pirate to stand on. Thousands of asteroids are known, but most are quite small; and given its vast size, the asteroid belt between Mars and Jupiter is mostly empty.

Nevertheless, we must consider the nature of the asteroids. In Chapter 16, we identified the asteroids as a characteristic of the solar system and concluded that they were the last remains of material that was unable to form a planet between Mars and Jupiter. We are now ready to examine these small worlds in more detail.

Properties of Asteroids

Asteroids are distant objects too small to study in detail with Earth-based telescopes. Yet astronomers have learned a surprising amount about these little worlds, and spacecraft have given us a few close-ups.

From the infrared radiation emitted by asteroids, astronomers can calculate their sizes. Ceres, the largest, is about 30 percent the diameter of our moon, and Pallas, next largest, is only 15 percent the diameter of the moon. Most are much smaller.

Because the brightness of the typical asteroid varies over periods of hours, astronomers concluded that most asteroids are not spherical. As their irregular shapes rotate, they reflect varying amounts of sunlight, and their brightness varies. Apparently, only the largest asteroids have enough gravity to overcome the strength of the rock they are made from and pull themselves into spherical shape.

The irregular shape of asteroids has been confirmed by photographs taken by the Galileo spacecraft on its way to Jupiter and by the NEAR spacecraft (Figure 19-5). The asteroids are irregular chunks of rock battered by craters. From the gravitational effects of the asteroids on the spacecraft, astronomers can compute their masses, and, dividing by their volumes, their densities. Although asteroids are clearly rocky, they have low densities. Mathilde, for example, has a density of just 1.3 g/cm³. It can't be solid rock but must be a collection of fragments only loosely held together by their mutual gravity.

In fact, some asteroids appear to be double objects. Radar observations of the asteroid Toutatis reveal that it is made up of two objects circling each other. Whether they orbit each other, are in contact with each other, or are fused into a single object is unknown. The asteroid Castalia is also a double asteroid. Photographs of the asteroid Ida, 52 km long, revealed that it is orbited by its own small moon, Dactyl, only 1.5 km in diameter.

Figure 19-5
Asteroids Mathilde, Gaspra, and Ida (left to right) were photographed by spacecraft. They are reproduced here to the same scale. Mathilde is 59 km in diameter. The brightness of the three images has been adjusted for easy viewing. Mathilde is actually much darker than Gaspra or Ida. *(Johns Hopkins Applied Physics Lab and NASA)*

Irregular shapes, craters, fragmented interiors, double asteroids, and asteroids with moons all serve to assure us that the asteroids have collided with each other often throughout their history. It should not surprise us that the inner solar system is filled with meteoroids from the asteroid belt.

Not all asteroids lie in the asteroid belt. Some 2000 objects larger than 1 km are believed to follow orbits that cross the orbit of Earth. Although a number of searches are beginning, only about 200 of these asteroids have been found. For example, LONEOS, Lowell Observatory Near Earth Object Search, will be able to survey the entire sky visible from Lowell Observatory three times a month and should be able to locate 1000 near-Earth asteroids over the next 10 years. A number of other projects are searching for these asteroids, not only because we want to understand the asteroids better, but because such asteroids must collide with Earth occasionally. Although such collisions occur very rarely, a single impact could cause planetwide devastation. We will discuss such impacts on Earth later in this chapter.

The color and spectra of asteroids help us understand their composition (Figure 19-6). From their bright, reddish colors, astronomers classify some asteroids as S types. These may be silicates mixed with metals, or they may resemble chondrite meteorites. C-type asteroids are very dark—about as bright as a lump of coal. They appear to be carbonaceous. M-type asteroids, bright but not red, appear to be mostly iron-nickel alloys. S types are common in the inner belt, and C types are common in the outer belt. That distribution is a clue to the origin of the asteroids.

Some asteroids have clearly evolved since they formed. Vesta, with a diameter of 525 km, is too small to image from Earth's surface, but Hubble Space Telescope images show that its surface is patchy, and spectra of these regions suggest solidified lavas. Furthermore, the best images reveal a giant crater 87 per-

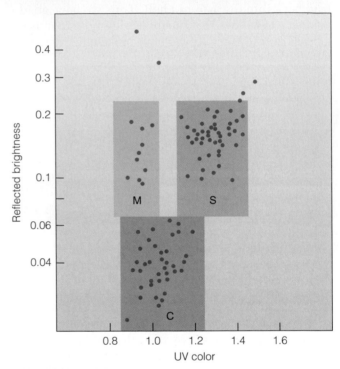

Figure 19-6
The three principal types of asteroids. In a diagram of albedo versus color (ultraviolet minus visual), the most reflective asteroids lie near the top, and the reddest at the right. The S types, which are mixed silicates and metals or perhaps chondrites, are clearly redder than the M types, which seem to be metallic. The darkest asteroids, the C types, are believed to be similar to carbonaceous chondrites. *(Diagram adapted from a figure by B. Zellner)*

cent of Vesta's diameter (Figure 19-7). A family of small asteroids appear to be fragments from Vesta, and a certain class of meteorites found on Earth have been identified spectrosopically as being from Vesta. Those meteorites are basalts typical of lava flows (Figure 19-7 inset). Thus there is strong evidence that Vesta has been at least partly resurfaced by lava flowing up from its interior. We must consider the evolution of asteroids as we think about their origin.

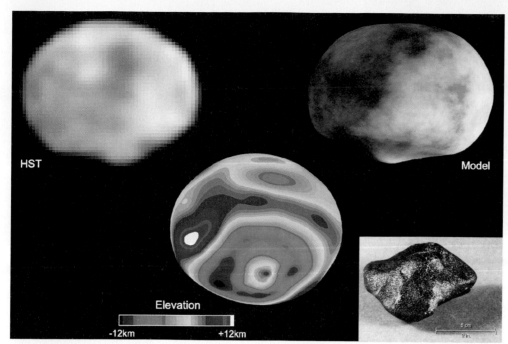

The Origin of the Asteroids

An old theory proposed that asteroids are the remains of a planet that exploded. Planet-shattering death rays may make for exciting science-fiction movies, but in reality planets do not explode. The gravitational field of a planet holds the mass tightly, and disrupting the planet would take tremendous energy. Shattering Earth, for example, would take all of the energy generated by the sun over a period of two weeks. In addition, the total mass of the asteroids is only about one-fifth the mass of the moon, hardly enough to be the remains of a planet.

Most astronomers now believe that the asteroids are the remains of material that was unable to form a planet at 2.8 AU from the sun because of the gravitational influence of Jupiter, the next planet outward. If this is true, then the asteroids are the remains of planetesimals fragmented by collisions with one another. This would explain why the C-type asteroids, which appear to be carbonaceous, are more common in the outer asteroid belt. It is cooler there, and the condensation sequence (see Chapter 16) predicts that carbonaceous material would form there more easily than in the inner belt.

Now we can understand the compositions of the meteorites. They are fragments of planetesimals, the largest of which developed molten cores, differentiated, may have suffered lava flows on their surfaces, and then cooled slowly. The largest asteroids we see today may be nearly unbroken planetesimals, but the rest are just the fragments produced by 4.6 billion years of collisions.

R E V I E W   Critical Inquiry

What is the evidence that asteroids have been fragmented?

First, we might note that the solar nebula theory of the formation of the solar system predicts that planetesimals collided and either stuck together or fragmented. This is suggestive, but it is not evidence. A theory can never be used as evidence to support some other theory or hypothesis. Evidence means observations or the results of experiments, so we must turn to observations of asteroids. The Galileo spacecraft photographs of Mathilde, Ida, and Gaspra show irregularly shaped little worlds heavily scarred by impact craters, and the radar images obtained so far of asteroids also show irregularly shaped bodies that appear to be the fragments of larger bodies. In fact, radar images of Castalia and Toutatis show what may be pairs of bodies in contact, and the Galileo image of Ida reveals its small satellite, Dactyl. Furthermore, meteorites appear to come from the asteroid belt, and a few have been linked to specific asteroids such as Vesta. All of this evidence suggests that the asteroids have suffered violent collisions in their history.

The impact fragmentation of asteroids has been important, but it has not erased all traces of the original planetesimals from which the asteroids formed. What evidence do we have to tell us what those planetesimals were like?

The asteroids are only the last crumbs left behind by failed planet building near Jupiter. Farther out in the solar system are other remains—ice shards from the age of planet formation.

a

b

19-3 Comets

Of all the fossils left behind by the solar nebula, comets are the most beautiful. Asteroids are dark, rocky worlds, and meteors are flitting specks of fire, but the comets move with the grace and beauty of a great ship at sea (Figure 19-8). Comet Hale-Bopp, for example, was visible for weeks in 1996 and 1997. Scientifically, comets are interesting because observations of the tail and head of a comet tell us about the small icy nucleus and about the ancient solar nebula from which we inherit our wealth of comets.

Properties of Comets

The head, or **coma,** of a comet is a vast cloud of gas and dust up to 100,000 km in diameter—seven times the diameter of Earth (Figure 19-9). The gas in the coma is made up of H_2O, CO_2, CO, H, OH, O, S, C, and so on. Mixed with the gases of the coma are tiny specks of dust, most smaller than the particles in cigarette smoke. The dirty snowball model, described in Chapter 16, proposes that all of the features of the coma are created by a central lump of dirty ices only a dozen kilometers in diameter. As the ices vaporize, dust is released, and larger molecules of gas are broken into the smaller fragments we see in the coma.

The tail of a comet springs from the coma and typically extends 10^7 to 10^8 km through space, with the longest reaching 1 AU (1.5×10^8 km). Seen from Earth, the tail of a large comet can span 30° across the sky.

A comet can have two kinds of tails (Figure 19-10). A type I tail, also called a *gas tail,* is straight and wispy and looks like a narrow trail of smoke. Its spectrum contains emission lines of ionized gases excited by the ultraviolet radiation from the sun and blown outward by the moving gas and the embedded magnetic field in the solar wind.

The ionized gases in the gas tail interact with the magnetic field in the solar wind to give the tail its long, thin shape. When a space probe named ICE plunged through the tail of a comet, it found the magnetic field draped over the nucleus like seaweed over a fishhook. Disturbances in the solar wind (caused, perhaps, by activity on the sun) can even produce kinks in the comet's tail.

The type II tail, also known as a *dust tail,* is immune to magnetic fields. Its spectrum is identical to

Figure 19-8
(a) Comet Ikeya-Seki was beautiful in the dawn sky in 1965. The long, graceful tail sprang from a nucleus of ice only a dozen kilometers in diameter. *(Stephen M. Larson)* (b) Comet Halley, like all comets, looked white to the eye, but computer enhancement and false-color graphics are often used to reveal subtle detail. This image from 1910 has been digitally enhanced to reveal subtle differences in intensity. *(NOAO)*

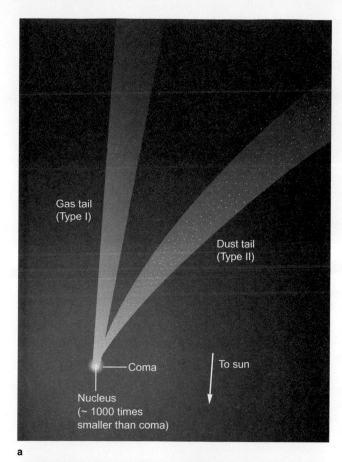

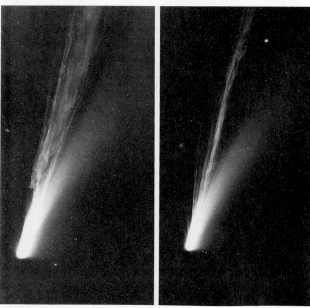

b

Figure 19-9
(a) The parts of a comet include an extended coma, or head, of gas and dust surrounding an icy nucleus only a few kilometers in diameter. Gas is carried away by the solar wind in a wispy gas tail, but dust moves more slowly and forms a curving dust tail. (b) Comet Mrkos in 1957 had pronounced gas and dust tails, which changed dramatically from night to night. *(Palomar Obs/Caltech)*

Figure 19-10
Comet Hale-Bopp was brilliant in the sky in early 1997. With an unusually large nucleus estimated to be 40 km in diameter, it produced large amounts of gas and dust. As the gas became ionized, it was carried away rapidly by the solar wind to form the straight, blue Type I tail. The dust was less affected by the solar wind. It was pushed outward by the pressure of sunlight and produced the broad, curving Type II tail. *(Dean Ketelsen)*

sunlight's, which means it is made up of solid bits of dust that reflect sunlight. The dust is evidently the dirt released by the dirty snowball as the ices vaporize. The Infrared Astronomy Satellite found that this dust fills the solar system, and it can be collected by spacecraft and high-flying aircraft (Figure 19-11).

The dust in the dust tail is not affected by magnetic fields but is pushed outward by the pressure of sunlight. It thus forms a long, curving tail with a much more uniform appearance than a gas tail. The dust tail curves because the comet moves along its orbit with a velocity that is comparable to the velocity of the dust. Like the stream of water from the moving nozzle of a lawn sprinkler, the dust tail appears

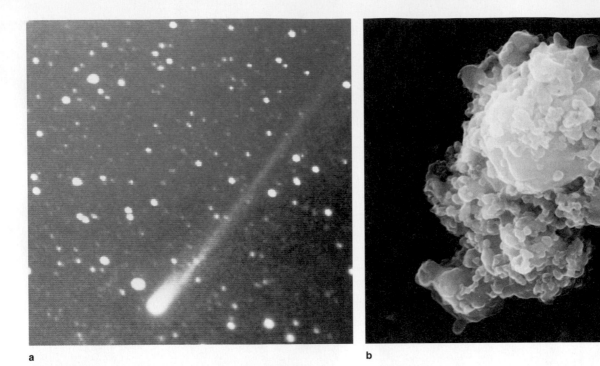

a b

Figure 19-11
(a) Comet Halley is a typical comet in that much of the light we receive from its tail is reflected by microscopic dust particles released from the head of the comet. *(NOAO)* (b) NASA aircraft flying at an altitude of 18 km (59,000 ft) can collect dust falling into Earth's atmosphere from space. This dust speck is only 5 by 6 micrometers. Comets continually release such dust throughout the inner solar system, and it can be detected by infrared telescopes in Earth orbit. *(NASA)*

curved. It lacks the wisps and twists of the gas tail because it is not affected by kinks and shifts in the magnetic field of the solar wind. Thus, the gas tail and the dust tail both point roughly away from the sun, but for slightly different reasons.

Comet Geology

The nucleus of a comet is quite small and cannot be resolved by any telescopes now in use. Nevertheless, astronomers are beginning to understand the geology of these peculiar worlds.

Of course, the nuclei of comets must contain water, but they must also contain compounds such as ammonia and carbon dioxide trapped in the ice. These ices sublime—change from a solid directly into a vapor—to produce the observed tails. Telescopes in space can detect vast clouds of hydrogen gas surrounding the heads of active comets (Figure 19-12a). The gases that comets release must come from these volatile compounds.

Comet Halley visited the inner solar system in 1985 and 1986, and five spacecraft flew past the nucleus. The photos sent back show a dark, irregularly shaped body venting gas and dust from active regions on its sunward side (Figure 19-12b). The surface is so dark it reflects only 4 percent of the light that hits it; a sooty lump of coal reflects 6 percent. The dark color

of this dust suggests the composition of the meteorites called carbonaceous chondrites.

From the gravitational effects the nucleus had on the spacecraft, astronomers can find the mass of the nucleus and thus its density, 0.1 to 0.25 g/cm^3, much less than the density of ice. From these observations we can conclude that the nucleus is not a solid ball of ice but must be a fluffy mixture of ices and a dark dusty crust.

Photographs of the comas of comets often show jets springing from the nucleus and being swept back by the pressure of sunlight and by the solar wind to form the tail (Figure 19-12c). Studies of the motions of these jets as the nucleus rotates and the photographs of the nucleus of Comet Halley reveal that the jets originate from active regions that may be faults or vents. As the rotation of a cometary nucleus carries an active region into sunlight, it begins venting gas and dust, and as it rotates into darkness it shuts down.

The irregular shape of the nucleus of Comet Halley suggests that the crust is a fluffy layer of dust left behind by the vaporizing ices. Most of the ice is hidden below this dark crust, and, as it sublimes from ice to vapor, the crust collapses into the resulting voids. Some regions of cometary nuclei seem extra rich in highly volatile ices, so the nuclei are probably not uniform mixtures.

The nucleus of Comet Halley was observed to be 16 by 8 by 7 km, and the nucleus of Comet Hale-Bopp

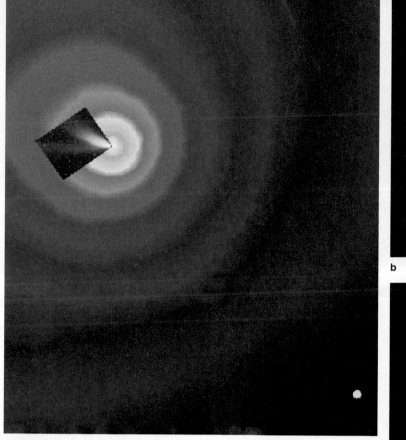

Figure 19-12

(a) An ultraviolet image of Comet Hale-Bopp reveals a vast cloud of hydrogen gas surrounding the visible comet (inset). Concentric bands in the cloud are produced by the digitizing process. At lower right is the sun to scale. *(SOHO/SWAN Consortium, ESA, and NASA)* (b) This image of the nucleus of Comet Halley taken by the Giotto spacecraft shows jets of dust and gas venting from the dark nucleus. *(©1986 Max-Planck Institute)* (c) In the inner coma of Comet Halley, jets from the nucleus are bent back to form the tail. The black dot is produced by the imaging process. The actual nucleus is too small to detect. *(Steven Larson)*

is believed to be 30 to 40 km in diameter. Each passage near the sun costs such an object many millions of tons of ices, so the nuclei slowly loose their ices until there is nothing left but dust and rock.

The Origin of Comets

Family relationships among the comets give us clues to their origin. Most comets have long, elliptical orbits with periods greater than 200 years. These are known as long-period comets. Their orbits are randomly inclined with comets falling into the inner solar system from all directions. As many circle the sun clockwise as counterclockwise.

In contrast, about 100 or so of the 600 well-studied comets have orbits with periods less than 200 years. These short-period comets follow orbits that lie within 30° of the plane of the solar system, and most revolve around the sun counterclockwise—the same direction the planets orbit. Comet Halley, with a period of 76 years, is a short-period comet.

Comets cannot survive long before the heat of the sun drives away their ices and reduces them to inactive bodies of rock and dust. A comet may last only 100 to 1000 orbits around the sun. The comets we see in our skies can't have survived 4.6 billion years since the formation of the solar system, so there must be a continuous supply of new comets. Where do they come from?

In the 1950s, Dutch astronomer Jan Oort proposed that the long-period comets are objects that fall in from the **Oort cloud,** a spherical cloud of icy bodies believed to extend from 10,000 to 100,000 AU from the sun (Figure 19-13). Astronomers estimate that the cloud contains several trillion icy bodies. Far from the sun, they are very cold, lack comas and tails, and are invisible. The gravitational influence of occasional passing stars could perturb a few of these objects to fall into the inner solar system, where the heat of the sun warms their ices and transforms them into comets. Because the Oort cloud is spherical, these long-period comets fall inward from random directions.

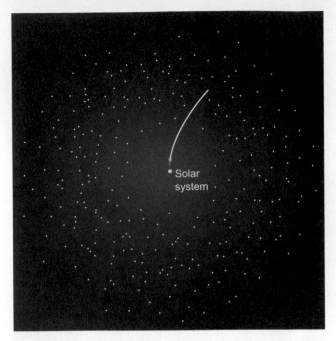

Figure 19-13
The long-period comets appear to originate in the Oort cloud. Objects that fall into the inner solar system from this cloud arrive from all directions.

The short-period comets, however, require a different explanation. In 1951, Dutch-American astronomer Gerard P. Kuiper proposed that the formation of the solar system should have left behind a belt of small, icy planetesimals beyond the Jovian planets and in the plane of the solar system. Known today as the **Kuiper belt,** this disk extends from 30 AU to as much as 100 AU from the sun. The entire Kuiper belt would be hidden in Figure 19-13 behind the yellow dot symbolizing the solar system. The short-period comets are visitors from the Kuiper belt.

Although the Kuiper belt was originally proposed as a theory, astronomers are now able to observe some of the Kuiper-belt objects. Searches for small icy bodies in the outer solar system have so far identified roughly 60 objects that are members of the Kuiper belt. Also, the physical and chemical similarities among the icy moons of the outer solar system such as Triton and Charon suggest these are large Kuiper-belt objects captured as moons. In fact, astronomers commonly think of Pluto not as a tiny planet but as the largest of the remaining Kuiper-belt objects. Working at the extreme limit of its capability, the Hubble Space Telescope has produced statistical evidence of a population of Kuiper-belt objects roughly the size of the nucleus of Comet Halley.

We can find more evidence that our solar system has a Kuiper belt by looking at other stars. A number of nearby stars such as Beta Pictoris (Chapter 16) are surrounded by disks of dust believed to be released by comets in the equivalent of Kuiper belts around these stars. Thus the detection of the dust implies the presence of Kuiper belts.

If comets are icy planetesimals from the Oort cloud and the Kuiper belt, how did those planetesimals form? The Kuiper-belt objects appear to have formed as icy planetesimals in the outer solar nebula not much farther from the sun than the outer planets. But the objects in the Oort cloud lie much farther from the sun, and they can't have formed out there. The solar nebula would have been too tenuous at such great distances. Also, we would expect objects that formed from the nebula to be confined to a disk and not distributed in a sphere. Rather, astronomers think the objects now in the Oort cloud formed in the outer solar system near the present orbits of Saturn and Uranus. As the Jovian planets grew more massive, they swept up some of these planetesimals and ejected others to form the Oort cloud. If this idea is true, then even the long-period comets are icy planetesimals.

Comet Impacts on Earth

For centuries, superstitious people have associated comets with doom, and that seems silly. Comets are graceful visitors from the icy fringes of the solar system. Of course, comets must hit planets now and then, so we might wonder just how dangerous such impacts would be.

Earthlings watched in awe during the summer of 1994 as the fragmented head of a comet (Figure 19-14a) slammed into Jupiter and produced impacts equaling millions of megatons of TNT (Figure 18-5). Such impacts on Jupiter probably occur every century or so, and we might expect similar impacts on Earth, smaller and with less gravitational power, to occur much less often. Nevertheless, these impacts do occur, and we can find chains of craters on moons that seem to have been formed by fragmented comets (Figure 19-14b).

Small meteorite impacts occur quite often, and a building is damaged by a falling meteorite every few years. Larger impacts are less common, and truly large impacts are rare. Earth is marked by many dozens of meteorite craters that illustrate the power of an impacting object (Figure 19-15 and Figure 17-8). A large impact could have devastating consequences.

Recent studies of sediments laid down all over the world 65 million years ago at the time of the extinction of the dinosaurs have found overabundances of the element iridium—common in meteorites but rare in Earth's crust. This finding suggests that the impact of a large meteorite or comet may have altered the atmosphere and climate on Earth so dramatically that the dinosaurs and over 75 percent of the species then on Earth became extinct.

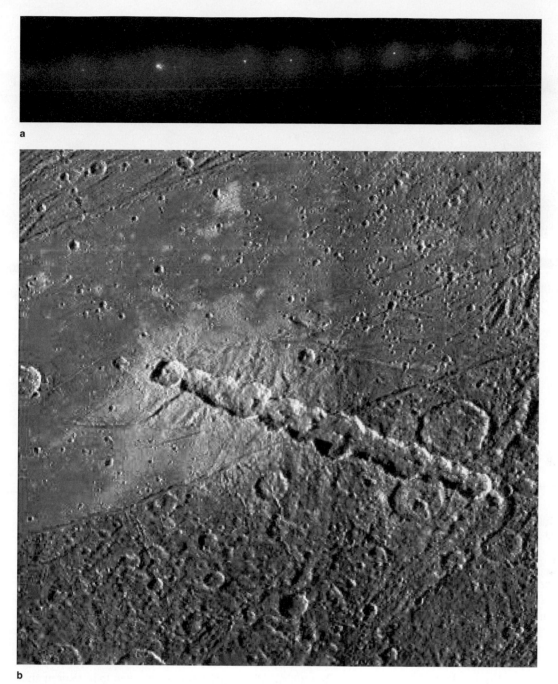

Figure 19-14
(a) The heart of Comet Shoemaker-Levy 9 came too close to Jupiter and was pulled apart. At least 21 fragments later smashed into Jupiter. (See Figure 18-5.) (b) This 140-km-long crater chain on Jupiter's moon Callisto was probably formed by the impact of a fragmented comet. The impacts that form such chains probably occur within a span of seconds. *(NASA)*

Mathematical models and observations of the impact of the comet fragments on Jupiter in 1994 have combined to create a plausible scenario of the events following a major impact on Earth. Of course, creatures living near the site of the impact would probably die in the initial shock, and an impact at sea would create tsunamis (tidal waves) many hundreds of meters high that would devastate coastal regions for many kilometers inland halfway around the world. But the worst effects would begin after the initial explosion. On land or sea, a major impact would excavate large amounts of pulverized rock, heat it to high temperatures, and loft it high above the atmosphere. As this material fell back, Earth's atmosphere would be turned into a glowing oven of red-hot rock streaming through the air as a rain of meteors, and the

a

b

Figure 19-15
(a) The Barringer Meteorite Crater (near Flagstaff, Arizona) is nearly a mile in diameter and was formed about 50,000 years ago by the impact of an iron meteorite roughly 90 m in diameter. It hit with energy equivalent to that of a 3-megaton hydrogen bomb. Notice the raised and deformed rock strata all around the crater. For scale, locate the brick building on the far rim at right. *(M. A. Seeds)* (b) Like all larger-impact features, the Barringer Meteorite Crater has a raised rim and scattered ejecta. *(USGS)*

heat would trigger massive forest fires around the world. Soot from such fires has been detected in the layers of clay laid down at the end of the Cretaceous period. Once the fire storms cooled, the remaining dust in the atmosphere would block sunlight and produce deep darkness for a year or more. At the same time, large amounts of carbon dioxide locked in limestone deposits would be released into the atmosphere, where it would produce acid rain. All of these consequences make it surprising that any life could have survived such an impact.

Geologists have located a crater at least 150 km in diameter centered near the village of Chicxulub in the northern Yucatán (Figure 19-16). Although the crater is totally covered by sediments, mineral samples show that it contains shocked quartz typical of impact sites and that it is the right age. A gigantic impact formed the crater about 65 million years ago, just when the dinosaurs and many other species died out, and many Earth scientists now believe that this is the scar of the impact that ended the Cretaceous period.

Could such an impact happen again? Earth gets hit by small meteorites every day and by larger objects less often. Impacts by large asteroids may happen many millions of years apart, but they continue to happen. In mid-March 1998, newspaper headlines announced, "Mile-Wide Asteroid to Hit Earth in October 2028." The frightening story was true except that the news media did not emphasize the uncertainty in the orbit. Within days, astronomers found the asteroid on old photographic plates, recalculated the orbit adding the new data, and concluded that the asteroid, known as 1997XF$_{11}$, would miss Earth by 600,000 miles. There will be no impact by this asteroid in 2028, but there are plenty more asteroids that haven't been discovered. It is just a matter of time.

A solar system is a dangerous place to put an inhabited planet. Comets, asteroids, and meteoroids constantly rain down on the planets, and Earth gets hit often. Chicxulub isn't the only large impact scar on Earth. About 150 are known, including giant craters buried under sediment in Iowa and another underlying most of Chesapeake Bay.

a

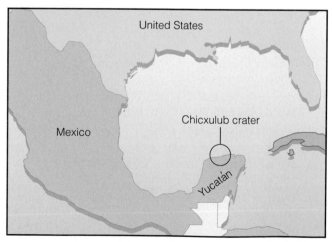

b

Figure 19-16

(a) The theory that the impact of one or more comets altered Earth's climate and drove dinosaurs to extinction has become so popular it appeared on this Hungarian stamp. The spacecraft shown (ICE) flew through the tail of Comet Giacobini-Zinner in 1985. Note the dead dinosaurs in the background. (b) The giant impact scar buried in Earth's crust near the village of Chicxulub in the northern Yucatán was formed about 65 million years ago. Such a large impact would have altered Earth's climate dramatically, and thus the Chicxulub impact is suspected of being the trigger that ended the Cretaceous period and destroyed the dinosaurs.

How do comets help us understand the formation of the planets?

According to the solar nebula hypothesis, the planets formed from planetesimals that accreted in a disk-shaped nebula around the forming sun. The planetesimals that formed in the inner solar nebula were warm and could not incorporate much ice. The asteroids may be the last remains of such bodies. In the outer solar nebula, it was colder, and the planetesimals would have contained large amounts of ices. Many of these planetesimals were destroyed when they fell together to make the planets, but some may have survived. The icy bodies of the Oort cloud and the Kuiper belt may be the last surviving icy planetesimals in our solar system. When these bodies fall into the inner solar system, we see them as comets, and the gases they release tell us that many of them are rich in volatile materials such as water and carbon dioxide. These are the ices we would expect to find in the icy planetesimals. Furthermore, comets are rich in dust, and the planetesimals must have included large amounts of dust frozen into the ices when they formed. Thus, the nuclei of comets seem to be frozen samples of the ancient solar nebula.

Nearly all of the mass of a comet is in the nucleus, but the light we see comes from the coma and the tail. What kind of spectra do we get from comets, and what does that tell us about the process that converts a dirty iceberg into a comet?

Life on Earth seems fragile and exposed. Are we just lucky to have survived so long? Does life exist on other worlds? If so, has it, too, survived long enough to develop intelligence? That is the story of the next chapter.

Summary

The solar nebula theory proposes that the solar system formed from a disk of gas and dust around the forming sun and that most of that nebula was blown away when the sun became luminous. The asteroids and comets appear to be rocky and icy debris left behind.

Meteorites can be classified according to their mineral content. Iron meteorites are mostly iron and nickel, and when sliced open, polished, and etched with acid, they show Widmanstätten patterns. These suggest the metal cooled from a molten state very slowly.

Stony meteorites can be classified as chondrites if they contain small, glassy particles called chondrules. If such a meteorite is rich in volatiles and carbon, it is called a carbonaceous chondrite. An achondrite is a stony meteorite that contains no chondrules. Achondrites appear to have been melted after they formed, but chondrites were never melted. Carbonaceous chondrites were never even warm enough to drive off volatiles.

This evidence suggests that the different kinds of meteorites were formed in bodies roughly 100 km in diameter. Such objects could have melted, differentiated, and cooled very slowly, and when they were later broken by collisions, their fragments would resemble the different kinds of meteorites. Also, we may suppose that the carbonaceous chondrites were formed farther from the sun, where it was cooler. This evidence suggests that meteorites are fragments of the asteroids.

Most of the meteors we see, however, are small, weak objects that never reach the ground. These bits of matter appear to be debris from comets, and we see meteor showers when Earth crosses the debris-laden orbit of a comet.

Most asteroids lie in the asteroid belt between Mars and Jupiter. Infrared studies show that asteroids in the outer belt are darker and redder than others. These may resemble the volatile-rich carbonaceous chondrites. Apparently, the asteroids are the remains of planetesimals that were unable to form a planet between Mars and Jupiter. The strong gravitational influence of Jupiter could have prevented the material from accumulating into a large body.

A comet is produced by a lump of dirty ices only a few dozen kilometers in diameter. In a long, elliptical orbit, the icy body stays frozen until the object draws close to the sun. Then the ices vaporize and release the imbedded dust and debris. The gas is caught in the solar wind and blown outward to form a type I, or gas, tail. The pressure of sunlight blows the dust away to form a type II, or dust, tail. The coma of a comet can be up to 100,000 km in diameter, and it contains jets issuing from the nucleus.

When spacecraft flew past Comet Halley in 1986, astronomers discovered that the nucleus was coated by a dark crust and that jets of vapor and dust were venting from active regions on the sunlit side. The low density of the nucleus showed that it was a fluffy mixture of ices and silicate dust.

Comets are believed to have formed as icy planetesimals in the outer solar system, and some of these objects remain as the Kuiper belt, the source of short-period comets. Some icy planetesimals were swept up by the formation of the Jovian planets, and others were ejected to form the Oort cloud, the source of the long-period comets.

New Terms

Widmanstätten pattern	meteor shower
chondrite	coma
chondrule	Oort cloud
carbonaceous chondrite	Kuiper belt
achondrite	

Review Questions

1. What do Widmanstätten patterns tell us about the history of iron meteorites?

2. What do chondrules tell us about the history of chondrites?

3. Why are there no chondrules in achondritic meteorites?

4. Why do astronomers refer to carbonaceous chondrites as "unmodified"?

5. How do observations of meteor showers reveal one of the sources of meteoroids?

6. How can most meteors be cometary if most, perhaps all, meteorites are asteroidal?

7. Why do we think the asteroids were never part of a planet?

8. What evidence do we have that the asteroids are fragmented?

9. What evidence do we have that some asteroids have differentiated?

10. What evidence do we have that some asteroids have had active surfaces?

11. How is the composition of meteorites related to the formation and evolution of asteroids?

12. What is the difference between a type I tail and a type II tail?

13. What evidence do we have to support the dirty snowball model of cometary nuclei?

14. Why do short-period comets tend to have orbits near the plane of the solar system?

15. How did the bodies in the Kuiper belt and the Oort cloud form?

Discussion Questions

1. Futurists suggest we may someday mine the asteroids for materials to build and supply space colonies. What kinds of materials could we get from asteroids? (*Hint:* What are S-, M-, and C-type asteroids made of?)

2. If cometary nuclei were heated by internal radioactive decay rather than by solar heat, how would comets differ from what we observe?

3. From what you know now, do you think the government should spend money to locate near-Earth asteroids? How serious is the risk?

Problems

1. Large meteorites are hardly slowed by Earth's atmosphere. Assuming the atmosphere is 100 km thick and that a large meteorite falls perpendicular to the surface, how long does it take to reach the ground? (*Hint:* About how fast do meteoroids travel?)

2. What is the orbital velocity of a meteoroid whose average distance from the sun is 2 AU? (*Hint:* Find the orbital period from Kepler's third law. See Table 4-1.)

3. If a single asteroid 1 km in diameter were fragmented into meteoroids 1 m in diameter, how many would it yield? (*Hint:* Volume of sphere $= \frac{4}{3}\pi r^3$.)

4. What is the orbital period of a typical asteroid? (*Hint:* Use Kepler's third law. See Table 4-1.)

5. If half a million asteroids each 1 km in diameter were assembled into one body, how large would it be? (*Hint:* Volume of sphere $= \frac{4}{3}\pi r^3$.)

6. What is the maximum angular diameter of Ceres as seen from Earth? Could Earth-based telescopes detect surface features? Could the Hubble Space Telescope? (*Hint:* See By the Numbers 3-1.)

7. If the velocity of the solar wind is about 400 km/sec and the visible tail of a comet is 10^8 km long, how long does it take an atom to travel from the nucleus to the end of the visible tail?

8. If you saw Comet Halley when it was 0.7 AU from Earth and it had a visible tail 5° long, how long was the tail in kilometers? Suppose that the tail was not perpendicular to your line of sight. Is your answer too large or too small? (*Hint:* See By the Numbers 3-1.)

9. What is the orbital period of a cometary nucleus in the Oort cloud? What is its orbital velocity? (*Hints:* Use Kepler's third law. The circumference of a circular orbit $= 2\pi r$.)

10. The mass of an average comet's nucleus is about 10^{12} kg. If the Oort cloud contains 200×10^9 cometary nuclei, what is the mass of the cloud in Earth masses? (*Hint:* Mass of Earth $= 6 \times 10^{24}$ kg.)

Critical Inquiries for the Web

1. Some nights are better for looking for meteors than others (see the Table of Meteor Showers—Table 19-1, page 401. We know these showers are associated with comets, but how are these associations made? There are several sites on the Internet that provide information on meteor showers, including historical data and information on parent comets. Pick a shower whose parent comet is known, and summarize how we came to know that the meteors and the comet are related.

2. The chances are small that you will be killed by an asteroid impact, but if there are objects out there that astronomers are not aware of whose orbits intersect Earth, we could be in for a surprise one day. Look for information on the Spacewatch project and other investigations into near-Earth asteroids. How many such objects have been discovered? What is the record for closest known passage of an asteroid to Earth?

 Go to the Brooks/Cole Astronomy Resource Center (www.brookscole.com/astronomy) for critical thinking exercises, articles, and additional readings from InfoTrac College Edition, Brooks/Cole's online student library.

CHAPTER 20

Life on Other Worlds

Did I solicit thee from darkness to

promote me?

John Milton
Paradise Lost

Guidepost

This chapter is either unnecessary or critical depending on our point of view. If we believe that astronomy is the study of the physical universe above the clouds, then this chapter does not belong here. But if we believe that astronomy is the study of our position in the universe, not only our physical position but also our role as living beings in the origin and evolution of the universe, then everything else in this book is just preparation for this chapter.

Astronomy is the only science that truly acts as a mirror. In studying the universe up there, we learn what we are down here. Astronomy is not really about stars, galaxies and planets; it is about us.

Science is a way of understanding the world around us, and the heart of that understanding is the explanations that science gives us for natural phenomena. Whether we call these explanations stories, histories, theories, or hypotheses, they are attempts to describe how nature works based on fundamental rules of evidence and intellectual honesty. While we may take these explanations as factual truth, we should understand that they are not the only explanations that satisfy the rules of logic.

A separate class of explanations involves religion, and those explanations can be quite logical. The Old Testament description of the creation of the world, for instance, does not fit scientific observations; but if we accept the existence of an omnipotent being, then the biblical explanation is internally logical and acceptable. Of course, it is not a scientific explanation; but religion is a matter of faith and not subject to the rules of evidence. Religious explanations follow their own logic, and we would be wrong to demand that they follow the rules of evidence that govern scientific explanations, just as we would be wrong to demand that scientific explanations accept certain religious beliefs on faith. Both kinds of explanations are logical, but the rules are different.

If scientific explanations are not the only logical explanations, then why do we give them such weight? First, we must notice the tremendous success of scientific explanations in producing technological advances in our daily lives. Smallpox is a disease of the past thanks to the application of scientific explanations to modern medicine. The power of science to shape our world can lead us to think its explanations are unique. But second, the process we call science depends on the use of evidence to test and perfect our explanations, and the logical rigor of this process gives us great confidence in our conclusions.

Scientific explanations have given us tremendous insight into the workings of nature, and consequently both scientists and nonscientists tend to forget there can be other kinds of explanations. The so-called conflict between science and religion has been symbolized for centuries by the trial of Galileo. That conflict is easier to understand when we consider the nature of scientific explanations and the role of evidence in testing scientific understanding.

As living things, we have been promoted from darkness. We are made of heavy atoms that could not have formed at the beginning of the universe. Successive generations of stars fusing light elements into heavier elements have built the atoms so important to our existence. When a dark cloud of interstellar gas enriched in these heavy atoms fell together to form our sun, a small part of the cloud gave birth to the planet we inhabit.

Are there intelligent beings living on other planets? That is the last and perhaps the most challenging question in our study of astronomy. We will try to answer it in three steps, each dealing with a different aspect of life.

First, we must decide what we mean by *life*. Life is not so much a form of matter as it is a behavior. Living matter extracts energy from its environment to modify itself and its surroundings so as to preserve itself and to create offspring. Thus, life is based on information, the recipe for the process of survival and reproduction.

Second, we must study the origin of life on Earth. If we can understand how life began on our planet, then we can try to estimate the likelihood that it has originated on other worlds as well.

Third, we must try to understand how the first primitive living organism on Earth could give rise to the tremendously diverse life forms that now inhabit our world. By understanding how living things evolve to fit their environment, we will see evidence that intelligence is a natural development in the evolution of life.

If life can originate on other worlds, and if intelligence is a natural result of the evolution of life forms, then we might expect that other intelligent races inhabit other worlds. Visits between worlds seem impossible, but perhaps we can detect their existence by radio.

Alien life forms could be quite different from us, but if they are alive, then they must share with us certain characteristics. Our goal in this chapter is to use our knowledge of science to explain the greatest of mysteries—the origin and evolution of life on Earth and on other worlds (Window on Science 20-1).

20-1 The Nature of Life

What is life? Philosophers have struggled with that question for thousands of years, so it is unlikely that we will answer it here. But we must agree on a working model of life before we can speculate on its occurrence on other worlds. To that end, we will identify in living things two important aspects: a physical basis and a unit of controlling information.

The Physical Basis of Life

On Earth, the physical basis of life is the carbon atom (Figure 20-1). Because of the way this atom bonds to

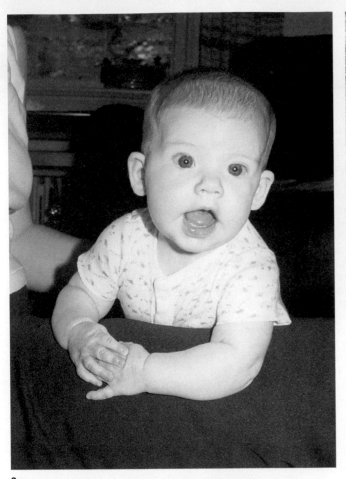
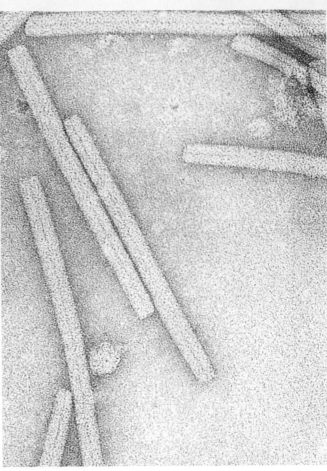

a

b

Figure 20-1

All living things on Earth are based on carbon chemistry. Even the long molecules that carry genetic information, DNA and RNA, have a framework defined by chains of carbon atoms. (a) Katie, a complex mammal, contains about 30 AU of DNA. *(Michael Seeds)* (b) Each rod of the tobacco mosaic virus contains a single spiral strand of RNA about 0.01 mm long. *(L. D. Simon)* All life on Earth stores its genetic information in such carbon-chain molecules.

other atoms, it can form long, complex, stable chains that are capable of extracting, storing, and utilizing energy. Other chemical bases of life may exist. Science-fiction stories and movies abound with silicon creatures, living things whose body chemistry is based on silicon rather than carbon. However, silicon forms weaker bonds than carbon does, and it cannot form double bonds as easily. Consequently, it cannot form the long, complex, stable chains that carbon can. Silicon is 135 times more common on Earth than carbon is, yet there are no silicon creatures among us. All Earth life is carbon based. Thus, the likelihood that distant planets are inhabited by silicon people seems small, but we cannot rule out life based on noncarbon chemistry.

In fact, nonchemical life might be possible. What is required is some mechanism capable of supporting the extraction and utilization of energy that we have identified as life. One could at least imagine life based on electromagnetic fields and ionized gas. No one has

ever met such a creature, but science-fiction writers conjure up all sorts.

Clearly, we could range far in space and time, theorizing about different bases for alien life, but to make progress we must discuss what we know best—carbon-based life on Earth. How can a lump of carbon-rich matter live? The answer lies in the information that guides its life processes.

Information Storage and Duplication

The key to understanding life is information—the information the organism uses to control its utilization of energy. We must discover how life stores and uses that information and how the information changes and thus preserves the species.

The unit of life on Earth is the cell, the self-contained factory capable of absorbing nourishment from its surroundings, maintaining its own existence, and, in multicellular organisms, performing its task

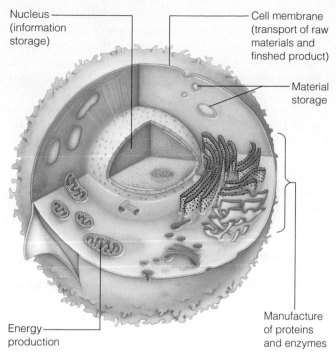

Nucleus (information storage)

Cell membrane (transport of raw materials and finshed product)

Material storage

Energy production

Manufacture of proteins and enzymes

Figure 20-2
A living cell is a self-contained factory that absorbs raw materials from its surroundings and uses them to maintain itself and manufacture finished products for the use of the organism as a whole.

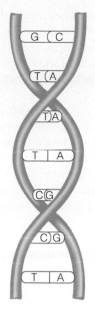

Figure 20-3
The DNA molecule consists of two rails of sugars and phosphates (dark) and rungs made of bases: adenine (A), cytosine (C), guanine (G), and thymine (T).

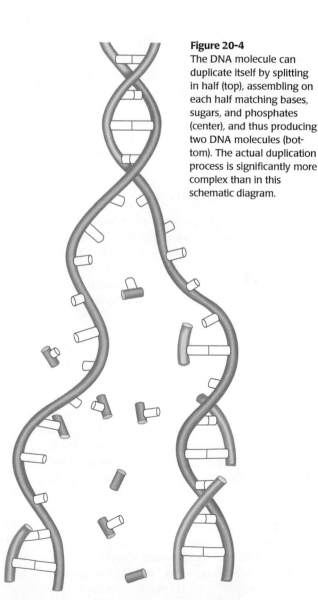

Figure 20-4
The DNA molecule can duplicate itself by splitting in half (top), assembling on each half matching bases, sugars, and phosphates (center), and thus producing two DNA molecules (bottom). The actual duplication process is significantly more complex than in this schematic diagram.

within the larger organism (Figure 20-2). The foundation of the cell's activity is a set of patterns that describe how it is to function. This information must be stored in the cell in some safe location, yet it must be passed on easily to new cells and be used readily to guide the cell's activity. To understand how matter can be alive, we must have an understanding of how the cell stores, reproduces, and uses this information.

The information is stored in long carbon-chain molecules called **DNA (deoxyribonucleic acid),** most of which reside in the cell nucleus. The structure of DNA resembles a long, twisted ladder. The rails of the ladder are made of alternating phosphates and sugars; the rungs are made of pairs of molecules called bases (Figure 20-3). Only four kinds of bases are present in DNA, and the order in which they appear on the DNA ladder represents the information the cell needs to function. One human cell stores about 1.5 m of DNA, containing about 4.5 billion pairs of bases. Thus 4.5 billion pieces of information are available to run a human cell. That is enough to record all of the works of Shakespeare over 200 times. Because the human body contains about 60×10^{12} cells, the total DNA in a single human adult would stretch 9×10^{13} m, about 600 AU.

Storing all this data in each cell does the organism no good unless the data can be reproduced and passed on to new cells. The DNA molecule is specially adapted for duplicating itself by splitting its ladder down the center of the rungs, producing two rails with protruding bases (Figure 20-4). These quickly bond with the proper bases, phosphates, and sugars to reconstruct the missing part of the molecule, and presto—the cell has two complete copies of the critical information. One set goes to each of the newly forming cells. Thus, the DNA is the genetic information passed from parent to offspring.

Segments of the DNA molecules are patterns for the production of **proteins.** Many proteins are structural molecules—the cell might make protein to repair its cell wall, for example. **Enzymes** are special proteins that control other processes—growth, for example. Thus, the DNA molecule contains the recipes to make all of the different molecules required in an organism.

Actually, the cell does not risk its precious DNA patterns by involving them directly in the manufacture of protein. The DNA stays safely in the cell nucleus, where it produces a copy of the patterns by assembling a long carbon-chain molecule called **RNA (ribonucleic acid).** This RNA carries the information out of the nucleus and then assembles the proteins from simple molecules called **amino acids,** the basic building blocks of protein. Thus the RNA acts as a messenger, carrying copies of the necessary plans from the central office to the construction site.

Although the information coded in the DNA must be preserved for the survival of the organism, it must be changeable or the species will not survive very long at all. To see why, we must study evolution, the process that rewrites the data in the DNA.

Modifying the Information

If living things are to survive for many generations, then the information stored in their DNA must change as the environment changes. Without change in DNA, a slight warming of the climate, for example, might kill a species of plant, in turn starving the rabbits, deer, and other plant eaters, and leaving the hawks, wolves, and mountain lions with no prey. If the information stored in DNA could never change, then environmental changes would quickly drive life forms to extinction. If life is to survive in a changing world, then the information in DNA must be changeable. Living things must evolve.

Species evolve by **natural selection.** Each time an organism reproduces, its offspring receive the data stored in the DNA, but some variation is possible. For example, most of the rabbits in a litter may be normal, but it is possible for one to get a DNA recipe that gives it stronger teeth. If it has stronger teeth, it may be able to eat something other than the plant the others depend on, and if that plant is becoming scarce, the rabbit with stronger teeth has a survival advantage. It can eat other plants and so will be healthier than its littermates and have more offspring. Some of these offspring may also

have stronger teeth, as the altered DNA data are handed down to the new generation. Thus nature selects and preserves those attributes that contribute to the survival of the species. Those that are unfit die. Natural selection is merciless to the individual, but it gives the species the best possible chance to survive.

The only way nature can obtain new DNA patterns from which to select the best is from DNA molecules that have changed. This can happen through chance mismatching of base pairs—errors—in the reproduction of the DNA molecule. Another way this can occur is through damage to reproductive cells from exposure to radioactivity. Cosmic rays or natural radioactivity in the soil might perform this function. In any case, an offspring born with altered DNA is called a **mutant.** Most mutations are fatal, and the individual dies long before it can have offspring of its own. But in rare cases, a mutation may give a species a new survival advantage. Then natural selection makes it likely that the new DNA message will survive and be handed down, making the species more capable of surviving.

REVIEW Critical Inquiry

Why can't the information in DNA be permanent?

The information stored in a creature's DNA provides the recipes that make the creature what it is. For example, the DNA in a starfish must contain all the recipes for making the various kinds of proteins needed to consume and digest food. That information must be passed on to offspring starfish, or they will be unable to survive. But the information must be changeable because our environment is changeable. Ice ages come and go, mountains rise, lakes dry up, and ocean currents shift. If the environment changes in some way, one or more of the recipes may no longer work. In our example, a change in the temperature of the ocean water may kill off the shellfish the starfish eat. If they can't digest other shellfish, the entire species will become extinct. Natural variation in DNA means that among all the infant starfish in any generation, some of the recipes are different; if the environment changes, all of the old-style starfish may die, but a few—those with the different DNA—can carry on.

The survival of life depends on this delicate balance between reliable reproduction and the introduction of small variations in DNA information. What are some of the ways these small changes in DNA can arise?

Life is based not only on information, but also on the duplication of information. Today, that process seems so complex that it is hard to imagine how it could have begun.

20-2 The Origin of Life

If life on Earth is based on the storage of information in these long, complex, carbon-chain molecules, how could it have ever gotten started? Obviously, 4.5 billion chemical bases didn't just happen to drift together to form the DNA formula for a human being. The key is evolution. Once a life form begins to reproduce itself, natural selection preserves the most advantageous traits. Over long periods of time spanning thousands, perhaps millions, of generations, the life form becomes more fit to survive. This nearly always means the life form becomes more complex. Thus life could have begun as a very simple process that gradually became more sophisticated as it was modified by evolution.

We begin our search on Earth, where fossils and an intimate familiarity with carbon-based life give us a glimpse of the first living matter. Once we discover how Earthly life could have begun, we can look for signs that life began on other planets in our solar system. Finally, we can speculate on the chances that other planets, orbiting other stars, have conditions that give rise to life.

The Origin of Life on Earth

The oldest fossils hint that life began in the oceans. The oldest easily identified fossils appear in sedimentary rocks that formed between 0.6 and 0.5 billion years ago—the **Cambrian period.** Such Cambrian fossils were simple ocean creatures, the most complex of which were trilobites (Figure 20-5), but there are no Cambrian fossils of land plants or animals. Evidently, land surfaces were totally devoid of life until only 400 million years ago.

Precambrian deposits contain no obvious fossils, but microscopes reveal microfossils that were the ancestors of the Cambrian creatures. Fig tree chert, a rock that resembles flint, in South Africa is 3.0 to 3.3 billion years old, and chert from the Pilbara Block in northwestern Australia is 3.5 billion years old. Both contain structures that appear to be microfossils of bacteria or simple algae such as those that live in water (Figure 20-6). Apparently, life was already active in Earth's oceans a billion years after the planet formed.

The key to the origin of this life may lie in an experiment performed by Stanley Miller and Harold Urey in 1952. This **Miller experiment** sought to reproduce the conditions on Earth under which life began. In a closed glass container, the experimenters placed water (to represent the oceans), the gases hydrogen, ammonia, and methane (to represent the primitive atmosphere), and an electric arc (to represent lightning bolts). The apparatus was sterilized, sealed, and set in operation (Figure 20-7).

Figure 20-5
Trilobites made their first appearance in the Cambrian oceans about 600 million years ago. This example, about the size of a human hand, lived 400 million years ago in an ocean floor that is now a limestone deposit in Pennsylvania. *(Grundy Observatory photograph)*

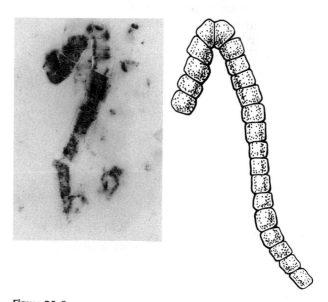

Figure 20-6
Among the oldest fossils known, this microscopic filament resembles modern bacterial forms (artist's reconstruction at right). This fossil was found in the 3.5-billion-year-old chert of the Pilbara Block in northwestern Australia. *(Courtesy J. William Schopf)*

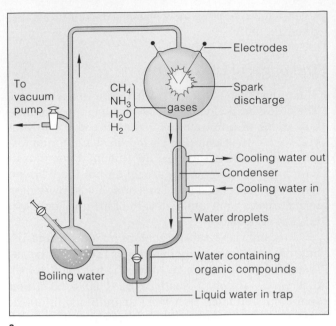

a

b

Figure 20-7
The Miller experiment (a) circulated gases through water in the presence of an electric arc. This simulation of primitive conditions on Earth produced amino acids, the building blocks of proteins. (b) Stanley Miller with a Miller apparatus. *(Courtesy Stanley Miller)*

After a week, Miller and Urey stopped the experiment and analyzed the material in the flask. Among the many compounds the experiment produced, they found four amino acids (building blocks of protein), various fatty acids, and urea, a molecule common to many life processes. Evidently, the energy from the electric arc had molded the atmospheric gases into some of the basic components of living matter. Other energy sources such as hot silica (to simulate hot lava spilling into the sea) and ultraviolet radiation (to simulate sunlight) give similar results.

Recent studies of the composition of meteorites and models of planet formation suggest that Earth's first atmosphere did not resemble the gases used in the Miller experiment. Earth's first atmosphere was probably composed of carbon dioxide, nitrogen, and water vapor. This finding, however, does not invalidate the Miller experiment. When such gases are processed in a Miller apparatus, a tarry gunk rich in organic molecules soon coats the inside of the chamber.

The Miller experiment did not create life, nor did it necessarily imitate the exact conditions on the young Earth. Rather, it is important because it shows that complex organic molecules form naturally in a wide variety of circumstances. The chemical deck is stacked to deal nature a hand of complex molecules. If we could travel back in time, we would probably find Earth's oceans filled with a rich mixture of organic compounds in what some have called the **primordial soup.**

The next step on the journey toward life is for the compounds dissolved in the oceans to link up and form larger molecules. Amino acids, for example, can link together to form proteins. This linkage occurs when amino acids join together end to end and release a water molecule (Figure 20-8). For many years, experts have assumed that this process must have happened in sun-warmed tidal pools where evaporation concentrated the broth. But recent studies suggest that the young Earth was subject to extensive volcanism and large meteorite impacts that periodically modified the climate enough to destroy any life forms exposed on the surface. Thus, the early growth of complex molecules likely took place among the hot springs along the midocean ridges. The heat from such springs could have powered the growth of long protein chains. Deep in the oceans, they would have been safe from climate changes.

Although these proteins might have contained hundreds of amino acids, they would not have been alive. Not yet. Such molecules would not have reproduced, but would have merely linked together and broken apart at random. Because some molecules are more stable than others, however, and because some molecules bond together more readily than others, this blind **chemical evolution** would have led to the concentration of the varied smaller molecules into the most stable larger forms. Eventually, somewhere in the oceans, a molecule took shape that could reproduce itself. At that point the chemical evolution of molecules became the biological evolution of living things.

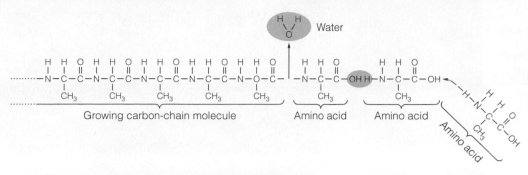

Figure 20-8
Amino acids can link together through the release of a water molecule to form long carbon-chain molecules. The amino acid in this hypothetical example is alanine, one of the simplest.

Figure 20-9
A sample of the Murchison meteorite, a carbonaceous chondrite that fell in 1969 near Murchison, Australia. Analysis of the interior of the meteorite revealed evidence of amino acids. Whether the first building blocks of life originated in space is unknown, but the amino acids found in meteorites illustrate how commonly amino acids and other complex molecules occur even in the absence of living things. *(Courtesy Chip Clark, National Museum of Natural History)*

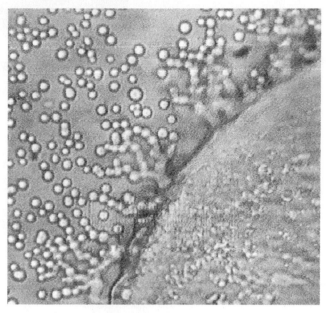

Figure 20-10
Single amino acids can be assembled into long proteinlike molecules. When such material cools in water, it can form microspheres, microscopic spheres with double-layered boundaries similar to cell membranes. Microspheres may have been an intermediate stage in the evolution of life between compex molecules and cells holding molecules reproducing genetic information. *(Courtesy Sidney Fox and Randall Grubbs)*

An alternative theory proposes that primitive living things such as reproducing molecules did not originate on Earth but came here in meteorites or comets. Radio astronomers have found a wide variety of organic molecules in the interstellar medium, and some studies have found similar compounds inside meteorites (Figure 20-9). Such molecules form so readily that we would be surprised if they were not present in space. A few investigators, however, have speculated that living, reproducing molecules originated in space and came to Earth as a cosmic contamination. If

this is true, every planet in the universe is contaminated with the seeds of life. However entertaining this theory may be, it is presently untestable, and an untestable theory is of little use in science. Experts studying the origin of life proceed on the assumption that life began as reproducing molecules in Earth's oceans.

Which came first, reproducing molecules or the cell? Because we think of the cell as the basic unit of life, this question seems to make no sense, but in fact the cell may have originated during chemical evolution. If a dry mixture of amino acids is heated, the acids form long, proteinlike molecules that, when poured into water, collect to form microscopic spheres that function in ways similar to cells (Figure 20-10). They have a thin membrane surface, they can absorb material from their surroundings, they grow in

Figure 20-11
A 3.5-billion-year-old fossil stromatolite from western Australia is one of the oldest known fossils (below). Stromatolites were formed, layer by layer, by mats of bacteria living in shallow water. Such life may have been common in shallow seas when Earth was young (right). Stromatolites are still being formed today in similar environments. *(Mural by Peter Sawyer; photo courtesy Chip Clark, National Museum of Natural History)*

size, and they can divide and bud just as cells do. They contain no large molecule that copies itself, however. Thus the structure of the cell may have originated first and the reproducing molecules later.

An alternative theory proposes that the replicating molecule developed first. Such a molecule would have been exposed to damage if it had been bare, so the first to manufacture or attract a protective coating of protein would have had a significant survival advantage. If this was the case, the protective cell membrane was a later development of biological evolution.

The first living things must have been single-celled organisms much like modern bacteria. Some of the oldest fossils known are **stromatolites,** structures produced by communities of photosynthesizing bacteria that grew in mats and, year by year, deposited layers of minerals that were later fossilized. One of the oldest such fossils known is 3.5 billion years old (Figure 20-11). If such bacteria were common when Earth was young, the early atmosphere may have contained a small amount of oxygen produced by the photosynthesis. Recent studies suggest that an oxygen abundance of only 0.1 percent would have been sufficient to provide an ozone screen that would protect organisms from the sun's ultraviolet radiation.

How evolution shaped creatures to live in the ancient oceans, to photosynthesize and respire, to become multicellular, and to reproduce sexually is a fascinating story, but we cannot explore it in detail here.

We can see that life could have begun through simple chemical reactions building complex molecules, and that once some DNA-like molecule formed, it protected its own survival with selfish determination. Over billions of years, the genetic information stored in living things kept those qualities that favored survival and discarded the rest. As Samuel Butler said, "The chicken is the egg's way of making another egg." In that sense, all living matter on Earth is merely the physical expression of DNA's mindless determination to continue its existence.

Perhaps this seems harsh. Human experience goes far beyond mere reproduction. *Homo sapiens* has art, poetry, music, philosophy, religion, science. Perhaps all of the great accomplishments of our intelligence represent more than mere reproduction of DNA. Nevertheless, intelligence, the ability to analyze complex situations and respond with appropriate action, must have begun as a survival mechanism. For example, a fixed escape strategy stored in the DNA is a disadvantage for a creature that frequently moves from one environment to another. A rodent that always escapes from predators by automatically climbing the nearest tree would be in serious jeopardy if it met a hungry fox in a treeless clearing. Even a faint glimmer of intelligence might allow the rodent to analyze the situation and, finding no trees, to choose running over climbing. Thus, intelligence, of which *Homo sapiens* is so proud, may have developed in ancient creatures as a way of making them more versatile.

Any discussion of the evolution of life seems to involve highly improbable coincidences until we consider how many years have passed in the history of Earth. We read the words—4.6 billion years—so easily, but it is in truth hard to grasp the meaning of such a long period of time.

Geologic Time

Humanity is a very new experiment on planet Earth. We can fit all of the evolution that leads from the primitive life forms in the oceans of the Cambrian period 600 million years ago to fishes, amphibians, reptiles, and mammals into a single chart such as Figure 20-12. We can take comfort in thinking that creatures like us have walked on Earth for roughly 3 million years, but when we add our history to the chart, we discover that the entire history of humanity makes up no more than a thin line at the top. In fact, if we tried to represent the entire 4.6-billion-year history of Earth on the chart, the portion describing the rise of life on the land would be an unreadably small segment.

One way to represent the evolution of life is to compress the 4.6-billion-year history of Earth into a 1-year-long videotape. In such a program, Earth forms as the video begins on January 1, and through all of January and February it cools and is cratered, and the first oceans form. Search as we might, we would find no trace of life in these oceans until sometime in March or early April, when the first living things develop. The slow development of these simplest of living forms grinds on slowly through the spring and summer of our videotape. The entire 4-billion-year history of Precambrian evolution lasts until the video reaches mid-November, when the primitive ocean life begins to evolve into more complex organisms such as trilobites.

While our year-long videotape plays on and on, we might amuse ourselves by looking at the land instead of the oceans, but we would be disappointed. The land is a lifeless waste with no plants or animals of any kind. Not until November 28 in our video does life appear on the land; but once it does, it evolves rapidly into a wide range of plants and animals. Dinosaurs, for example, appear about December 12th and vanish by Christmas evening as mammals and birds flourish.

Throughout the 1-year run of our video there have been no humans; and even during the last days of the year, as the mammals rise to dominate the landscape, there are no people. In the early evening of December 31, vaguely human forms move through the grasslands, and by late evening they begin making stone tools. The Stone Age lasts till about 11:45 P.M.,

and the first signs of civilization, towns and cities, do not appear until 11:54 P.M. The Christian era begins only 14 seconds before the New Year, and the Declaration of Independence is signed with but 1 second to spare.

By converting the history of Earth into a year-long videotape, we have placed the rise of life in perspective. Tremendous amounts of time were needed for the first simple living things to evolve in the oceans, and even more time was needed for the evolution of complex creatures that could colonize the land. As life became more complex, it evolved and diversified faster and faster, as if evolution were drawing on a growing library of solutions that had been previously invented with great effort to solve earlier problems. The burst of diversity on land led slowly to the rise of intelligent creatures like us, a process that has taken 4.6 billion years.

If life could originate on Earth and develop into intelligent creatures, perhaps the same thing could have happened on other planets. This raises three questions. First, could life originate on another world if conditions were suitable? Second, if life begins on a planet, will it evolve toward intelligence? The answer to both questions seems to be yes. The process of chemical and biological evolution is based on survival, which should lead to versatility and intelligence. But what of the third question: Are suitable conditions so rare that life almost never gets started? The only way to answer that is to search for life on other planets. We begin, in the next section, with the other planets in our solar system.

Life in Our Solar System

Though life based on something other than carbon chemistry may be possible, we must limit our discussion to life as we know it and the conditions it requires. The most important condition is the presence of liquid water, not only as part of chemical reactions but also as a medium to transport nutrients and wastes within the organism. This means the temperature must be moderate. Thus our search for life in the solar system must look for a planet where liquid water could exist.

The water requirement automatically eliminates a number of worlds. Our moon and Mercury are airless, and thus liquid water could not exist on their surfaces. Venus has some water vapor, but it is much too hot for liquid water, and in the outer regions of the solar system, the temperature is much too low.

The Jovian planets themselves have deep hydrogen-rich atmospheres but no real surfaces. At some level in the atmospheres, the temperature might be right for the formation of complex carbon molecules, and a few people have speculated about living things evolving to survive in the clouds of the Jovian planets. Convection in

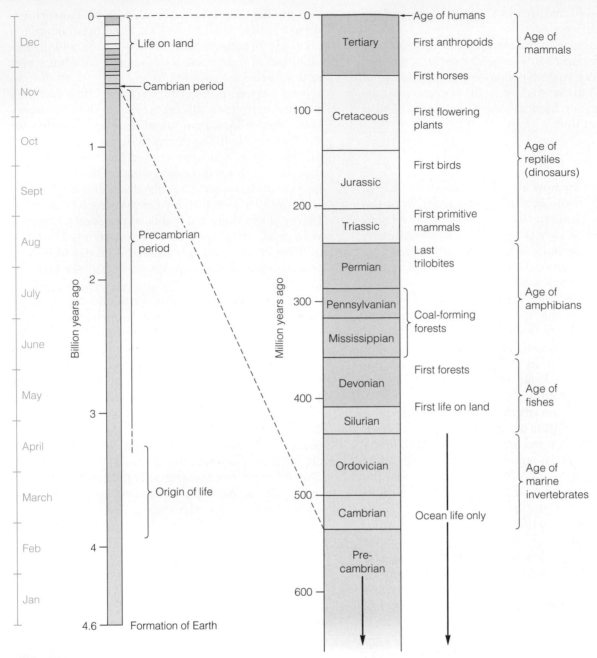

Figure 20-12
Complex life has developed on Earth only recently. If the entire history of Earth were represented in a time line (left), we would have to magnify the end of the line to see details such as life leaving the oceans and dinosaurs appearing. The age of humans would still be only a thin line at the top of our diagram. If the history of Earth were a year-long videotape, humans would not appear until the last hours of December 31.

these atmospheres, however, would carry newly formed molecules into regions too cold or too hot for them to survive, so life on the Jovian planets themselves seems very unlikely.

A few of the satellites of the Jovian planets might have conditions that could support life. Jupiter's moon Europa seems to have a liquid-water ocean below its icy crust, but we know little about its long-term history. Saturn's moon Titan may have methane oceans on its surface and an organic tarry goo settling out of its atmosphere. That sounds ideal, but the surface temperature is −179°C (−290°F), and that seems too cold for life-based chemical reactions.

Mars is the most likely place for life in our solar system. The evidence, however, is not encouraging. In 1976, two robotic spacecraft, Viking 1 and Viking 2, landed on the Martian surface, scooped up soil samples, and subjected them to tests for the presence of

a

b

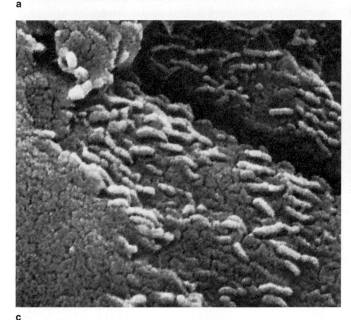

c

Figure 20-13

(a) ALH84001 is one of a dozen meteorites known to have originated on Mars. (b) Microscopic carbonate globules found inside the meteorite resemble similar deposits made by Earthly bacteria. (c) Concentrations of features that resemble Earthly bacteria have been found inside the meteorite. *(NASA)*

living organisms. The results were negative. The samples tested did not seem to contain living things. But the Viking landers only tested two places on Mars and were not able to look for traces of past life.

Meteorite ALH84001 (Figure 20-13a) was found on the Antarctic ice in 1984 and only recently recognized as one of 12 meteorites believed to have come from Mars. How do we know it is from Mars? About a dozen of these SNC-class meteorites are known, and one of them contains gases trapped in small glassy particles. Analysis of these gases show that they have the same abundance as the gases in the atmosphere of Mars as measured by the two Viking spacecraft that landed there in 1976. The abundances are like fingerprints and they match exactly, so it seems very likely that the SNC meteorites are rocks that were blasted off of the surface of Mars at some point in the past when a comet or an asteroid struck the planet. From traces in ALH84001, scientists conclude it left the surface of Mars about 16 million years ago, orbited the sun for a long time, and then fell to Earth in Antarctica about 13,000 years ago.

In 1996, NASA scientists announced that they had found three kinds of evidence that life once existed in ALH84001. Part of the evidence consisted of carbonate deposits that resemble those produced by some Earthly bacteria (Figure 20-13b). A second part of the evidence was complex organic molecules that could be the remains of ancient bacteria. The third piece of evidence was revealed in electron microscope images that show what seem to be small fossils of living things (Figure 20-13c).

Scientists, being professionally skeptical (see Window on Science 16-2), began testing these results immediately, and in many cases the tests did not confirm the conclusion that life once existed in the meteorite. There is evidence that the water that deposited the carbonates was much too hot for living things to survive. Also, the organic molecules found in the meteorite may have formed from nonbiological processes, or they may be contamination. There is conclusive evidence that the interior of the meteorite has been contaminated by chemical compounds such as amino acids that exist in the Antarctic ice. Furthermore, the microscopic fossils appear to be natural mineral features of the rock samples and not the remains of living things. Although studies of ALH84001 continue, it does not give us conclusive evidence that life once existed on Mars.

We are left to conclude that, so far as we know, our solar system is bare of life except for the Earth. Consequently our search for life in the universe takes us to other planetary systems.

Life in Other Planetary Systems

Might life exist in other solar systems? To consider this question, let us try to decide how common planets are and what conditions a planet must fulfill for life to originate and evolve to intelligence. The first question is astronomical; the second is biological. Our ability to discuss the problem of life outside our solar system is severely limited by our lack of experience.

In Chapter 16 we concluded that planets form as a natural by-product of star formation, and we learned that a number of extra-solar planets have been found circling nearby stars. From this we can conclude that planetary systems are very common.

If a planet is to become a suitable home for life, it must have a stable orbit around its sun. This is simple in a solar system like our own, but in a binary system most planetary orbits are unstable. Most planets in such systems would not last long before they were swallowed up by one of the stars or ejected from the system.

Thus, single stars are the most likely to have planets suitable for life. Because our galaxy contains about 10^{11} stars, half of which are single, there should be roughly 5×10^{10} planetary systems in which we might look for life.

A few million years of suitable conditions does not seem to be enough time to originate life. On our planet it took at least 0.5 to 1 billion years for the first cells to evolve and 4.6 billion years for intelligence to evolve. Clearly, conditions on a planet must remain acceptable over a long time. This eliminates massive stars that remain stable on the main sequence for only a few million years. If it takes a few billion years for life to originate and evolve to intelligence, no star hotter than about F5 will do. This is not really a serious restriction, because upper-main-sequence stars are rare anyway.

In previous sections, we decided that life, at least as we know it, requires liquid water. That requirement defines a **life zone** (or ecosphere) around each star, a region within which a planet has temperatures that permit the existence of liquid water.

The size of the life zone depends on the temperature of the star (Figure 20-14). Hot stars have larger life zones because the planets must be more distant to remain cool. But the short main-sequence lives of these stars make them unacceptable. M stars have small life zones because they are extremely cool—only planets very near the star receive sufficient warmth. However, planets that are close to a star would probably become tidally coupled, keeping the same side toward the star. This might cause the water and atmosphere to freeze in the perpetual darkness of the planet's night side and end all chance of life. Also, M stars are subject to sudden flares that might destroy life on a planet close to the star. Thus, the life zone restricts our search for life to main-sequence G and K stars.

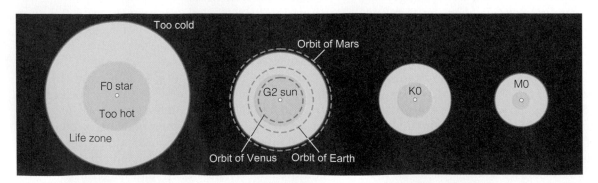

Figure 20-14
The life zone around a star (green) is the region where a planet would have a moderate temperature. M stars are not good candidates because they are cool and their life zone is small. F stars have a large life zone but evolve too quickly to allow the rise of complex life. K and G stars like our sun seem the best candidates in a search for intelligent life.

Even a star on the main sequence is not perfectly stable. Main-sequence stars gradually grow more luminous as they convert their hydrogen to helium, and thus the life zone around a star gradually moves outward. A planet might form in the life zone and life might begin and evolve for billions of years only to be destroyed as the slowly increasing luminosity of its star moves the life zone outward, evaporates the planet's oceans, and drives off its atmosphere. If a planet is to remain in the life zone for 4 to 5 billion years, it must form on the outer edge of the zone. This may be the most serious restriction we have yet discussed.

If all these requirements are met, will life begin? Early in this chapter we decided that life could begin through simple chemical reactions, so perhaps we should change our question and ask, What could prevent life from beginning? Given what we know about life, it should arise whenever conditions permit, and our galaxy should be filled with planets that are inhabited with living creatures.

REVIEW Critical Inquiry

What evidence do we have that life is at least possible on other worlds?

Our evidence is limited almost entirely to Earth, but it is promising. The fossils we find show that life originated in Earth's oceans almost 4 billion years ago, and biologists have proposed relatively simple chemical processes that could have created these first reproducing molecules. The fossils show that life developed from a very slow beginning into more and more complex creatures that filled the oceans. The pace of evolution quickened dramatically about 600 million years ago, the beginning of the Cambrian period, when life began taking on complex forms; and later, when life emerged onto the surface of the land, it evolved rapidly to produce the tremendous diversity we see around us. Human intelligence has been a very recent development; it is only a few million years old.

If this process occurred on Earth, then it seems reasonable that it could have occurred on other worlds as well. We also have exciting but limited evidence of ancient life on Mars. Thus, the evidence suggests we can expect that life might begin and evolve to intelligent forms on any world where conditions are right. What are the conditions we should expect of other worlds that host life?

It is both easy and fun to speculate about life on other worlds, but it leads to a simple question: Is there really life beyond Earth? If we can't leave Earth and visit other worlds, then the only life we can detect will be beings intelligent enough to communicate with us. Why haven't we heard from them?

20-3 Communication with Distant Civilizations

If other civilizations exist, perhaps we can communicate with them in some way. Sadly, travel between the stars is more difficult in real life than in science fiction—and may in fact be impossible. If we can't physically visit, perhaps we can communicate by radio. Again, nature places restrictions on such conversations, but the restrictions are not too severe. As we will see, the real problem lies with the life expectancy of civilizations.

Travel Between the Stars

Practically speaking, roaming among the stars is tremendously difficult because of three limitations: distance, speed, and fuel. The distances between stars are almost beyond comprehension. It does little good to explain that if we use a golf ball in New York City to represent the sun, the nearest star would be another golf ball in Chicago. It is only slightly better to note that the fastest commercial jet would take about 4 million years to reach the nearest star.

The second limitation is a speed limit—we cannot travel faster than the speed of light. Though science fiction writers invent hyperspace drives so their heroes can zip from star to star, the speed of light is a natural and unavoidable limit that we cannot exceed. This, combined with the large distances between stars, makes interstellar travel very time consuming.

The third limitation is that we can't even approach the speed of light without using a fantastic amount of fuel. Even if we ignore the problem of escaping from Earth's gravity, we must still use energy stored in fuel to accelerate to high speed and to decelerate to a stop when we reach our destination. To return to Earth, assuming we wish to, we have to repeat the process. These changes in velocity require a tremendous amount of fuel. If we flew a spaceship as big as a large yacht to a star 5 light-years (1.5 pc) away and wanted to get there in only 10 years, we would use 40,000 times as much energy as the United States consumes in a year.

Travel for a few individuals might be possible if we accept very long travel times. That would require some form of suspended animation (currently unknown) or colony ships that carry a complete, though small, society in which people are born, live, and die generation after generation. Whether the occupants of such a ship would retain the social characteristics of humans over a long voyage is questionable.

These three limitations not only make it difficult for us to leave our solar system, they would also make it difficult for aliens to visit Earth. Reputable scientists

If we discuss life on other worlds, then we might be tempted to use UFO sightings and supposed visits by aliens from outer space as evidence to test our hypotheses. We don't do so for two reasons, both related to the reliability of these observations.

First, the reputation of UFO sightings and alien encounters does not give us confidence that these data are reliable. Most people hear of such events in the grocery store tabloids, daytime talk shows, or sensational "specials" on viewer-hungry cable networks. We must take note of the low reputation of the media that report UFOs and space aliens. Most of these reports are simply made up for the sake of sensation, and we cannot use them as reliable evidence.

Second, the remaining UFO sightings, those not simply made up, do not survive careful examination. Most are mistakes and unconscious misinterpretations of natural events made by honest people. In short, there is no dependable evidence that Earth has been visited by aliens from space.

That's too bad. A confirmed visit by intelligent creatures from beyond our solar system would answer many of our questions. It would be exciting, enlightening, and, like any real adventure, a bit scary. But none of the UFO sightings are dependable, and we are left with no direct evidence of life on other worlds.

have studied "unidentified flying objects" (UFOs) and related phenomena and have never found any evidence that Earth is being visited or has ever been visited by aliens from other worlds (Window on Science 20-2). Thus, humans are unlikely ever to meet an alien face to face. The only way we can communicate with other civilizations is via radio.

Radio Communication

Nature places two restrictions on our ability to communicate with distant societies by radio. One has to do with simple physics, is well understood, and merely makes the communication difficult. The second has to do with the fate of technological civilizations, is still unresolved, and may severely limit the number of societies we can detect by radio.

Radio signals are electromagnetic waves that travel at the speed of light. Because even the nearest civilizations must be a few light-years away, this limits our ability to carry on a conversation with distant beings. If we ask a question of a creature 10 light-years away, we will have to wait 20 years for a reply. Clearly, the give-and-take of normal conversation will be impossible.

Instead, we could simply broadcast a radio beacon of friendship to announce our presence. Such a beacon would have to consist of a pattern of pulses obviously designed by intelligent beings, to distinguish it from natural radio signals emitted by nebulae, pulsars, and so on. For example, pulses counting off the first dozen prime numbers would do. In fact, we are already broadcasting a recognizable beacon. Short-wavelength radio signals, such as TV and FM, have been leaking into space for the last 50 years or so. Any civilization within 50 light-years might already have detected us.

If we intentionally broadcast a signal, we can anticode it. That is, we can arrange it to make it easy to decode. We could transmit pulses to represent 1s and gaps to represent 0s. A message counting through the first few prime numbers would distinguish our signal from natural sources of radio noise. We could even transmit a picture by sending a string of 1s and 0s that can be arranged in only two ways. One way produces nonsense, but the other way produces a meaningful picture (Figure 20-15).

In 1974, at the dedication of the 1000-ft radio telescope at Arecibo, radio astronomers transmitted such a signal toward the globular cluster M13, which is located 26,000 ly from Earth. When the signal finally arrives, any aliens who detect it will be able to arrange its 1679 pulses in only two ways, as 23 rows of 79 or 79 rows of 23. The second arrangement will form a picture that describes life on Earth (Figure 20-16).

It took only minutes to transmit the Arecibo message. If more time were taken, a more detailed picture could be sent, and if we were sure our radio telescope was pointed at a listening civilization, we could send a long series of pictures. With pictures we could teach aliens our language and tell them all about our life, our difficulties, and our accomplishments.

If we can think of sending such signals, aliens can think of it too. If we point our radio telescopes in the right direction and listen at the right wavelength, we might hear other intelligent races calling out to one another. This raises two questions: Which stars are the best candidates, and what wavelengths are most likely? We have already answered the first question. Main-sequence G and K stars have the most favorable characteristics. But the second question is more complex.

Only certain wavelengths are useful for communication. We cannot use wavelengths longer than about

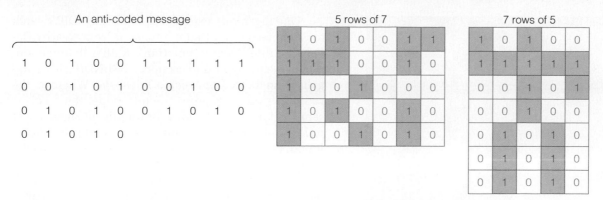

An anti-coded message

1	0	1	0	0	1	1	1	1	1	1
0	0	1	0	1	0	0	1	0	0	
0	1	0	1	0	0	1	0	1	0	
0	1	0	1	0						

5 rows of 7

1	0	1	0	0	1	1
1	1	1	0	0	1	0
1	0	0	1	0	0	0
1	0	1	0	0	1	0
1	0	0	1	0	1	0

7 rows of 5

1	0	1	0	0
1	1	1	1	1
0	0	1	0	1
0	0	1	0	0
0	1	0	1	0
0	1	0	1	0
0	1	0	1	0

Figure 20-15
An anticoded message is designed for easy decoding. Here a string of 35 radio pulses, represented as 1s and 0s, can be arranged in only two ways, as 5 rows of 7 or 7 rows of 5. The second way produces a friendly message. Any number of pulses can be used so long as it is the product of two prime numbers. Then the pulses can be arranged in only two ways.

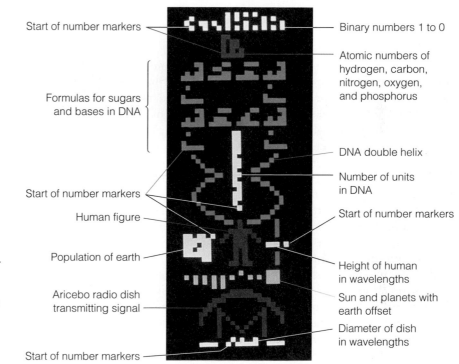

Figure 20-16
The Arecibo message to M13 begins by counting from 1 to 10 and goes on to describe our solar system and the biochemistry of life on Earth (color added for clarity). Binary numbers give the height of the human figure (1110) and the diameter of the telescope dish (1001011111100) in units of the wavelength of the signal, 12.3 cm. *(NASA)*

30 cm because the signal would be lost in the background radio noise from our galaxy. Nor can we go to wavelengths much shorter than 1 cm because of absorption within our atmosphere. Thus only a certain range of wavelengths, a radio window, is open for communication.

This communications window is very wide, so a radio telescope would take a long time to tune over all the wavelengths searching for intelligent signals. Nature may have given us a way to narrow the search, however. Within the communications window lie the 21-cm line of neutral hydrogen and the 18-cm line of OH. The interval between these two lines has been dubbed the **water hole** because the combination of H

and OH yields water (H_2O). Water is the fundamental solvent in our life form, so it might seem natural for similar water creatures to call out to each other at wavelengths in the water hole (Figure 20-17). But even silicon creatures would be familiar with the 21-cm line of hydrogen.

This is not idle speculation. A number of searches for extraterrestrial radio signals have been made, and some major searches are now under way. The field has become known as **SETI**, Search for Extra-Terrestrial Intelligence, and it has generated heated debate. Some scientists and philosophers argue that life on other worlds can't possibly exist, so it is a waste of money to search. Others argue that life on other worlds is

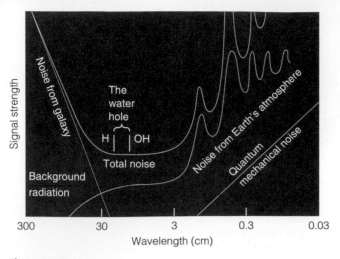

Figure 20-17
Radio noise from various sources makes it difficult to detect distant signals at wavelengths longer than 30 cm or shorter than 1 cm. In this range, radio emission from H atoms and from OH mark a small wavelength range dubbed the water hole, which may be a likely place for communication.

common and could be detected with present technology. Congress funded a NASA search for a short time but then ended support because political leaders feared public reaction. In fact, the annual cost of a major search is only about as much as a single Air Force attack helicopter. The controversy may spring in part from the theological and philosophical controversy that would result from the discovery of intelligent life on another world.

In spite of the controversy, some searches have been made, and others are continuing. The first was Project Ozma in 1960. It listened to two stars at a wavelength of 21 cm. Since then, numerous small-scale searches have been conducted, but in order to have some assurance of success a search must scan the entire sky at millions of closely spaced frequency bands. Only recently has technology made that possible, and as technology improves, the searches can scan more sky and more frequencies.

Since 1985, META, Megachannel Extra-Terrestrial Assay, has been searching the entire sky at millions of frequency bands in the water hole. Funded by the Planetary Society, the search uses radio antennas at Harvard and near Buenos Aires, Argentina. The project has instituted a second, even more efficient search called BETA. Since it began, META has detected dozens of candidate signals. That is, it has found dozens of radio signals that satisfy its most basic criteria. Unfortunately, none of these signals has proved to be continuous, and no signal has been detected when the radio telescopes were redirected at any of the candidate stars. Many of these signals are probably unusual noise from sources on Earth, but the candidate signals are still under study.

One ingenious search is called SERENDIP, Search for Extraterrestrial Radio Emission from Nearby Developed Intelligent Populations. Rather than monopolize an entire radio telescope, SERENDIP rides piggyback on the 305-m Arecibo Telescope. Wherever the radio astronomers point the telescope, the receiver samples the signal, looking for intelligent signals over millions of frequency bands. If candidate signals are found, the receivers note the position of the radio telescope for future investigation.

The NASA search originally called SETI was renamed High Resolution Microwave Survey to reduce criticism, but Congress cut off all of its funds in the fall of 1993. The search continues, however, with private funding as Project Phoenix at the SETI Institute. It uses an 8.4-million-channel receiver on major radio telescopes and plans to survey the entire sky in 10 years. In addition, it will listen carefully to 800 nearby solar-type stars. The search officially began on October 12, 1992, 500 years after the landing of Columbus in the New World.

What should we expect if signals are found? By international agreement, those searching for signals have agreed to share candidate signals and obtain reliable confirmation before making public announcements. False alarms would be embarrassing. The chances of success depend on the number of inhabited worlds in our galaxy, and that number is difficult to estimate.

How Many Inhabited Worlds?

The technology exists, and given enough time the searches will find other inhabited worlds, assuming there are at least a few out there. If intelligence is common, then we should find the signals soon — in the next few decades — but if intelligence is rare in the universe, it may be a very long time before we confirm that we are not alone.

Simple arithmetic can give us an estimate of the number of technological civilizations with which we might communicate. The formula for the number of communicative civilizations in a galaxy, N_c, is:

$$N_c = N^* \cdot f_P \cdot n_{LZ} \cdot f_L \cdot f_I \cdot F_S$$

N^* is the number of stars in a galaxy, and f_P represents the probability that a star has planets. If all single stars have planets, f_P is about 0.5. The factor n_{LZ} is the average number of planets in a solar system suitably placed in the life zone, f_L is the probability that life will originate if conditions are suitable, and f_I is the probability that the life form will evolve to intelligence. These factors can be roughly estimated, but the remaining factor is much more uncertain.

F_S is the fraction of a star's life during which the life form is communicative. Here we assume that a star lives about 10 billion years. If a society survives

Table 20-1 The Number of Technological Civilizations per Galaxy

	Variables	Estimates Pessimistic	Estimates Optimistic
N^*	Number of stars per galaxy	2×10^{11}	2×10^{11}
f_P	Fraction of stars with planets	0.01	0.5
n_{LZ}	Number of planets per star that lie in life zone for longer than 4 billion years	0.01	1
f_L	Fraction of suitable planets on which life begins	0.01	1
f_I	Fraction of life forms that evolve to intelligence	0.01	1
F_S	Fraction of star's life during which a technological society survives	10^{-8}	10^{-4}
N_c	Number of communicative civilizations per galaxy	2×10^{-5}	10×10^6

at a technological level for only 100 years, our chances of communicating with it are small. But a society that stabilizes and remains technological for a long time is much more likely to be in the communicative phase at the proper time to signal to us. If we assume that technological societies destroy themselves in about 100 years, F_S is 100 divided by 10 billion, or 10^{-8}. But if societies can remain technological for a million years, then F_S is 10^{-4}. The influence of the factors in the formula is shown in Table 20-1.

If the optimistic estimates are true, there may be a communicative civilization within a few dozen light-years of us, and we could locate it by searching through only a few thousand stars. On the other hand, if the pessimistic estimates are correct, we may be the only planet in our galaxy capable of communication. We may never know until we understand how technological societies function.

REVIEW Critical Inquiry

Why does the number of inhabited worlds we might hear from depend on how long civilizations survive at a technological level?

When we turn radio telescopes toward the sky and scan millions of frequency bands, we take a snapshot of the universe at a particular time when we are living and able to build radio telescopes. To detect other civilizations, they must be in a similar technological stage so they will be broadcasting either intentionally or accidentally. If we search for decades and detect no other signal, it may mean not that life is rare but rather that civilizations do not survive at a technological level for very long. Only a few decades ago, our civilization was threatened by nuclear war, and now we are threatened by the pollution of our environment. If nearly all civilizations in our galaxy are either on the long road up from primitive life forms in oceans or on the long road down from nuclear war or environmental collapse, there may be no one transmitting during the short interval when we are capable of building radio telescopes with which to listen.

Radio communication between inhabited worlds is limited because it requires very fast computers to search many frequency intervals. Why must we search so many frequencies when we suspect that the water hole would be a good place to listen?

Are we the only thinking race? If we are, we bear the sole responsibility to understand and admire the universe. Then we are the sole representatives of that state of matter called intelligence. The mere detection of signals from another civilization would demonstrate that we share the universe with others. Although we might never leave our solar system, such communication would end the self-centered isolation of humanity and stimulate a reevaluation of the meaning of our existence. We may never realize our full potential as humans until we communicate with nonhuman intelligent life.

Summary

To discuss life on other worlds, we must first understand something about life in general, life on Earth, and the origin of life. In general, we can identify three properties of living things: a process, a physical basis, and a controlling unit of information. The process must extract energy from the surroundings, maintain the organism, and modify the surroundings to promote the organism's survival. The physical basis is the arrangement of matter and energy that implements the life process. On Earth, all life is based on carbon chemistry. The controlling information is the data necessary to maintain the organism's function. Data for Earth life are stored in long carbon-chain molecules called DNA.

The DNA molecule stores information in the form of chemical bases linked together like the rungs of a ladder. When these patterns are copied by RNA molecules, they can direct the manufacture of proteins and enzymes. Thus the DNA information is the chemical formulas the cell needs to function. When a cell divides, the DNA molecule splits lengthwise and duplicates itself so that each of the new cells has a copy of the information. Errors in the duplication or damage to the DNA molecule can produce mutants, organisms that contain new DNA information and have new properties. Natural selection determines which of these new organisms are best suited to survive, and the species evolves to fit its environment.

The Miller experiment duplicated conditions in Earth's primitive environment and suggests that energy sources such as lightning could have caused amino acids and other complex molecules to form. Chemical evolution would have connected these together in larger and more complex, but not yet living, molecules. When a molecule acquired the ability to produce copies of itself, natural selection perfected the organism through biological evolution. Though this may have happened in the first billion years, life did not become diverse and complex until the Cambrian period, about 0.6 billion years ago. Life emerged from the oceans about 0.4 billion years ago, and humanity developed only a few million years ago.

It seems unlikely that there is now life on other planets in our solar system. Most of the planets are too hot or too cold; Mars may have been suitable in the past, but it is now a cold, dry desert. Nevertheless, evidence in a meteorite originally from Mars suggests that life may have gotten started on Mars some 3.6 billion years ago.

To find life, we must look beyond our solar system. Because we suspect that planets form from the leftover debris of star formation, we suspect that most stars have planets. The rise of intelligence may take billions of years, however, so short-lived massive stars and binary stars with unstable planetary orbits must be discarded. The best candidates are G and K main-sequence stars.

The distances between stars are too large to permit travel, but communication by radio could be possible. A certain wavelength range called a radio window is suitable, and a small range between the radio signals of H and OH, the so-called water hole, is especially likely.

We now have the technology to search for other intelligent life in the universe. Although such searches are controversial, a number of radio astronomers are now searching for radio signals from extraterrestrial civilizations. If life is common in the universe, success seems inevitable.

New Terms

DNA (deoxyribonucleic acid)

protein

enzyme

RNA (ribonucleic acid)

amino acid

natural selection

mutant

Cambrian period

Miller experiment

primordial soup

chemical evolution

stromatolite

life zone

water hole

SETI

Review Questions

1. If life is based on information, what is that information?

2. What would happen to a life form if the information handed down to offspring was always the same? How would that endanger the future of the life form?

3. How does the DNA molecule produce a copy of itself?

4. Give an example of natural selection acting on new DNA patterns to select the most advantageous characteristics.

5. Why do we believe that life on Earth began in the sea?

6. Why do we think that liquid water is necessary for the origin of life?

7. What is the difference between chemical evolution and biological evolution?

8. What was the significance of the Miller experiment?

9. How does intelligence make a creature more likely to survive?

10. What role did the controls play in the analysis of the Martian meteorite?

11. What evidence supports the claim for traces of ancient life in the Martian meteorite?

12. Why are upper-main-sequence stars unlikely sites for intelligent civilizations?

13. Why do we suspect that travel between stars is nearly impossible?

14. How does the stability of technological civilizations affect the probability that we can communicate with them?

15. What is the water hole, and why would it be a good place to look for other civilizations?

Discussion Questions

1. What would you change in the Arecibo message if humanity lived on Mars instead of Earth?

2. What do you think it would mean if decades of careful searches for radio signals for extraterrestrial intelligence turned up nothing?

Problems

1. A single human cell encloses about 1.5 m of DNA containing 4.5 billion base pairs. What is the spacing between these base pairs in nanometers? That is, how far apart are the rungs on the DNA ladder?

2. If we represent the history of the Earth by a line 1 m long, how long a segment would represent the 400 million years since life moved onto the land? How long a segment would represent the 3-million-year history of human life?

3. If a human generation, the time from birth to childbearing, is 20 years, how many generations have passed in the last million years?

4. If a star must remain on the main sequence for at least 5 billion years for life to evolve to intelligence, how massive could a star be and still harbor intelligent life on one of its planets? (*Hint:* See By the Numbers 9-1.)

5. If there are about 1.4×10^{-4} stars like the sun per cubic light-year, how many lie within 100 light-years of Earth? (*Hint:* The volume of a sphere is $\frac{4}{3} \pi r^3$.)

6. Mathematician Karl Gauss suggested planting forests and fields in a gigantic geometric proof to signal to possible Martians that intelligent life exists on Earth. If Martians had telescopes that could resolve details no smaller than 1 second of arc, how large would the smallest element of Gauss's proof have to be? (*Hint:* See By the Numbers 3-1.)

7. If we detected radio signals with an average wavelength of 20 cm and suspected that they came from a civilization on a distant planet, roughly how much of a change in wavelength should we expect to see because of the orbital motion of the distant planet? (*Hint:* See By the Numbers 6-2.)

8. Calculate the number of communicative civilizations per galaxy from your own estimates of the factors in Table 20-1.

Critical Inquiries for the Web

1. The popular movie *Contact* focused interest on the SETI program by profiling the work of a radio astronomer dedicated to the search for extraterrestrial intelligence. Visit Web sites that give information about the movie, SETI programs, and radio astronomy, and discuss how realistic the movie was in capturing how such research is done.

2. Where outside the solar system would you look for habitable planets? NASA has increasingly focused its interest on this question. Look for information online about programs dedicated to detecting which planets might support life as we know it. What criteria are used to choose targets for the planned searches? What methods will be used to carry out the searches?

Go to the Brooks/Cole Astronomy Resource Center (www.brookscole.com/astronomy) for critical thinking exercises, articles, and additional readings from InfoTrac College Edition, Brooks/Cole's online student library.

Supernatural is a null word.

Robert A. Heinlein
The Notebooks of Lazarus Long

Afterword

Our journey is over, but before we part company, there is one last thing to discuss—the place of humanity in the universe. Astronomy gives us some comprehension of the workings of stars, galaxies, and planets, but its greatest value lies in what it teaches us about ourselves. Now that we have surveyed astronomical knowledge, we can better understand our own position in nature.

To some, the word *nature* conjures up visions of furry rabbits hopping about in a forest glade dotted with pastel wildflowers. To others, nature is the blue-green ocean depths filled with creatures swirling in a mad struggle for survival. Still others think of nature as windswept mountaintops of gray stone and glittering ice. As diverse as these images are, they are all Earthbound. Having studied astronomy, we can view nature as a beautiful mechanism composed of matter and energy interacting according to simple rules to form galaxies, stars, planets, mountaintops, ocean depths, and forest glades.

Perhaps the most important astronomical lesson is that we are a small but important part of the universe. Most of the universe is lifeless. The vast reaches between the galaxies appear to be empty of all but the thinnest gas, and the stars, which contain most of the mass, are much too hot to preserve the chemical bonds that seem necessary to allow life to survive and develop. Only on the surfaces of a few planets, where temperatures are moderate, could atoms link together to form living matter.

If life is special, then intelligence is precious. The universe must contain many planets devoid of life, planets where the wind has blown unfelt for billions of years. There may also be planets where life has developed but has not become complex, planets on which the wind stirs wide plains of grass and rustles dark forests. On some planets, insects, fish, birds, and animals may watch the passing days unaware of their own existence. It is intelligence, human or alien, that gives meaning to the landscape.

Science is the process by which intelligence tries to understand the universe. Science is not the invention of new devices or processes. It does not create home computers, cure the mumps, or manufacture plastic spoons—that is engineering and technology, the adaptation of scientific understanding for practical purposes. Science is understanding nature, and astronomy is understanding nature on the grandest scale. Astronomy is the science by which the universe, through its intelligent lumps of matter, tries to understand its own existence.

As the primary intelligent species on this planet, we are the custodians of a priceless gift—a planet filled with living things. This is especially true if life is rare in the universe. In fact, if Earth is the only inhabited planet, our responsibility is overwhelming. In any case, we are the only creatures who can take action to preserve the existence of life on Earth, and ironically, it is our own actions that are the most serious hazards.

The future of humanity is not secure. We are trapped on a tiny planet with limited resources and a population growing faster than our ability to produce food. In our efforts to survive, we have already driven some creatures to extinction and now threaten others. If our civilization collapses because of starvation, or if our race destroys itself somehow, the only bright spot is that the rest of the creatures on Earth will be better off for our absence.

But even if we control our population and conserve and recycle our resources, life on Earth is doomed. In 5 billion years, the sun will leave the main sequence and swell into a red giant, incinerating Earth. Earth will be lifeless long before that, however. Within the next few billion years, the growing luminosity of the sun will first alter Earth's climate and then boil away its atmosphere and oceans. Our Earth is, like everything else in the universe, only temporary.

To survive, humanity must leave Earth and search for other planets. Colonizing the moon and other planets of our solar system will not save us, because they face the same fate as Earth when the sun dies. But travel to other stars is tremendously difficult and may be impossible with the limited resources we have in our small solar system. We and all the living things that depend on us for survival may be trapped.

This is a depressing prospect, but a few factors are comforting. First, everything in the universe is temporary. Stars die; galaxies die; perhaps the entire universe will fall back in a "big crunch" and die. That our distant future is limited only assures us that we are a part of a much larger whole. Second, we have a few billion years to prepare, and a billion years is a very long time. Only a few million years ago, our ancestors were learning to walk erect and communicate with one another. A billion years ago our ancestors were microscopic organisms living in the primeval oceans. To suppose that a billion years hence we humans will still be human, or that we will still be the dominant species on Earth, or that we will still be the sole intelligence on Earth, is the ultimate conceit.

Our responsibility is not to save our race for all eternity, but to behave as dependable custodians of our planet, preserving it, admiring it, and trying to understand it. That will call for drastic changes in our behavior toward other living things and a revolution in our attitude toward our planet's resources. Whether we can change our ways is debatable—humanity is far from perfect in its understanding, abilities, or intentions. We must not imagine, however, that we and our civilization are less than precious. We have the gift of intelligence, and that is the finest thing this planet has ever produced.

We must not cease from exploration and the end of all our exploring will be to arrive where we began and to know the place for the first time.

—T. S. Eliot

Units and Astronomical Data

Introduction

The metric system is used worldwide as the system of units not only in science but also in engineering, business, sports, and daily life. Developed in 18th-century France, the metric system has gained acceptance in almost every country in the world because it simplifies computations.

A system of units is based on the three fundamental units for length, mass, and time. Other quantities such as density and force are derived from these fundamental units. In the English (or British) system of units (commonly used in the United States, although no longer in Britain!) the fundamental unit of length is the foot, composed of 12 inches. The metric system is based on the decimal system of numbers, and the fundamental unit of length is the meter, composed of 100 centimeters.

To see the advantage of having a decimal-based system, try computing the volume of a bathtub that is 5′9″ long, 1′2″ deep, and 2′10″ wide. In the metric system the length of the tub is 1 m and 75 cm or 1.75 m. The other dimensions are 0.35 m and 0.86 m, and the volume is just $1.75 \times 0.35 \times 0.86$ cm^3. To make the computation in English units, we must first convert inches to feet by dividing by 12. We can convert centimeters to meters by the simpler process of moving the decimal point. Thus, the computation is much easier if we measure the tub in meters and centimeters instead of feet and inches.

Because the metric system is a decimal system, it is easy to express quantities in larger or smaller units as is convenient. We can express distances in centimeters, meters, kilometers, and so on. The prefixes specify the relation of the unit to the meter. Just as a cent is $\frac{1}{100}$ of a dollar, a centimeter is $\frac{1}{100}$ of a meter. A kilometer is 1000 m, and a kilogram is 1000 g. The meanings of the commonly used prefixes are given in Table A-1.

The SI Units

Any system of units based on the decimal system would be easy to use, but by international agreement, the preferred set of units, known as the *Système International d'Unités* (SI units) is based on the meter, kilogram, and second. These three fundamental units define the rest of the units as given in Table A-2.

The SI unit of force is the newton (N), named after Isaac Newton. It is the force needed to accelerate a 1 kg mass by 1 m/sec^2, or the force roughly equivalent to the weight of an apple at Earth's surface. The SI unit of energy is the joule (J), the energy produced by a force of 1 N acting through a distance of 1 m. A joule is roughly the energy in the impact of an apple falling off a table.

Exceptions

Units help us in two ways. They make it possible to make calculations, and they help us to conceive of certain quantities. For calculations the metric system is far superior, and we use it for our calculations throughout this book.

But Americans commonly use the English system of units, so for conceptual purposes we can express quantities in English units. Instead of saying the average person would weigh 133 N on the moon, we could express the weight as 30 lb. Thus, the text commonly gives quantities in metric form followed by the English form in parentheses: the radius of the moon is 1738 km (1080 miles).

In SI units, density should be expressed as kilograms per cubic meter, but no human can enclose a cubic meter in his or her hand, so this unit does not help us grasp the significance of a given density. This book refers to density in grams per cubic centimeter. A gram is roughly the mass of a paperclip, and a cubic centimeter is the size of a small sugar cube, so we can conceive of a density of 1 g/cm^3, roughly the density of water. This is not a bothersome departure from SI units because we will not make complex calculations using density.

Table A-1 Metric Prefixes

Prefix	Symbol	Factor
Mega	M	10^6
Kilo	k	10^3
Centi	c	10^{-2}
Milli	m	10^{-3}
Micro	μ	10^{-6}
Nano	n	10^{-9}

Table A-2 SI Metric Units

Quantity	SI Unit	English Unit
Length	Meter (m)	Foot
Mass	Kilogram (kg)	Slug (sl)
Time	Second (sec)	Second (sec)
Force	Newton (N)	Pound (lb)
Energy	Joule (J)	Foot-pound (fp)

Conversions

To convert from one metric unit to another (from meters to kilometers, for example), we have only to look at the prefix. However, converting from metric to English or English to metric is more complicated. The conversion factors are given in Table A-3.

Example: The radius of the moon is 1738 km. What is this in miles? Table A-3 indicates that 1 mile equals 1.609 km, so:

$$1738 \text{ km} \times \frac{1 \text{ mile}}{1.609 \text{ km}} = 1080 \text{ miles}$$

Temperature Scales

In astronomy, as in most other sciences, temperatures are expressed on the Kelvin scale, although the centigrade (or Celsius) scale is also used. The Fahrenheit scale commonly used in the United States is not used in scientific work.

Temperatures on the Kelvin scale are measured from absolute zero, the temperature of an object that contains no extractable heat. In practice, no object can be as cold as absolute zero, although laboratory apparatuses have reached temperatures less than 10^{-6} K. The scale is named after the Scottish mathematical physicist William Thomson, Lord Kelvin (1824–1907).

The centigrade scale refers temperatures to the freezing point of water (0°C) and to the boiling point of water (100°C). One degree centigrade is 1/100th the temperature difference between the freezing and boiling points of water. Thus the prefix *centi*. The centigrade scale is also called the Celsius scale after its inventor, the Swedish astronomer Anders Celsius (1701–1744).

The Fahrenheit scale fixes the freezing point of water at 32°F and the boiling point at 212°F. Named after the German physicist Gabriel Daniel Fahrenheit (1686–1736), who made the first successful mercury thermometer in 1720, the Fahrenheit scale is used only in the United States.

It is easy to convert temperatures from one scale to another using the information given in Table A-4.

Powers of 10 Notation

Powers of 10 make writing very large numbers much simpler. For example, the nearest star is about 43,000,000,000,000 km from the sun. Writing this number as 4.3×10^{13} km is much easier.

Very small numbers can also be written with powers of 10. For example, the wavelength of visible light is about 0.0000005 m. In powers of 10 this becomes 5×10^{-7} m.

The powers of 10 used in this notation appear in the list below. The exponent tells us how to move the decimal point. If the exponent is positive, we move the decimal point to the right. If the exponent is negative, we move the decimal point to the left. Thus, 2×10^3 equals 2000.0, and 2×10^{-3} equals 0.002.

$$\vdots$$

10^5	=	100,000
10^4	=	10,000
10^3	=	1,000
10^2	=	100
10^1	=	10
10^0	=	1
10^{-1}	=	0.1
10^{-2}	=	0.01
10^{-3}	=	0.001
10^{-4}	=	0.0001

$$\vdots$$

If you use scientific notation in calculations, be sure you correctly enter numbers into your calculator. Not all calculators accept scientific notation, but those that can have a key labeled EXP, EEX, or perhaps EE that allows you to enter the exponent of ten. To enter a number such as 3×10^8 we press the keys 3 EXP 8. To enter a number with a negative exponent, we must use the change-sign key, usually labeled +/− or CHS. To enter the number 5.2×10^{-3} we press the keys 5.2 EXP +/− 3. Try a few examples.

To read a number in scientific notation from a calculator we must read the exponent separately. The number 3.1×10^{25} may appear in a calculator display as 3.1 25 or on some calculators as 3.1 10^{25}. Examine your calculator to determine how such numbers are displayed.

Astronomy, and science in general, is a way of learning about nature and understanding the universe we live in. To test hypotheses about how nature works, scientists use observations of nature. The tables that follow contain some of the basic observations that support our best understanding of the astronomical universe. Of course, these data are expressed in the form of numbers, not because science reduces all understanding to mere numbers, but because the struggle to understand nature is so demanding we must use every tool available. Quantitative thinking, reasoning mathematically, is one of the most powerful tools ever invented by the human brain. Thus the tables that follow are not nature reduced to mere number, but number supporting humanity's growing understanding of the natural world around us.

Table A-3 Conversion Factors

1 inch = 2.54 centimeters	1 centimeter = 0.394 inch
1 foot = 0.3048 meter	1 meter = 39.36 inches = 3.28 feet
1 mile = 1.6093 kilometers	1 kilometer = 0.6214 mile
1 slug = 14.594 kilograms	1 kilogram = 0.0685 slug
1 pound = 4.4482 newtons	1 newton = 0.2248 pound
1 foot-pound = 1.35582 joules	1 joule = 0.7376 foot-pound
1 horsepower = 745.7 joules/sec	1 joule/sec = 1 watt

Table A-4 Temperature Scales

	Kelvin (K)	Centigrade (°C)	Fahrenheit (°F)
Absolute zero	0 K	−273°C	−459°F
Freezing point of water	273 K	0°C	32°F
Boiling point of water	373 K	100°C	212°F

Conversions:

$$K = °C + 273$$
$$°C = \frac{5}{9}(°F - 32)$$
$$°F = \frac{9}{5}°C + 32$$

Table A-6 Constants

astronomical unit (AU)	$= 1.495979 \times 10^{11}$ m
parsec (pc)	$= 206{,}265$ AU
	$= 3.085678 \times 10^{16}$ m
	$= 3.261633$ ly
light-year (ly)	$= 9.46053 \times 10^{15}$ m
velocity of light (c)	$= 2.997925 \times 10^{8}$ m/sec
gravitational constant (G)	$= 6.67 \times 10^{-11}$ N·m^2/kg^2
mass of Earth ($M_\oplus$)	$= 5.976 \times 10^{24}$ kg
Earth equatorial radius ($R_\oplus$)	$= 6378.164$ km
mass of sun ($M_\odot$)	$= 1.989 \times 10^{30}$ kg
radius of sun ($R_\odot$)	$= 6.9599 \times 10^{8}$ m
solar luminosity ($L_\odot$)	$= 3.826 \times 10^{26}$ J/sec
mass of moon	$= 7.350 \times 10^{22}$ kg
radius of moon	$= 1738$ km
mass of H atom	$= 1.67352 \times 10^{-27}$ kg

Table A-5 Units Used in Astronomy

1 angstrom (Å)	$= 10^{-8}$ cm
	$= 10^{-10}$ m
1 astronomical unit (AU)	$= 1.495979 \times 10^{11}$ m
	$= 92.95582 \times 10^{6}$ miles
1 light-year (ly)	$= 6.3240 \times 10^{4}$ AU
	$= 9.46053 \times 10^{15}$ m
	$= 5.9 \times 10^{12}$ miles
1 parsec (pc)	$= 206{,}265$ AU
	$= 3.085678 \times 10^{16}$ m
	$= 3.261633$ ly
1 kiloparsec (kpc)	$= 1000$ pc
1 megaparsec (Mpc)	$= 1{,}000{,}000$ pc

Table A-7 The Nearest Stars

Name	Absolute Magnitude (M_v)	Distance (ly)	Spectral Type	Apparent Visual Magnitude (m_v)
Sun	4.83		G2	−26.8
Proxima Cen	15.45	4.28	M5	11.05
α Cen A	4.38	4.37	G2	0.1
B	5.76	4.37	K5	1.5
Barnard's Star	13.21	5.9	M5	9.5
Wolf 359	16.80	7.6	M6	13.5
Lalande 21185	10.42	8.1	M2	7.5
Sirius A	1.41	8.6	A1	−1.5
B	11.54	8.6	white dwarf	7.2
Luyten 726-8A	15.27	8.9	M5	12.5
B (UV Cet)	15.8	8.9	M6	13.0
Ross 154	13.3	9.4	M5	10.6
Ross 248	14.8	10.3	M6	12.2
ε Eri	6.13	10.7	K2	3.7
Luyten 789–6	14.6	10.8	M7	12.2
Ross 128	13.5	10.8	M5	11.1
61 CYG A	7.58	11.2	K5	5.2
B	8.39	11.2	K7	6.0
ε Ind	7.0	11.2	K5	4.7
Procyon A	2.64	11.4	F5	0.3
B	13.1	11.4	white dwarf	10.8
Σ 2398 A	11.15	11.5	M4	8.9
B	11.94	11.5	M5	9.7
Groombridge 34 A	10.32	11.6	M1	8.1
B	13.29	11.6	M6	11.0
Lacaille 9352	9.59	11.7	M2	7.4
τ Ceti	5.72	11.9	G8	3.5
BD + 5° 1668	11.98	12.2	M5	9.8
L 725–32	15.27	12.4	M5	11.5
Lacaille 8760	8.75	12.5	M0	6.7
Kapteyn's Star	10.85	12.7	M0	8.8
Kruger 60 A	11.87	12.8	M3	9.7
B	13.3	12.8	M4	11.2

Table A-8 Properties of Main-Sequence Stars

Spectral Type	Absolute Visual Magnitude (M_v)	Luminosity*	Temp. (K)	λ max (nm)	Mass*	Radius*	Average Density (g/cm³)
05	−5.8	501,000	40,000	72.4	40	17.8	0.01
B0	−4.1	20,000	28,000	100	18	7.4	0.1
B5	−1.1	790	15,000	190	6.4	3.8	0.2
A0	+0.7	79	9900	290	3.2	2.5	0.3
A5	+2.0	20	8500	340	2.1	1.7	0.6
F0	+2.6	6.3	7400	390	1.7	1.4	1.0
F5	+3.4	2.5	6600	440	1.3	1.2	1.1
G0	+4.4	1.3	6000	480	1.1	1.0	1.4
G5	+5.1	0.8	5500	520	0.9	0.9	1.6
K0	+5.9	0.4	4900	590	0.8	0.8	1.8
K5	+7.3	0.2	4100	700	0.7	0.7	2.4
M0	+9.0	0.1	3500	830	0.5	0.6	2.5
M5	+11.8	0.01	2800	1000	0.2	0.3	10.0
M8	+16	0.001	2400	1200	0.1	0.1	63

*Luminosity, mass, and radius are given in terms of the sun's luminosity, mass, and radius.

Table A-9 The Brightest Stars

Star	Name	Apparent Visual Magnitude (m_v)	Spectral Type	Absolute Visual Magnitude (M_v)	Distance (ly)
α CMa A	Sirius	−1.47	A1	1.4	8.7
α Car	Canopus	−0.72	F0	−3.1	98
α Cen	Rigil Kentaurus	−0.01	G2	4.4	4.3
α Boo	Arcturus	−0.06	K2	−0.3	36
α Lyr	Vega	0.04	A0	0.5	26.5
α Aur	Capella	0.05	G8	−0.6	45
β Ori A	Rigel	0.14	B8	−7.1	900
α CMi A	Procyon	0.37	F5	2.7	11.3
α Ori	Betelgeuse	0.41	M2	−5.6	520
α Eri	Achernar	0.51	B3	−2.3	118
β Cen AB	Hadar	0.63	B1	−5.2	490
α Aql	Altair	0.77	A7	2.2	16.5
α Tau A	Aldebaran	0.86	K5	−0.7	68
α Cru	Acrux	0.90	B2	−3.5	260
α Vir	Spica	0.91	B1	−3.3	220
α Sco A	Antares	0.92	M1	−5.1	520
α PsA	Fomalhaut	1.15	A3	2.0	22.6
β Gem	Pollux	1.16	K0	1.0	35
α Cyg	Deneb	1.26	A2	−7.1	1600
β Cru	Beta Crucis	1.28	B0.5	−4.6	490

Table A-10 Greatest Elongations of Mercury

Evening Sky	Morning Sky
Oct. 24, 1999	Dec. 3, 1999
Feb. 15, 2000	March 28, 2000
June 9, 2000	July 27, 2000
Oct. 6, 2000	Nov. 15, 2000
Jan. 28, 2001	March 11, 2001
May 22, 2001	July 9, 2001
Sept. 18, 2001	Oct. 29, 2001*
Jan. 11, 2002	Feb. 21, 2002
May 4, 2002	June 21, 2002
Sept. 1, 2002	Oct. 13, 2002*
Dec. 26, 2002	Feb. 4, 2003
April 16, 2003*	June 3, 2003
Aug. 14, 2003	Sept. 27, 2003*
Dec. 9, 2003	Jan. 17, 2004

*Most favorable elongations.

Table A-11 Greatest Elongations of Venus

Evening Sky	Morning Sky
Jan. 17, 2001	June 8, 2001
Aug. 22, 2002	Jan. 11, 2003
March 29, 2004	Aug. 17, 2004
Nov. 3, 2005	March 25, 2006
June 9, 2007	Oct. 28, 2007
Jan. 14, 2009	June 5, 2009
Aug. 20, 2010	Jan. 8, 2011

Table A-12 Meteor Showers

Shower	Dates	Hourly Rate	Radiant R.A.	Radiant Dec.	Associated Comet
Quadrantids	Jan. 2–4	30	15^h24^m	50°	
Lyrids	April 20–22	8	18^h4^m	33°	1861 I
η Aquarids	May 2–7	10	22^h24^m	0°	Halley?
δ Aquarids	July 26–31	15	22^h36^m	−10°	
Perseids	Aug. 10–14	40	3^h4^m	58°	1982 III
Orionids	Oct. 18–23	15	6^h20^m	15°	Halley?
Taurids	Nov. 1–7	8	3^h40^m	17°	Encke
Leonids	Nov. 14–19	6	10^h12^m	22°	1866 I Temp
Geminids	Dec. 10–13	50	7^h28^m	32°	

Table A-13 Properties of the Planets

PHYSICAL PROPERTIES (EARTH = $\oplus$)

Planet	Equatorial Radius (km)	Equatorial Radius ($\oplus = 1$)	Mass ($\oplus = 1$)	Average Density (g/cm^3)	Surface Gravity ($\oplus = 1$)	Escape Velocity (km/sec)	Sidereal Period of Rotation	Inclination of Equator to Orbit
Mercury	2439	0.382	0.0558	5.44	0.378	4.3	58.646^d	0°
Venus	6052	0.95	0.815	5.24	0.903	10.3	244.3^d	177°
Earth	6378	1.00	1.00	5.497	1.00	11.2	23^{h}56^{m}04.1^s	23°27′
Mars	3398	0.53	0.1075	3.94	0.379	5.0	24^{h}37^{m}22.6^s	23°59′
Jupiter	71,494	11.20	317.83	1.34	2.54	61	9^{h}50^{m}30^s	3°5′
Saturn	60,330	9.42	95.147	0.69	1.16	35.6	10^{h}13^{m}59^s	26°24′
Uranus	25,559	4.01	14.54	1.19	0.919	22	17^{h}14^m	97°55′
Neptune	24,750	3.93	17.23	1.66	1.19	25	16^{h}3^m	28°48′
Pluto	1151	0.18	0.0022	2.0	0.06	1.2	6^{d}9^{h}21^m	122°

ORBITAL PROPERTIES

Planet	(AU)	(10^6 km)	(y)	(days)	Average Orbital Velocity (km/sec)	Orbital Eccentricity	Inclination to Ecliptic
Mercury	0.3871	57.9	0.24084	87.969	47.89	0.2056	7°0′16″
Venus	0.7233	108.2	0.61515	224.68	35.03	0.0068	3°23′40″
Earth	1	149.6	1	365.26	29.79	0.0167	0°
Mars	1.5237	227.9	1.8808	686.95	24.13	0.0934	1°51′09″
Jupiter	5.2028	778.3	11.867	4334.3	13.06	0.0484	1°18′29″
Saturn	9.5388	1427.0	29.461	10,760	9.64	0.0560	2°29′17″
Uranus	19.18	2869.0	84.013	30,685	6.81	0.0461	0°46′23″
Neptune	30.0611	4497.1	164.793	60,189	5.43	0.0100	1°46′27″
Pluto	39.44	5900	247.7	90,465	4.74	0.2484	17°9′3″

Table A-14 Satellites of the Solar System

Planet	Satellite	Radius (km)	Distance from Planet (10³ km)	Orbital Period (days)	Orbital Eccentricity	Orbital Inclination
Earth	Moon	1738	384.4	27.322	0.055	5°8′43″
Mars	Phobos	14 × 12 × 10	9.38	0.3189	0.018	1°.0
	Deimos	8 × 6 × 5	23.5	1.262	0.002	2°.8
Jupiter	Metis	20	126	0.29	0.0	0°.0
	Adrastea	12 × 8 × 10	128	0.294	0.0	0°.0
	Amalthea	135 × 100 × 78	182	0.4982	0.003	0°.45
	Thebe	50	223	0.674	0.0	1°.3
	Io	1820	422	1.769	0.000	0°.3
	Europa	1565	671	3.551	0.000	0°.46
	Ganymede	2640	1071	7.155	0.002	0°.18
	Callisto	2420	1884	16.689	0.008	0°.25
	Leda	~8	11,110	240	0.146	26°.7
	Himalia	~85	11,470	250.6	0.158	27°.6
	Lysithea	~20	11,710	260	0.12	29°
	Elara	~30	11,740	260.1	0.207	24°.8
	Ananke	15	21,200	631	0.169	147°
	Carme	22	22,350	692	0.207	163°
	Pasiphae	35	23,300	735	0.40	147°
	Sinope	20	23,700	758	0.275	156°
Saturn	Pan	10	133.570	0.574	0.000	0°
	Atlas	20 × 15 × 15	137.7	0.601	0.002	0°.3
	Prometheus	70 × 40 × 50	139.4	0.613	0.003	0°.0
	Pandora	55 × 35 × 50	141.7	0.629	0.004	0°.05
	Epimetheus	70 × 50 × 50	151.42	0.694	0.009	0°.34
	Janus	110 × 80 × 100	151.47	0.695	0.007	0°.14
	Mimas	196	185.54	0.942	0.020	1°.5
	Enceladus	250	238.04	1.370	0.004	0°.0
	Tethys	530	294.67	1.888	0.000	1°.1
	Calypso	17 × 11 × 12	294.67	1.888	0.0	~1°.?
	Telesto	12	294.67	1.888	0.0	~1°?
	Dione	560	377	2.737	0.002	0°.0
	Helene	20 × 15 × 15	377	2.74	0.005	0°.15
	Rhea	765	527	4.518	0.001	0°.4
	Titan	2575	1222	15.94	0.029	0°.3
	Hyperion	205 × 130 × 110	1484`	21.28	0.104	~0°.5
	Iapetus	720	3562	79.33	0.028	14°.72
	Phoebe	110	12,930	550.4	0.163	150°
Uranus	Cordelia	20	49.8	0.3333	~0	~0°
	Ophelia	15	53.8	0.375	~0	~0°
	Bianca	25	59.1	0.433	~0	~0°
	Cressida	30	61.8	0.462	~0	~0°
	Desdemona	30	62.7	0.475	~0	~0°
	Juliet	40	64.4	0.492	~0	~0°
	Portia	55	66.1	0.512	~0	~0°
	Rosalind	30	69.9	0.558	~0	~0°
	Belinda	30	75.2	0.621	~0	~0°

Table A-14 (continued)

Planet	Satellite	Radius (km)	Distance from Planet (10^3 km)	Orbital Period (days)	Orbital Eccentricity	Orbital Inclination
Uranus	Puck	85 ± 5	85.9	0.762	~0	~0°
(continued)	Miranda	242 ± 5	129.9	1.414	0.017	3°.4
	Ariel	580 ± 5	190.9	2.520	0.003	0°
	Umbriel	595 ± 10	266.0	4.144	0.003	0°
	Titania	805 ± 5	436.3	8.706	0.002	0°
	Oberon	775 ± 10	583.4	13.463	0.001	0°
Neptune	Naiad	30	48.2	0.296	~0	~0°
	Thalassa	40	50.0	0.312	~0	~0°
	Despina	90	52.5	0.333	~0	~0°
	Galatea	75	62.0	0.396	~0	~0°
	Larissa	95	73.6	0.554	~0	~0°
	Proteus	205	117.6	1.121	~0	~0°
	Triton	1352	354.59	5.875	0.00	160°
	Nereid	170	5588.6	360.125	0.76	27.7°
Pluto	Charon	596	19.7	6.38718	~0	122°

Table A-15 The Greek Alphabet

A, α	alpha	H, η	eta	N, ν	nu	T, τ	tau
B, β	beta	Θ, θ	theta	Ξ, ξ	xi	Y, υ	upsilon
Γ, γ	gamma	I, ι	iota	O, o	omicron	Φ, ϕ	phi
Δ, δ	delta	K, κ	kappa	Π, π	pi	X, χ	chi
E, ε	epsilon	Λ, λ	lambda	P, ρ	rho	Ψ, ψ	psi
Z, ζ	zeta	M, μ	mu	Σ, σ	sigma	Ω, ω	omega

Table A-16 Periodic Table of the Elements

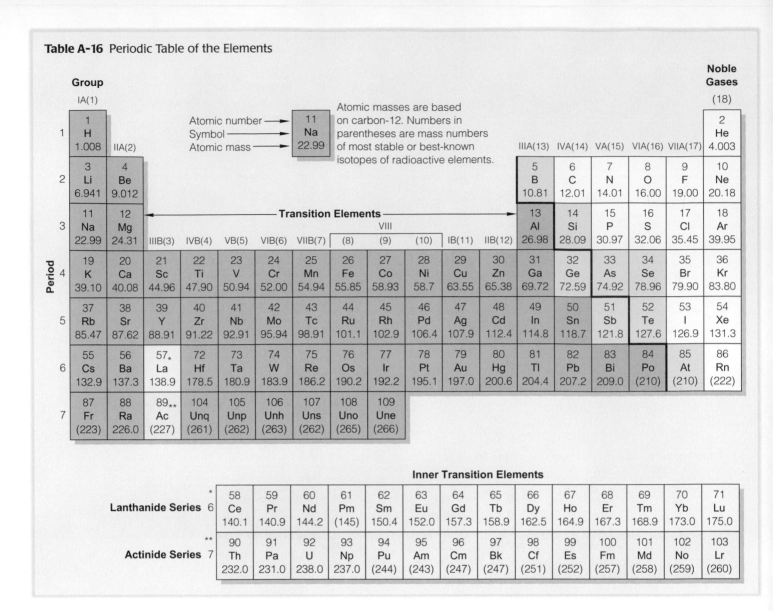

Group

Atomic masses are based on carbon-12. Numbers in parentheses are mass numbers of most stable or best-known isotopes of radioactive elements.

Atomic number → 11
Symbol → Na
Atomic mass → 22.99

Group																	Noble Gases
IA(1)												IIIA(13)	IVA(14)	VA(15)	VIA(16)	VIIA(17)	(18)
1 H 1.008	IIA(2)																2 He 4.003
3 Li 6.941	4 Be 9.012											5 B 10.81	6 C 12.01	7 N 14.01	8 O 16.00	9 F 19.00	10 Ne 20.18
11 Na 22.99	12 Mg 24.31	IIIB(3)	IVB(4)	VB(5)	VIB(6)	VIIB(7)	(8)	(9)	(10)	IB(11)	IIB(12)	13 Al 26.98	14 Si 28.09	15 P 30.97	16 S 32.06	17 Cl 35.45	18 Ar 39.95
19 K 39.10	20 Ca 40.08	21 Sc 44.96	22 Ti 47.90	23 V 50.94	24 Cr 52.00	25 Mn 54.94	26 Fe 55.85	27 Co 58.93	28 Ni 58.7	29 Cu 63.55	30 Zn 65.38	31 Ga 69.72	32 Ge 72.59	33 As 74.92	34 Se 78.96	35 Br 79.90	36 Kr 83.80
37 Rb 85.47	38 Sr 87.62	39 Y 88.91	40 Zr 91.22	41 Nb 92.91	42 Mo 95.94	43 Tc 98.91	44 Ru 101.1	45 Rh 102.9	46 Pd 106.4	47 Ag 107.9	48 Cd 112.4	49 In 114.8	50 Sn 118.7	51 Sb 121.8	52 Te 127.6	53 I 126.9	54 Xe 131.3
55 Cs 132.9	56 Ba 137.3	57* La 138.9	72 Hf 178.5	73 Ta 180.9	74 W 183.9	75 Re 186.2	76 Os 190.2	77 Ir 192.2	78 Pt 195.1	79 Au 197.0	80 Hg 200.6	81 Tl 204.4	82 Pb 207.2	83 Bi 209.0	84 Po (210)	85 At (210)	86 Rn (222)
87 Fr (223)	88 Ra 226.0	89** Ac (227)	104 Unq (261)	105 Unp (262)	106 Unh (263)	107 Uns (262)	108 Uno (265)	109 Une (266)									

Transition Elements

VIII (8) (9) (10)

Period

Inner Transition Elements

Lanthanide Series 6*	58 Ce 140.1	59 Pr 140.9	60 Nd 144.2	61 Pm (145)	62 Sm 150.4	63 Eu 152.0	64 Gd 157.3	65 Tb 158.9	66 Dy 162.5	67 Ho 164.9	68 Er 167.3	69 Tm 168.9	70 Yb 173.0	71 Lu 175.0
Actinide Series 7**	90 Th 232.0	91 Pa 231.0	92 U 238.0	93 Np 237.0	94 Pu (244)	95 Am (243)	96 Cm (247)	97 Bk (247)	98 Cf (251)	99 Es (252)	100 Fm (257)	101 Md (258)	102 No (259)	103 Lr (260)

APPENDIX B

Observing the Sky

Observing the sky with the naked eye is of no more importance to modern astronomy than picking up pretty pebbles is to modern geology. But the sky is a natural wonder unimaginably bigger than the Grand Canyon, the Rocky Mountains, or any other natural wonder that tourists visit every year. To neglect the beauty of the sky is equivalent to geologists neglecting the beauty of the minerals they study. This supplement is meant to act as a tourist's guide to the sky. We analyzed the universe in the regular chapters, but here we will admire it.

The brighter stars in the sky are visible even from the centers of cities with their air and light pollution. But in the countryside only a few miles beyond the cities, the night sky is a velvety blackness strewn with thousands of glittering stars. From a wilderness location, far from the city's glare, and especially from high mountains, the night sky is spectacular.

Using Star Charts

The constellations are a fascinating cultural heritage of our planet, but they are sometimes a bit difficult to learn because of Earth's motion. The constellations above the horizon change with the time of night and the seasons.

Because Earth rotates eastward, the sky appears to rotate around us westward. A constellation visible in the southern sky soon after sunset will appear to move westward, and in a few hours it will disappear below the horizon. Other constellations will rise in the east, so the sky changes gradually through the night.

In addition, Earth's orbital motion makes the sun appear to move eastward among the stars. Each day the sun moves about twice its own diameter eastward along the ecliptic, and each night at sunset, the constellations are about 1° farther toward the west.

Orion, for instance, is visible in the southern sky in January, but as the days pass, the sun moves closer to Orion. By March, Orion is difficult to see in the southwest sky soon after sunset. By June, the sun is so close to Orion it sets with the sun and is invisible. Not until late July is the sun far enough past Orion for

Figure B-1
To use the star charts in this book, select the appropriate chart for the date and time. Hold it overhead, and turn it until the direction at the bottom of the chart is the same as the direction you are facing.

the constellation to become visible rising in the eastern sky just before dawn.

Clearly Earth's rotation and orbital motion require that we use more than one star chart to map the sky. Which chart we select depends on the time of night and the time of year. The charts given here show the evening sky for each month.

The constellations we see in the sky also depend on where we are, so two sets of charts are provided. Those labeled Northern Hemisphere Sky show the sky as seen from a latitude representative of most of the United States. People living in Earth's southern hemisphere should use the charts labeled Southern Hemisphere Sky. Wherever you live, look at both sets of charts and notice the differences.

To use the charts, select the chart for the appropriate month and hold it overhead as shown in Figure B-1. If you face south, turn the chart until the words *Southern Horizon* are at the bottom. If you face other directions, turn the chart appropriately.

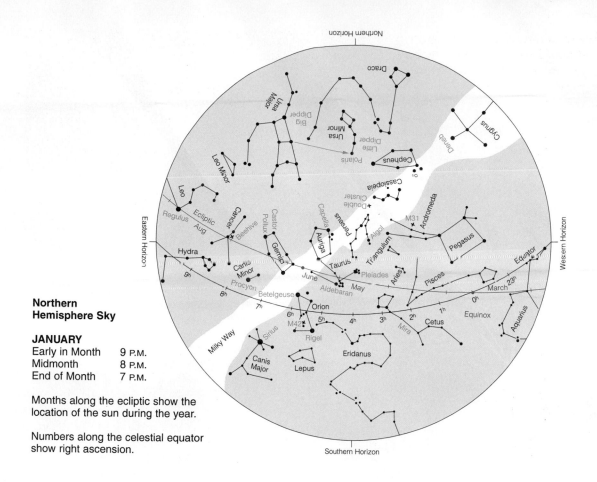

Northern Hemisphere Sky

JANUARY

Early in Month   9 P.M.
Midmonth   8 P.M.
End of Month   7 P.M.

Months along the ecliptic show the location of the sun during the year.

Numbers along the celestial equator show right ascension.

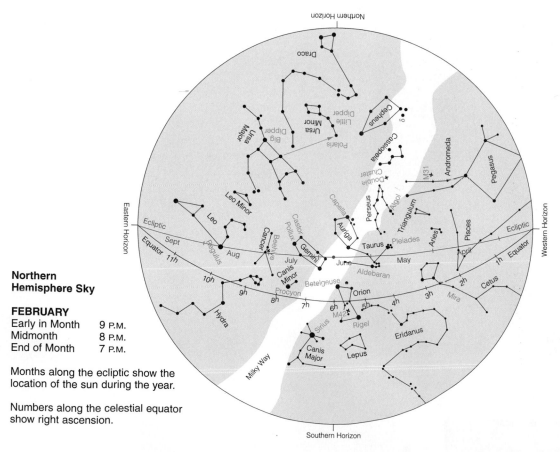

Northern Hemisphere Sky

FEBRUARY

Early in Month   9 P.M.
Midmonth   8 P.M.
End of Month   7 P.M.

Months along the ecliptic show the location of the sun during the year.

Numbers along the celestial equator show right ascension.

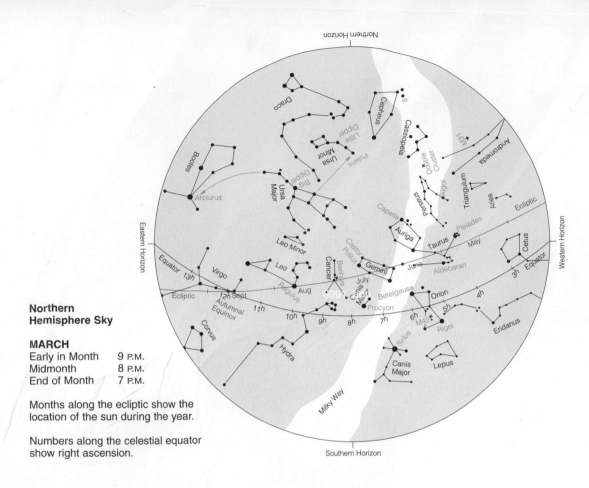

Northern Hemisphere Sky

MARCH

Early in Month	9 P.M.
Midmonth	8 P.M.
End of Month	7 P.M.

Months along the ecliptic show the location of the sun during the year.

Numbers along the celestial equator show right ascension.

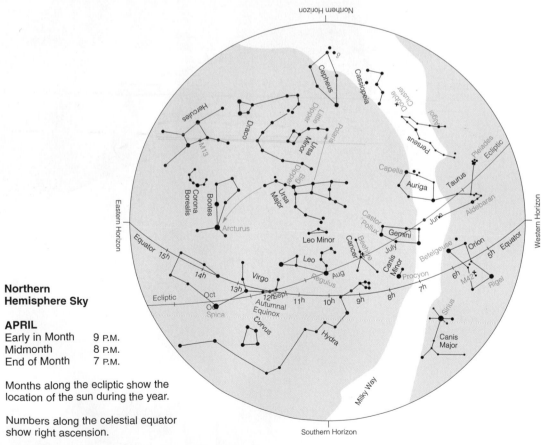

Northern Hemisphere Sky

APRIL

Early in Month	9 P.M.
Midmonth	8 P.M.
End of Month	7 P.M.

Months along the ecliptic show the location of the sun during the year.

Numbers along the celestial equator show right ascension.

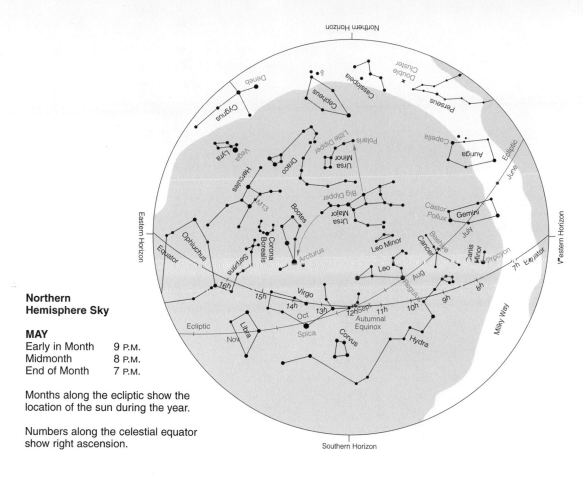

Northern Hemisphere Sky

MAY

Early in Month	9 P.M.
Midmonth	8 P.M.
End of Month	7 P.M.

Months along the ecliptic show the location of the sun during the year.

Numbers along the celestial equator show right ascension.

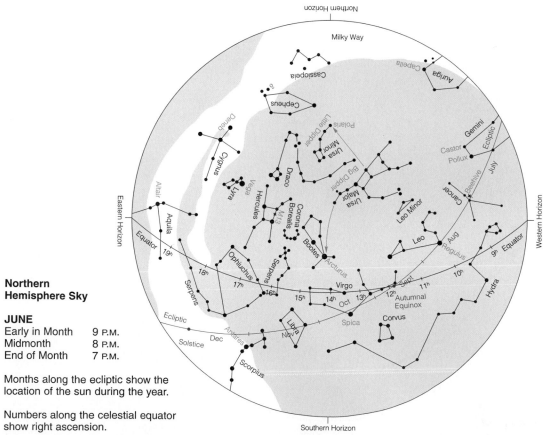

Northern Hemisphere Sky

JUNE

Early in Month	9 P.M.
Midmonth	8 P.M.
End of Month	7 P.M.

Months along the ecliptic show the location of the sun during the year.

Numbers along the celestial equator show right ascension.

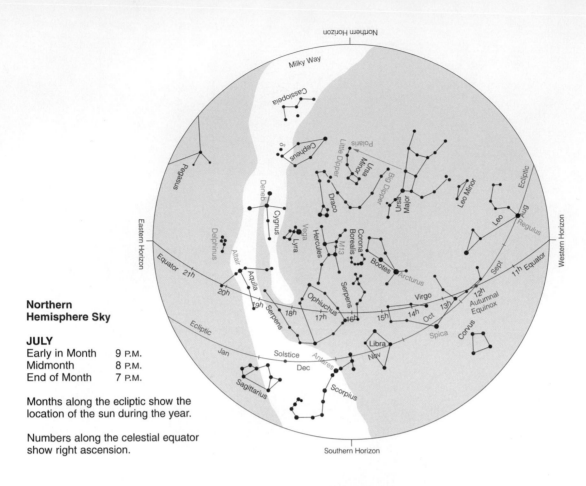

Northern Hemisphere Sky

JULY

Early in Month	9 P.M.
Midmonth	8 P.M.
End of Month	7 P.M.

Months along the ecliptic show the location of the sun during the year.

Numbers along the celestial equator show right ascension.

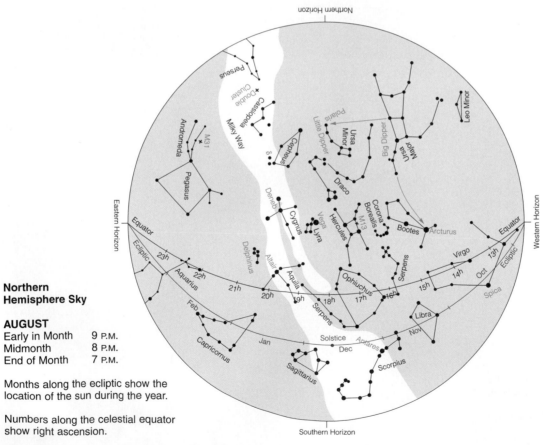

Northern Hemisphere Sky

AUGUST

Early in Month	9 P.M.
Midmonth	8 P.M.
End of Month	7 P.M.

Months along the ecliptic show the location of the sun during the year.

Numbers along the celestial equator show right ascension.

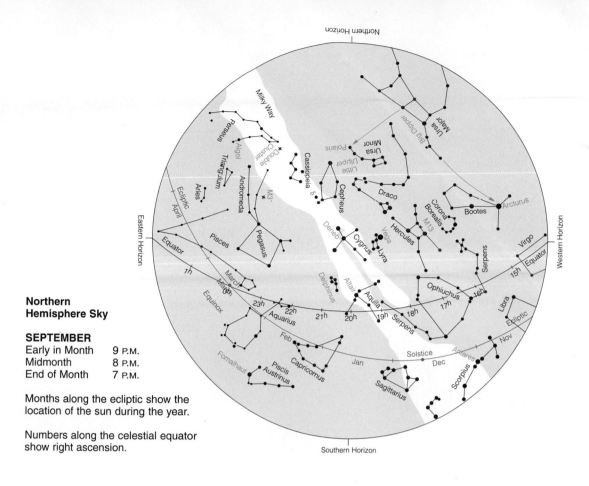

Northern Hemisphere Sky

SEPTEMBER

Early in Month	9 P.M.
Midmonth	8 P.M.
End of Month	7 P.M.

Months along the ecliptic show the location of the sun during the year.

Numbers along the celestial equator show right ascension.

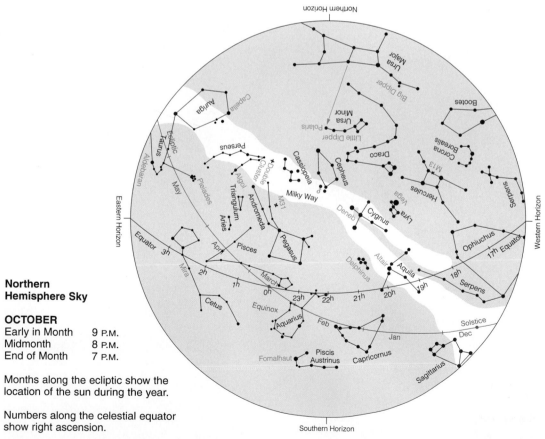

Northern Hemisphere Sky

OCTOBER

Early in Month	9 P.M.
Midmonth	8 P.M.
End of Month	7 P.M.

Months along the ecliptic show the location of the sun during the year.

Numbers along the celestial equator show right ascension.

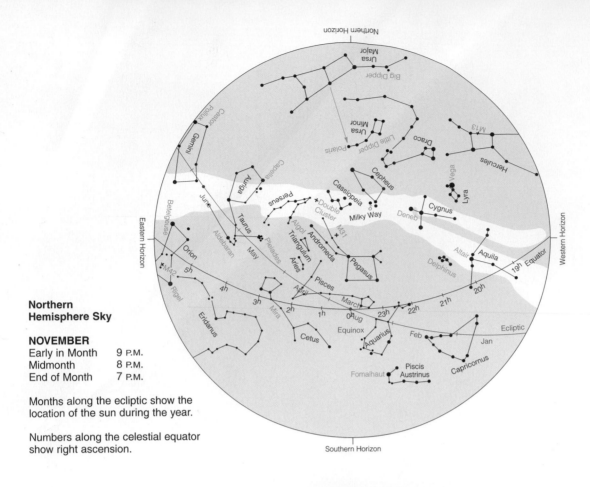

Northern Hemisphere Sky

NOVEMBER

Early in Month   9 P.M.
Midmonth   8 P.M.
End of Month   7 P.M.

Months along the ecliptic show the location of the sun during the year.

Numbers along the celestial equator show right ascension.

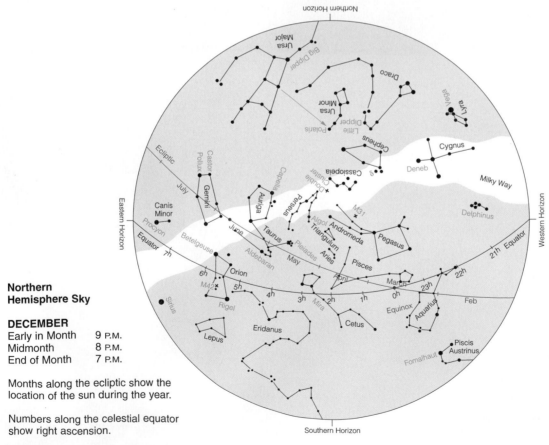

Northern Hemisphere Sky

DECEMBER

Early in Month   9 P.M.
Midmonth   8 P.M.
End of Month   7 P.M.

Months along the ecliptic show the location of the sun during the year.

Numbers along the celestial equator show right ascension.

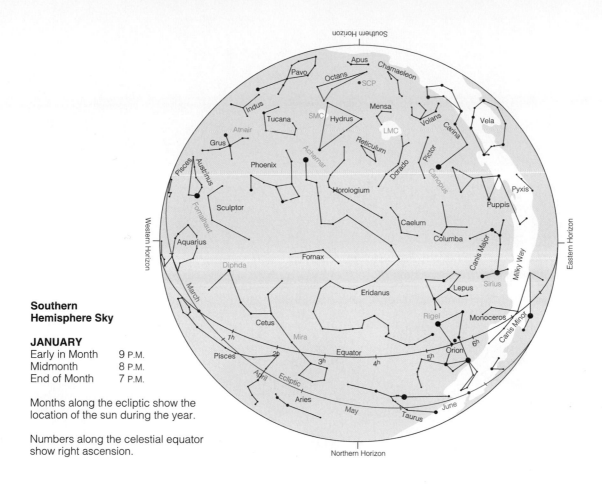

Southern Hemisphere Sky

JANUARY

Early in Month	9 P.M.
Midmonth	8 P.M.
End of Month	7 P.M.

Months along the ecliptic show the location of the sun during the year.

Numbers along the celestial equator show right ascension.

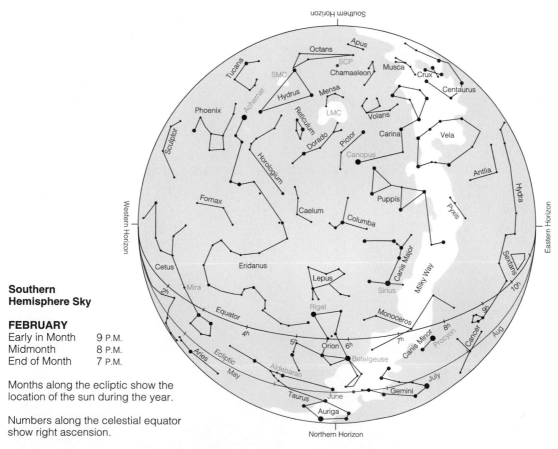

Southern Hemisphere Sky

FEBRUARY

Early in Month	9 P.M.
Midmonth	8 P.M.
End of Month	7 P.M.

Months along the ecliptic show the location of the sun during the year.

Numbers along the celestial equator show right ascension.

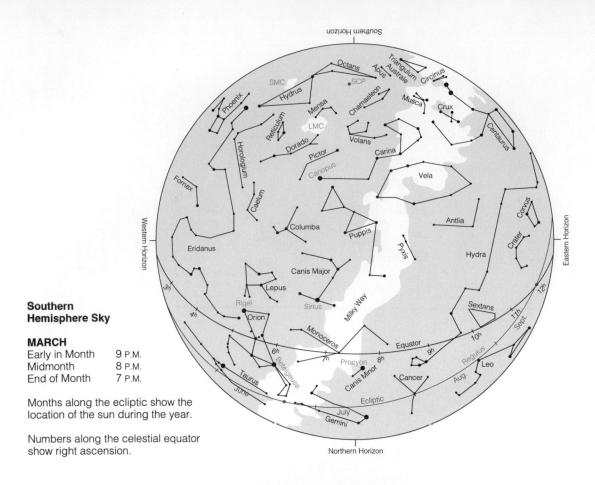

Southern Hemisphere Sky

MARCH

Early in Month	9 P.M.
Midmonth	8 P.M.
End of Month	7 P.M.

Months along the ecliptic show the location of the sun during the year.

Numbers along the celestial equator show right ascension.

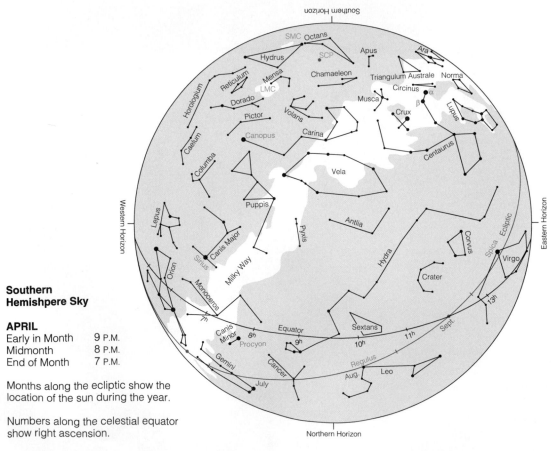

Southern Hemishpere Sky

APRIL

Early in Month	9 P.M.
Midmonth	8 P.M.
End of Month	7 P.M.

Months along the ecliptic show the location of the sun during the year.

Numbers along the celestial equator show right ascension.

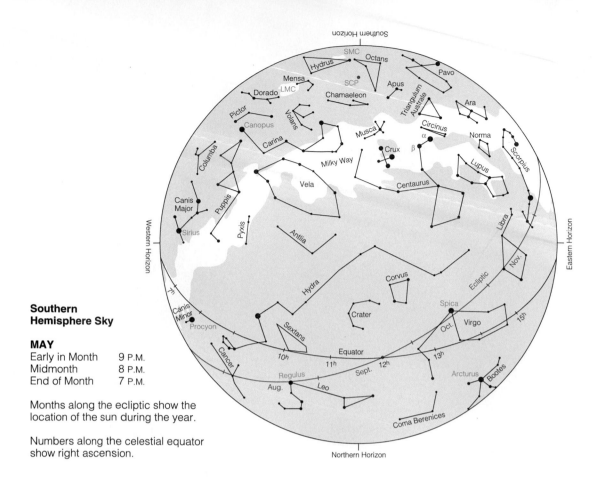

Southern Hemisphere Sky

MAY

Early in Month	9 P.M.
Midmonth	8 P.M.
End of Month	7 P.M.

Months along the ecliptic show the location of the sun during the year.

Numbers along the celestial equator show right ascension.

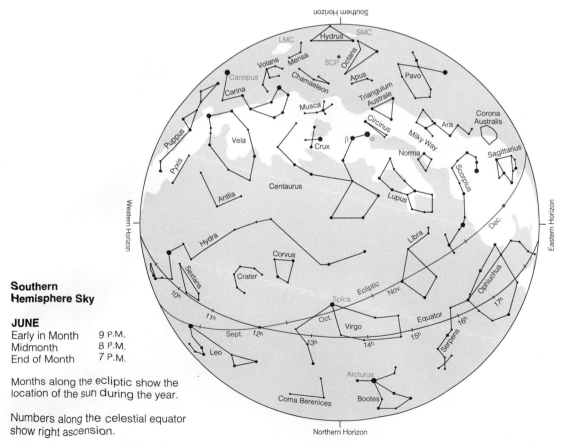

Southern Hemisphere Sky

JUNE

Early in Month	9 P.M.
Midmonth	8 P.M.
End of Month	7 P.M.

Months along the ecliptic show the location of the sun during the year.

Numbers along the celestial equator show right ascension.

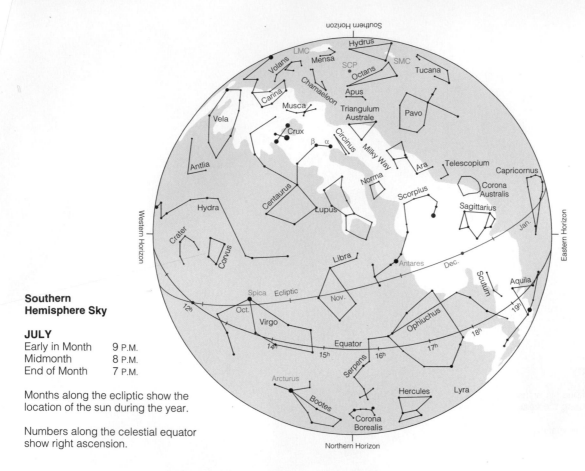

Southern Hemisphere Sky

JULY

Early in Month   9 P.M.
Midmonth   8 P.M.
End of Month   7 P.M.

Months along the ecliptic show the location of the sun during the year.

Numbers along the celestial equator show right ascension.

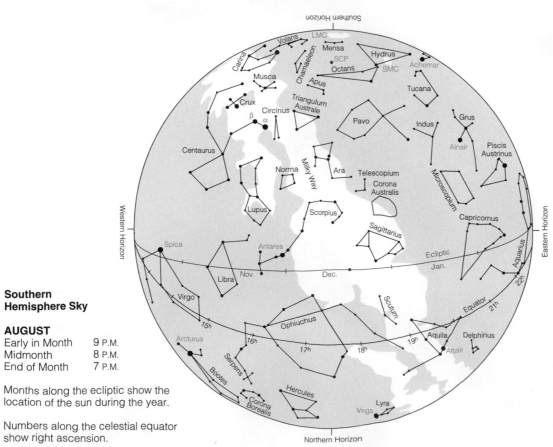

Southern Hemisphere Sky

AUGUST

Early in Month   9 P.M.
Midmonth   8 P.M.
End of Month   7 P.M.

Months along the ecliptic show the location of the sun during the year.

Numbers along the celestial equator show right ascension.

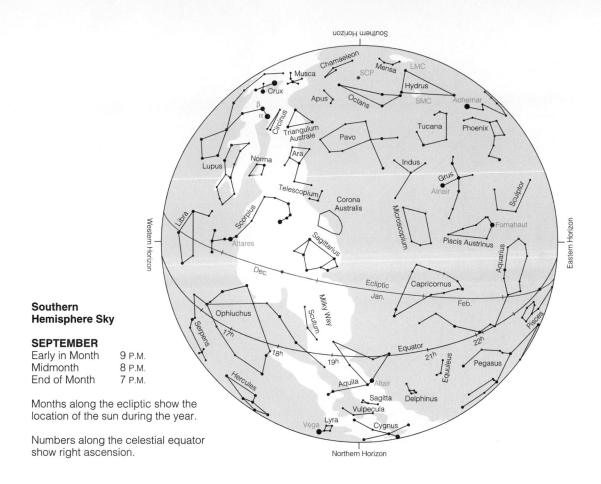

Southern Hemisphere Sky

SEPTEMBER

Early in Month	9 P.M.
Midmonth	8 P.M.
End of Month	7 P.M.

Months along the ecliptic show the location of the sun during the year.

Numbers along the celestial equator show right ascension.

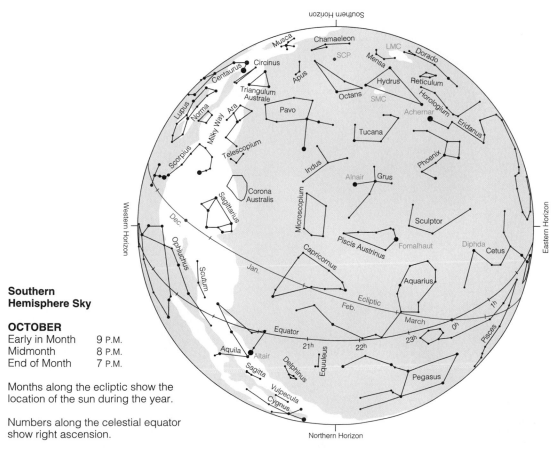

Southern Hemisphere Sky

OCTOBER

Early in Month	9 P.M.
Midmonth	8 P.M.
End of Month	7 P.M.

Months along the ecliptic show the location of the sun during the year.

Numbers along the celestial equator show right ascension.

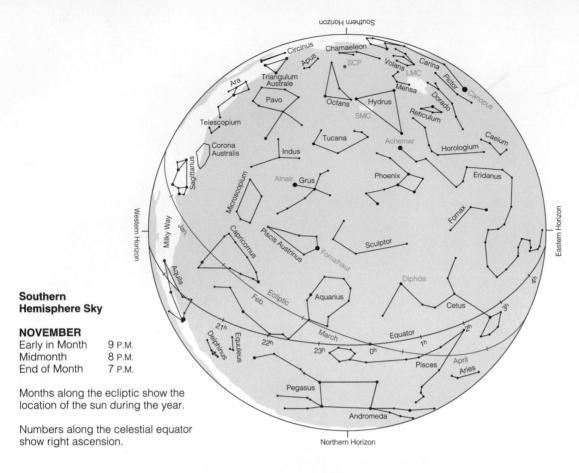

Southern Hemisphere Sky

NOVEMBER

Early in Month	9 P.M.
Midmonth	8 P.M.
End of Month	7 P.M.

Months along the ecliptic show the location of the sun during the year.

Numbers along the celestial equator show right ascension.

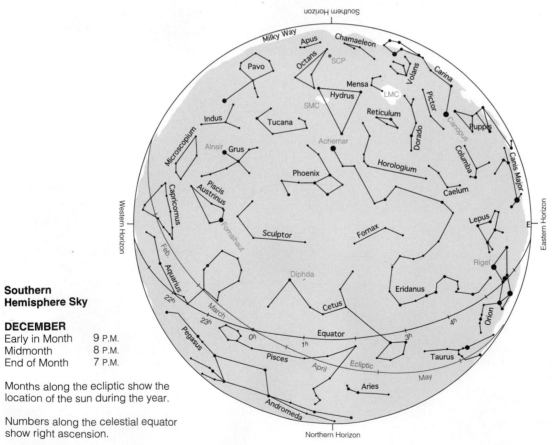

Southern Hemisphere Sky

DECEMBER

Early in Month	9 P.M.
Midmonth	8 P.M.
End of Month	7 P.M.

Months along the ecliptic show the location of the sun during the year.

Numbers along the celestial equator show right ascension.

Glossary

Numbers in parentheses refer to the page where the term is first discussed in the text.

absolute visual magnitude (M_v) (130) Intrinsic brightness of a star. The apparent visual magnitude the star would have if it were 10 pc away.

absolute zero (95) The theoretical lowest possible temperature at which a material contains no extractable heat energy. Zero on the Kelvin temperature scale.

absorption line (98) A dark line in a spectrum. Produced by the absence of photons absorbed by atoms or molecules.

absorption spectrum (dark-line spectrum) (98) A spectrum that contains absorption lines.

accretion (330) The sticking together of solid particles to produce a larger particle.

accretion disk (200) The whirling disk of gas that forms around a compact object such as a white dwarf, neutron star, or black hole as matter is drawn in.

achondrite (ā·kŏn′drīt) (400) Stony meteorite containing no chondrules or volatiles.

achromatic lens (70) A telescope lens composed of two lenses ground from different kinds of glass and designed to bring two selected colors to the same focus and correct for chromatic aberration.

active galactic nuclei (AGN) (272) The centers of active galaxies that are emitting large amounts of excess energy. See also **active galaxy.**

active galaxy (272) A galaxy whose center emits large amounts of excess energy, often in the form of radio emission. Active galaxies are suspected of having massive black holes in their centers into which matter is flowing.

active optics (77) Thin telescope mirrors that are controlled by computers to maintain proper shape as the telescope moves.

active region (118) A magnetic region on the solar surface that includes sunspots, prominences, flares, etc.

adaptive optics (78) A computer-controlled optical system used to partially correct for seeing in an astronomical telescope.

albedo (344) The ratio of the light reflected from an object divided by the light that hits the object. Albedo equals 0 for perfectly black and 1 for perfectly white.

alt-azimuth mounting (72) A telescope mounting that allows the telescope to move in altitude (perpendicular to the horizon) and in azimuth (parallel to the horizon). See also **equatorial mounting.**

amino acid (ŭ·mē′nō) (420) Carbon-chain molecule that is the building block of protein.

angstrom (Å) (ăng′strŭm) (68) A unit of distance. 1 Å = 10^{-10} m. Commonly used to measure the wavelength of light.

angular diameter (16) The angle formed by lines extending from the observer to opposite sides of an object.

angular distance (15) The angle formed by lines extending from the observer to two locations.

angular momentum (200) A measure of the tendency of a rotating body to continue rotating. Mathematically, the product of mass, velocity, and radius.

annular eclipse (32) A solar eclipse in which the solar photosphere appears around the edge of the moon in a bright ring, or annulus. The corona, chromosphere, and prominences cannot be seen.

anorthosite (ăn·ôr′thŭ·sīt) (344) Rock of aluminum and calcium silicates found in the lunar highlands.

antimatter (301) Matter composed of antiparticles, which upon colliding with a matching particle of normal matter annihilate and convert the mass of both particles into energy. The antiproton is the antiparticle of the proton, and the positron is the antiparticle of the electron.

aphelion (ŭ·fē′le·ŭn) (25) The orbital point of greatest distance from the sun.

apogee (34) The point farthest from Earth in the orbit of a body circling Earth.

apparent visual magnitude (m_v) (13) The brightness of a star as seen by human eyes on Earth.

association (155) Group of widely scattered stars (10 to 100) moving together through space. Not gravitationally bound into clusters.

asterism (11) A named grouping of stars that is not one of the recognized constellations, such as the Big Dipper or the Pleiades.

asteroid (324) Small, rocky world. Most asteroids lie between Mars and Jupiter in the asteroid belt.

astronomical unit (AU) (4) Average distance from Earth to the sun; 1.5×10^8 km, or 93×10^6 mi.

atmospheric window (69) Wavelength region in which our atmosphere is transparent—at visual, infrared, and radio wavelengths.

aurora (ô·rôr′ŭ) (122) The glowing light display that results when a planet's magnetic field guides charged particles toward the north and south magnetic poles, where they strike the upper atmosphere and excite atoms to emit photons.

autumnal equinox (23) The point on the celestial sphere where the sun crosses the celestial equator going southward. Also, the time when the sun reaches this point and autumn begins in the northern hemisphere—about September 22.

Babcock model (119) A model of the sun's magnetic cycle in which the differential rotation of the sun winds up and tangles the solar magnetic field in a 22-year cycle. This is thought to be responsible for the 11-year sunspot cycle.

Balmer series (99) Spectral lines in the visible and near-ultraviolet spectrum of hydrogen produced by transitions whose lowest energy level is the second.

barred spiral galaxy (253) A spiral galaxy with an elongated nucleus resembling a bar from which the arms originate.

basalt (340) Dark igneous rock characteristic of solidified lava.

belt–zone circulation (372) The atmospheric circulation typical of Jovian planets. Dark belts and bright zones encircle the planet parallel to its equator.

big bang (298) The theory that the universe began in a high-density, high-temperature state from which the expanding universe of galaxies formed.

binary stars (136) Pairs of stars that orbit around their common center of mass.

binding energy (93) The energy needed to pull an electron away from its atom.

bipolar flow (158) Jets of gas flowing away from a central object in opposite directions. Usually applied to protostars.

birth line (157) In the H-R diagram, the line above the main sequence where protostars first become visible.

black body radiation (95) Radiation emitted by a hypothetical perfect radiator. The spectrum is continuous, and the wavelength of maximum emission depends on the body's temperature.

black dwarf (189) The end state of a white dwarf that has cooled to low temperature.

black hole (214) A mass that has collapsed to such a small volume that its gravity prevents the escape of all radiation. Also, the volume of space from which radiation may not escape.

blazars See **BL Lac objects.**

BL Lac objects (278) Objects that resemble quasars. Thought to be highly luminous cores of distant active galaxies.

blue shift (104) A Doppler shift toward shorter wavelengths caused by a velocity of approach.

Bok globule (158) Small, dark cloud only about 1 ly in diameter that contains 10 to 1000 solar masses of gas and dust. Believed to be related to star formation.

breccia (brĕch′ē·ŭ) (344) Rock composed of fragments of earlier rocks bonded together.

bright-line spectrum See **emission spectrum.**

brown dwarf (171) A star whose mass is too low to ignite nuclear fusion. Heated by contraction.

calibration (228) The establishment of the relationship between a parameter that is easily determined and a parameter that is more difficult to determine. For example, the periods of Cepheid variables have been calibrated to reveal absolute magnitudes, which can then be used to find distance. Thus astronomers say Cepheids have been calibrated as distance indicators.

Cambrian period (kăm′brē·ŭn) (421) A geological period 0.6 to 0.5 billion years ago during which life on Earth became diverse and complex. Cambrian rocks contain the oldest easily identifiable fossils.

capture hypothesis (349) The theory that Earth's moon formed elsewhere in the solar nebula and was later captured by Earth.

carbonaceous chondrite (kär·bŭ·nā′-shŭs kŏn′drīt) (400) Stony meteorite that contains both chondrules and volatiles. These chondrites may be the least-altered remains of the solar nebula still present in the solar system.

carbon detonation (191) The explosive ignition of carbon burning in some giant stars. A possible cause of some supernova explosions.

carbon–nitrogen–oxygen (CNO) cycle (162) A series of nuclear reactions that use carbon as a catalyst to combine four hydrogen atoms to make one helium atom plus energy. Effective in stars more massive than the sun.

Cassegrain telescope (kăs′ŭ·grăn) (71) A reflecting telescope in which the secondary mirror reflects light back down the tube through a hole in the center of the objective mirror.

CCD See **charge-coupled device.**

celestial equator (15) The imaginary line around the sky directly above Earth's equator.

celestial pole (north or south) (14) One of the two points on the celestial sphere directly above Earth's poles.

celestial sphere (14) An imaginary sphere of very large radius surrounding Earth and to which the planets, stars, sun, and moon seem to be attached.

center of mass (137) The balance point of a body or system of masses. The point about which a body or system of masses rotates in the absence of external forces.

Cepheid (sĕ·fē′ĭd) (225) Variable star with a period of 60 days. Period of variation related to luminosity.

Chandrasekhar limit (shăn′drä·sä′-kär) (189) The maximum mass of a white dwarf, about 1.4 solar masses. A white dwarf of greater mass cannot support itself and will collapse.

charge-coupled device (CCD) (79) An electronic device consisting of a large array of light-sensitive elements used to record very faint images.

chemical evolution (422) The chemical process that led to the growth of complex molecules on primitive Earth. This did not involve the reproduction of molecules.

chondrite (kŏn′drīt) (400) A stony meteorite that contains chondrules.

chondrule (kŏn′dro͞ol) (400) Round, glassy body found in some stony meteorites. Believed to have solidified very quickly from molten drops of silicate material.

chromatic aberration (krō·măt′ĭk) (70) A distortion found in refracting telescopes because lenses focus different colors at slightly different distances. Images are consequently surrounded by color fringes.

chromosphere (krō′mŭ·sfîr) (32) Bright gases just above the photosphere of the sun. Responsible for the emission lines in the flash spectrum.

circular velocity (62) The velocity an object needs to stay in orbit around another object.

circumpolar constellation (north or south) (17) A constellation so close to one of the celestial poles that it never sets or never rises as seen from a particular latitude.

closed universe (296) A model universe in which the average density is great enough to stop the expansion and make the universe contract.

cluster method (260) The method of determining the masses of galaxies based on the motions of galaxies in a cluster.

CNO cycle See **carbon–nitrogen–oxygen cycle.**

coma (406) The glowing head of a comet.

comet (325) One of the small, icy bodies that orbit the sun and produce tails of gas and dust when they approach the sun.

comparative planetology (338) The study of planets in relation to one another.

comparison spectrum (81) A spectrum of known spectral lines used to identify unknown wavelengths in an object's spectrum.

condensation (329) The growth of a particle by addition of material from surrounding gas, atom by atom.

condensation hypothesis (348) The theory that Earth and the moon condensed from the same cloud of material in roughly their present orbital relationship.

condensation sequence (329) The sequence in which different materials condense from the solar nebula as we move outward from the sun.

conservation of energy (167) One of the basic laws of stellar structure. The amount of energy flowing out of the top of a shell must equal the amount coming in at the bottom plus whatever energy is generated within the shell.

conservation of mass (167) One of the basic laws of stellar structure. The total mass of the star must equal the sum of the masses of the shells, and the mass must be distributed smoothly through the star.

constellation (10) One of the stellar patterns identified by name, usually of mythological gods, people, animals, or objects. Also, the region of the sky containing that star pattern.

continuous spectrum (97) A spectrum in which there are no absorption or emission lines.

convection (112) Circulation in a fluid driven by heat. Hot material rises and cool material sinks.

corona (32, 232, 353) On the sun, the faint outer atmosphere composed of low-density, high-temperature gas. On Venus, large, round geological faults in the crust caused by the intrusion of magma below the crust.

coronal hole (122) An area of the solar surface that is dark at X-ray wavelengths. Thought to be associated with divergent magnetic fields and the source of the solar wind.

cosmological constant (307) A constant in Einstein's equations of space and time that represents a force of repulsion.

cosmological principle (293) The assumption that any observer in any galaxy sees the same general features of the universe.

cosmology (291) The study of the nature, origin, and evolution of the universe.

Coulomb barrier (ko͞o·lôn) (162) The electrostatic force of repulsion between bodies of like charge. Commonly applied to atomic nuclei.

Coulomb force (93) The electrostatic force of repulsion or attraction between charged bodies.

critical density (302) The average density of the universe needed to make its curvature flat.

dark-line spectrum See **absorption spectrum.**

dark halo (232) The low-density extension of the halo of our galaxy believed to be composed of dark matter.

dark matter (232) The undetected matter that some astronomers believe makes up the missing mass of the universe.

deferent (dĕf′ŭr·ŭnt) (44) In the Ptolemaic theory, the large circle around Earth along which the center of the epicycle was thought to move.

degenerate matter (183) Extremely high-density matter in which pressure no longer depends on temperature due to quantum mechanical effects.

density wave theory (240) Theory proposed to account for spiral arms as compressions of the interstellar medium in the disk of the galaxy.

diamond ring effect (32) During a total solar eclipse, the momentary appearance of a spot of photosphere at the edge of the moon, producing a brilliant glare set in the silvery ring of the corona.

differential rotation (118, 231) The rotation of a body in which different parts of the body have different periods of rotation. This is true of the sun, the Jovian planets, and the disk of the galaxy.

differentiation (331) The separation of planetary material according to density.

diffraction fringe (73) Blurred fringe surrounding any image, caused by the wave properties of light. Because of this, no image detail smaller than the fringe can be seen.

dirty snowball model (325) The theory that the nuclei of comets are kilometer-sized chunks of ice with imbedded silicates in long, elliptical orbits around the sun.

disk component (229) All material confined to the plane of the galaxy.

distance indicator (256) Object whose luminosity or diameter is known. Used to find the distance to a star cluster or galaxy.

DNA (deoxyribonucleic acid) (419) The long carbon-chain molecule that records information to govern the biological activity of the organism. DNA carries the genetic data passed to offspring.

Doppler effect (104) The change in the wavelength of radiation due to relative radial motion of source and observer.

double-exhaust model (272) The theory that double radio lobes are produced by pairs of jets emitted in opposite directions from the centers of active galaxies.

double-lobed radio galaxy (272) A galaxy that emits radio energy from two regions (lobes) located on opposite sides of the galaxy.

dynamo effect (118) The theory that Earth's magnetic field is generated in the conducting material of its molten core.

east point (15) One of the four cardinal directions. The point on the horizon directly east.

eccentricity, e (53) A number between 1 and 0 that describes the shape of an ellipse.

eclipsing binary system (140) A binary star system in which the stars eclipse each other.

ecliptic (22) The apparent path of the sun around the sky.

ejecta (344) Pulverized rock scattered by meteorite impacts on a planetary surface.

electromagnetic radiation (67) Changing electric and magnetic fields that travel through space and transfer energy from one place to another; examples are light or radio waves.

electron (92) Low-mass atomic particle carrying a negative charge.

ellipse (53) A closed curve around two points called the foci such that the total distance from one focus to the curve and back to the other focus remains constant.

elliptical galaxy (251) A galaxy that is round or elliptical in outline and contains little gas and dust, no disk or spiral arms, and few hot, bright stars.

emission line (98) A bright line in a spectrum caused by the emission of photons from atoms.

emission nebula (152) A cloud of glowing gas excited by ultraviolet radiation from hot stars.

emission spectrum (bright-line spectrum) (98) A spectrum containing emission lines.

energy level (94) One of a number of states an electron may occupy in an atom, depending on its binding energy.

energy transport (168) Flow of energy from hot regions to cooler regions by one of three methods: conduction, convection, or radiation.

enzyme (420) Special protein that controls processes in an organism.

epicycle (ĕp′ŭ·sī·kŭl) (44) The small circle followed by a planet in the Ptolemaic theory. The center of the epicycle follows a larger circle (the deferent) around Earth.

equant (ē′kwŭnt) (44) In the Ptolemaic theory, the point off center in the deferent from which the center of the epicycle appears to move uniformly.

equatorial mounting (72) A telescope mounting that allows motion parallel to and perpendicular to the celestial equator.

escape velocity (215) The initial velocity an object needs to escape from the surface of a celestial body.

evening star (26) Any planet visible in the sky just after sunset.

event horizon (215) The boundary of the region of a black hole from which no radiation may escape. No event that occurs within the event horizon is visible to a distant observer.

evolutionary track (156) The path a star follows in the H-R diagram as it gradually changes its surface temperature and luminosity.

excited atom (94) An atom in which an electron has moved from a lower to a higher energy level.

extra–solar planet (317) A planet orbiting a star other than the sun.

eyepiece (69) A short-focal-length lens used to enlarge the image in a telescope. The lens nearest the eye.

false-color image (79) A representation of graphical data with added or enhanced color to reveal detail.

filtergram (112) A photograph (usually of the sun) taken in the light of a specific region of the spectrum—for example, an H-alpha filtergram.

fission hypothesis (348) The theory that the moon and Earth formed when a rapidly rotating protoplanet split into two pieces.

flare (121) A violent eruption on the sun's surface.

flatness problem (305) In cosmology, the peculiar circumstance that the early universe must have contained almost exactly the right amount of matter to make space–time flat.

flat universe (296) A model of the universe in which space–time is not curved.

flocculent galaxy (241) A galaxy whose spiral arms have a woolly or fluffy appearance.

flux (130) A measure of the flow of energy out of a surface. Usually applied to light.

focal length (69) The focal length of a lens is the distance from the lens to the point where it focuses parallel rays of light.

folded mountain chain (341) A long range of mountains formed by the compression of a planet's crust.

forward scattering (375) The optical property of finely divided particles to preferentially direct light in the original direction of the light's travel.

galactic cannibalism (264) The theory that large galaxies absorb smaller galaxies.

galactic corona (232) The extended, spherical distribution of low-luminosity matter believed to surround the Milky Way and other galaxies.

galaxy (6) A large system of stars, star clusters, gas, dust, and nebulae orbiting a common center of mass.

galaxy seeds (310) The first irregularities in the universe, which stimulated the formation of the galaxies.

Galilean satellites (găl·ŭ·lē′ŭn) (324) The four largest satellites of Jupiter, named after their discoverer Galileo.

geocentric universe (44) A model universe with Earth at the center, such as the Ptolemaic universe.

giant star (134) Large, cool, highly luminous star in the upper right of the H-R diagram. Typically 10 to 100 times the diameter of the sun.

globular star cluster (225) A star cluster containing 100,000 to 1 million stars in a sphere about 75 ly in diameter. Generally old, metal-poor, and found in the spherical component of the galaxy.

gossamer rings (375) Jupiter's largest and most tenuous rings of dust.

grand unified theories (GUTs) (305) Theories that attempt to unify (describe in a similar way) the electromagnetic, weak, and strong forces of nature.

granulation (111) The fine structure of bright grains covering the sun's surface.

grating (80) A piece of material in which numerous microscopic parallel lines are scribed. Light encountering a grating is dispersed to form a spectrum.

gravitational collapse (328) The process by which a forming body such as a planet gravitationally captures gas from its surroundings.

gravitational lens (284) The focusing of light from a distant galaxy or quasar by an intervening galaxy to produce multiple images of the distant body.

gravitational red shift (217) The lengthening of the wavelength of a photon due to its escape from a gravitational field.

greenhouse effect (342) The process by which a carbon dioxide atmosphere traps heat and raises the temperature of a planetary surface.

grooved terrain (377) Regions of the surface of Ganymede consisting of parallel grooves. Believed to have formed by repeated fracture and refreezing of the icy crust.

ground state (94) The lowest permitted electron energy level in an atom.

GUTs See **grand unified theories.**

half-life (326) The time required for half of the atoms in a radioactive sample to decay.

halo (230) The spherical region of a spiral galaxy, containing a thin scattering of stars, star clusters, and small amounts of gas.

head–tail radio galaxy (275) A radio galaxy with a contour consisting of a head and a tail. Believed to be caused by the motion of an active galaxy through the intergalactic medium.

heat (95) Energy stored in a material as agitation among its particles.

heat of formation (331) In planetology, the heat released by in-falling of matter during the formation of a planetary body.

heavy bombardment (334) The intense cratering during the first 0.5 billion years in the history of the solar system.

heliocentric universe (46) A model of the universe with the sun at the center, such as the Copernican universe.

helioseismology (114) The study of the interior of the sun by the analysis of its modes of vibration.

helium flash (184) The explosive ignition of helium burning that takes place in some giant stars.

Herbig-Haro objects (157) Small nebulae that vary irregularly in brightness. Believed to be associated with star formation.

Hertzsprung-Russell (H-R) diagram (hĕrt′sprŭng·rŭs·ŭl) (132) A plot of the intrinsic brightness versus the surface temperature of stars. It separates the effects of temperature and surface area on stellar luminosity. Commonly plotted as absolute magnitude versus spectral type but also as luminosity versus surface temperature or color.

homogeneity (293) The assumption that on the large scale, matter is uniformly spread through the universe.

horizon (14) The circular boundary between the sky and Earth.

horizon problem (305) In cosmology, the circumstance that the primordial background radiation seems much more isotropic than can be explained by the standard big bang theory.

hot spot (272) In geology, a place on Earth's crust where volcanism is caused by a rising convection cell in the mantle below. In radio astronomy, a bright spot in a radio lobe.

H-R diagram See **Hertzsprung-Russell diagram.**

H II region (257) A region of ionized hydrogen around a hot star.

Hubble constant (H) (258) A measure of the rate of expansion of the universe. The average value of velocity of recession divided by distance. Presently believed to be between 50 and 100 km/sec/Mpc.

Hubble law (258) The linear relation between the distances to galaxies and their velocity of recession.

Hubble time (306) The age of the universe, equivalent to 1 divided by the Hubble constant. The Hubble time is the age of the universe if it has expanded since the big bang at a constant rate.

hydrostatic equilibrium (167) The balance between the weight of the material pressing downward on a layer in a star and the pressure in that layer.

hypothesis (54) A conjecture, subject to further tests, that accounts for a set of facts.

inflationary universe (305) A version of the big bang theory that includes a rapid expansion when the universe was very young. Derived from grand unified theories.

infrared cirrus (153) Wispy network of cold dust clouds discovered by the Infrared Astronomy Satellite.

instability strip (225) The region of the H-R diagram in which stars are unstable to pulsation. A star passing through this strip becomes a variable star.

interstellar medium (151) The gas and dust distributed between the stars.

interstellar reddening (152) The process in which dust scatters blue light out of starlight and makes the stars look redder.

inverse square relation (61) A rule that the strength of an effect (such as gravity) decreases in proportion as the distance squared increases.

ion (92) An atom that has lost or gained one or more electrons.

ionization (92) The process in which atoms lose or gain electrons.

irregular galaxy (253) A galaxy with a chaotic appearance, large clouds of gas and dust, and both population I and II stars, but without spiral arms.

isotopes (92) Atoms that have the same number of protons but a different number of neutrons.

isotropy (ī·sŏt′rŭ·pē) (293) The assumption that in its general properties the universe looks the same in every direction.

joule (J) (jool) (97) A unit of energy equivalent to a force of 1 newton acting over a distance of 1 m. One joule per second equals 1 watt of power.

Jovian planet (321) Jupiterlike planet with a large diameter and low density.

Kelvin temperature scale (95) A temperature scale using Celsius degrees and based on zero at absolute zero.

kiloparsec (kpc) (239) A unit of distance equal to 1000 pc or 3260 ly.

Kirchoff's laws (98) A set of laws that describe the absorption and emission of light by matter.

Kuiper belt (410) The collection of icy planetesimals believed to orbit in a region from just beyond Neptune out to 100 AU or more.

Lagrangian point (199) Points of stability in the orbital plane of a planet or moon. One is located 60° ahead and one 60° behind the orbiting bodies.

large-impact hypothesis (349) The theory that the moon formed from debris ejected during a collision between Earth and a large planetesimal.

Large Magellanic Cloud (253) An irregular galaxy that is a satellite of our Milky Way galaxy. It is visible in the southern sky.

L dwarf (104) A main-sequence star cooler than an M star.

life zone (428) A region around a star within which a planet can have temperatures that permit the existence of liquid water.

light curve (141) A graph of brightness versus time commonly used in analyzing variable stars and eclipsing binaries.

light-gathering power (72) The ability of a telescope to collect light. Proportional to the area of the telescope's objective lens or mirror.

lighthouse theory (207) The description of a pulsar as a neutron star producing pulses of radiation by sweeping radio beams around the sky as it spins.

light pollution (74) The illumination of the night sky by waste light from cities and outdoor lighting, which prevents the observation of faint objects.

light-year (ly) (5) A unit of distance. The distance light travels in one year.

liquid metallic hydrogen (323) A form of liquid hydrogen that is a good electrical conductor, found in the interiors of Jupiter and Saturn.

lobate scarp (lō′bāt·skärp) (351) A curved cliff such as those found on Mercury.

local group (262) The small cluster of a few dozen galaxies that contains our Milky Way Galaxy

look-back time (257) The amount by which we look into the past when we look at a distant galaxy. A time equal to the distance to the galaxy in light-years.

luminosity (L) (131) The total amount of energy a star radiates in 1 second.

luminosity class (135) A category of stars of similar luminosity. Determined by the widths of lines in their spectra.

lunar eclipse (29) The darkening of the moon when it moves through Earth's shadow.

Lyman series (lī′měn) (99) Spectral lines in the ultraviolet spectrum of hydrogen produced by transitions whose lowest energy level is the ground state.

MACHOs (304) Massive Compact Halo Objects. Low-luminosity objects such as planets and brown dwarfs that contribute to the mass of the halo.

Magellanic clouds (măj·ŭ·lăn′ĭk) (253) Small, irregular galaxies that are companions to the Milky Way. Visible in the southern sky.

magnetar (214) A class of neutron star having very strong magnetic fields.

magnetic carpet (113) The network of small magnetic loops that covers the solar surface.

magnetosphere (măg·nē′tō·sfîr) (372) The volume of space around a planet within which the motion of charged particles is dominated by the planetary magnetic field rather than the solar wind.

magnifying power (74) The ability of a telescope to make an image larger.

magnitude scale (12) The astronomical brightness scale. The larger the number, the fainter the star.

main sequence (133) The region of the H-R diagram running from upper left to lower right, which includes roughly 90 percent of all stars.

mantle (340) The layer of dense rock and metal oxides that lies between the molten core and Earth's surface. Also, similar layers in other planets.

mare (sea) (mä′rā) (344) One of the lunar lowlands filled by successive flows of dark lava.

mass (61) A measure of the amount of matter making up an object.

mass–luminosity relation (143) The more massive a star is, the more luminous it is.

Maunder butterfly diagram (môn′dŭr) (117) A graph showing the latitude of sunspots versus time. First plotted by W. W. Maunder in 1904.

Maunder minimum (môn′dŭr) (117) A period of less numerous sunspots and other solar activity between 1645 and 1715.

megaparsec (Mpc) (256) A unit of distance equal to 1,000,000 pc.

metals (233) In astronomical usage, all atoms heavier than helium.

meteor (326) A small bit of matter heated by friction to incandescent vapor as it falls into Earth's atmosphere.

meteorite (326) A meteor that survives its passage through the atmosphere and strikes the ground.

meteoroid (326) A meteor in space before it enters Earth's atmosphere.

meteor shower (400) A multitude of meteors that appear to come from the same region of the sky. Believed to be caused by comet debris.

midocean rift (340) Chasms that split the midocean rises where crustal plates move apart.

midocean rise (340) One of the undersea mountain ranges that push up from the seafloor in the center of the oceans.

Milankovitch hypothesis (36) Suggestion that Earth's climate is determined by slow periodic changes in the shape of its orbit, the angle of its axis, and precession.

Milky Way (6) The hazy band of light that circles our sky. Produced by the glow of our galaxy.

Milky Way Galaxy (6) The spiral galaxy containing our sun. Visible in the night sky as the Milky Way.

Miller experiment (421) An experiment that reproduced the conditions under which life began on Earth and manufactured amino acids and other organic compounds.

millisecond pulsar (210) A pulsar with a pulse period of only a few milliseconds.

model (15) See scientific model.

molecular cloud (154) A dense interstellar gas cloud in which atoms are able to link together to form molecules such as H_2 and CO.

molecule (92) Two or more atoms bonded together.

morning star (26) Any planet visible in the sky just before sunrise.

mutant (420) Offspring born with altered DNA.

nanometer (68) A unit of distance equaling one billionth of a meter (10^{-9} m).

natural law (54) A theory that is almost universally accepted as true.

natural selection (420) The process by which the best traits are passed on, allowing the most able to survive.

neap tide (28) Ocean tide of low amplitude occurring at first- and third-quarter moon.

nebula (152) A glowing cloud of gas or a cloud of dust reflecting the light of nearby stars.

neutrino (163) A neutral, massless atomic particle that travels at the speed of light.

neutron (92) An atomic particle with no charge and about the same mass as a proton.

neutron star (205) A small, highly dense star composed almost entirely of tightly packed neutrons. Radius about 10 km.

Newtonian telescope (71) A reflecting telescope in which a diagonal mirror reflects light out the side of the telescope tube for easier access.

node (35) The points where an object's orbit passes through the plane of Earth's orbit.

north celestial pole (14) The point on the celestial sphere directly above Earth's North Pole.

north circumpolar constellations (17) Constellations located so close to the north celestial pole they do not rise or set.

north point (15) One of the four cardinal directions. The point on the horizon directly north.

nova (181, 200) From the Latin, meaning "new," a sudden brightening of a star making it appear as a new star in the sky. Believed to be associated with eruptions on white dwarfs in binary systems.

nuclear bulge (230) The spherical cloud of stars that lies at the center of spiral galaxies.

nuclear fission (161) Reactions that break the nuclei of atoms into fragments.

nuclear fusion (161) Reactions that join the nuclei of atoms to form more massive nuclei.

nucleus (of an atom) (92) The central core of an atom containing protons and neutrons. Carries a net positive charge.

objective lens (69) In a refracting telescope, the long-focal-length lens that forms an image of the object viewed. The lens closest to the object.

objective mirror (69) In a reflecting telescope, the principal mirror (reflecting surface) that forms an image of the object viewed.

oblateness (380) The flattening of a spherical body. Usually caused by rotation.

observable universe (293) The part of the universe that we can see from our location in space and in time.

Olbers' paradox (ôl'bŭrs) (292) The conflict between observation and theory about why the night sky should or should not be dark.

Oort cloud (ōrt) (409) The hypothetical source of comets. A swarm of icy bodies believed to lie in a spherical shell 50,000 AU from the sun.

opacity (168) The resistance of a gas to the passage of radiation.

open star cluster (225) A cluster of 100 to 1,000 stars with an open, transparent appearance. The stars are not tightly grouped. Usually relatively young and located in the disk of the galaxy.

open universe (296) A model of the universe in which the average density is less than the critical density needed to halt the expansion.

oscillating universe theory (302) The theory that the universe begins with a big bang, expands, is slowed by its own gravity, and then falls back to create another big bang.

outgassing (331) The release of gases from a planet's interior.

ovoid (388) The oval features found on Miranda, a satellite of Uranus.

paradigm (49) A commonly accepted set of scientific ideas and assumptions.

parallax (*p*) (42) The apparent change in position of an object due to a change in the location of the observer. Astronomical parallax is measured in seconds of arc.

parsec (*pc*) (pär'sĕk) (129) The distance to a hypothetical star whose parallax is 1 second of arc. 1 pc = 206,265 AU = 3.26 ly.

Paschen series (pä'shŭn) (99) Spectral lines in the infrared spectrum of hydrogen produced by transitions whose lowest energy level is the third.

penumbra (pǐ·nŭm′brŭ) (30) The portion of a shadow that is only partially shaded.

perigee (32) The point closest to Earth in the orbit of a body circling Earth.

perihelion (pĕr·ŭ·hē′lē·ŭn) (25) The orbital point of closest approach to the sun.

period–luminosity relation (225) The relation between period of pulsation and intrinsic brightness among Cepheid variable stars.

permitted orbit (93) One of the energy levels in an atom that an electron may occupy.

photon (67) A quantum of electromagnetic energy. Carries an amount of energy that depends inversely on its wavelength.

photosphere (32) The bright visible surface of the sun.

planet (4) A small, nonluminous body formed by accretion in a disk around a protostar.

planetary nebula (187) An expanding shell of gas ejected from a star during the latter stages of its evolution.

planetesimal (plăn·ŭ·tĕs′ŭ·mŭl) (329) One of the small bodies that formed from the solar nebula and eventually grew into protoplanets.

plastic (340) A material with the properties of a solid but capable of flowing under pressure.

plate tectonics (340) The constant destruction and renewal of Earth's surface by the motion of sections of crust.

polar axis (72) In an equatorial telescope mounting, the axis that is parallel to Earth's axis.

poor galaxy cluster (262) An irregularly shaped cluster that contains fewer than 1000 galaxies, many spiral, and no giant ellipticals.

population I (233) Stars rich in atoms heavier than helium. Nearly always relatively young stars found in the disk of the galaxy.

population II (233) Stars poor in atoms heavier than helium. Nearly always relatively old stars found in the halo, globular clusters, or the nuclear bulge.

precession (18) The slow change in the direction of Earth's axis of rotation. One cycle takes nearly 26,000 years.

primary lens (69) In a refracting telescope, the largest lens.

primary mirror (69) In a reflecting telescope, the largest mirror.

prime focus (71) The point at which the objective mirror forms an image in a reflecting telescope.

primeval atmosphere (342) Earth's first air.

primordial background radiation (299) Radiation from the hot clouds of the big bang explosion. Because of its large red shift, it appears to come from a body whose temperature is only 2.7 K.

primordial soup (422) The rich solution of organic molecules in Earth's first oceans.

prominence (32) Eruption on the solar surface. Visible during total solar eclipses.

proper motion (226) The rate at which a star moves across the sky. Measured in seconds of arc per year.

protein (420) Complex molecule composed of amino acid units.

proton (92) A positively charged atomic particle contained in the nucleus of atoms. The nucleus of a hydrogen atom.

proton–proton chain (162) A series of three nuclear reactions that builds a helium atom by adding together protons. The main energy source in the sun.

protoplanet (330) Massive object resulting from the coalescence of planetesimals in the solar nebula and destined to become a planet.

protostar (156) A collapsing cloud of gas and dust destined to become a star.

pulsar (206) A source of short, precisely timed radio bursts. Believed to be spinning neutron stars.

P wave (339) A pressure wave. A type of seismic wave produced in Earth by the compression of the material.

quantum mechanics (93) The study of the behavior of atoms and atomic particles.

quasar (quasi-stellar object, or QSO) (kwā′zär) (281) Small, powerful sources of energy believed to be the active cores of very distant galaxies.

radial velocity (V_r) (104) That component of an object's velocity directed away from or toward Earth.

radiation pressure (333) The force exerted on the surface of a body by its absorption of light. Small particles floating in the solar system can be blown outward by the pressure of the sunlight.

radio galaxy (275) A galaxy that is a strong source of radio signals.

radio interferometer (82) Two or more radio telescopes that combine their signals to achieve the resolving power of a larger telescope.

rays (344) Ejecta from meteorite impacts forming white streamers radiating from some lunar craters.

recombination (301) The stage within 1 million years of the big bang, when the gas became transparent to radiation.

reconnection (121) On the sun, the merging of magnetic fields to release energy in the form of flares.

red dwarf (145) A faint, cool, low-mass, main-sequence star.

red shift (104) A Doppler shift toward longer wavelengths caused by a velocity of recession.

reflecting telescope (69) A telescope that uses a concave mirror to focus light into an image.

reflection nebula (152) A nebula produced by starlight reflecting off dust particles in the interstellar medium.

refracting telescope (69) A telescope that forms images by bending (refracting) light with a lens.

relativistic red shift (283) The red shift due to the Doppler effect for objects traveling at speeds near the speed of light.

resolving power (73) The ability of a telescope to reveal fine detail. Depends on the diameter of the telescope objective.

retrograde motion (44) The apparent backward (westward) motion of planets as seen against the background of stars.

revolution (22) Orbital motion about a point located outside the orbiting body. See also **rotation.**

rich galaxy cluster (262) A cluster containing over 1000 galaxies, mostly elliptical, scattered over a volume about 3 MPc in diameter.

rift valley (341) A long, straight, deep valley produced by the separation of crustal plates.

ring galaxy (264) A galaxy that resembles a ring around a bright nucleus. Believed to be the result of a head-on collision of two galaxies.

RNA (ribonucleic acid) (420) Long carbon-chain molecules that use the information stored in DNA to manufacture complex molecules necessary to the organism.

Roche limit (rōsh) (375) The minimum distance between a planet and a satellite that holds itself together by its own gravity. If a satellite's orbit brings it within its planet's Roche limit, tidal forces will pull the satellite apart.

Roche lobe (198) The volume of space a star controls gravitationally within a binary system.

rotation (22) Motion around an axis passing through the rotating body. See also **revolution.**

rotation curve (231) A graph of orbital velocity versus radius in the disk of a galaxy.

RR Lyrae variable star (är·är·lī′rē) (225) Variable star with periods of from 12 to 24 hours. Common in some globular clusters.

Sagittarius A* (244) The powerful radio source located at the core of the Milky Way Galaxy.

Saros cycle (sě′rōs) (36) An 18-year, 11-day period after which the pattern of lunar and solar eclipses repeats.

Schmidt-Cassegrain telescope (71) Cassegrain telescope that uses a thin corrector plate at the entrance to the tube. A popular design for small telescopes.

Schwarzschild radius (R_s) (schwôrts′ shēld) (216) The radius of the event horizon around a black hole.

scientific model (15) A tentative description of a phenomenon for use as an aid to understanding.

scientific notation (4) The system of recording very large or very small numbers by using powers of 10.

seeing (75) Atmospheric conditions on a given night. When the atmosphere is unsteady, producing blurred images, the seeing is said to be poor.

self-sustaining star formation (242) The process by which the birth of stars compresses the surrounding gas clouds and triggers the formation of more stars. Proposed to explain spiral arms.

semimajor axis, a (53) Half of the longest diameter of an ellipse.

SETI (431) The Search for Extra-Terrestrial Intelligence.

Seyfert galaxy (sē′fûrt) (275) An otherwise normal spiral galaxy with an unusually bright, small core that fluctuates in brightness. Believed to indicate the core is erupting.

Shapley–Curtis Debate (229) A 1920 debate between Harlow Shapley and Heber Curtis on the nature of spiral nebulae. Curtis argued they are other galaxies and Shapley argued they are internal to our own galaxy.

shepherd satellite (382) A satellite that, by its gravitational field, confines particles to a planetary ring.

shield volcano (353) Wide, low-profile volcanic cone produced by highly liquid lava.

shock wave (121) A sudden change in pressure that travels as an intense sound wave.

sidereal drive (sī·dîr′ē·ŭl) (71) The motor and gears on a telescope that turn it westward to keep it pointed at a star.

sidereal period (26) The time a celestial body takes to turn once on its axis or revolve once around its orbit relative to the stars.

singularity (215) The object of zero radius into which the matter in a black hole is believed to fall.

Small Magellanic Cloud (253) An irregular galaxy that is a satellite of our

Milky Way galaxy. It is visible in the southern sky.

solar eclipse (31) The event that occurs when the moon passes directly between Earth and the sun, blocking our view of the sun.

solar nebula theory (317) The theory that the planets formed from the same cloud of gas and dust that formed the sun.

solar system (4) The sun and its planets, asteroids, comets, and so on.

solar wind (114) Rapidly moving atoms and ions that escape from the solar corona and blow outward through the solar system.

south celestial pole (15) The point on the celestial sphere directly above Earth's South Pole.

south circumpolar constellations (17) Constellation located so close to the south celestial pole they do not rise or set.

south point (15) One of the four cardinal directions. The point on the horizon directly south.

spectral class or type (101) A star's position in the temperature classification system O, B, A, F, G, K, M. Based on the appearance of the star's spectrum.

spectral sequence (101) The arrangement of spectral classes (O, B, A, F, G, K, M) ranging from hot to cool.

spectrograph (80) A device that separates light by wavelengths to produce a spectrum.

spectroscopic binary system (138) A star system in which the stars are too close together to be visible separately. We see a single point of light, and only by taking a spectrum can we determine that there are two stars.

spectroscopic parallax (136) The method of determining a star's distance by comparing its apparent magnitude with its absolute magnitude as estimated from its spectrum.

spherical component (230) The part of the galaxy including all matter in a spherical distribution around the center (the halo and nuclear bulge).

spicule (spĭk′yōōl) (112) A small, flamelike projection in the chromosphere of the sun.

spiral arm (229) Long spiral pattern of bright stars, star clusters, gas, and dust. Spiral arms extend from the center to the edge of the disk of spiral galaxies.

spiral galaxy (252) A galaxy with an obvious disk component containing gas; dust; hot, bright stars; and spiral arms.

spiral tracer (238) Object used to map the spiral arms—for example, O and B associations, open clusters, clouds of ionized hydrogen, and some types of variable stars.

spring tide (28) Ocean tide of high amplitude that occurs at full and new moon.

star (4) A globe of gas held together by its own gravity and supported by the internal pressure of its hot gases, which generate energy by nuclear fusion.

starburst galaxy (267) A galaxy undergoing a rapid burst of star formation.

steady state theory (300) The theory (now generally abandoned) that the universe does not evolve.

stellar model (164) A table of numbers representing the conditions in various layers within a star.

stellar parallax (*p*) (128) A measure of stellar distance. See also **parallax.**

stromatolite (424) A layered fossil formation caused by ancient mats of algae or bacteria, which build up mineral deposits season after season.

strong force (161) One of the four forces of nature. The strong force binds protons and neutrons together in atomic nuclei.

subduction zone (340) A region of a planetary crust where a tectonic plate slides downward.

summer solstice (23) The point on the celestial sphere where the sun is at its most northerly point. Also, the time when the sun passes this point, about June 22, and summer begins in the northern hemisphere.

sunspot (115) Relatively dark spot on the sun that contains intense magnetic fields.

supercluster (309) A cluster of galaxy clusters.

supergiant star (134) Exceptionally luminous star whose diameter is 10 to 1000 times that of the sun.

supergranule (112) Very large convective features in the sun's surface.

supernova (181) The explosion of a star in which it increases it brightness by a factor of about a million.

supernova remnant (194) The expanding shell of gas marking the site of a supernova explosion.

S wave (339) A shear wave. A type of seismic wave produced in Earth by the lateral motion of the material.

synchrotron radiation (sǐn′krǔ·trŏn) (195) Radiation emitted when high-speed electrons move through a magnetic field.

synodic period (sǐ·nŏd′ǐk) (27) The time a solar system body takes to orbit the sun once and return to the same orbital relationship with Earth. That is, orbital period referenced to Earth.

temperature (95) A measure of the agitation among the atoms and molecules of a material. The intensity of heat.

terrestrial planet (321) An Earthlike planet—small, dense, rocky.

theory (54) A system of assumptions and principles applicable to a wide range of phenomena that have been repeatedly verified.

tidal heating (378) The heating of a planet or satellite because of friction caused by tides.

time dilation (217) The slowing of moving clocks or clocks in strong gravitational fields.

transition (99) The movement of an electron from one atomic energy level to another.

triple-alpha process (165) The nuclear fusion process that combines three helium nuclei (alpha particles) to make one carbon nucleus.

T Tauri stars (tôrē) (157) Young stars surrounded by gas and dust. Believed to be contracting toward the main sequence.

tuning fork diagram (250) A system of classification for elliptical, spiral, and irregular galaxies.

turnoff point (197) The point in an H-R diagram at which a cluster's stars turn off of the main sequence and move toward the red-giant region, re-

vealing the approximate age of the cluster.

umbra (ŭm′brŭ) (29) The region of a shadow that is totally shaded.

uncompressed density (329) The density a planet would have if its gravity did not compress it.

uniform circular motion (44) The classical belief that the perfect heavens could only move by the combination of uniform motion along circular orbits.

universality (293) The assumption that the physical laws observed on Earth apply everywhere in the universe.

variable star (225) A star whose brightness changes periodically.

velocity dispersion method (260) A method of finding a galaxy's mass by observing the range of velocities within the galaxy.

vernal equinox (23) The place on the celestial sphere where the sun crosses the celestial equator moving northward. Also, the time of year when the sun crosses this point, about March 21, and spring begins in the northern hemisphere.

vesicular basalt (vǔ·sǐk′yǔ·lǔr) (344) A porous rock formed by solidified lava with trapped bubbles.

visual binary system (138) A binary star system in which the two stars are separately visible in the telescope.

water hole (431) The interval of the radio spectrum between the 21-cm hydrogen radiation and the 18-cm OH radiation. Likely wavelengths to use in the search for extraterrestrial life.

wavelength (67) The distance between successive peaks or troughs of a wave. Usually represented by λ.

wavelength of maximum intensity (λ_{max}) (96) The wavelength at which a perfect radiator emits the maximum amount of energy. Depends only on the object's temperature.

weak force (161) One of the four forces of nature. The weak force is responsible for some forms of radioactive decay.

west point (15) One of the four cardinal directions. The point on the horizon directly west.

white dwarf star (134) Dying star that has collapsed to the size of Earth and is slowly cooling off. At the lower left of the H-R diagram.

Widmanstätten pattern (wĭd'mŭn·stä·tŭn) (400) Bands in iron meteorites due to large crystals of nickel–iron alloys.

winter solstice (23) The point on the celestial sphere where the sun is farthest south. Also the time of year when the sun passes this point, about December 22, and winter begins in the northern hemisphere.

Zeeman effect (117) The splitting of spectral lines into multiple components when the atoms are in a magnetic field.

zenith (14) The point on the sky directly above the observer.

Answers to Even-Numbered Problems

Chapter 1
2. 1.3 sec
4. 1.8×10^7 km

Chapter 2
2. 2800
4. The sun is 400,000 times brighter than the full moon.

Chapter 3
2. 1.9×10^5 cycles
4. (a) new, (b) full, (c) first quarter, (d) third quarter
6.

Total solar eclipse — Sun, Moon, Earth

Total lunar eclipse — Sun, Earth, Moon

Partial lunar eclipse — Sun, Earth, Moon

Annular eclipse — Sun, Moon, Earth

8. July 21, 1990. The eclipse was total in the northern part of the eastern Pacific near Alaska.

Chapter 4
2. Ratio of maximum to minimum distance is 6. This is not compatible with the Ptolemaic universe, because the orbit of Venus does not change this much in one epicycle.
4. 8 years
6. 39.4 AU
8. Yes, because Venus has a larger orbital velocity than Earth (1.6 times greater) and Saturn has a smaller orbital velocity than Earth (0.3 times smaller).

Chapter 5
2. 3 cm, microwave region
4. The Keck telescopes are 10 m and gather 1.6×10^6 more light than the eye.
6. $\alpha = 5.8$ seconds of arc. No, Galileo could not resolve the two stars.
8. 13 mm
10. Assuming each person is approximately ½ m across, a 45 cm diameter telescope is needed.

Chapter 6
2. 150 nm
4. 16
6. 420 nm
8. (a) B, (b) F, (c) M, (d) K
10. 0.58 nm

Chapter 7
2. 730 km
4. The surface is 3.6 times brighter.
6. 2.5×10^9 megatons of TNT
8. 3.9×10^9 sec or 124 years

Chapter 8

2. 28.2 m

4.

m	M	d(pc)	P (sec of arc)
7	7	10	0.1
11	1	1000	0.001
1	-2	40	0.025
4	2	25	0.040

6. B

8. 160 pc

10. a, c, c, c, d

Chapter 9

2. 0.097 pc

4. 100 K

6. 0.21 year

8. In each diagram, there are four input hydrogen atoms (^{1}H), which are used to create one output helium atom (^{1}He). Even though six hydrogen atoms are initially used in Figure 9-15, two are left over at the end of the reaction leaving a net of four. The gamma rays, positrons, and neutrinos represent energy production in the diagrams.

10. 1/1024 solar lifetimes or 9.77×10^6 years

Chapter 10

2. 1.75 ly

4. 32,000 years

6. 1.3×10^6

8. 50 km/s

10. 1/32 solar lifetime or 3.1×10^8 years

Chapter 11

2. 0.00096 AU (140,000 km) if the total mass is 2 M$_\odot$

4. 3 m, 1.1×10^{-25} m

6. 8.1 minutes

8. 3×10^8 m/s; no massive particles can travel this fast.

Chapter 12

2. 4%

4. 5 kpc

6. 25 pc

8. 21

10. 1500 K

Chapter 13

2. 2.6 Mpc

4. 93 km/s

6. 28.6 Mpc

8. 165 million years

Chapter 14

2. 7.6 million years

4. 172 Mpc

6. -28.5

8. 0.3 c or 90,000 km/s

10. 0.16

Chapter 15

2. 17.2 billion years, 11.5 billion years

4. 1.6×10^{-30} gm/cm^3

6. 76 km/sec/Mpc

8. 16.3 billion years, 10.9 billion years

Chapter 16

2. 43 billion times fainter (26.6 magnitudes), 22.6 mag

4. 3.9 billion years

6. rocky core of iron, nickel, and silicates with outer layers of liquid or gaseous ammonia, methane, and water

8. 134

Chapter 17

2. The ratio of surface area to volume is inversely proportional to r. The bigger a planet is, the smaller the ratio of surface area to volume and the more difficult it is to radiate away internal heat. Hence, small planets (with small radii) cool faster than large planets.

4. We can be sure there are no rifts or subduction zones on Earth's moon because it has been mapped at high resolution from lunar orbit.

6. 690 sec; at its farthest point from Earth, Mercury would be on the other side of the sun. We can't send radio signals through the sun.

8. 380 km

10. 6.4×10^{23} kg

Chapter 18

2. 9.7 degrees of arc

4. 33.6 km/sec

6. 7.3 sec

8. 60 μm

10. 1.46×10^{22} kg

Chapter 19

2. 21 km/sec

4. 4.7 years

6. 0.8 seconds of arc, no (unless there was extremely good seeing), yes

8. 9.2×10^6 km, too small

10. 0.033 Earth masses

Chapter 20

2. 8 cm, 60 μm

4. 1.3 M$_\odot$

6. At opposition, 380 km

Index

Page numbers in bold refer to definitions.

primordial soup, **422**
Principia (Newton), 60
Project Ozma, 432
prominences and flares, **32,**
       120–122
proper motions, **226**
proteins, **420**
proton–proton chain, **162**
protons, **92**
protoplanets, growth, **330**–332
protostars
       defined, **156**
       formation of, 155–156
Proxima Centauri, 5
Prutenic Tables, 47, 50
PSR 1257 + 12, 212–213
PSR 1013 + 10, 210
PSR 1937 + 21, 210
Ptolemaic model, 44
Ptolemy, Claudius, 43–45
pulsars
       binary, 210–212
       defined, **206**
       discovery of, 206–207
       evolution of, 208–209
       millisecond, **210**
       model, 207–208
       planets, 212–214
Puppis, 85
P waves, **339**

Q

quantum mechanics, **93,** 183
quantum universe theory, 305–306
quasar 0957 + 561, 285
quasar 1059 + 730, 285
quasars
       defined, **281**
       discovery of, 282–283
       distances, 283–284
       evidence of in distant, 284–285
       model, 285–288
QZ Vul, 219

R

radial velocity, **104**
radiation
       black body, **95**–96, 97
       electromagnetic, **67,** 68–69
       from a heated objects, 95–96
       light as a wave and a particle,
          67–68
       pressure, **333**
       primordial background, **299**–300
       synchrotron, **195,** 272
       ultraviolet, 85
radio communication, 430–432
radio galaxies. *See* active galaxies
radio interferometer, **82**
radio telescopes, 81–84

rays, **344**
recombination, **301**–302
reconnections, **121**–122
red dwarfs, **145,** 172, 185–186
red shift, **104,** 258
       gravitational, **217**
       relativistic, **283**–284
reflecting telescope, **69**–70, 71
reflection nebula, **152**
refracting telescope, **69,** 70
relativistic red shift, **283**–284
resolving power, **73,** 76
retrograde motion, **44,** 47
revolution
       defined, **22**
       solar system, 320–321
rich galaxy clusters, **262**
rift valley, **341**
Rigel, 135
ring galaxies, **264**
Ring Nebula, 187
rings
       Jupiter, 374–375
       Neptune, 390–392
       Saturn, 382
       Uranus, 386, 388
RNA (ribonucleic acid), **420**
Roche limit, **375**
Roche lobe, **198**–199
rocky core, **371**
rotation
       curve, **231**–232, **260**
       defined, **22**
       differential, **118, 231**
       Earth, 3, 15, 18
       Milky Way, 231–232
       moon, 26
       solar system, 320–321
RR Lyrae variable stars, **225**
Rudolphine Tables (Tycho Brahe),
       51, 52, 55
Russell, Henry Norris, 132

S

Sagittarius, 22, 99, 226, 227, 228
Sagittarius A*, **244**–246
Sagittarius Dwarf, 262, 263, 264
Saros cycle, **36**
satellites. *See* moon; moons
Saturn
       atmosphere, 380, 381
       distance to the sun, 4
       history of, 384–385
       interior, 380
       as a Jovian planet, **321**–324
       moons, 382–384
       motion, 25
       rings, 382
       statistics, 380, 445
scale, 2–8
Schmidt, Maarten, 282

Schmidt-Cassegrain telescopes, **71**
Schwabe, Heinrich, 115
Schwarzschild, Karl, 215
Schwarzschild black holes, 215–217
Schwarzschild radius, **216**
science
       birth of modern, and Galileo,
          55–58
       separating facts from theories,
          157
       statistical evidence, 277
scientific arguments, 279
scientific model, **15**
scientific notation, **4,** 439
scientific study, 144
seasons, 24–25
       affect of Earth's orbit on, 28
seeing, **75**
self-sustaining star formation, **242**
semimajor axis, **53**
SERENDIP (Search for Extraterres-
       trial Radio Emission from
       Nearby Developed Intelligent
       Populations), 432
SETI (Search for Extra-Terrestrial
       Intelligence), **431**–432
Seyfert galaxies, **275**–278
Shapley, Harlow, 225–229
Shapley-Curtis debate, **229**
shepherd satellites, **382**
shield volcanoes, **353**
shock wave, **121,** 154–155
sidereal drive, **71**
sidereal period, **26**
Sidereus Nuncius (Galileo), 56, 57
singularity, **215**
Sirius, 11, 12–13, 188–189
SI units, 439
sky, celestial sphere model, **13**–19
small-angle formula, 31
Small Magellanic Cloud (SMC),
       253, 264
SN1987A supernova, 193–194, 210
SOHO (Solar and Heliospheric
       Observatory), 113, 118
solar eclipses, **31**–34
solar flares, 120–122, **121**
Solar Maximum Mission satellite,
       114
solar nebula
       chemical composition of, 328
       clearing the, 333–334
       hypothesis, 332
       theory, **317**
solar neutrino problem, 163–164
solar system
       age of, 326–327
       defined, **4**
       explaining the characteristics of,
          332–333
       life in, 425–428
       origin of, 315–319